SPSS/PC+™ V2.0 Base Manual

for the IBM PC/XT/AT and PS/2

SPSS Inc.
444 N. Michigan Avenue
Chicago, Illinois 60611
312.329.3300

SPSS Europe B.V.
P.O. Box 115
4200 AC Gorinchem
The Netherlands
Tel. +31.1830.36711
Twx.: 21019 (SPSS NL)

For more information about the SPSS/PC+™ V3.0 system and other software produced and distributed by SPSS Inc., please write or call

Marketing Department
SPSS Inc.
444 North Michigan Avenue
Chicago, IL 60611
312/329-3500

In Europe and the Middle East, please write or call

SPSS International B.V.
P.O. Box 115
4200 AC Gorinchem
The Netherlands
+31.1830.36711
Twx: 21019 (SPSS nl)
Fax: +31.1830.35839

4 5 6 7 8 9 0   91 90 89

ISBN 0-918469-55-4

Library of Congress Catalog Card Number: 87-062353

Preface

SPSS/PC+™ V2.0 continues the SPSS line of powerful statistical and information analysis systems running on a wide selection of mainframe and personal computers. As the latest product for the PC-DOS/MS-DOS operating system, it brings a working environment of unequaled flexibility to the popular SPSS/PC+ product, while maintaining full compatibility with SPSS/PC+ command and data files. New features include an optional menu mode for building commands, a context-sensitive help facility, and an online glossary of statistical and related terms. Like its predecessors, the SPSS/PC+ V2.0 system is built around the reliable, tested routines developed for SPSS mainframe software systems over a period of over twenty years.

The manual for SPSS/PC+ has been highly acclaimed. In updating it for V2.0 we have provided a new Introduction containing an explanation of the different ways you can use the system: from a command-line environment, from a text-editor environment of REVIEW, or from the Menu and Help system built into the new REVIEW. As before, statistical overviews guide the new user in the rationale behind statistical procedures, their use, and the interpretation of their output. The comprehensive Command Reference remains an essential guide for the experienced user.

This manual describes the base system of SPSS/PC+. Add-on enhancements available at the release of V2.0 include SPSS/PC+ Advanced Statistics™, a set of seven procedures offering more sophisticated statistical analysis (now including reliability analysis); SPSS/PC+ TABLES™, a procedure that generates publication-quality tabulations and offers great flexibility in table layout and contents; SPSS/PC+ Trends™ for comprehensive and convenient analysis of time series; SPSS Data Entry II™ for data entry and validation; SPSS/PC+ Graphics™ featuring Microsoft® Chart; SPSS/PC+ Mapping™ featuring Map-Master™ from Ashton-Tate®; and SPSS/PC+ Graph-in-the-Box™ from New England Software®.

Compatibility

SPSS Inc. warrants that SPSS/PC+ and enhancements are designed for personal computers in the IBM PC and IBM PS/2™ lines with a 10MB or larger hard disk. These products also function on many closely IBM-compatible machines. Contact SPSS Inc. for details about specific IBM-compatible hardware.

Serial Numbers

Your serial number is your identification number with SPSS Inc. You will need this serial number when you call SPSS Inc. for information regarding support, payment, a defective diskette, or an upgraded system.

The serial number can be found on the diskette labeled B1 in your system. Before using the system, please copy this number to the **registration card.**

Registration Card

STOP! Before continuing on, *fill out and send us your registration card.* Until we receive your registration card, you have an unregistered system. Even if you have previously sent a card to us, please fill out and return the card enclosed in your SPSS/PC+ package.

Registering your system entitles you to

- Technical support on our customer hot-line.
- Favored customer status.
- *Keywords*—the SPSS user newsletter.
- New product announcements.

Of course, unregistered systems receive none of the above, so *don't put it off—send your registration card now!*

Replacement Policy

System Diskettes. Call the Micro Software Department at 312/329-3300 to report a defective diskette. You must provide us with the serial number of your system. (The normal installation procedure will detect any damaged diskettes.) SPSS Inc. will ship replacement diskettes the same day we receive notification from you. Please return the defective diskettes to the Micro Software Department, SPSS Inc., 444 North Michigan Avenue, Chicago, IL 60611.

Shipping List

The shipping list for SPSS/PC+ V2.0 is on a separate sheet in the package.

Training Seminars

SPSS Inc. provides both public and onsite training seminars for SPSS/PC+. There is a two-day introductory course to familiarize users with the basics of SPSS/PC+. In addition there is an advanced course, also two days, that deals with more sophisticated aspects of the program. Additional seminars treat specialized topics such as data entry, graphics, report writing, and time series analysis. All seminars feature hands-on workshops.

SPSS/PC+ seminars will be offered in major U.S. and European cities on a regular basis. For further information on these seminars or to schedule an onsite seminar, call the SPSS Inc. Training Department at 312/329-3557.

Additional Documentation

Additional copies of all SPSS product manuals may be purchased separately. To order additional manuals, just fill out the Documentation Card included with your system and send it to SPSS Inc. Documentation Sales, 444 N. Michigan Avenue, Chicago, IL, 60611.

Note: In Europe, additional copies of documentation can be purchased by site-licensed customers only. Please contact the European office at the address listed on the copyright page for more information.

Techline

The SPSS technical hot-line is available to registered customers of SPSS/PC+. Customers may call the Techline for assistance in using SPSS products or for installation help for one of the warranted hardware environments.

To reach a Techline consultant, call 312/329-3410, 9:00 a.m. to 5:00 p.m. CST. Be prepared to identify yourself, your organization, and the serial number of your system.

If you are a Value Plus or Customer EXPress customer, use the priority 800 number you received with your materials. Priority technical help is available from 7:00 a.m. until 9:00 p.m. CST, Monday through Friday. For information on subscribing to the Value Plus or Customer EXPress plan, call SPSS Inc. at 312/329-3313.

Lend Us Your Thoughts

Your comments are important. So send us a letter and let us know about your experiences with SPSS products. We especially like to hear about new and interesting applications using the SPSS/PC+ system. Write to SPSS Inc. Marketing Department, Attn: Micro Software Products Manager, 444 N. Michigan Avenue, Chicago, IL, 60611.

Contents

Introduction

Contents

About This Manual

This is a large manual, in part because it documents a large and versatile software system, and in part because it seeks to meet the differing needs of a diverse body of users. For those who have limited experience with statistics, computing, or the IBM PC environment, introductory material and statistical overviews are available. For those who are already familiar with statistical computing, a reference section presents SPSS/PC+ commands without extensive examples. You are likely to find different parts of the manual most valuable as your experience with SPSS/PC+ grows.

If you have just received the system, you should turn first to the Installation Instructions that accompany the system. You need to install SPSS/PC+ only once, unless you remove it from your system. (You can, however, remove and reinstall portions of the SPSS/PC+ system; for information, refer to SPSS MANAGER in Part C.)

Introduction. The Introduction (Part A) helps you become familiar with running SPSS/PC+ and using REVIEW, the SPSS/PC+ text editor. *Running SPSS/PC+* describes the system's command-based operation and gives overviews of the three environments available for entering commands. This includes a discussion of the Menu and Help system, a new facility in SPSS/PC+ V2.0 that gives you help in building commands by describing each one and guiding you through command syntax. *REVIEW: The SPSS/PC+ Editor* provides an example that shows you step-by-step how to edit files using the fully integrated SPSS/PC+ text editor, REVIEW.

Statistics Guide. The Statistics Guide (Part B) is a complete user's guide to the data definition and analysis facilities of SPSS/PC+. If you are new to computer data analysis, you can start at the beginning of the Guide and progress through introductions to coding and entering data, defining the data to SPSS/PC+, managing it, tabulating it, and analyzing it with the many statistical procedures that SPSS/PC+ makes available. If you are more experienced, you can turn to the Guide whenever you want a more guided approach to certain SPSS/PC+ facilities than you find in the Command Reference or when you want to further your understanding of the statistics calculated by certain procedures. The Guide does not discuss at length the SPSS/PC+ commands that control such things as the destination of output files or the format of the output. For that information consult Running SPSS/PC+ in the Introduction.

Command Reference. The Command Reference (Part C) is a detailed reference to the syntax and operations of each SPSS/PC+ command. Its opening section, Universals, documents the general characteristics of the system. Following that section, the individual commands are presented in alphabetical order. For each command, the Command Reference provides complete syntax rules plus details of operations.

Examples. The examples presented in Part D illustrate typical uses of SPSS/PC+ analytical procedures. The annotated input and output are arranged not to imitate the progress of an interactive SPSS/PC+ session but to demonstrate a set of commands that carry out a complete data analysis task. You may find that these examples, with their interpretative commentary, extend your understanding of the logic of SPSS/PC+ command structure.

Glossary. The Glossary (Part E) defines terms used in this manual that may be unfamiliar to many users. It does not attempt to cover the full vocabulary of DOS operating manuals or the statistical terms discussed in the Statistics Guide.

Communications. If you wish to transfer data files, portable system files, command files, or any other information between computers (particularly between your IBM PC and a mainframe computer), you will be interested in Kermit, a communications facility included with your system and available on many mainframes. *Transferring Files with Kermit,* in Part F, describes the use of Kermit with SPSS/PC+.

Appendixes. If your computer has an extended or expanded memory board and RAM-disk software, *Using SPSS/PC+ with RAM Disks* tells you how to use this capability to best advantage in SPSS/PC+. For users making the transition from earlier versions of SPSS/PC+, *Help for Old Friends* provides a list of important changes. *Files Created by Other Software* gives general suggestions for anyone wanting to use data or command files created by a text editor or other software.

Running SPSS/PC+

OVERVIEW

SPSS/PC+ is a system that accesses and analyzes data. The way you communicate with SPSS/PC+ is through the SPSS/PC+ command language. Commands are simply English-like statements. They can be simple, like

```
LIST.
```

They can contain additional specifications, such as

```
LIST CASES FROM 1 TO 100 BY 5.
```

Or commands can contain complex combinations of subcommands, such as

```
FREQUENCIES VARIABLES = SEX JOBCAT /BARCHART /STATISTICS ALL.
```

The command language is easy to understand, and SPSS/PC+ gives you a lot of help in both learning and using it.

You use commands to do three things, usually in the following order:

1 Convert data into a form that can be used by the computer and bring the data into SPSS/PC+. Your data may exist on paper, or they may already exist in some electronic form used by other software (including SPSS^X, the mainframe version of SPSS/PC+). *Data definition* commands like DATA LIST, TRANSLATE, and GET (among others) tell SPSS/PC+ where and how to read data.

2 Modify data. Once the data are in SPSS/PC+, you might decide, for example, to convert rainfall measured in centimeters to rainfall measured in inches, for which you could use a *data manipulation* command like

```
COMPUTE RAINFALL = RAINFALL/2.54.
```

3 Process data. When your data are in the form you want, *procedure* commands like FREQUENCIES, MEANS, and REGRESSION tell SPSS/PC+ to do something with your data, such as perform a statistical analysis, produce a report, listing, or plot, sort your cases into a different order, or save your data to a file.

During the course of an SPSS/PC+ session you can repeat any of the three steps above, to use different data, arrange the data differently, or run different analyses on them.

For example, a typical series of commands might resemble the following:

```
GET FILE = 'WEATHER.SYS'.
COMPUTE RAINFALL = RAINFALL/2.54.
MEANS RAINFALL BY MONTH REGION.
```

- The GET FILE command retrieves a *system file* named WEATHER.SYS. (A system file is a special type of SPSS/PC+ file that contains both data and information describing the data.)
- The COMPUTE command converts rainfall in inches to rainfall in centimeters.
- The MEANS command computes the average rainfall for each month and for each region.

SPSS/PC+ can execute these commands one at a time or as a group. The next sections tell you about ways to enter commands and have SPSS/PC+ execute them.

In addition to procedure and data definition and manipulation commands, SPSS/PC+ has *operation* commands like DISPLAY and SET that you use to find out about and control the SPSS/PC+ environment. For example, notice in the example above that each command must end with a period. If you want to use a different symbol to indicate your commands are complete, you can do so using the SET command.

Table 1 summarizes the commands used in SPSS/PC+.

Table 1 SPSS/PC+ command summary

Function	Commands
Operation commands	
Provide assistance	SHOW, DISPLAY
Specify options for operations and output	SET
Submit SPSS/PC+ commands from a file	INCLUDE
Edit a file	REVIEW
Access DOS or other facilities	DOS, EXECUTE
Data definition and manipulation commands	
Read data	DATA LIST, BEGIN DATA, END DATA, IMPORT GET, TRANSLATE
Transform data	RECODE, COMPUTE, IF, COUNT
Define missing data	MISSING VALUE
Select and weight cases	SELECT IF, PROCESS IF, N, SAMPLE, WEIGHT
Provide labels and formats	TITLE, SUBTITLE, *, VARIABLE LABELS, VALUE LABELS, FORMAT
Procedure commands	
Data display	LIST, PLOT, REPORT, TABLES*
Descriptive statistics	DESCRIPTIVES, FREQUENCIES
Categorical statistics	CROSSTABS, HILOGLINEAR*
Group comparisons	T-TEST, ONEWAY, MEANS, ANOVA
Multivariate statistics	CORRELATION, REGRESSION, CLUSTER*, QUICK CLUSTER*, FACTOR*, DSCRIMINANT*, MANOVA*, RELIABILITY*
Nonparametric statistics	NPAR TESTS
Time series analysis*	ACF, AREG, ARIMA, CASEPLOT, CCF, CURVEFIT, EXSMOOTH, FIT, NPPLOT, PACF, RMV, SPECTRA
Utilities	WRITE, AGGREGATE, SORT CASES, JOIN, EXPORT, SAVE
Graphics	GRAPH*, MAP*, FASTGRAF*

*Available only in SPSS/PC+ options.

Entering Commands into SPSS/PC+

There are three ways to enter commands into SPSS/PC+: at the command prompt, from the integrated text editor REVIEW, or using the Menu and Help system.

- The SPSS/PC+ command prompt. With this method, the system prompts you to enter a command; you enter it; the system evaluates and executes it. This method does not give you a clue as to which command to enter; nor does it provide a great deal of opportunity to edit a command or to change your mind once you've entered it. However, if you already know exactly which commands you want to use, you may prefer to work at the command prompt.
- The REVIEW text editor. This method lets you type your commands into a file instead of at a prompt, and then submit them for execution. This method gives you unlimited opportunity to edit and revise the commands before execution. Once you have a set of commands, making minor changes in them to achieve a different analysis can be the fastest way to work.
- The Menu and Help System. REVIEW includes a *Menu and Help system* that allows you to select commands from menus and paste them into the text editor. Along with the menus are descriptions of each command to help you figure out which one to

select. The advantage of using the menus is that you don't have to remember the command names or their specifications.

When you first enter SPSS/PC+, you will be in the Menu and Help system.

Getting into SPSS/PC+

Before you can run SPSS/PC+ for the first time, you have to install it, using the installation instructions that accompany the system. Next you should change your current directory to the one from which you want to execute SPSS/PC+ and in which you want to keep the data and other files used by the system. You may also want to change other DOS environmental parameters. For information about DOS and directories, see *Important DOS Concepts* at the end of Part A, or see your DOS manual.

Once the preliminaries are done, you enter SPSS/PC+ by typing

```
SPSSPC
```

at the DOS prompt, and pressing ↵. This displays the SPSS/PC+ logo screen. Usually, the system then takes you straight into the Menu and Help system. You can change the starting environment by changing the SET commands in your profile. (See the SET command in Part C, and "The SPSSPROF.INI File" in the next chapter, *REVIEW: The SPSS/PC+ Editor.*)

WORKING WITH MENUS

The menu system contains all SPSS/PC+ commands and their keywords. A help system describing the function of commands is built in. You use the Menu and Help system together with the text editor REVIEW to produce your commands.

The menus do not control SPSS/PC+ directly. When you select a command (or a piece of a command) from a menu, it is simply pasted into REVIEW's scratch pad (the scratch pad is discussed below). To execute the commands, you must tell SPSS/PC+ to do so. You can do this whenever you like—you can build a lengthy set of commands and execute them all at once, or you can run them one at a time.

You can take over from the Menu and Help system at any point and use REVIEW to type in commands yourself or to modify commands that you have pasted in. From REVIEW, you can easily recall the Menu and Help system. It appears with information showing and explaining the appropriate options for the command you were typing in REVIEW.

When you enter SPSS/PC+, you are automatically in the Menu and Help system. Your screen looks like the following:

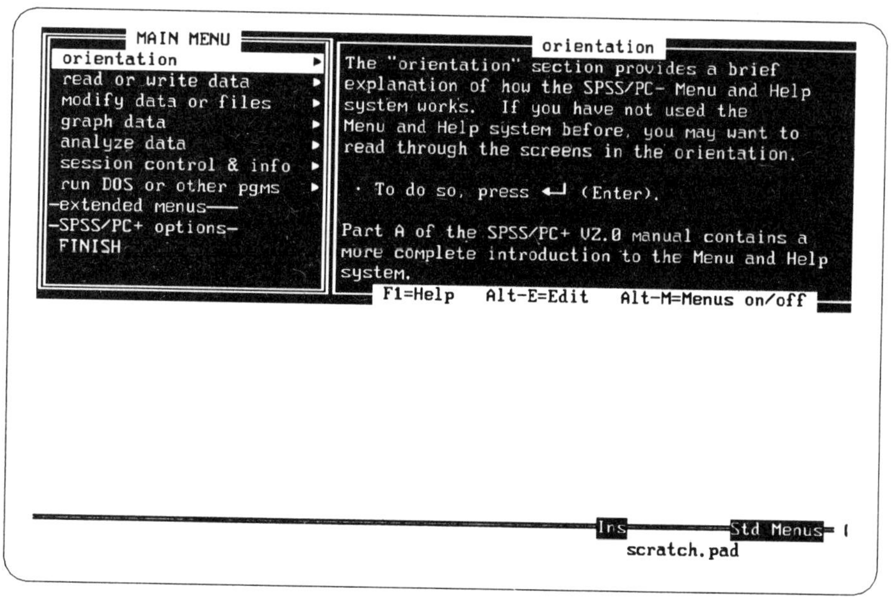

Important things to notice are:

- The bottom half of the screen is the *scratch pad*, where REVIEW works as a text editor.
- The top left part of the screen shows a *menu*, in this case the Main Menu, which is the "top" level of the Menu and Help system. The arrowhead characters to the right of each item on the menu indicate that the item has a lower-level menu that you can access to see more detail.
- The top right part of the screen shows a *Help window*. The Help window always describes the highlighted menu item.

When you first enter the system, the top half of the screen—that is, the Menu and Help system—is active.

To see the commands you need to get around in REVIEW and the Menu and Help system, press F1. A "mini-menu" appears at the bottom of the screen. The information you want is **Review help.** Since this item is highlighted, you can select it simply by pressing ↵. You can also make the selection by pressing the letter that is capitalized for that item, R, instead of ↵. The following screen will be displayed:

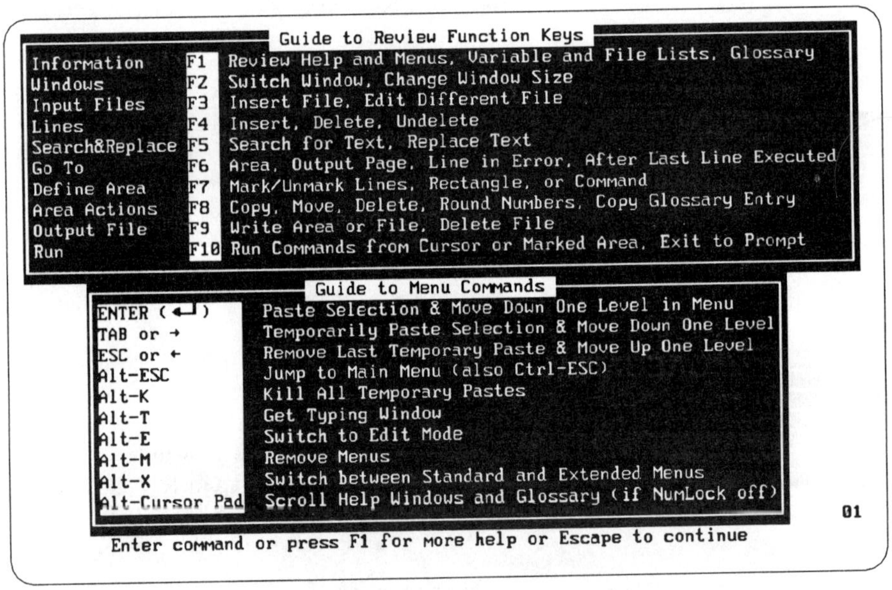

The commands you need to get around the Menu and Help system are shown at the bottom of the screen. As you can see, most of these commands use the Alt key, which you hold down while you press the corresponding command key:

- Alt M (*M* for *M*enu) clears away the Menu and Help system (leaving you in the REVIEW environment) or brings it back if it is not on the screen.
- Alt E allows you to edit the scratch pad but leaves the menus on screen, "frozen," until you press Alt E again (or Esc) to return to the Menu system.

You can also recall the menu system at any time by pressing F1 followed by M.

Exploring the Menus

The easiest way to explore the Menu and Help system is with the cursor arrows on your keyboard.

- ↑ and ↓ move the highlighting up and down the menu. As each menu item is highlighted, the Help window describes what that item does, or how to use it.
- If a menu is too long to fit, up or down arrowheads at the right edge of the menu indicate the presence of items offscreen. As you move the highlighting up and down,

the menu scrolls automatically to let you highlight these items. You can also use (PgUp) and (PgDn) to scroll the menus.

- (Home) and (End) take you to, respectively, the first and last items on the menu.
- (→) moves you "deeper" into the menus. For example, if the item **orientation** is highlighted when you press (→), the Orientation menu appears. You know that there is an Orientation menu because of the arrowhead beside **orientation** back on the Main Menu.
- (←) brings you back out of the menus, one level at a time, until you reach the Main Menu.
- If the text in the Help window is too long to fit on the screen, up or down arrowheads at the right edge of the Help window indicate the presence of offscreen text. Hold down the (Alt) key and use (↑), (↓), (PgUp), (PgDn), (Home), or (End) to scroll the text in the Help window (make sure that the (NumLck) key is not active).

Pasting Menu Selections

To use the menu system in SPSS/PC+, you need to do more than just browse around and read it. You use the menus by pasting commands, keywords, variable names, and filenames into REVIEW's scratch pad. The difference between exploring the menus with the cursor arrows, as discussed above, and pasting commands, is simple:

- To *explore* a menu selection without affecting the scratch pad, press (→). The only thing that happens is you move to the lower-level menu associated with the current selection. (The menu selection is held for you, in case you decide to accept it after all.)
- To accept a menu selection and also (if it is part of a command that you are building) to *paste* it into the scratch pad, press (↵). The selection appears in the scratch pad, and again you move to the lower-level menu associated with the current selection, if there is one.

Menu selections that are in upper case are either commands or parts of commands. They may be preceded on the menu by either a slash (/) or an exclamation point (!). The slash is a separator that SPSS/PC+ expects between different *subcommands* (parts of commands). The exclamation point indicates a selection that is required: SPSS/PC+ cannot execute the command without that specification.

- You can always paste an upper-case menu selection into the scratch pad by pressing (↵).
- Some other selections (such as the symbols +, −, *, and /) will also paste if you select them with (↵).
- Selections in lower case do not paste into the Edit window, regardless of whether you use (→) or (↵) to select them. You do move to the lower-level menu if there is one.
- **When you paste a selection that needs to be "filled in," such as a pair of parentheses ( ) or apostrophes ' ', REVIEW automatically asks you to type the contents, unless selections on a lower-level menu constitute the contents.**

You can move to the scratch pad in the bottom half of the screen at any point by simply pressing (Alt) (E). From there, you can

- Edit the commands you've pasted.
- Run them. We'll see below how you pass commands to SPSS/PC+ for execution.
- Save them into a file on disk, to use later. See the next chapter, *REVIEW: The SPSS/PC+ Editor,* for an explanation of saving commands.

To get back to the Menu and Help system, press (Esc).

Pasting after Exploring

If you explore a menu selection by pressing (→) and then decide to accept and paste a selection on a lower-level menu, REVIEW goes back to fill in the intermediate step(s) before pasting the lower-level selection.

For example,

- You select the **FREQUENCIES** command and the **/VARIABLES** subcommand, and choose the variables for which you want frequency tables. All this is pasted into the scratch pad.
- You aren't sure whether you want to change the default format, so you explore **/FORMAT** by pressing [→]. The FORMAT subcommand is not pasted into the scratch pad, but it is moved into a holding area or *buffer* at the bottom of the screen, in case it is needed later:

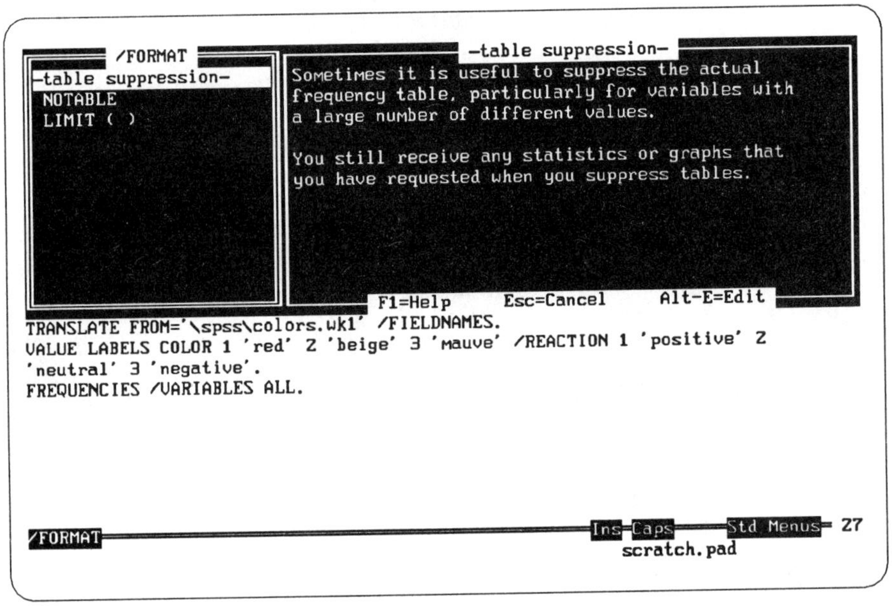

- After reading the Help windows for the various FORMAT specifications, you decide that you don't need any of them, so you "back out" by pressing [←] (or [Esc], if you prefer). The buffer area and /FORMAT disappear from the bottom of the screen, and the FREQUENCIES menu reappears.
- You move the cursor to **/STATISTICS** and explore it by pressing [→]. The STATISTICS menu appears, and /STATISTICS shows up in the buffer. After looking through the various choices on the STATISTICS menu, you decide that you want them all—so you highlight **ALL** and press [↵].
- REVIEW first pastes the subcommand /STATISTICS from the buffer, and *then* the keyword ALL that you explicitly pasted.

In this example, REVIEW held the specifications that you explored (/FORMAT and /STATISTICS) in case you decided that you needed them. When you backed out of the /FORMATS menu, it discarded that specification. When you chose a statistics keyword, it filled in /STATISTICS so SPSS/PC+ would know that ALL referred to the STATISTICS subcommand.

- You can clear this holding buffer at any time by pressing [Alt] [K].
- If you don't remember [Alt] [K] (or any other REVIEW command), press [F1], use the cursor arrows to highlight **Review help,** and press [↵].

Editing Pasted Commands

It's easy to add to or modify the commands you paste into the REVIEW scratch pad.

- Press [Alt] [E] to jump to the scratch pad, or [Alt] [M] to clear the menus completely.
- Then fix the commands. Usually the [Ins] and [Del] keys will be all you need. See the next chapter, *REVIEW: The SPSS/PC+ Editor,* for a discussion of more advanced editing functions.

To return to the menus, press (Alt) (E) again, or, if you cleared the menus completely, press (Alt) (M) again. (You can also press (F1), highlight **Menus** with the cursor arrows, and press (↵).) The Menu and Help window for the specific command or subcommand you're working on in the scratch pad will be displayed. You can paste items from the menu, or you can press (Alt) (E) to continue editing with the Help window to guide you.

Extended Menus

The standard menus do not show all of the keywords you can use to build SPSS/PC+ commands. Seldom-used keywords and options are omitted to reduce the complexity of the menus. To switch to the extended menus that show all available specifications, press (Alt) (X). The message at the right-hand side of the bottom of the screen then changes from *Std Menus* to *Ext Menus*.

Once you are familiar with the system you may want to place the command

```
SET MENUS EXTENDED.
```

in your automatic profile SPSSPROF.INI (see the section on SPSSPROF.INI in the next chapter, *REVIEW: The SPSS/PC+ Editor*). This will cause extended windows to be displayed automatically.

Executing Commands

When you have built a command (or several commands), you can submit those lines to SPSS/PC+ for execution.

1 If you haven't done so already, press (Alt) (E) to move to the REVIEW scratch pad (or (Alt) (M) to clear away the menu system entirely).
2 Put the cursor on the first command, if it's not there already. (You don't have to do these first two steps if you want to execute a single command and the cursor is already on that command in the scratch pad.)
3 Press (F10). On the mini-menu that appears at the bottom of the screen, **run from Cursor** is highlighted, so just press (↵). This tells SPSS/PC+ to run all the commands starting at the one containing the cursor and below.

The screen clears, and your commands appear on the screen one at a time as they are executed. The word **MORE** appears in the corner of the screen when SPSS/PC+ is about to display something else and wants to make sure that you've had time to read what's already there. Press the space bar to continue. After all of the commands and their results have displayed, the system returns you to REVIEW and the Menu and Help system. (You can see the results again whenever you like in the *listing file,* described below.)

Quitting

The FINISH command is at the bottom of the Main Menu. It takes you out of both REVIEW and SPSS/PC+, back to DOS. You can paste and execute it just like any other command.

To leave the Menu and Help system and REVIEW but remain in SPSS/PC+, press (F10) and use the cursor arrows to highlight **Exit to prompt.** Press (↵). (Alternatively, you can just press (F10) followed by (E)). This takes you to the SPSS/PC+ command prompt (see "Working from the Command Prompt," below).

WORKING IN REVIEW

You can also use REVIEW without the menu system. With this method, you type in your commands directly instead of pasting them in from the menus.

To clear away the Menu and Help system after you enter SPSS/PC+, press (Alt) (M). Alternatively, you can go straight to the REVIEW scratch pad when you enter SPSS/PC+ by putting the command

```
SET AUTOMENU OFF
```

in your automatic profile SPSSPROF.INI.

There are several advantages to using REVIEW alone:

- Some commands, particularly DATA LIST and the data transformation commands, are much quicker to type than to select in pieces from the menus.
- When you are repeating or modifying commands, REVIEW gives you the best possible raw material to work with: your previous commands, and their results. It does this through three files: the scratch pad, the listing file, and the log file.

The Scratch Pad

By default, REVIEW's bottom window contains the *scratch pad,* a place where you can play with commands until you're ready to run them. When you leave REVIEW to run commands, this file is saved for you, so that when you return to REVIEW (assuming you didn't leave SPSS/PC+) all your commands are still there. If you want to run them again, perhaps with some alterations to try a different approach or correct errors, they're easy to edit (see the next chapter, *REVIEW: The SPSS/PC+ Editor*).

The Listing File

The *listing file* contains a copy of all the output produced by your commands. The main use for the listing file is to review your work or to edit the output. When the session is over, you can print the listing file from DOS (it's named SPSS.LIS by default).

The listing file is displayed in REVIEW's upper window, which is initially covered by the Menu and Help system. When you clear the menus by pressing Alt M, you see the listing file in the top half of the screen. (If you haven't yet executed any commands, you won't have any output, so the top half of the screen will be blank.)

While you're in SPSS/PC+, it's often helpful to examine your commands vs. the output they produce by getting the scratch pad or the log file (described below) in bottom window and the listing file in the top window. You can scroll whichever window is active (whichever one contains the cursor). To switch the cursor from one window to the other, press F2 and then press ↵ to select the highlighted item **Switch windows** from the mini-menu at the bottom of the screen.

If you want to see the Menu and Help system while using the listing file, you can call the menus by pressing F1, using the cursor arrows to highlight **Menus,** and pressing ↵ (or just by pressing F1 followed by M). The menus appear in the bottom half of your screen:

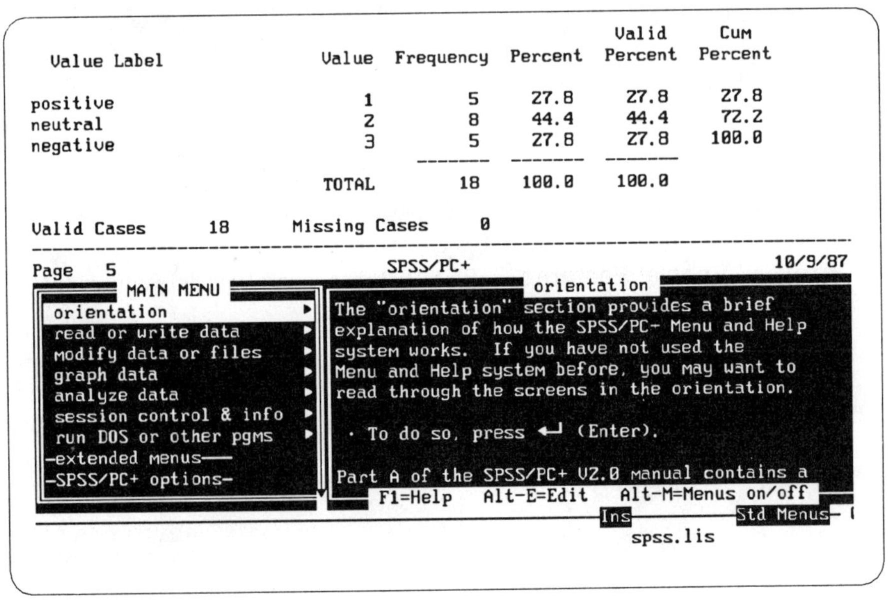

When you switch back to editing the bottom window, the Menu and Help system reappears in the top window.

The Log File

The log file contains a copy of all the commands that SPSS/PC+ has executed during your current session, in the order in which they were submitted. From within REVIEW, it's easy to access the log file:

1 Press (F3), and then press (↵) to select the highlighted item *Edit different file* from the mini-menu that appears at the bottom of the screen (or just press (F3) followed by (E)).

2 Type the name of the file to edit, SPSS.LOG, in the box at the bottom of the screen, and press (↵).

The log file is a useful record of what you've already done since, in addition to the commands you've run, it also contains

- Error messages about commands that couldn't be executed.
- Notes that direct you to the location of output in the listing file.

Note that the logfile contains a record of all the commands you submit in the session. For example, if you run a set of commands twice, the second time with modifications, the log file will contain both sets. The scratch pad, however, will only contain the modified set.

Working Faster

The following features can help you work in REVIEW more quickly:

- To select an item from a mini-menu, you can press the letter capitalized for that item, instead of using the cursor arrows to highlight the item and pressing (↵). For example, to delete a line, you can just press (D) after pressing (F4). For commonly used selections, if you know the letter for item you want, you can bypass the mini-menu altogether by simply holding down the (Alt) key while pressing the letter. For example, to insert a line, you can just press (Alt) (I) rather than pressing (F4) and then selecting **Insert after.** For the functions for which this shortcut is available, see REVIEW in Part C. Or, if you know that the selection you want is the highlighted one, you can just press the function key followed by (↵).
- When you press (F1) and select **Menus,** REVIEW displays the menu for the command under the cursor, rather than the Main Menu.
- When the Menu and Help system is active, you can use incremental search to move the cursor to a menu item—just type in the first letter or letters of the item you want to select to make the cursor jump to that item.
- When you are in the scratch pad, you can quickly place the cursor at specified locations by pressing (F6) and selecting the location you want— **Error line,** or **after executed Line.** (Note: For **Error line** to be displayed, you must set RUNREVIEW to AUTO; see SET in Part C.)
- When you are in the listing file, you can quickly move to the output page you want by pressing (F6), selecting **Output page** from the mini-menu, and entering the number of the page you want.

Special Features

REVIEW has several special features that make your work easier. These features are available even when you aren't using the Menu and Help system:

- To see lists of files, press (F1), move the cursor arrows to highlight **File list,** press (↵) and then either specify particular types of files or press (↵) again to see all files in your current directory. SPSS/PC+ then displays a list of the files you requested. In the list, one of the files is highlighted; more detailed information about this file appears in the narrow box below the list. If you press (↵), the name will be pasted into the scratch pad. Or you can use the cursor arrows to highlight another file.
- After you have brought data into your computer with a command like DATA LIST or GET, you can see a list of the variables in the active file by pressing (F1), highlighting **Var list,** and pressing (↵). SPSS/PC+ then displays a list of the currently

defined variables. In the list, one of the variables is highlighted; more detailed information about this variable appears in the narrow box below the list. If you press ↵ again, the variable name will be pasted into the scratch pad. You can also highlight a group of variables by pressing F7 at the first variable and moving the cursor arrows over the other variables you want. To paste the highlighted group into the edit window, press ↵.

- SPSS/PC+ contains an online glossary that defines the terms used in commands and their output. To look up a term in the glossary, press F1, use the cursor arrows to highlight *Glossary,* and press ↵. The system displays a box containing the term currently under the cursor. To look up this term, just press ↵; to look up a different term, type it in the box and then press ↵. The system then displays the definition you requested. For more information about the Glossary, see "Getting Help," below.

Quitting

To leave REVIEW, press F10, use the cursor arrows to highlight **Exit to prompt,** and press ↵. This takes you to the SPSS/PC+ prompt. To leave SPSS/PC+, at the prompt type

```
FINISH
```

and press ↵. You will be returned to DOS.

WORKING FROM THE COMMAND PROMPT

If you like the speed of working directly with commands and you know the command you want to use, you can ignore both REVIEW and the Menu and Help system altogether and work from the SPSS/PC+ command prompt. To go to the command prompt as soon as you enter SPSS/PC+, change the SET RUNREVIEW command in the profile SPSSPROF.INI to MANUAL. To reach the command prompt when you are already in REVIEW in SPSS/PC+, press F10, use the cursor arrows to highlight **Exit to prompt,** and press ↵.

The command prompt works like this:

- SPSS/PC+ prompts for a command.
- You enter a command.
- SPSS/PC+ responds to your command.
- SPSS/PC+ prompts for another command.

This process continues until you end your SPSS/PC+ session.

The SPSS/PC+ Command Prompt and Terminator

SPSS/PC+ tells you it is ready for a command by displaying a command prompt. The default command prompt is **SPSS/PC:**. Whenever SPSS/PC+ displays the prompt, you can begin entering a command on that line.

You must end every command with a command terminator to tell SPSS/PC+ that the command is complete. The default command terminator is a period (.). *Do not type any other characters after the command terminator.* To submit the command to SPSS/PC+, just press ↵.

If a command is too long to fit on one line, type whatever fits on the line and press ↵. As long as you do not include a command terminator, SPSS/PC+ will respond with a continuation prompt. This signals you to continue the command on that line. The default continuation prompt is seven spaces followed by a colon.

Use as many lines as each command requires, pressing ↵ to submit each line of the command. SPSS/PC+ gives continuation prompts until you enter the command terminator. When the command specifications are complete, enter the terminator and press ↵. SPSS/PC+ responds with the output for the command, if any, and prompts for another command.

If you enter a complete command and forget to include the command terminator, simply enter the terminator in response to the continuation prompt. You can also terminate the command by simply pressing ↵ again—that is, entering a completely empty line. (Failing to end a command with a period, or whatever character you have specified as the command terminator, is a common oversight when just beginning with SPSS/PC+. Be alert for a continuation prompt when you think you have finished entering a command. Enter a period, or an empty line, to complete the command.)

You can use the SET command to change the default prompts and terminator (see SET in Part C).

SPSS/PC+ Error Detection

As you enter each command, SPSS/PC+ checks it for proper syntax. If you misspell or improperly specify a command or subcommand, SPSS/PC+ immediately displays an error message. It also checks to make sure that you do not use variables that are not defined.

For example, the VARIABLE LABELS command below is mistyped as *vriable labels*. SPSS/PC+ displays an appropriate error message and prompts for a new command.

```
                                                              MORE

DATA LIST FILE 'phys1.dat'/phys 1-3 studied 5 avstay 7-10(1) avstayZ 12-15(1).
VRIABLE LABELS PHYS'' /STUDIED 'Total Patients Studied' /AVSTAY 'Average Stay'

ERROR       1, Text: VRIABLE LABELS
INVALID COMMAND--Check spelling.  If it is intended as a continuation of a
previous line, the terminator must not be specified on the previous line.
If a DATA LIST is in error, in-line data can also cause this error.
This command not executed.
```

Since you are allowed to abbreviate SPSS/PC+ keywords to their first three characters, you will not get an error message if you misspell a keyword after the first three characters.

Correcting Errors

Several special keys on the keyboard help you correct simple errors you make in entering commands. If you have made an error on the line you are typing, just backspace over the incorrect characters and type the rest of the line correctly before you press the enter key.

If you want to cancel an entire line, press the escape key. SPSS/PC+ then displays a backslash (\) at the end of the line and places the cursor on the next line. You can then retype the line correctly and continue with your session by pressing (↵). After you enter the corrected line, SPSS/PC+ prompts you for the next command. If you were in the middle of entering a multi-line command, you would retype only the line that you canceled with the escape key and then continue typing the command.

For information on using function keys to edit input lines, look up "Editing Keys" in the index of your DOS manual. On the IBM and most compatible machines, the function keys (F1), (F2), and (F3) work at the SPSS/PC+ command prompt as described in the DOS manual.

Running Commands

As described at the beginning of this chapter, there are three basic types of SPSS/PC+ commands:

- Operation commands.
- Data definition and manipulation commands.
- Procedure commands.

SPSS/PC+ executes an operation command as soon as you enter it. As you enter a data definition or manipulation command, SPSS/PC+ checks that it follows the language rules for the command. However, SPSS/PC+ does not actually execute any data definition commands until you enter a *procedure* command, which reads the data.

When you enter a procedure command, SPSS/PC+ first executes all preceding data definition and manipulation commands. These commands construct the active data file of your cases. The active file contains the data you tell SPSS/PC+ to read, the results of any transformations you request, and a dictionary of information you have provided about each variable (names, labels, missing values, and so forth), all in a format that the computer can understand.

The SPSS/PC+ system is divided into several interlinked modules, each of which contain a small number of procedures. When you enter a procedure command, SPSS/PC+ checks whether that procedure is part of the current module. If not, SPSS/PC+ displays the message **MODULE SWAP** in the status area. Then it automatically loads the module containing the procedure. This can take several seconds.

As SPSS/PC+ processes cases, it displays a case counter in the status area in the upper right-hand corner of the screen, as shown in the following:

```
                                                              CASE     552









VARIABLE LABELS PHYS'' /STUDIED 'Total Patients Studied' /AVSTAY 'Average Stay'
/AVSTAYZ 'Average Stay for Matching Patients'.
CROSSTABS /TABLES PHYS BY AVSTAY.

 ***** Given WORKSPACE allows for  5554 Cells with
     2 Dimensions for CROSSTAB problem *****
```

When the procedure is complete, SPSS/PC+ displays the message **MORE** in the status area. Press any key to display the procedure results.

You can specify different types of analyses for the same active file, and you can modify the active file. For example, you can use a CROSSTABS command after a FREQUENCIES command to explore the same variables with different analyses. Or you might follow a FREQUENCIES command with a RECODE command and another FREQUENCIES command to compare the results of the same analysis on a recoded variable.

Quitting

To exit from SPSS/PC+, at the command prompt simply type

```
FINISH
```

and press (↵). This returns you to the DOS prompt.

GETTING HELP

SPSS/PC+ provides two kinds of online assistance: a contextual help system, and an online glossary.

The Help System

Contextual help for building commands is supplied by the Menu and Help system. Any time you activate the menus, REVIEW attempts to identify the command on which the cursor is located and to display the appropriate menu. If you are typing commands into REVIEW and your memory fails you, activate the menus to see what your options are.

The Online Glossary

SPSS/PC+ contains an online glossary to help you understand the terms used in commands and in their output.

To use the glossary, you must first be in a file in REVIEW (not in the Menu and Help system).

1 Place the cursor under the term you want to look up. (To look up a command or keyword from the Menu system, first paste it into the scratch pad.)

2 Press (F1), use the cursor arrows to highlight **Glossary** on the mini-menu, and press (↵). The system then displays a search string box at the bottom of the screen, filled in with the word that's under the cursor.

3 If the displayed word isn't the one you want to look up, type the desired word over the one that appears in the box. Otherwise, just press (↵).

If the word is in the glossary, the system then displays its definition. If the word isn't in the glossary, the system displays the closest alphabetic match.

- The bottom line of the display shows the terms that precede and follow the displayed one in the glossary. Press (Ctrl) (PgUp) or (Ctrl) (PgDn) to display these definitions.
- If you don't understand a word used in the definition, you can *paste* the whole definition into the editing window, put the cursor on the problem word, and ask for *its* definition. (F8) copies the definition into the active editing window. The definition appears in the window as a marked area, so that you can easily copy or move it to another place.
- You can also paste the definition into the listing file to annotate your output for later reference. (Remember to save the listing file if you make editing changes like this.)

MANAGING FILES

SPSS/PC+ writes six types of files: the scratch pad, listing files, log files, system files, portable files, and results from procedures.

- The *scratch pad* contains the commands you type or paste into it. It's name is SCRATCH.PAD. This file is saved automatically whenever you leave REVIEW, and when you leave SPSS/PC+.
- The *listing file* contains your display output. By default this output is sent to the screen and to the SPSS.LIS file on disk. You can specify a different disk file. You can also have this output sent to the printer.
- The *log file* contains a log of the commands executed by SPSS/PC+, together with messages about commands that caused errors, and about the location of command output in the listing file. The default log file is SPSS.LOG.
- A *system file* saves data and a data dictionary for use in subsequent SPSS/PC+ sessions, after you use the SAVE command. The default system file, when you use the SAVE command without specifying another filename, is SPSS.SYS.
- The *portable file* is used to transport data and a data dictionary across machines, when you use the EXPORT command. There is no default portable file.

- The *results file* includes results (matrix materials, new data, etc.) from specific commands: WRITE, CORRELATION, CLUSTER, QUICK CLUSTER, FACTOR, REGRESSION, ONEWAY, DSCRIMINANT, and MANOVA. The default results file is SPSS.PRC.

At the beginning of each session, the default files SCRATCH.PAD, SPSS.LIS, and SPSS.LOG are reinitialized. This means that the scratch pad, listing, and log files from any previous session are lost unless they were directed to a file other than the default (see Command Reference: SET) or unless they have been renamed using the DOS *RENAME* command (see "DOS Commands" in your DOS manual). The default system file SPSS.SYS and results file SPSS.PRC are not automatically reinitialized. However, each time you write to these files, any existing contents are *replaced* by the new material. To avoid this, you can specify a name explicitly on the SAVE command (OUTFILE='filename') or you can specify SET RESULTS='filename'. Alternatively, you can use the DOS *RENAME* command at the end of a session to give these files a new name that will not be overwritten.

SPSS/PC+ reads from and writes to the current directory unless you specify otherwise. To specify a file in a different directory, you must follow DOS conventions for directory names (see "Using Tree-Structured Directories" in your DOS manual). Portable files and the results file must always be in the current directory.

REVIEW: The SPSS/PC+ Editor

REVIEW is the SPSS/PC+ text editor. As explained in the previous chapter, it is particularly designed to give you special help for entering SPSS/PC+ commands and for viewing their output, by means of the Menu and Help system, and the scratch pad, listing, and log files. But it is also very useful simply for standard editing tasks, such as creating, editing, or browsing through *any* text file. This section shows you how to perform standard text-editing tasks with REVIEW.

Note that not all REVIEW commands are described in this chapter. For a complete list, see REVIEW in Part C.

WORKING WITH A DATA FILE

A text file can serve a variety of purposes. One common use of a text file is to hold the data that you want to analyze with SPSS/PC+. For this session, we'll enter data from a fictitious survey of people's reactions to various colors for the SPSS/PC+ manual cover.

Getting into REVIEW

When you want to use REVIEW simply as a text editor, you can get into it directly from DOS, without using SPSS/PC+ at all. To do this, at the DOS prompt you simply type SPSSPC/RE followed by a space and the name of the file you want to edit. If you want to edit two files—one in the top window and one in the bottom—you simply enter both file names; the file entered first appears in the top window. (You can also get into REVIEW from within SPSS/PC+ by typing the word REVIEW followed by the file name(s) *in apostrophes* at the SPSS/PC+ command prompt. This overrides the default scratch pad and listing files, and gives you the file(s) you named.)

For this example, at the DOS prompt type:

```
SPSSPC/RE COLORS.DAT
```

When you press ⏎, you see the SPSS/PC+ logo screen and then the REVIEW screen, with the cursor positioned at the top. (The screen is initially blank, since we are creating a new file.) Since we're only editing one file, REVIEW shows only one window, which takes up the whole screen.

Entering Data

The data we want to enter in this file consist simply of numbered codes representing colors and respondents' reactions to them. We'll put the color codes in the first column of each line, and the reaction codes in the third column of each line. Type the data shown below. Press ⏎ to start each new line. If you make mistakes, ignore them for now; we'll see in a minute how to make corrections.

```
1 1
1 1
1 2
1 2
1 3
1 3
2 2
2 2
2 2
2 2
2 2
2 2
3 1
3 1
3 1
3 3
3 3
3 3
```

Moving Around the Screen

When you have entered the data (mistakes and all), you can experiment with moving the cursor around the screen. To do this, use the cursor-movement keys on the right side of your keyboard: (Home), (End), (↑), (↓), (←), and (→). These keys will operate correctly only when they are not locked into *numerical mode,* a mode that is indicated by the appearance of **Num** in the bottom-right corner of your screen. To get out of numerical mode, press (NumLck); the **Num** message then disappears from your screen.

As you move the cursor, notice that the column number in the lower-right corner of the screen changes to correspond to the cursor location. Notice also that the cursor will not move right of the last character in any line. If your file had lines of unequal length and you moved the cursor up or down, you would find that the cursor always tried to stay in the column in which it started—but if the line it was moving onto had fewer columns, it would move left to the last column in that line and then move back to the right when it reached a longer line.

The following keys allow quick cursor movement. (Since your file isn't very large at the moment, you won't be able to experiment with all of them now, but you'll certainly find them useful later.)

- (Tab) moves the cursor right to the next tab stop on the line. A tab stop occurs every eight characters. However, the cursor won't move to a tab stop that's beyond the last character in a line.
- (⇧) (Tab) moves the cursor left to the preceding tab stop on the line.
- (Ctrl) (→) moves the cursor to the end of the line.
- (Ctrl) (←) moves the cursor to the beginning of the line.
- (↵) moves the cursor to the start of the next line when you are in *overtype mode* (see "Insert Mode versus Overtype Mode," below).
- (PgDn) moves the cursor down one screen.
- (PgUp) moves the cursor up one screen.
- (Ctrl) (Home) moves the cursor to the top of the file.
- (Ctrl) (End) moves the cursor to the bottom of the file.

Making Changes

It's easy to make changes or corrections in your file using REVIEW. The following section shows you a variety of ways of doing this.

Deleting Characters and Joining Lines

To delete unwanted characters, you either put the cursor on the character and press the (Del) key, or put the cursor immediately to the right of the character and press the backspace key. Try this on anything you mistyped while entering the data. (If you didn't make any mistakes, type some characters at the end of the file.) You'll find that the character under or to the left of the cursor disappears, and the rest of the text on the line moves over one space to fill in.

If you press (Del) when the cursor is on the space at the end of a line, the next line jumps up to join the line the cursor is on. Try this by putting the cursor at the end of the first line and pressing (Del). (We'll see below how to split these lines again.)

Insert Mode versus Overtype Mode

There are two modes of editing in REVIEW: insert mode and overtype mode. In *insert mode,* the text you type is inserted to the left of the current character, and the text to the right of the cursor moves over to make room for the new text. In *overtype mode,* the text you type replaces text in the current line.

When you are in insert mode, the message **Ins** displays in the lower-right corner of your screen. In overtype mode, the **Ins** message disappears. Press (Ins) to switch from one editing mode to the other.

It's easy to see how this works. Press (Home) to move to the top of the screen. Make sure you're in insert mode; then type

```
jjjj
```

The numeric codes that were already there move over to make room for the letters.

Now press (Ctrl) (←) to get back to the beginning of the line. Press (Ins) to get into overtype mode; then type

```
ssss
```

Because you're in overtype mode, the *s*'s replace the *j*'s.

Only one editing key behaves differently depending on the mode. In overtype mode, (↵) moves the cursor to the beginning of the next line. In insert mode, (↵) splits the line the cursor is currently on, at the cursor location. To try this, make sure you're still in overtype mode, and press (↵); the cursor simply moves down to the next line. Now put the cursor under the **1** on the first line, and get into insert mode by pressing (Ins). This time when you press (↵), the line splits, with the part that was to the right of the cursor moving down to become the next line. Use this same method to split the line joined above with the (Del) key, putting the second set of codes back on their own line.

While it is impossible to enter more than 80 characters on a line, you can create lines longer than 80 characters by joining two lines together. However, the only way to view the part of the line that extends past the 80th column is to split the line again. It is good practice to avoid such long lines. SPSS/PC+ does not read commands past the 80th column.

Editing Functions

You perform many editing functions in REVIEW by pressing a function key and then making a selection from a mini-menu that appears at the bottom of the screen. For example, when you press (F4) you get a small menu that says:

```
lines: Insert after   insert Before   Delete   Undelete
```

- The most likely choice is highlighted. Just press (↵) to select and execute it. To select a different command on the menu, use the cursor arrows to move the highlighting, and then press (↵).
- To avoid executing any of the commands on a mini-menu, press (Esc).
- Only relevant commands appear in the mini-menu.
- One letter (usually the first) of each menu item is capitalized. Instead of selecting an item by highlighting it and pressing (↵), you can select it by just pressing its capitalized letter. For example, to insert a line after the one where the cursor is, press (I); to delete the current line, press (D), and so on.
- As alternatives to the mini-menus themselves, some commands let you use a combination of the (Alt) key and the selection key from the mini-menu. For example, to insert a line, you can hold down (Alt) and press (I), instead of pressing (F4) and selecting **Insert after** from the menu. See REVIEW in Part C for these commands and key combinations.
- If you are an experienced SPSS/PC+ user, you can still use all the old key combinations if you have learned them. For example, (Ctrl) (F4) deletes the current line without going through the mini-menu. Only the unmodified function keys call up the mini-menus.
- Function keys that perform editing functions (such as (F4)) are not available when the Menu and Help system is active. Press (Alt) (E) before using one of these function keys if you need to.

Put the cursor anywhere you like and try some editing functions. Note that if you accidentally delete a line you wanted to keep, REVIEW gives you a chance to get it back—by pressing (F4) and selecting **Undelete.** However, you can only restore the most recently deleted line. If you accidentally delete several lines, you can only restore the last one deleted.

Working with Areas

Some REVIEW editing functions operate on entire areas of lines or parts of lines. In this section we'll illustrate how to perform editing tasks—like copying, moving, and deleting—on areas.

To mark an area for editing, you press (F7) at one boundary of the area, select the type of area you want to work with, and press (F7) again at the other boundary. It doesn't matter whether you start at the beginning or ending location.

For example, let's work with an area consisting of the last six lines of your data file.

1 Press (End) to move the cursor to the last line.
2 Press (F7) followed by (↵) to select the highlighted item, *Lines,* marking this line as one boundary of the area. The line begins to flash. On IBM PCs and some other computers, it will also be highlighted. The message

   ```
   Waiting for second line mark
   ```

   appears at the bottom of the screen.
3 Move the cursor up six lines to the first line that contains **3 1**.
4 Press (F7) to mark this as the first line of the area.

The message **Area marked - 6 lines** appears at the bottom of the screen. If your computer has highlighting, the entire six-line area will be highlighted. We can now invoke any of the REVIEW commands that operate on areas.

Copying Areas. First let's put a copy of the area at the end of the file. Move the cursor to the last line of the file by pressing (Ctrl) (End). To insert the copy of the area starting on the line following the cursor, press (F8) followed by (↵) to select **Copy.** The copy then appears. Since the copy took more lines than were left on the screen, the screen instantly scrolls to make room for them.

Notice that the original area is still marked. Another area command would affect this original area, not the copy.

Moving Areas. Now let's move the original area to another location, at the top of the file. First move the cursor to the top of the file by pressing (Ctrl) (Home). Then move the block to the line following the cursor by pressing (F8), using the right cursor arrow to highlight **Move,** and pressing (↵). The original marked area moves from its location and reappears just below the first line of the file. The copy that we made is still at the end of the file.

Note that the original area is still marked, even after its move to the top of the file. We can therefore continue to work with this same area in its new location.

Deleting Areas. This time, let's try deleting it: to do this, press (F8), highlight **Delete,** and press (↵). Notice that, while the original area at the top of the file disappeared, the copy that we made at the end of the file did not. The Delete Area command, like all the area commands, applies only to the currently marked area, not to copies.

Unmarking Areas. When you're done working with an area, press (F7) a third time. When a marked area already exists, (F7) unmarks it. You can try this by marking any area of lines you like (press (F7) followed by (↵) at one line, and (F7) again at the other line); then press (F7) once again. The highlighting of the marked area disappears and the message **Area cancelled** appears at the bottom of the screen.

Rectangular Areas. The area that you mark does not need to consist of complete lines. Instead, it can be a rectangle made up only of selected columns within lines. Marking a rectangle is similar to marking lines—you place the cursor at one corner of the rectangle, press (F7) and select **Rectangle.** Then place the cursor at the diagonally opposite corner of the rectangle and press (F7) again.

All of the area commands work for rectangles. To copy or move a rectangle, you place the cursor at the *top left* corner of the area to which you want to copy or move it, press (F8) and then select **Copy** or **Move**. If REVIEW is in *insert* mode, copying or moving the rectangle pushes existing text to the right. If REVIEW is in *overtype* mode, copying or moving overlays existing text, replacing it. If the copy or move location is to the right of existing lines, REVIEW adds spaces as needed to the ends of existing lines to keep the rectangle aligned.

Working with Two Files

When you are editing two files, the copy and move functions—including those for areas—can operate between the files. That is, you can copy or move an area from one file into the other. The procedure is the same as working within one file—mark the area in one file, position the cursor in the other file (by pressing [F2] followed by [↵] to select **Switch windows**), and press the appropriate function key.

If you want to see more text in one window than in the other, you can change the size of the windows. Press [F2], highlight **Change window size**, and press [↵]. Then fill in the number of lines you want to see in the upper window (the smallest number allowed is 7). After you press [↵] again, the system adjusts both windows accordingly.

The Help Displays

If you don't remember which function key does what, you can press [F1] and select **Review help** to see the following help screen:

```
                       Guide to Review Function Keys
Information    F1  Review Help and Menus, Variable and File Lists, Glossary
Windows        F2  Switch Window, Change Window Size
Input Files    F3  Insert File, Edit Different File
Lines          F4  Insert, Delete, Undelete
Search&Replace F5  Search for Text, Replace Text
Go To          F6  Area, Output Page, Line in Error, After Last Line Executed
Define Area    F7  Mark/Unmark Lines, Rectangle, or Command
Area Actions   F8  Copy, Move, Delete, Round Numbers, Copy Glossary Entry
Output File    F9  Write Area or File, Delete File
Run            F10 Run Commands from Cursor or Marked Area, Exit to Prompt

                       Guide to Menu Commands
ENTER (↵)          Paste Selection & Move Down One Level in Menu
TAB or →           Temporarily Paste Selection & Move Down One Level
ESC or ←           Remove Last Temporary Paste & Move Up One Level
Alt-ESC            Jump to Main Menu (also Ctrl-ESC)
Alt-K              Kill All Temporary Pastes
Alt-T              Get Typing Window
Alt-E              Switch to Edit Mode
Alt-M              Remove Menus
Alt-X              Switch between Standard and Extended Menus
Alt-Cursor Pad     Scroll Help Windows and Glossary (if NumLock off)
                                                                      01
Enter command or press F1 for more help or Escape to continue
```

The top half of this display shows the functions associated with each key. You can call up this screen whenever you aren't sure which function key to press. (For full descriptions of all of REVIEW's functions, see REVIEW in Part C.)

Commands are grouped by function so that related commands are all accessed with the same function key. For example, all the line commands (inserting, deleting, and undeleting lines) are accessed with [F4].

If you press [F1] again while viewing the help screen, you see a second help display showing the keys that allow you to move around in REVIEW.

A number of commands prompt you for additional information when you issue them. These prompts always appear at the bottom of your screen. You can press [F1] in response to any prompt to get a fuller description of the prompt. REVIEW will tell you exactly what information it is requesting and will list possible responses.

Entering Extended ASCII Characters

You can enter extended ASCII characters by activating [NumLck] and holding down the [Alt] key while you type the character's ASCII numeric code on the numeric key pad. The character appears when you release the [Alt] key.

Saving the File

After you've entered the data and made any corrections, you can save the file. (If you were to leave REVIEW without saving the file, everything you've typed would be lost.) To save the file, press (F9) and select **write Whole file.** REVIEW asks for confirmation that you want to save the file under the same name you started with; to do so just press (↵). (If you wanted to give the file a different name, you could just type it in.) REVIEW then saves the file and displays a message when it's done.

OTHER FILES

In this session we've used REVIEW to create and edit a data file. However, REVIEW is also useful for editing *any* medium-sized (up to several hundred lines) file, including:

- Files that hold commands, like the scratch pad and log files, or any other file into which you've entered commands. However, when you run REVIEW outside of SPSS/PC+ you can't *execute* commands in the file, and you can't see a variable list, since there is no active file.
- The listing file, which holds SPSS/PC+ output.

The SPSSPROF.INI File

Another file you may want to create and edit with REVIEW is the file called SPSSPROF.INI, which is the *automatic profile*. This file contains commands that are automatically executed whenever you get into SPSS/PC+. Usually these are SET commands, which control the way SPSS/PC+ operates. Since these commands are executed automatically when you enter the system, they allow you to begin with the environment you prefer. You can learn about SET commands in Part C or under the Main Menu selection **session control & info**.

To edit the automatic profile, specify its name as you get into REVIEW, and then use any of REVIEW's editing functions to add or change the commands you want, or delete the ones you don't want.

File Limitations

In general, REVIEW can edit ASCII files that are small enough to fit into the available memory (RAM) of your computer. Note that REVIEW itself—and any other software currently loaded into memory—occupies some of your computer's RAM. REVIEW *cannot* edit

- Binary files. If a file has graphics characters in it when you use the DOS *TYPE* command to display it on your screen, it is probably a binary file. (SPSS/PC+ system files are binary files.)
- Files too large to fit in the working memory (RAM) available to REVIEW. REVIEW tells you if it does not have enough memory. (Operations such as scrolling may slow down noticeably as REVIEW runs out of memory.)

Important DOS Concepts

SPSS/PC+ is specifically tailored to the IBM PC/XT, PC/AT, and closely compatible computers running Release 2.0 or later of DOS. To use such a computer effectively you must acquire a basic familiarity with the concepts and commands of DOS. You should not rely on this manual for such information. Read at least the chapters in your DOS manual on files and file names, and on using directories. Browse through the descriptions of important commands such as *COPY, DEL, RENAME, TYPE, BACKUP,* and *DISKCOPY.* After reading about directories, note how to work with them using the *MKDIR* (or *MD*) and *CHDIR* (or *CD*) commands. The *CHKDSK* command provides useful information about both disk space and memory (RAM).

In this chapter, you will find information on DOS as it is used with SPSS/PC+. This discussion is *not* adequate to make you proficient in the use of your computer. You do not need to be an expert on DOS to use SPSS/PC+, but a little time invested in learning basic concepts from the DOS manual will make your work easier and more efficient. If you are already familiar with DOS, you may wish to skim this section to see how the SPSS/PC+ system fits in.

DOS is an operating system, a control program that manages the hardware functions of your computer such as accepting commands from the keyboard, writing to the screen, and managing the creation and use of files on disk. When you use an application program such as SPSS/PC+, the most important concepts to understand are those involving disk files.

DOS FILENAMES

The files that contain information stored on your hard disk, or on floppy diskettes, are referred to by name. When you create a file yourself, you assign a name to it. When SPSS/PC+ creates a file it assigns a name, either according to your specifications or using a built-in default. A complete file specification can contain any or all of the following parts, although you rarely have to specify them all.

drive name A single letter indicating which of the disk drives connected to your computer holds the disk on which the file is stored. When you specify this, follow it immediately by a colon (:) to show that it is a drive name. Drive names vary, but by far the most common convention is for A: (and perhaps B:) to be the names of floppy-diskette drives, and for C: to be the drive containing the hard disk.

pathname A specification for the directory containing the file. Pathnames are discussed in "The Path Command," below.

filename A name from one to eight characters that identifies the file. This is also called the "primary filename," since you must always include the extension (perhaps using a wildcard) when specifying a file whose name has an extension. Letters and numbers are allowed in filenames. Blank spaces are not allowed *in the middle* of a filename. See your DOS manual for a complete discussion of the characters that are permitted in filenames.

extension A specification from one to three characters, typically used to indicate what kind of information the file contains. The extension is always separated from the filename by a period. For example, SPSS.LIS has filename SPSS and extension LIS.

In fact, most references to files use only the filename and extension. This is understood to mean that the file is on your *default drive* and in your *current directory,* as explained below.

Wildcards

You can often use DOS "wildcard" characters when specifying filenames or extensions. Use of these characters, the asterisk and the question mark, is explained in the DOS manual. Basically, the asterisk represents *any* character or characters at the end of a filename or extension, and the question mark represents any single character in a filename or extension. Thus

```
erase *.bak
```

erases all files with the extension BAK in the current directory, and

```
copy a:*.* c:
```

copies all files (any name, any extension) from the floppy diskette in the A: drive onto the hard disk in C:. Consult the DOS manual for more information on these shortcuts.

THE DEFAULT DRIVE

When you are running DOS or a DOS program (including SPSS/PC+), a "default drive" is always assigned to you. DOS displays the letter associated with this drive in its command prompt and assumes that any file for which you do not explicitly specify a drive resides on the default drive. If your default drive is A: then the DOS command prompt may look like this (perhaps with some additional information):

```
A>
```

To make another drive the default, simply type in its letter followed by a colon, and DOS will confirm what you have done by using the new drive in its command prompt.

```
A> c:

C>
```

DIRECTORIES AND PATHS

A DOS directory is simply a subset of the contents of a disk. Directories are particularly convenient on hard disks, which can contain hundreds of files. They are less often used on floppy diskettes, although they are perfectly legal. Since directories can be nested inside other directories, you have to specify a "path," *either explicitly or implicitly,* to tell DOS where to find a particular directory. Advantages of using directories include:

- A directory listing of a group of related files will often fit on a single screen, while a listing of all the files on a hard disk would be too extensive to be useful.
- You can use the same name (for example, SPSS.LIS) for different files as long as they are in different directories.
- Different people can share a machine without getting in each other's way by establishing individual directories to hold their files.
- Valuable software, such as SPSS/PC+, can be saved in a directory that is never used as anyone's default directory, to reduce the likelihood of accidental damage. To delete a file that is not in your default directory, you have to include a pathname on the DEL or ERASE command.

Directories can contain not only files but also other directories (which can contain other directories, and so on).

The Current Directory

Just as you always have a default drive in DOS, you always have a *current directory.* If you name a file without specifying a directory, DOS assumes that the file is in your current directory. When you start up a system that initially assigns the hard disk as your default drive, you will be in the its "root directory," which is the main directory of the whole disk. Operating from the root directory, you do not even need to be aware that directories exist. When your disk contains a large number of files, as all hard disks eventually do, you should create and work from one or more smaller directories.

It is a good idea always to be aware of your current directory. If you enter the DOS command,

```
prompt $p$g
```

then the DOS command prompt will always remind you of your current directory. See "Using Batch Files to Avoid DOS," below, to find out how to have this command entered for you automatically so that you don't have to remember it.

DOS Commands for Directories

The most important DOS commands for using directories are

MKDIR or MD *Make directory.* Use this command to create a new directory. The short form, *MD,* is easier to spell and works just as well.

CHDIR or CD *Change directory.* Use this command anytime you want to change your current directory. Unlike other directory commands, *CD* is typically used at least once in a session.

RMDIR or RD *Remove directory.* This command deletes a directory. It can only be used after everything in the directory has been deleted.

Example. Here the DOS prompts are in upper case, while commands entered by the user are shown in lower case.

```
A>c:

C>prompt $p$g

C:\>md myfiles

C:\>cd myfiles

C:\MYFILES>
```

- The first command sets the default drive to C:.
- The *PROMPT* command changes the command prompt to show the current directory. Since the user is still in the root directory, the current directory is simply represented by a backslash (\), as discussed in the next section.
- The *MD* command creates a new directory named MYFILES.
- The *CD* command establishes MYFILES as the current directory. Notice that the final prompt from DOS displays the current directory.

Paths

A path specification is simply a way to indicate a particular directory. You use path specifications on the directory commands discussed above, or whenever you need to name a file that is not in your current directory. The basic rules are simple, but you should consult your DOS manual for more information.

- A path specification consists of one or more directory names, separated by backslashes (\). If you name more than one directory they are nested, with the one named first including them all.
- The directories in a path specification must already exist (except, of course, on the *MD* command that creates them).
- A path specification does not include any blank spaces.
- If the first character of the path specification is a backslash, DOS understands the path to begin at the root directory.
- If the first character of the path specification is not a backslash, DOS understands the path to begin in your current directory (see the example below). This means that the path can only indicate directories *inside* your current directory. (If your current directory is the root, a backslash in the first character of a path specification is optional.)
- If you are using the path specification to identify a particular *file,* rather than a directory, put a backslash after the name of the directory that contains the file and follow this immediately with the filename and the extension, if any.

Example. This example first shows the *DEL* (delete) command using a complete file specification, including drive, path, primary filename, and extension. As before, the DOS prompt is shown in upper case, while the command is shown in lower case. The file is in a directory named DATA, which in turn is in a directory named WORK. Since this is a complete specification, it means the same thing regardless of what your default drive or current directory is:

```
A>del c:\work\data\employ.dat
```

If the default drive is C: but the root is still the current directory, the specification is:

```
C:\>del work\data\employ.dat
```

If, instead, the current directory is WORK, the same command can be entered like this:

```
C:\WORK>del data\employ.dat
```

Finally, suppose that the current directory is the DATA directory inside the WORK directory:

```
C:\WORK\DATA>del employ.dat
```

As long as you are working with files in your current directory on your default drive, you do not need to specify anything but the filename and extension. For most people this is the typical situation.

THE PATH COMMAND

The *PATH* command enables you to dispense with the effort of specifying (and remembering) the paths to commonly used files in other directories. Often the most convenient way to issue this command is through the AUTOEXEC.BAT file discussed below in "Using Batch Files to Avoid DOS."

Normally, if you enter the name of a program, of a batch file, or of one of the DOS disk-based commands, DOS searches for it only in your current directory and will not find it if it is somewhere else. (The error message is "Bad command or filename.")

One solution to this problem is to specify paths always; another is to copy the SPSSPC.COM file, along with the disk-based DOS command files and other programs and batch files that you use frequently, to your working directory so that you can access them directly.

A better solution to the problem is to use the DOS *PATH* command, which permits you to define other directories that should be searched to find programs, batch files, or disk-based DOS commands that do not exist in your current directory. You can enter paths to several directories, separated by semicolons, on the *PATH* command. (If you enter a second *PATH* command, the directories listed on it completely replace the list of directories on the first *PATH* command.) Directories are searched in the order you list them on the *PATH* command. For example,

```
path \;\spss;\dos
```

will permit you to execute programs, batch files, and disk-based DOS commands that are in the root directory, the \SPSS directory, or the \DOS directory, without explicitly specifying a path.

THE SPSS/PC+ SYSTEM DIRECTORY

The SPSS/PC+ system must be installed in a specific directory on your system. You must specify which directory to use when you install the system. The installation procedure then creates this directory and saves the program modules into it.

Normally, you should use a directory named \SPSS. If you choose a directory other than \SPSS, you must issue the DOS command

```
SET SPSS=path
```

to indicate where the modules can be found. (This is a DOS system environment command, not an SPSS/PC+ command.) For example, to indicate that SPSS/PC+ is installed in a directory named \PCPLUS, specify

```
SET SPSS=\PCPLUS
```

(You can place such a command into your AUTOEXEC.BAT file. You might also wish to put a path to this directory on your *PATH* command. See the next section.)

If the SPSS/PC+ system is on a drive other than your default drive, you must indicate this also:

```
SET SPSS=E:\PCPLUS
```

This DOS command specifies that the SPSS/PC+ system is on the E: drive, in a directory named \PCPLUS.

USING BATCH FILES TO AVOID DOS

You cannot avoid DOS entirely. The more you learn about DOS, the easier your work will be. However, there is always a point when remembering command syntax becomes more trouble than it is worth. Some people reach this point quickly when studying operating-system commands. Batch files, which contain DOS commands or groups of DOS commands that can be invoked with a single word, provide a convenient way to avoid memorizing the names or syntax of DOS commands.

You can create a batch file with any editor or word processor, for example EDLIN, which comes with DOS and is described in the DOS manual, or REVIEW, which comes with SPSS/PC+ and is described in this manual. You simply enter some commands into a file just as you would enter them directly; or if you like, you can use some of the simple programming facilities that DOS supports in batch files.

After you have saved these commands into an ASCII file (a file containing ordinary letters, numbers, and punctuation, rather than special binary codes used only by computer programs), you can execute them by simply typing the filename. The file extension for a batch file must be BAT. You do not need to type this extension when invoking a batch file.

AUTOEXEC.BAT

One batch file, AUTOEXEC.BAT, is particularly useful in conjunction with SPSS/PC+. If you create an AUTOEXEC.BAT file in the root directory of your initial default disk, it is executed automatically whenever you start up your system. As explained in the DOS manual, there are a number of useful commands you can put into this file. Two commands are especially helpful when you are running SPSS/PC+.

PROMPT PG As explained above, this command sets the DOS command prompt to include the path to your current directory. The specification PG is not particularly easy to remember, so this command is well placed in this batch file, where you do not need to remember it.

PATH \SPSS The SPSS/PC+ system is stored in a directory that is usually named \SPSS. If you include this directory on a PATH command in your AUTOEXEC.BAT file, you will be able to run both SPSS/PC+ and REVIEW from any directory, without entering a pathname to it. You may wish to define paths to other directories also; if so, enter all of the paths on the same *PATH* command, separated by semicolons.

Since AUTOEXEC.BAT is searched for and executed immediately after you start up (or "boot") your system, it must be on the startup disk from which DOS is loaded. In most systems with built-in hard disks, this is the hard disk. Systems that require a floppy diskette to be inserted at startup must have the AUTOEXEC.BAT file on that floppy diskette. AUTOEXEC.BAT must always be in the root directory, since that is the current directory at system startup.

SETTING SYSTEM PARAMETERS

When you boot or reboot your system, DOS reads a special file named CONFIG.SYS (if it exists) to customize the system configuration. You can set a number of parameters by creating a CONFIG.SYS file, as explained in your DOS manual. To run SPSS/PC+, you should set up this file to include at least the following two parameters:

FILES Specify FILES=20. This parameter controls the maximum number of open files allowed. The default number, 8, is insufficient when using SPSS/PC+.

BUFFERS Specify BUFFERS=8 to increase the number of file buffers that the system allocates. This can significantly speed up the performance of SPSS/PC+.

Use any editor, such as EDLIN or REVIEW, to create the CONFIG.SYS file. After creating it, reboot the system (hold the Ctrl and Alt keys down and press Del), so that the new parameters will take effect.

Statistics Guide

Contents

Chapter 1 The SPSS/PC+ System

Many purchases come with suggestions for use. Children's blocks, microwave ovens, and woodcutting tools are all accompanied by colorful booklets describing results that can be obtained with them. What should a book accompanying a statistical software package contain? Although the results of statistical analyses are not as photogenic as block structures, gourmet meals, and fine furniture, data analysis is a creative process that can result in important contributions to many different undertakings. Increased profits in business, improved treatments for disease, as well as insights into social phenomena, are often attributable to the careful acquisition and analysis of data.

In this part of your SPSS/PC+ manual, we illustrate the application of various statistical procedures to solve a variety of real problems. The problems and their solutions range from the simple—counting the number of people who die on Mondays, to the complex—searching for salary discrimination. The goal is to introduce the building blocks that can be used alone or in many combinations to analyze and display data.

Before proceeding to a detailed discussion of the hows and whys of data analysis with SPSS/PC+, let's take a quick overview of the types of analyses that can be produced. The data we will use is from a recent *Consumer Reports* evaluation of 35 beers. The beers were rated on overall quality and a variety of other attributes, such as price, calories, sodium, and alcohol content. Each type of analysis discussed in this chapter is explained further in Chapters 7 through 18.

1.1 PREPARING A REPORT

One of the first steps in examining the beer data may be to prepare a report that contains detailed information about each of the beers. Figure 1.1 is an excerpt from output produced by the SPSS/PC+ REPORT procedure. Reports may contain additional information, such as summary statistics. Chapter 18 describes the REPORT procedure in detail.

Figure 1.1 Excerpt from a report produced by REPORT

CONSUMER REPORTS BEER RATING - JULY 1983

RATING	BEER	6-PACK PRICE	AVAILABLE IN U.S.	PRICE CLASS
VERY GOOD	MILLER HIGH LIFE	2.49	NATIONAL	PREMIUM
	BUDWEISER	2.59	NATIONAL	PREMIUM
	SCHLITZ	2.59	NATIONAL	PREMIUM
	LOWENBRAU	2.89	NATIONAL	SUPER-PREMIUM
	MICHELOB	2.99	NATIONAL	SUPER-PREMIUM
	HENRY WEINHARD	3.65	REGIONAL	SUPER-PREMIUM
	ANCHOR STEAM	7.19	REGIONAL	SUPER-PREMIUM
MEAN		$3.48		
GOOD	OLD MILWAUKEE	1.69	REGIONAL	POPULAR
	SCHMIDTS	1.79	REGIONAL	POPULAR
	PABST BLUE RIBBON	2.29	NATIONAL	PREMIUM
	AUGSBERGER	2.39	REGIONAL	SUPER-PREMIUM
	STROHS BOHEMIAN STYLE	2.49	REGIONAL	PREMIUM
	COORS	2.65	REGIONAL	PREMIUM
	OLYMPIA	2.65	REGIONAL	PREMIUM
MEAN		$2.28		
FAIR	BLATZ	1.79	REGIONAL	POPULAR
	ROLLING ROCK	2.15	REGIONAL	PREMIUM
	HAMMS	2.59	REGIONAL	PREMIUM
	HEILEMANS OLD STYLE	2.59	REGIONAL	PREMIUM
	TUBORG	2.59	REGIONAL	PREMIUM
MEAN		$2.34		

1.2 DESCRIBING THE DATA

A simple report just displays the data values. It does not attempt to organize or summarize the data. Several SPSS/PC+ procedures are designed especially for summarizing data. For example, a frequency table contains counts of the number of times a response occurs—the number of men and women in a sample, the number of children in families, or the number of visits to the dentist in a year by the head of the household. Figure 1.2a, which was produced by the SPSS/PC+ FREQUENCIES procedure, is a frequency table of the number of beers rated very good, good, and fair by the *Consumer Reports* panel. Of the 35 beers, 11 (31.4%) were rated very good, 14 (40%) good, and the remaining 10 (28.6%) fair.

Figure 1.2a Frequency table from FREQUENCIES

RATING

Value Label	Value	Frequency	Percent	Valid Percent	Cum Percent
VERY GOOD	1	11	31.4	31.4	31.4
GOOD	2	14	40.0	40.0	71.4
FAIR	3	10	28.6	28.6	100.0
	TOTAL	35	100.0	100.0	

The information presented in a frequency table can also be displayed in a bar chart. Figure 1.2b shows a bar chart of the beer ratings. Each value in the table is represented by a bar whose length is proportional to the number of times the value occurs in the data. The FREQUENCIES procedure can also be used to produce bar charts as well as various statistics that are useful for describing data. FREQUENCIES is described in Chapter 7.

Figure 1.2b Bar chart from FREQUENCIES

```
RATING
                     I
                   1 -------------------------------+
           VERY GOOD I                         11 I
                     -------------------------------+
                     I
                   2 ----------------------------------------+
                GOOD I                                  14 I
                     ----------------------------------------+
                     I
                   3 ----------------------------+
                FAIR I                      10 I
                     ----------------------------+
                     I
                     I.........I.........I.........I.........I.........I
                     0         4         8        12        16        20
                                             Frequency
Valid Cases      35       Missing Cases      0
```

1.3 COUNTING COMBINATIONS OF RESPONSES

A frequency table just counts the number of times various responses occur to a single item. Often, however, it is useful to count the number of times certain combinations of responses occur. For example, you might want to know how many men and how many women answered yes, no, or maybe to a survey question. Or you might want to know the number of fatal, serious, or minor accidents involving standard, compact, and subcompact cars.

Figure 1.3 shows a table that tabulates two items together—the rating of the beer and whether it was light or not. From this table one can see that none of the light beers were rated as very good, 4 were rated as good, and 3 were rated as fair. This type of table, known as a crosstabulation or contingency table, is available with the SPSS/PC+ procedure CROSSTABS. CROSSTABS can also compute a variety of percentages and statistics that indicate how closely two (or more) variables are related. The CROSSTABS procedure is described in Chapter 9.

Figure 1.3 Crosstabulation from CROSSTABS

```
Crosstabulation:        LIGHT
                     By RATING

                        RATING
                Count  I
                       IVERY      GOOD      FAIR         Row
                       IGOOD                             Total
                       I       1I        2I        3I
LIGHT          --------+--------+---------+---------+
                     0 I    11  I    10   I     7   I     28
  NO                   I        I         I         I   80.0
                       +--------+---------+---------+
                     1 I        I     4   I     3   I      7
  YES                  I        I         I         I   20.0
                       +--------+---------+---------+
                Column     11        14        10         35
                 Total   31.4      40.0      28.6      100.0

Number of Missing Observations =          0
```

1.4 SUMMARIZING RESPONSES

Both frequency tables and crosstabulation tables summarize the data by counting the number of times each response occurs. When a response can have many possible values—age, weight, or income—counting the number of times each possible individual response occurs may result in very large tables which are not very useful. Instead of looking at all responses, you might want to group values that are close to one another and see how often such groups of values (people in their twenties, individuals over 250 pounds) occur.

Figure 1.4 shows the distribution of alcohol content for the 35 beers. Each row of asterisks represents a range of alcohol values. As in the bar chart, the length of the row is proportional to the number of times the values occur. For example, the longest row corresponds to alcohol values between 4.25 and 4.75 and represents 17 beers. Such figures, called histograms, can be obtained from the SPSS/PC+ FREQUENCIES procedure.

Figure 1.4 Histogram of alcohol content from FREQUENCIES

```
ALCOHOL  ALCOHOL BY VOLUME (IN %)

   COUNT   MIDPOINT    ONE SYMBOL EQUALS APPROXIMATELY   .40 OCCURRENCES

       1       2.50  ***
       1       3.00  ***
       1       3.50  ***
       3       4.00  ********
      17       4.50  *******************************************
      11       5.00  ****************************
       1       5.50  ***
                    I....+....I....+....I....+....I....+....I
                    0         4         8        12        16        20
                              HISTOGRAM FREQUENCY
```

1.5 SUMMARY STATISTICS

It is possible to summarize the information contained in a histogram even further by calculating single numbers that represent an average or typical value and the amount of spread or variability in the data. Figure 1.5a contains the mean, the mode (the most frequently occurring value), the median (the value above which half the values fall), the variance (a measure of how spread out the values are), and the smallest and largest values for the alcohol content of the beers in the survey. These statistics and many others are calculated in the SPSS/PC+ FREQUENCIES and MEANS procedures.

Figure 1.5a Some summary statistics available from FREQUENCIES

```
Mean          4.577
Mode          4.700
Median        4.700
Variance       .364
Minimum       2.300
Maximum       5.500
```

Although it is informative to know that the average alcohol content of all the beers is 4.58%, you may also want to see if alcohol content is similar for the three beer rating groups. Figure 1.5b shows the average alcohol content for the three ratings. Beers rated as very good had the highest alcohol content (4.9%), while those rated fair had the lowest (4.2%). The good beers were in the middle with an average alcohol content of 4.6%. This type of table, which shows the means of a variable for subgroups of cases, can be obtained from the SPSS/PC+ MEANS procedure (see Chapter 10).

Figure 1.5b Table from MEANS

```
Summaries of   ALCOHOL     ALCOHOL BY VOLUME (IN %)
By levels of   RATING

Variable        Value  Label                          Mean    Std Dev    Cases

For Entire Population                                4.5771      .6030       35

RATING              1  VERY GOOD                     4.9000      .1789       11
RATING              2  GOOD                          4.5786      .4300       14
RATING              3  FAIR                          4.2200      .8954       10

  Total Cases =        35
```

1.6 PLOTTING THE DATA

When you want to examine the relationship between two variables, both of which can have many values, plotting the two variables may be helpful. Figure 1.6a, which was produced by the SPSS/PC+ PLOT procedure, is a plot of the price of the beer and the alcohol content. Each point is also identified by its *Consumer Reports* rating. Note that there does not appear to be a strong relationship between price and alcohol content since there is beer in various price ranges for the values of alcohol content. No "pattern" between cost and alcohol content appears to exist.

Examining plots of several variables together is a valuable step in many analyses. Plots of sales with advertising expenditures, blood pressure with weight, and birth rates with GNP all reveal interesting relationships between the two variables. The PLOT procedure produces a variety of plots that can be used to examine relationships among variables and is described in Chapter 12.

Figure 1.6a Sample output from PLOT

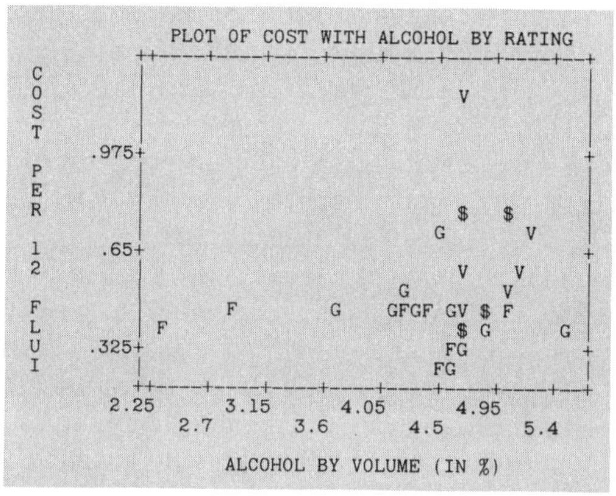

Another way to examine the strength of association between two variables is through indexes such as the correlation coefficient. Figure 1.6b contains correlation coefficients between price, alcohol content, and sodium content. Since these values are small (close to 0), there appears to be no linear association between the pairs of variables. (A linear association is one in which points cluster around a straight line.) The SPSS/PC+ CORRELATION procedure can be used to calculate correlation coefficients and various associated statistics. CORRELATION is described in Chapter 13.

Figure 1.6b Correlation coefficients available from CORRELATION

```
Correlations:  PRICE       ALCOHOL     SODIUM

 PRICE          1.0000       .1961      -.0897
               (    0)     (   35)     (   35)
               P= .        P= .129     P= .304

 ALCOHOL         .1961      1.0000       .2050
               (   35)     (    0)     (   35)
               P= .129     P= .        P= .119

 SODIUM         -.0897       .2050      1.0000
               (   35)     (   35)     (    0)
               P= .304     P= .119     P= .

(Coefficient / (Cases) / Significance)

" . " is printed if a coefficient cannot be computed
```

1.7 TESTING HYPOTHESES AND BUILDING MODELS

SPSS/PC+ also provides statistical tests for evaluating the likelihood of actual differences in a population based on observed differences in a sample. For example, we may test the hypothesis that beers rated very good and fair differ in price or alcohol content—not just for beers in the survey but for all beers. Or we may hypothesize that beers in the three rating categories differ in sodium content. The SPSS/PC+ T-TEST procedure (Chapter 11) can be used to test the hypothesis that two population means are equal, while the ONEWAY and ANOVA procedures (Chapters 14 and 15) are useful for testing hypotheses about several population means.

Since beers are known to vary considerably in cost, it may be interesting to try to predict the cost of a beer based on variables such as alcohol content, number of calories, availability, and origin. The SPSS/PC+ REGRESSION procedure (Chapter 17) is used to develop a model that examines the relationship between a dependent variable, in this case cost, and a set of independent variables. Special facilities for selecting variables to be included in a model as well as testing the adequacy of fit of the model are also available.

The procedures available in SPSS/PC+ Advanced Statistics (HILOGLINEAR, FACTOR, CLUSTER, QUICK CLUSTER, DSCRIMIN-ANT, and MANOVA) offer more sophisticated analytical techniques. You may want to study the associations among categorical variables, that is, variables whose values are categories—such as beer rating, class, and availability. The SPSS/PC+ procedure available for this type of analysis is HILOGLINEAR, a hierarchical loglinear technique. Loglinear models are a class of statistical techniques that model the number of cases in a cell of a multidimensional crosstabulation as a function of the variables used for classification.

Suppose that a number of individuals rated the beer on a number of characteristics, such as lightness, body, color, head, taste when drunk from a bottle, can, or glass, packaging, sizes available, and so on. You would expect that the responses to different items would be correlated. One explanation for the observed correlations is the idea that the items are related because they tap some of the same dimensions on which people rate beer. These dimensions may be quality, value, and accessibility. Factor analysis is one of the statistical techniques that is used to search for these underlying common dimensions, called *factors*. The SPSS/PC+ FACTOR procedure provides several methods for estimating the factors and for making their interpretation easier.

Although there are many different brands of beer in the world, one may wonder whether they are all that different from one another or whether the beers can be lumped into several fairly homogeneous categories. Since there are many attributes which can be used to describe the beers, any subset of these might serve as the basis for looking for similarities. The SPSS/PC+ CLUSTER procedure can be used to calculate "distances" between pairs of beers and then, based on these distances, group the beers into similar categories, called clusters. Several different ways of defining distances between items and forming clusters are available.

When you know in advance the number of clusters to expect, you can use the SPSS/PC+ QUICK CLUSTER procedure instead. QUICK CLUSTER is less flexible than CLUSTER. However, it can process large files efficiently, while the number of cases you can process with CLUSTER is limited by the amount of memory available.

Sometimes you already know the way your cases are grouped, but you want to know why the groupings came out that way. The beers were rated as "very good,' "good," or "fair" in the *Consumer Reports* evaluation. Is there any way to predict these ratings from the other variables available? The SPSS/PC+ DS-CRIMINANT procedure can be used to predict how the beers were rated in quality, on the basis of their price, alcohol content, sodium content, and so on. If the prediction works well for the beers actually rated, you could then use the DSCRIMINANT procedure to try and predict how other beers would have been rated if they had been included in the evaluation.

Sometimes the tests available in procedures ANOVA and ONEWAY are not general enough for a complex analysis. The SPSS/PC+ MANOVA procedure lets you test a wide variety of hypotheses about the effects of categorical or continuous variables on one or more dependent variables. It can handle most types of linear models, including "repeated measures" analyses in which dependent variable(s) have been measured on several occasions.

For more information on HILOGLINEAR, FACTOR, CLUSTER, QUICK CLUSTER, DSCRIMINANT, and MANOVA, refer to the *SPSS/PC+ Advanced Statistics* manual.

1.8 SUMMARY

The SPSS/PC+ product contains many facilities for analyzing and reporting data. For most problems, you will probably want to explore your data using several different techniques. Although it may be tempting to run a lot of different procedures with which you are not very familiar in hope of making sense of the data, this is not a very good tactic. You should instead think about the problem you want to solve, spend some time considering the statistical techniques that may be helpful in arriving at a solution, and only then proceed with the analysis. You should also keep in mind the caveat that the most complicated procedure is not necessarily the best. A little common sense and thought will not only save time but give better results as well.

Contents

Chapter 2 Preparing Data for Analysis

Before information can be analyzed by SPSS/PC+, it must be entered into a disk file. This entails two steps—arranging the data into a suitable format and entering the data into the computer. You can use an editor program to create the file and enter the data. If you are not already familiar with an editor, you may wish to use REVIEW, which is included with your system (see Part A: REVIEW). It is also possible to enter data directly into an SPSS/PC+ command file or to bring in a file that already exists on a mainframe computer. In this chapter we will consider only the first step—taking data which are stored in some form that a computer cannot read and preparing it for analysis.

2.1 CASES, VARIABLES, AND VALUES

Consider Table 2.5a, which contains an excerpt from the *Consumer Reports* report on beers discussed in Chapter 1. Each line in the table represents a *case,* or observation, for which *values* are available for a set of *variables.*

For the first case, MILLER HIGH LIFE beer, the value of the cost variable is 42 cents, and of the alcohol variable, 4.7%. For each beer, the same variables—rating, origin, availability, price, cost, calories, sodium, alcohol content, class, and light (type of beer)—are recorded. What differs are the actual values of the variables. Each case has one and only one value for each variable. "Unknown" and "missing" are acceptable values for a variable, although these values require special treatment during analysis.

The case is the basic unit for which measurements are taken. In this analysis, the case is a brand of beer. In studies of political opinion or brand preference, the case is most likely the individual respondent to a questionnaire. A case may be a larger unit, such as a school, county, or nation; it may be a time period, such as a year or month in which measurements are obtained; or it may be an event, such as an auto accident.

For any single analysis, the cases must be the same. If the unit of analysis is a county, all cases are counties, and the values of each variable are for individual counties. If the unit is a state, then all cases are states and the values for each variable are for states.

2.2 Identifying Important Variables

A critical step in any study is the selection of variables to be included. For example, an employee can be described using many variables, such as place of residence, color of hair and eyes, years of education, work experience, and so forth. The variables that are relevant to the problem under study must be chosen from the vast array of information available. If important variables are excluded from the data file, the results will be of limited use. This point may seem obvious, but it is all too easy to overlook an important variable until you need it for analysis, when it is too late to get the information. For example, if a variable such as years of work experience is excluded from a study of salary discrimination, few—if any—correct conclusions can be drawn. All potentially relevant variables should be included in the study since it is much easier to exclude unnecessary variables from analysis than to gather additional information.

2.3 Recording the Data

Once the variables have been selected, you must decide how they will be recorded. Do you need to record the actual date of birth or can you simply record the age in years? Is it sufficient to know if someone is a high-school or college graduate or do you need to know the actual number of years of education? It is usually a good idea to record the data in as much detail as possible. For example, if you record actual ages, cases can be grouped later into age categories. But if you just record each case as over 50 years or under 50 years of age, you can never analyze your data using any other age categories.

2.4 Coding the Variables

One way to simplify data entry is to assign numbers or symbols to represent responses. This is known as *coding* the data. For example, instead of typing "light" or "regular" as the values for the type-of-beer variable, the codes *1* and *0* can be used. If only numbers are included in a coding scheme it is called *numeric*. If letters or a mixture of numbers, letters, and special symbols are chosen, the code is termed *alphanumeric* or *string*. By coding, you substantially decrease the number of symbols that you need to type, especially for variables whose values are originally recorded as words (such as class of beer). If you want the coded values to be labeled on the output, a few instructions in SPSS/PC+ will take care of it.

Coding schemes are arbitrary by their very nature. The type-of-beer variable could also be coded *R* for regular and *L* for light. All that is necessary is that each possible response have a distinct code. For example, coding the states by their first letter is unacceptable since there are many states that begin with the same letter. Maine, Massachussetts, Michigan, Maryland, Minnesota, Mississippi, Missouri, and Montana would be indistinguishable.

It is usually helpful to have one variable that uniquely identifies each case. For the beer data, that variable is the name of the beer. Sometimes it is useful to identify cases with an ID number. This identifier can help you easily locate the data lines for cases with unusual values or missing information.

2.5 An Example

Table 2.5a shows a portion of the uncoded data from the beer study. A possible coding scheme for this data is shown in Table 2.5b. Figure 2.5a contains data for the first three beers coded according to this scheme. Once the data are coded, a format for arranging the data in a computer file must be determined. Each data line (usually entered from a terminal) is also known as a *record*. Each line is composed of columns in which the numbers or characters are stored. Two decisions that must be made are how many lines will be needed for each case and in what column locations each variable will be stored.

Table 2.5a Excerpt from uncoded data for the beer study

Rating	Beer	Origin	Avail	Price	Cost
Very good	MILLER HIGH LIFE	USA	National	2.49	.42
Very good	BUDWEISER	USA	National	2.59	.43
Very good	SCHLITZ	USA	National	2.59	.43
Good	DOS EQUIS	Mexico	Regional	4.22	.70
Fair	PABST EXTRA LIGHT	USA	National	2.29	.38

Calories	Sodium	Alchohol	Class	Light
149	17	4.7	Premium	Regular
144	15	4.7	Premium	Regular
151	19	4.9	Premium	Regular
145	14	4.5	Not given	Regular
68	15	2.3	Not given	Light

Table 2.5b Coding scheme for beer data form

Variable	Coding scheme
RATING	1=Very good 2=Good 3=Fair
BEER	Actual name of the beer
ORIGIN	1=USA 2=Canada 3=France 4=Holland 5=Mexico 6=Germany 7=Japan
AVAIL	1=National 2=Regional
PRICE	price per six pack of 12 ounce containers
COST	cost per 12 fluid ounces
CALORIES	calories per 12 fluid ounces
SODIUM	sodium per 12 fluid ounces in mg
ALCOHOL	alcohol by volume (in %)
CLASS	0=Not given 1=Super-premium 2=Premium 3=Popular
LIGHT	0=Regular 1=Light

Figure 2.5a Coded data

```
RATING    BEER              ORIGIN AVAIL PRICE COST CALORIES SODIUM ALCOHOL CLASS LIGHT

   1   MILLER HIGH LIFE      1     1     249   42     149     17      47      2     0
   1   BUDWEISER             1     1     259   43     144     15      47      2     0
   1   SCHLITZ               1     1     259   43     151     19      49      2     0
```

Figure 2.5b One-record file

```
         1         2         3         4         5
12345678901234567890123456789012345678901234567890  Columns

1 MILLER HIGH LIFE        1 1 249  42 149 17 47 2 0
1 BUDWEISER               1 1 259  43 144 15 47 2 0
1 SCHLITZ                 1 1 259  43 151 19 49 2 0
```

Figure 2.5b shows a listing of a file in which one line is used for each case. The column locations for the variables are also indicated. Rating is in column 1; the name of the beer in columns 3 through 22; origin in column 25, the availability of the beer in column 27; the price in columns 29–31; the cost in columns 33–35; calories in columns 37–39; the sodium content in columns 41–42; the alcohol content in columns 44–45; class in column 47, and the light or regular designation in column 49. The numbers are positioned in each field so that the last digit is in the last column of the field for the variable. For example, a calorie count of 72 would have the number 7 in column 38; leading blanks or zeros occupy the beginning columns. This is known as *fixed-column format.* (Freefield input is discussed in Section 2.6.) The decimal points for the price, cost, and alcohol variables are not included in the file. The decimal point does not need to be included since SPSS/PC+ commands can be used to indicate its location. If the decimal point is included, it occupies a column like any other symbol.

When there are many variables for each case, more than one line may be necessary to store the information. For example, if your screen width is 80, you may prefer to enter information that requires more than 80 columns on two or more lines. It is usually recommended that you enter an identification number for each case and a record number onto each line if it takes more than one line to record the data for a case. You can then easily locate missing or out-of-order data lines.

It is important to allocate a sufficient number of columns for each variable. For example, if only two columns are used to record a weight variable, only weights less than 100 pounds will fit. Always allocate the maximum number of columns that you might need. Don't worry if your observed data do not actually require that many columns.

All data files considered in this manual are *rectangular*. That is, all cases have the same variables and the same number of lines per case. Some data files are not rectangular. For instance, every case may not have the same variables recorded. So, in a study of adverse drug reactions, cases that are alive will not have a data line detailing autopsy findings. Another nonrectangular file might not define all cases as the same unit, as in a file containing some lines with data about families and some lines with data about individual members within families. Currently, SPSS/PC+ does not contain facilities for handling these kinds of files.

2.6 Freefield Format

Sometimes it is more convenient not to have to worry about arranging variables in particular column locations. Instead, for each case, variables are entered in the same order with at least one blank separating values. Figure 2.6 shows how a freefield data file for the first three cases of the beer data might look.

Figure 2.6 Beer data in freefield format

```
1 'MILLER HIGH LIFE' 1 1 2.49 .42 149 17 4.7 2 0
1 BUDWEISER 1 1 2.59 .43 144 15 4.7 2 0
1 SCHLITZ 1 1 2.59 .43 151 19 4.9 2 0
```

Figure 2.5b differs from Figure 2.6 in several ways. Whenever there is a blank within the name, the name of the beer is enclosed in apostrophes (or quotation marks). This indicates that the blanks are part of the value. Decimal points must be included in the data. Freefield data are discussed in greater detail in the Command Reference under DATA LIST FREE.

2.7 DESIGNING FORMS

When a study is based on data already gathered, there is not much that can be done about the forms on which data reside or how the information is recorded. For example, if education is recorded in categories, the actual number of years cannot be entered into the data file. However, when a study is planned in advance, special forms can be designed that indicate both the type of information to be collected and where it will reside on the computer file. This type of form makes data entry much easier. You can enter the information directly from the form onto a disk using a terminal.

Sometimes data collection forms are designed with space for miscellaneous comments. These comments can be analyzed only if they are coded. For example, if undergraduate major is listed in the comments section it must be coded into a variable. A coding scheme such as 1=physical sciences, 2=social sciences, 3=humanities, 4=engineering, and so forth could be used. Unless the comment section has specific codable information, it cannot be analyzed in any reasonable manner.

2.8 THE DATA FILE

The data file is the most crucial component of any analysis. Unless the data have been carefully gathered, recorded, and entered, all subsequent analyses will be of limited use. Always try to obtain as much of the necessary information as possible for all of the cases that are to be included in a study. A special code standing for missing information should be reserved only for cases where it is impossible to ascertain a certain value. Once the data have been coded and entered, make sure to check the values. Any suspicious values should be confirmed. They may be the result of coding or data-entry errors. Subsequent chapters show how you can use SPSS/PC+ to help locate errors in a data file.

2.9 THE SPSS/PC+ SESSION

After you have prepared your data, you are ready to run the SPSS/PC+ program. The way in which you start the program and enter commands to read, modify, and analyze your data, or to obtain help about the system itself, were described in Running SPSS/PC+, earlier in this manual.

The SPSS/PC+ commands for carrying out analyses follow a simple progression. At the start of a session you need to tell the system how to interpret (and perhaps where to find) your data. You might want to add some labels and print formats to make the output more readable and identify values that stand for missing data. These commands are discussed in Chapter 3. Next you might wish to make some modifications to the data file. For example, if you have recorded age in years (as you were advised to do above), you might want to create a new variable that gives age categories. The commands for this type of operation are in Chapter 4. For a given analysis, you might want to select a particular subset of cases or perhaps a random sample. Commands for selecting and sampling cases are discussed in Chapter 5. Finally, you can use one of many procedure commands to produce a report or a statistical analysis. Chapters 7 through 18 contain detailed information about the procedures available in SPSS/PC+. You can continue modifying the data and running procedures until you end the session with the FINISH command.

Contents

Chapter 3 Defining Data

The data definition commands in SPSS/PC+ answer the following questions:

- Where is the collection of data stored on your machine?
- How many lines are there for each case?
- What are the names of the variables, and where are they located on the data file?
- What labels should be attached to variables and values?
- What values are used to represent missing information?

3.1 DESCRIBING THE DATA FILE

The SPSS/PC+ commands in Figure 3.1 define and produce a listing of data from the beer data. Assume that the data are entered in fixed format. That is, variables are stored in the same column locations for all of the cases. Section 3.7 discusses data definition when data are entered using freefield format.

The first data definition command is DATA LIST, which tells SPSS/PC+ where to find the data and how to read it.

Figure 3.1 Command file for the beer data

```
DATA LIST /RATING 1 BEER 3-22(A) ORIGIN 25 AVAIL 27
PRICE 29-31(2) COST 33-35(2) CALORIES 37-39 SODIUM 41-42
ALCOHOL 44-45(1) CLASS 47 LIGHT 49.

VARIABLE LABELS AVAIL 'AVAILABILITY IN THE U.S.'
/PRICE 'PRICE PER 6-PACK'
/COST 'COST PER 12 FLUID OUNCES'
/CALORIES 'CALORIES PER 12 FLUID OUNCES'
/SODIUM 'SODIUM PER 12 FLUID OUNCES IN MG'
/ALCOHOL 'ALCOHOL BY VOLUME (IN %)'
/CLASS 'PRICE CLASS'.

VALUE LABELS RATING 1 'VERY GOOD' 2 'GOOD' 3 'FAIR'
/ORIGIN 1 'USA' 2 'CANADA' 3 'FRANCE' 4 'HOLLAND' 5 'MEXICO'
6 'GERMANY'  7 'JAPAN'
/AVAIL 1 'NATIONAL' 2 'REGIONAL'
/CLASS 0 'NOT GIVEN' 1 'SUPER-PREMIUM' 2 'PREMIUM' 3 'POPULAR'
/LIGHT 0 'REGULAR' 1 'LIGHT'.

MISSING VALUE CLASS(0).
BEGIN DATA.
1  MILLER HIGH LIFE      1 1 249  42 149 17 47 2 0
1  BUDWEISER             1 1 259  43 144 15 47 2 0
1  SCHLITZ               1 1 259  43 151 19 49 2 0
 .....      Remainder of cases not shown
END DATA.

LIST VARIABLES=RATING TO PRICE CALORIES ALCOHOL /CASES=10.
FINISH.
```

3.2 Locating the Data

You can enter data along with your SPSS/PC+ commands or read data from a separate file. If the data are in a file other than the SPSS/PC+ command file, name the file in which the data are stored with the FILE subcommand, as in

```
DATA LIST FILE='BEER.DAT'
```

BEER.DAT is the name SPSS/PC+ uses to locate the file on which the data are stored. You specify the name of the file in apostrophes (or quotes). If the file is not stored in the current directory, you can give the path name within the apostrophes (see Part A: Running SPSS/PC+). If you enter the data in the same file as the SPSS/PC+ commands, you do not need to use a FILE subcommand (see Section 3.14).

3.3 Choosing Variable Names

After you have identified the data file, you assign names to each of the variables and give their location on the file. You use the assigned variable name to refer to a variable throughout the SPSS/PC+ session. For example, a variable that describes father's occupation might be named PAOCCUP. Keep in mind the following rules when you name variables:

- The name must begin with a letter or the @ symbol The remaining characters in the name can be any letter, any digit, a period, or the symbols _, $, or @.
- The length of the name cannot exceed eight characters.
- Blanks and special symbols such as &, !, ?, /, ', cannot occur in a variable name.
- Each variable must have a unique name—duplication is not allowed.
- The reserved keywords in Table 3.3 cannot be used as variable names since they have special meaning in SPSS/PC+.

The following are all valid variable names: LOCATION, LOC@5, X_1, and OVER$500.

You can create a set of variable names by using the keyword TO. When you are assigning new names, as in DATA LIST specifications, ITEM1 TO ITEM5 is equivalent to five names: ITEM1, ITEM2, ITEM3, ITEM4, and ITEM5. The prefix can be any valid name and the numbers can be any integers, so long as the first number is smaller than the second, and the full variable name, including the number, does not exceed eight characters.

Table 3.3 SPSS/PC+ reserved keywords

ALL	AND	BY	EQ	GE	GT	LE
LT	NE	NOT	OR	TO	WITH	

It is a good idea to assign names that help you identify the variables. You could give the names X and Z to variables for age and sex, but the names AGE and SEX give you a much better idea of the nature of each variable. The variable names assigned to the beer data include RATING for the rating of the beer, ALCOHOL for the alcohol content, PRICE for the price of a six-pack, and CALORIES for caloric content.

3.4 Indicating Column Locations

Along with a variable's name, you specify its column location on the data file. All variables on the same line are identified at the same time. For example, the command

```
DATA LIST FILE='BEER.DAT'
/RATING 1 BEER 3-22(A) ORIGIN 25 AVAIL 27.
```

describes four variables. Variable definition begins with the first slash. The numbers after the variable names give their column locations. For example, RATING is in column 1, and BEER is in columns 3 through 22 (and is alphanumeric, see Section 3.8).

Although variables from the same data line must be defined together, they do not need to be defined in any particular sequence within that line. That is, variables at the end of a line can be defined before those at the beginning of the same line. It is the order in which you define variables that determines their order on your SPSS/PC+ active file, not necessarily their original order on your file.

If several variables are recorded in adjacent columns of the same line and have the same width and format type (numeric or string), you can use an abbreviated format to define them on DATA LIST. List all of the variable names followed by the beginning column location of the first variable in the list, a dash, and the ending column location of the last variable in the list. For example, in the command

```
DATA LIST FILE='HUB.DAT'
/DEPT82 19 SEX 20 MOHIRED YRHIRED 12-15/.
```

MOHIRED and YRHIRED form a list of variables, and 12–15 is the column specification for both. (The second slash is needed for this file to skip a second line of data not being defined, see Section 3.6). The DATA LIST command divides the total number of columns specified equally among the variables in the list. Thus, MOHIRED is in columns 12–13 and YRHIRED is in columns 14–15. Be careful to use variables of equal width when defining data this way. If you use variables of different widths, and SPSS/PC+ can divide the number of columns by the number of variables equally, your data will be read incorrectly. If the total number of columns is not an even multiple of the number of variables listed, SPSS/PC+ displays an error message and does not read the file.

3.5 Establishing Display Formats

Whenever you see the values of a variable displayed, SPSS/PC+ knows what format to use because it knows the variable's width and type from the DATA LIST specifications. This information, along with the variable name, labels, and missing values (see Sections 3.10 and 3.11) form the *dictionary* portion of your SPSS/PC+ active file. Any time that you want to change the format of a numeric variable (string variable formats cannot be changed), use the FORMAT command (see Command Reference: FORMAT).

3.6 Specifying Data Recorded on Multiple Lines

Sometimes your data are located on more than one line or record for each case. To read more than one line for each case, enter a slash and define the variables recorded on the first line, and then enter a slash followed by the the variable definitions for the next data line. Repeat this procedure until you have defined all lines for each case in your data file. For example, the following DATA LIST command defines a personnel file that was entered with two lines per case:

```
DATA LIST FILE='HUB.DAT'
/DEPT82 19 SEX 20 MOHIRED YRHIRED 12-15
/SALARY82 21-25.
```

This DATA LIST reads variables DEPT82, SEX, MOHIRED, and YRHIRED from the first line and SALARY82 from the second line.

3.7 Freefield Data Input

With freefield format, successive data values are simply separated by one or more blanks or one comma. Variables must be in the same order for each of the cases, but they need not be in the same columns (see Chapter 2). If you choose this manner of entering data, specify the keyword FREE after the DATA LIST command. In this case, column locations are not specified after the variable names. However, you must indicate the length of long string variables using the A notation (see Section 3.8). It is probably a good idea to also give the length of short strings so the dictionary format will be correct (the default for short strings is A8). For example,

```
DATA LIST FREE
/RATING BEER (A20) ORIGIN AVAIL PRICE COST CALORIES
SODIUM ALCOHOL CLASS LIGHT.
```

can be used to define the variables for the beer example. Note that if values of string variables include blanks (such as Miller High Life for the BEER variable), they must be enclosed within apostrophes in the data file. Otherwise, the blanks are read as indicating a new variable.

The advantage of freefield format is obvious: data entry is much simpler since variables do not have to be put in particular locations. The major disadvantage of freefield data entry is that if you inadvertently omit a data value, all values for subsequent variables and cases are incorrect. For example, if the rating variable is omitted for the second case, the value for the type of beer is taken as the rating (which will cause an error since it is a string value), and everything that follows is wrong. A similar problem can arise if you mistakenly enter an extra value. Therefore, it is particularly important to list and check the data values after input with freefield format.

Another disadvantage of freefield input is that all numeric variables are assigned dictionary formats of width eight and two decimal places. However, you can use the FORMAT command to assign proper formats following the DATA LIST command.

3.8 Types of Variables

You can define two types of variables with SPSS/PC+: numeric and string (alphanumeric). A numeric variable contains only numbers. Numeric variables can be either decimals (such as 12.345) or integers (such as 1234). A string variable can contain a combination of letters, numbers, and special characters. There are two types of string variables—short strings and long strings. A string variable whose values contain eight characters or less is considered a short string. The variable SEX, coded as F or M, is a short string. In the beer data example, the name of the beer is a long string. The difference is that short strings can be used in several data transformation and procedure commands where long strings cannot. String variables are identified with the letter A in parentheses following the column specification on the DATA LIST command, as in

```
DATA LIST FILE='BEER.DAT' /BEER 3-22(A).
```

where variable BEER is defined as a string variable.

When using freefield format, you should also indicate the width of the string variable, as in BEER (A20). Use the maximum string-value length for a variable as the width. Count all characters and blanks in calculating the width. For example, "Miller High Life" has a width of 16 with blanks included.

3.9 Indicating Decimal Places

By default, DATA LIST assumes that the data format type is numeric and that the numbers are integers, or that any decimal points are explicitly coded. To indicate noninteger values when the decimal point is not actually coded in the data, specify the number of *implied* decimal places by enclosing the intended number in parentheses following the column specification. The specification

```
DATA LIST FILE='BEER.DAT' /ALCOHOL 44-45(1).
```

locates the variable that measures alcohol content in columns 44 through 45. The last digit of ALCOHOL is stored as a decimal position.

For example, if the number 47 is stored in columns 44–45, the specification ALCOHOL 44–45 (1) results in the number 4.7. The specification ALCOHOL 44–45 (2) results in the number 0.47. The dictionary format is also affected by the implied decimal. The two-column designation 44–45 (1) results in a three-column dictionary format in order to accommodate a decimal point. If the number is stored in the data file with the decimal point, the decimal point overrides the DATA LIST format specification (but the dictionary format might have to be adjusted). Implied decimals can only be used with fixed-format data.

3.10 VARIABLE AND VALUE LABELS

The VARIABLE LABELS and VALUE LABELS commands supply information that is used for labeling the output of SPSS/PC+ sessions. These labels are optional, but using them often makes the output more readable. Some variables that have many values, such as age or weight, do not need value labels since the values themselves are meaningful.

The VARIABLE LABELS command assigns variables an extended descriptive label. Specify the variable name, followed by at least one comma or blank, and the label enclosed in apostrophes or quotation marks. Multiple label specifications are optionally separated by slashes, as in

```
VARIABLE LABELS AVAIL 'AVAILABILITY IN THE U.S.'
/PRICE 'PRICE PER 6-PACK'
/COST 'COST PER 12 FLUID OUNCES'
/CALORIES 'CALORIES PER 12 FLUID OUNCES'
/SODIUM 'SODIUM PER 12 FLUID OUNCES IN MG'
/ALCOHOL 'ALCOHOL BY VOLUME (IN %)'
/CLASS 'PRICE CLASS'.
```

This command assigns variable labels to the variables AVAIL through CLASS. A variable label applies to only one variable. The variable must have been previously defined on a DATA LIST, GET, or IMPORT command, or on one of the transformation commands that create new variables. The label can be up to 40 characters long and can include blanks and any other characters.

To use an apostrophe as part of a label, enclose the label in quotation marks, as in

```
VARIABLE LABELS SALARY82 "EMPLOYEE'S 1982 SALARY".
```

Quotation marks are entered in a label in the same manner.

The VALUE LABELS command assigns descriptive labels to values. The VALUE LABELS command is followed by a variable name, or variable list, and a list of values with associated labels. The command

```
VALUE LABELS RATING 1 'VERY GOOD' 2 'GOOD' 3 'FAIR'
/ORIGIN 1 'USA' 2 'CANADA' 3 'FRANCE' 4 'HOLLAND' 5 'MEXICO'
6 'GERMANY' 7 'JAPAN'
/AVAIL 1 'NATIONAL' 2 'REGIONAL'
/CLASS 0 'NOT GIVEN' 1 'SUPER-PREMIUM' 2 'PREMIUM' 3 'POPULAR'
/LIGHT 0 'REGULAR' 1 'LIGHT'.
```

assigns labels to the values for the variables RATING, ORIGIN, AVAIL, CLASS, and LIGHT. The labels for each variable are separated from the labels for the preceding variable by a slash. You can assign labels for values of any variable already defined. If the variable is a string, the value must be enclosed in apostrophes. Value labels can be up to 20 characters long and can contain any characters, including blanks.

3.11 IDENTIFYING MISSING VALUES

Sometimes information for a particular variable is not available for a case. When information about the value of a variable is unknown, a special code is used to indicate that the value is missing. For example, if an patient's age is not known, this can be indicated by a code such as −1 to indicate that the information is missing.

The MISSING VALUE command identifies the value that represents missing information. Specify the variable name or variable list and the specified missing value in parentheses, as in

```
MISSING VALUE CLASS(0).
```

This command assigns the value 0 as missing for variable CLASS.

User-defined missing values specified on the MISSING VALUE command are distinguished from the *system-missing* value (which is indicated on output by a period). SPSS/PC+ assigns the system-missing value when it encounters a value other than a number for a variable declared as numeric on the DATA LIST command. For example, blanks are set to system-missing for numeric variables. An alternative to entering a special value, then, is to leave a field blank. However, you will find that assigning a user-defined missing value gives you more control in tables and other results from SPSS/PC+.

System-missing values are also assigned when new variables created with data transformation commands are undefined, as when an attempt is made to divide by 0 or when a case is missing a value for a variable used in computing the new variable.

3.12 THE ACTIVE FILE

The DATA LIST command defines an *active file*. This is the file you work with during your session. The active file exists only temporarily, unless you use the SAVE command. It consists of a dictionary of variable names and labels, value labels, and missing-value specifications; and also the actual data, whether you entered them interactively or read them in from another file. At any time in your session you can modify the labels (by entering new labels) or the data (with the commands discussed in the next chapter). You can use the SAVE command to write a copy of the active file to disk for later use. When saved to disk in this way, it is called a *system file*.

In addition to modifying the data values or the labels in your active file, you can completely replace it with another active file, or you can combine it with system files that you have previously saved to disk. Any command which defines a new active file will replace the existing active file. (DATA LIST, GET, IMPORT, JOIN, and sometimes AGGREGATE do this.) Use the JOIN command to combine the active file with existing system files.

3.13 LISTING DATA

Once you have defined the data file, you are ready to specify an SPSS/PC+ procedure. SPSS/PC+ procedures are used to tabulate the data, to calculate statistics, and to generate reports and plots. The job in Figure 3.1 requests a listing of the data values for the first 10 cases. The LIST command for this job is

```
LIST VARIABLES=RATING TO PRICE CALORIES ALCOHOL /CASES=10.
```

The subcommand VARIABLES indicates which of the variables are to be displayed (the default is to list all variables). The CASES subcommand indicates the number of observations for which the values are to be listed (the default is to list all cases). LIST uses the dictionary formats to display the variable's values. Therefore, it is a good check on whether a new variable has the proper width and number of decimal places.

Figure 3.13 shows the listing of the first 10 cases. This listing is useful for spotting errors in data entry or data definition. For example, if the wrong columns have been given for the AVAIL variable, strange values will probably appear in the listing. From this listing, you might decide to assign a DOLLAR format to PRICE (see Command Reference: FORMAT).

Figure 3.13 Output from LIST

RATING	BEER	ORIGIN	AVAIL	PRICE	CALORIES	ALCOHOL
1	MILLER HIGH LIFE	1	1	2.49	149	4.7
1	BUDWEISER	1	1	2.59	144	4.7
1	SCHLITZ	1	1	2.59	151	4.9
1	LOWENBRAU	1	1	2.89	157	4.9
1	MICHELOB	1	1	2.99	162	5.0
1	LABATTS	2	2	3.15	147	5.0
1	MOLSON	2	2	3.35	154	5.1
1	HENRY WEINHARD	1	2	3.65	149	4.7
1	KRONENBOURG	3	2	4.39	170	5.2
1	HEINEKEN	4	1	4.59	152	5.0

NUMBER OF CASES READ = 10 NUMBER OF CASES LISTED = 10

You can use the TO convention on a procedure command to refer to variables that are adjacent in your SPSS/PC+ active file (the order in which they were defined on your DATA LIST command). Thus, if you specify VARA TO VARD on a procedure, VARA, VARD, and any variables that have been defined between VARA and VARD are analyzed. For example, the command

```
LIST VARIABLES=RATING TO PRICE CALORIES ALCOHOL.
```

requests a listing for variables RATING, BEER, ORIGIN, AVAIL, and PRICE, plus variables CALORIES and ALCOHOL. See Figure 3.1 for the DATA LIST command that defines these variables.

3.14 INLINE DATA

Sometimes, instead of keeping your data in an external file (that you refer to with the FILE subcommand), you may prefer to enter your data along with your SPSS/PC+ commands. When this is the case, separate the *inline data* from the other lines in the command file with the BEGIN DATA and END DATA commands. The BEGIN DATA command follows the data-definition commands and precedes the data, and the END DATA command follows the last line of the data (see Figure 3.1).

3.15 READING MATRICES

For the SPSS/PC+ procedures FACTOR, REGRESSION, ONEWAY, and CLUSTER, instead of reading the original cases you can enter certain summary statistics such as means, sample sizes, correlations, covariances, or distance coefficients. This results in a considerable decrease in processing time. All statistical computations are based on the summary statistics. (The results you get are the same as if you had entered the original cases, since all of the necessary information is contained in the summary statistics.)

Reading intermediate values instead of the actual cases is useful when you have used SPSS/PC+ procedures to write a file with summary results or when the summary results are available from some other source such as journals.

When you enter summary statistics, special specifications are required on the DATA LIST command (see Command Reference: DATA LIST—Matrix Materials).

3.16 USING AN SPSS/PC+ SYSTEM FILE

Once you have defined your data file in SPSS/PC+, you do not need to repeat the data definition process. Information from the data definition commands described in this chapter can be permanently saved along with the data on specially formatted files called the SPSS/PC+ *system file* and the *portable file.* Variables created or altered by data transformations and the descriptive information for these variables can also be saved on these files.

The system file is used in subsequent SPSS/PC+ sessions without requiring respecification of variable locations, formats, missing values, or variable and value labels. You can update the system file, altering the descriptive information or modifying the data, and you can save the updated version in a new system file. See the SAVE and GET commands in the Command Reference.

The portable file is used to transport your data plus definitions between SPSS/PC+ and SPSS^X on a mainframe computer without having to redefine them each time. See the EXPORT and IMPORT commands in the Command Reference.

3.17 USING SPREADSHEET OR DATABASE FILES

If your data are already in a spreadsheet or database file, you may be able to use the TRANSLATE command to read them directly into SPSS/PC+. TRANSLATE reads

- Spreadsheet files from Lotus 1-2-3 or Symphony.
- Spreadsheet files saved in SYLK format by Multiplan or other programs.
- Database files from dBase-II, dBase-III, or dBase-III Plus.

TRANSLATE can also write your data into a spreadsheet or database file.

It is inefficient to translate the file format each time you use the data, so you should save it in the form of an SPSS/PC+ system file if you plan to analyze the same data repeatedly.

See Command Reference: TRANSLATE for details on using the TRANSLATE command.

Contents

Chapter 4 Data Transformations

In the beer data (see Chapter 1), 35 brands of beer are rated on a three-point scale with the categories "very good,' "good," and "fair." Suppose you want to compare the beers on their ratings, but are only interested in making a dichotomous distinction between ratings of "good" and "fair." To do this, you would want to collapse the rating categories of "very good" and "good" into a single category.

Or suppose you have done a survey of political attitudes, and you have five "yes" or "no" questions on the topic of women's rights. You might want to create a new variable that counts the total number of "yes" responses to the five items.

Operations such as these, where you take existing variables and alter their values or use them to create new variables, are called *data transformations.*

There are four commands in SPSS/PC+ that allow you to perform a wide variety of data transformations. Use the RECODE command to alter the values of an existing variable. Typical reasons for recoding variables include combining several values into one, rearranging the order of categories, or carrying out simple data checks. You would use RECODE to collapse the rating categories in the beer data.

The COMPUTE command creates new variables through numeric transformations of existing ones. For instance, the beer data contains a variable that is the number of calories in 12 ounces of beer. You might want to use COMPUTE to figure out how many calories there would be in an eight-ounce glass of each brand.

The COUNT command creates a new variable that, for each case, counts the occurrences of certain values across a list of variables. You would use COUNT to add the "yes" responses to the women's rights questions on the political survey.

You can use the IF command to transform data differently for subsets of cases. For example, a company may award vacation time on the basis of length of employment. The IF command could be used to calculate vacation time for employees who have been with the company for varying lengths of time.

4.1 RECODING VALUES OF VARIABLES

The RECODE command tells SPSS/PC+ to make specified changes in a variable's values as the data are being read. Take the availability variable from the beer data as an example. The command

```
RECODE AVAIL (1=2)(2=1).
```

reverses values 1 and 2 for variable AVAIL (i.e., if a case has value 1, it is changed to 2, and vice versa).

To be recoded, a variable must already exist in your data file. The variable's name precedes the list of value specifications on the RECODE command. You can create as many new values for a variable as you wish, as long as each specification is enclosed in parentheses. A single specification can be used to recode several values as one new value. Thus, the command

```
RECODE RATING (1,2=1)(3,4=2).
```

changes RATING variable values to 1 for cases with original ratings of 1 or 2, and to 2 for cases originally rated as 3 or 4.

You cannot list more than one *new* value in a single value specification. Thus, specifications like (2,3=0,1) or (1=5,6) cannot be used.

The value specifications on a RECODE command are evaluated from left to right, and the value of a case is recoded only once in a single RECODE command. For example, if a case has the value 0 for the variable SEX, the command

```
RECODE SEX (0=1)(1=0)(3=99).
```

recodes SEX as 1 for that case. This value is not recoded back to 0 by the second value specification. Variable values that you do not mention on a RECODE command are left unchanged.

If you want to recode several variables in the same way, you can use a single RECODE command to do so, as in the command

```
RECODE SEX RACE SURGERY (0=1)(1=0)(3=99).
```

In addition, you can use one RECODE command to perform different recodes for different variables by separating the variable names and their specifications with a slash, as in the command

```
RECODE SODIUM (10 THRU 15=1)(15 THRU 20=2)
/COST (.28 THRU .39=1)(.40 THRU .50=2)(.51 THRU 1.20=3).
```

You can use the TO keyword to refer to several consecutive variables in the file. For example, the command

```
RECODE SCORE1 TO SCORE5 (5=1)(6=2)(7=3).
```

recodes SCORE1, SCORE5, and all the variables between them in the file.

4.2 Recoding Numeric Variables

Several keywords are available to facilitate recoding of numeric variables. Use the keywords THRU, LOWEST, and HIGHEST to recode a range of variables. Thus, the command

```
RECODE CALORIES (68 THRU 100=1)(101 THRU 170=2).
```

recodes all the values between 68 and 100 (inclusive) as 1 and all the values between 101 and 170 (inclusive) as 2 for the CALORIES variable. The LOWEST (or LO) keyword specifies the lowest value of a variable, while HIGHEST (or HI) specifies the highest value. The command

```
RECODE CALORIES (LO THRU 100=1)(101 THRU HI=2).
```

is equivalent to the previous command. When you use LOWEST or HIGHEST to specify a range, user-missing values in the range are recoded, but system-missing values are not changed.

You can use the ELSE keyword to recode all values not previously mentioned into a single category. Thus, the command

```
RECODE CALORIES (LO THRU 100=1)(ELSE=2).
```

is equivalent to the previous two commands. ELSE should be the last specification for a variable, since RECODE will ignore subsequent specifications for that variable. ELSE does recode system-missing values.

You can also use ELSE as a data-cleaning device. For example, if the variables SCORE1 to SCORE5 have only 5, 6, and 7 as legitimate values, you might use the command

```
RECODE SCORE1 TO SCORE5 (5=1)(6=2)(7=3)(ELSE=SYSMIS).
```

to recode the valid values to new values and recode the nonvalid values as system-missing.

4.3 Recoding Missing Values

The MISSING keyword is useful for recoding all missing values (user- or system-missing) to a single value. For example, if −99 is the missing value you have declared for the AGE variable, the command

```
RECODE AGE (MISSING=-1).
```

recodes −99 and any system-missing values for AGE to −1 while leaving the other AGE values unchanged.

You can use the SYSMIS keyword as either an input or output specification. The command

```
RECODE AGE (SYSMIS=-1).
```

recodes the system-missing value to −1.

The SYSMIS keyword as an output specification recodes specified values to system-missing, as in the commands

```
RECODE AGE (MISSING=SYSMIS).
```

and

```
RECODE AGE (-99=SYSMIS).
```

Both commands recode the AGE missing values to system-missing.

You *cannot* use the MISSING keyword as an output specification on RECODE to recode values as user-missing. Thus, SPSS/PC+ does not accept a recode specification like (17=MISSING) for the variable AGE. To classify value 17 for AGE as missing, use the MISSING VALUE command, as in

```
MISSING VALUE AGE(17).
```

When you use the MISSING and SYSMIS keywords to recode missing values to a single new value, that value is not automatically considered a missing value. The MISSING VALUE command is required to define it as such, as in the commands

```
RECODE AGE (MISSING=-1).
MISSING VALUE AGE(-1).
```

4.4 Recoding Continuous Value Ranges

If a variable has noninteger values, some values may escape recoding unless you make certain they are included in a value range. For example, the command

```
RECODE AGE (0 THRU 17=1)(18 THRU 65=2)(66 THRU 99=3).
```

does not recode values between 17 and 18 and between 65 and 66. Thus, values like 17.2 and 65.8 would be left unchanged. You can avoid this problem by using overlapping endpoint values in the specifications, as in the command

```
RECODE AGE (66 THRU 99=3)(18 THRU 66=2)(0 THRU 18=1).
```

Note that the order of the recode specifications has been reversed, since a value is recoded only once into the first specification it meets. Thus, the value 66 is coded as a 3 and is not altered further, even though it serves as an endpoint on the following specification.

4.5 Recoding String Variables

You can use the RECODE command to recode string variables. Only short strings (those containing eight or fewer characters) can be recoded. The keywords LOWEST or LO, HIGHEST or HI, THRU, SYSMIS and MISSING do not apply to recoding string variables.

When recoding string variables, you must enclose all values in apostrophes (or quotation marks). For example, the command

```
RECODE LIGHT ('Y'='A')('N'='B').
```

recodes *Y* into *A* and *N* into *B* for the string variable LIGHT.

When recoding string variables, the values in a specification must be of equal length and must have the same length that you have defined for the alphanumeric variable being recoded. Use blanks to specify the exact string-value length. For example, the command

```
RECODE GRADES ('ABC '='ABCD').
```

recodes the value *ABC* into *ABCD* for the four-character string variable GRADES.

4.6 COMPUTING NEW VARIABLES

The COMPUTE command creates new variables through numeric transformations of already existing variables. COMPUTE names the variable you want to create (the *target variable*) followed by an *expression* defining the variable. For example, the command

```
COMPUTE TOTSCORE=MIDTERM+FINAL+HOMEWORK.
```

defines the new variable TOTSCORE as the sum of the variables MIDTERM, FINAL, and HOMEWORK.

The target variable can be a variable that already exists or a new variable. If the target variable already exists, its values are replaced with those produced by the specified transformation. If it is a new variable, it is added to the end of the dictionary in your active file.

The expression on the COMPUTE command can use existing numeric variables, constants, arithmetic operators (such as + and −), numeric functions such as SQRT (square root) and TRUNC (truncate), the missing value function (VALUE), the cross-case function (LAG), random-number functions, and the date function (YRMODA). For example, the command

```
COMPUTE GRADESCR=.35*MIDTERM+.45*FINAL+.2*HOMEWORK.
```

creates a new variable, GRADESCR, that is the weighted average of the variables MIDTERM, FINAL, and HOMEWORK.

4.7 Transforming String Variables

You can use COMPUTE to create or modify short string variables. A variable can be set equal to an existing string variable or to a string constant, as in the command

```
COMPUTE STATE='IL'.
```

which creates variable STATE with the value *IL* for all cases.

String values and constants must be enclosed in apostrophes or quotation marks.

When you create a new string variable by setting a variable name equal to a string constant, the new string variable is assigned a dictionary format equal to the width of the string constant. Thus, the previous command creates STATE as a two-character string variable. When you create a string variable by setting a variable name equal to an existing string variable, the new variable's dictionary format is the same as that of the original variable.

Leading or trailing blanks must be specified. Once you have created a string variable, all subsequent transformations on the variable must use the width first specified.

4.8 Specifying Arithmetic Operations

The following arithmetic operators are available for transforming numeric variables with COMPUTE:

- \+ *Addition.*
- − *Subtraction.*
- * *Multiplication.*
- / *Division.*
- ** *Exponentiation.*

Arithmetic operators must be explicitly specified. You cannot, for example, write (PROPTAX)(100) instead of (PROPTAX)*100.

You can include blanks in an arithmetic expression to improve readability, as in the command

```
COMPUTE TAXTOTAL = PROPTAX + FICA + STATETAX + FEDTAX.
```

Since fairly complex expressions are possible, it is important to keep in mind the order in which operations are performed. Functions (see Sections 4.9 through 4.10) are evaluated first, then exponentiation, then multiplication and division, and, finally, addition and subtraction. Thus, if you specify

```
COMPUTE NEWRATE=SQRT(RATE1)/SQRT(RATE1)+SQRT(RATE3).
```

the square roots (SQRT) are calculated first, then the division is performed, and the addition is performed last.

You can control the order in which operations are performed by enclosing the operation you want executed first in parentheses. Thus, the command

```
COMPUTE NEWRATE=SQRT(RATE1)/(SQRT(RATE1)+SQRT(RATE3)).
```

produces different results than the previous command, since addition is performed before division.

Operations at the same level, as far as order of execution is concerned, are evaluated from left to right. For example, the command

```
COMPUTE SCORE=( A/B * C ).
```

results in a different value than the command

```
COMPUTE SCORE=( A/ (B * C) ).
```

since in the first command *A* is divided by *B*, and the resulting quantity multiplied by *C*, while the second command first multiplies *B* times *C* and then divides *A* by the resulting quantity.

If you are uncertain about the order of execution, you should use parentheses to make the order you want explicit.

4.9 Specifying Numeric Functions

You can specify numeric functions such as square roots, logarithms, and trigonometric functions in a COMPUTE expression. The quantity to be transformed by such a function is called the *argument* and is specified in parentheses after the function keyword. For example, in the command

```
COMPUTE TOTLCOST=RND(COST * 6).
```

the function RND (round to the nearest integer) acts on the argument COST * 6 to create the new variable TOTLCOST.

The argument can be a variable name, a number, or an expression involving several variables. A numeric function can have only one argument.

The following numeric function keywords are available:

ABS *Absolute value.* For example, ABS(−4.7) is 4.7; ABS(4.7) is 4.7.

RND *Argument rounded to the nearest integer.* For example, RND(−4.7) is −5.

TRUNC *Argument truncated to its integer part.* For example, TRUNC(−4.7) is −4.

MOD10 *Remainder resulting when the argument is divided by 10.* For example, MOD(198) is 8.

SQRT *Square root.*

EXP *Exponential. e* is raised to the power of the argument.

LG10 *Base 10 logarithm.*

LN *Natural or Naperian logarithm.*

ARTAN *Arctangent.*

SIN *Sine.* The argument must be in radians.

COS *Cosine.* The argument must be in radians.

For example, in the command

```
COMPUTE LOGINCOM=LN(INCOME+1).
```

LOGINCOM is the natural logarithm of the expression INCOME+1.

4.10 Other Functions

Also available on COMPUTE are two random-number functions (UNIFORM and NORMAL), the cross-case function (LAG), and the date function (YRMODA).

UNIFORM(arg) *A uniform pseudo-random number.* The number is drawn from a distribution having values uniformly distributed between zero and the value of the argument.

NORMAL(arg) *A normal pseudo-random number.* The number is drawn from a normal distribution with a mean of zero and a standard deviation equal to the argument.

LAG(arg) *Get the value of the previous case for the variable named.*

YRMODA(arg list) *Convert the year, month, and day in the argument list into a day number.* The year, month, and day are specified in that order. The number computed is the number of days since October 15, 1582 (the first day of the Gregorian calendar).

All of these functions can be used with numeric variables. Only the LAG function is available for short string variables.

4.11 The YRMODA Function

The YRMODA function converts a given date into the number of days since October 15, 1582. For example the expression YRMODA(1582,10,15) returns a value of 1. YRMODA(1800,1,1) returns 79337, indicating that January 1, 1800 is 79,336 days after the beginning of the Gregorian calendar.

The time interval between two dates can be calculated by converting each of the dates to day numbers and then subtracting the earlier day from the later one. For example, to calculate an individual's age in years on July 4, 1982, specify

```
AGE=(YRMODA(1982,7,4) - YRMODA(BYR,BMO,BDAY)) /365.25
```

where BYR is the year, BMO the month, and BDAY the day of birth.

The YRMODA function has three arguments, which can be variables, constants, or expressions that result in integer values.

- The first argument can be any year from 1582 to 47516. If you specify a number between 00 and 99, SPSS/PC+ will interpret it to mean 1900 to 1999.
- The second argument is the month, coded from 1 to 13. Month 13 refers to the first month of the subsequent year. For example, YRMODA(84,13,1) specifies January 1, 1985.
- The third argument is a day from 0 through 31. Day 0 specifies the last day of the previous month, regardless of whether it was 28, 29, 30, or 31. Thus, (84,2,0) refers to the last day of January in 1984. This is equivalent to (84,1,31) since January has 31 days.

4.12 Using Functions in Complex Expressions

You can specify more than one function in an argument as well as combine functions with arithmetic operators. Such arguments will be evaluated in the order described in Section 4.8 or in the order specified by parentheses. For example, if the command

```
COMPUTE PCTTAXES=RND((TAXES/INCOME)*100).
```

is used, TAXES is first divided by INCOME, the result is multiplied by 100, and this result is rounded off to the nearest integer to get the new variable PCTTAXES.

4.13 Missing Values

If a case has missing values for any of the variables used in a COMPUTE expression, the case is assigned the system-missing value for the computed variable. For example, if the command

```
COMPUTE AGECUBE=AGE**3.
```

is used, the AGECUBE variable will not be computed for any case with a missing value for AGE.

A case is also assigned the system-missing value for a computed variable when the specified operation is not defined for that case. For example, if the command

```
COMPUTE PCTTAXES=(TAXES/INCOME)*100.
```

is used, a case with the value 0 for INCOME is assigned the system-missing value for PCTTAXES because division by 0 is not defined. If the result of an expression cannot be represented on the computer (even when valid values are used in the expression itself), the system-missing value is assigned to the new variable. The following errors will result in assignment of the system-missing value:

**	A negative number to a noninteger power.
/	A divisor of 0.
SQRT	A negative argument.
EXP	An argument that produces a result too large to be represented on the computer.
LG10	A negative or 0 argument.
NORMAL	A negative or 0 argument.
YRMODA	Arguments that do not form a valid date.

Some arithmetic operations involving 0 produce the same result whether any of the arguments have missing values or not. These operations are shown in Table 4.13a.

Table 4.13a Missing-value exceptions in numeric expressions

Expression	Computed value
0 * missing	= 0
0 / missing	= 0
missing ** 0	= 1
0 ** missing	= 0

SPSS/PC+ tries to evaluate a function using all the information it has, assigning the system-missing value only when there is insufficient information to compute the new variable. Table 4.13b summarizes the ways in which the system-missing value is assigned.

Table 4.13b Missing values in arguments to functions

Function	Assigns system-missing value if
ABS (x)	x is missing
ARTAN (x)	x is missing
COS (x)	x is missing
EXP (x)	x is missing
LG10 (x)	x is missing
LN (x)	x is missing
NORMAL (x)	x is missing
RND (x)	x is missing
SIN (x)	x is missing
SQRT (x)	x is missing
TRUNC (x)	x is missing
UNIFORM (x)	x is missing
VALUE (x)	x is system-missing
YRMODA (x1,x2,x3)	any x is missing
MOD (x)	x is missing
LAG (x)	x is missing

4.14 Including User-Missing Values

The VALUE function on COMPUTE allows you to include user-missing values in a transformation.

VALUE *Ignore user-missing values for the variable specified.* The argument must be a variable name.

Thus, the command

```
COMPUTE TOTAL=VALUE(SCORE1)+VALUE(SCORE2)+VALUE(SCORE3).
```

includes user-missing values in computing TOTAL.

4.15 COUNTING VALUES ACROSS VARIABLES

Use the COUNT command to create a variable that records, for each case, the number of times some value or list of values occurs in a list of variables. For example, the command

```
COUNT FEMINISM=ERA JOBEQUAL POLEQUAL (1).
```

creates the variable FEMINISM, which indicates the number of times in a case the value 1 occurs for ERA, JOBEQUAL, and POLEQUAL. Thus, the value of FEMINISM is 0, 1, 2, or 3. You can count across more than one variable list and more than one value as in the command

```
COUNT FEMINISM=ERA JOBEQUAL POLEQUAL(1) VOTE CAMPAIGN(3,4).
```

which counts the number of times the value 1 occurs in the variables ERA, JOBEQUAL, and POLEQUAL, and the values 3 or 4 in VOTE and CAMPAIGN.

The criterion variable list can include both string and numeric variables, provided they have separate value specifications. String values must be enclosed in apostrophes.

You can specify adjacent variables with the TO keyword and ranges of numeric values with the LOWEST, HIGHEST, and THRU keywords. (You cannot specify any keywords with string variables). More than one variable can be created with a single COUNT command by separating the specifications with a slash, as in the command

```
COUNT PSYCHTIC=PTEST1 TO PTEST10(51 THRU HIGHEST)
/SCHIZPHR=STEST1 TO STEST10(LOWEST THRU 20).
```

The COUNT command counts user-missing values. For example, the command

```
COUNT RACISM=SCALE1 TO SCALE12(LOWEST THRU 5).
```

counts the value of −99 even though it has been defined by the user as missing.

If you want to count missing values, in order to deduct them from your previous sum of the values, specify the SYSMIS or MISSING keyword in parentheses after the numeric variable list. For example, the command

```
COUNT PHYSMISS=AGE WEIGHT HEIGHT(SYSMIS).
```

creates the new variable PHYSMISS which records the number of system-missing values each case has for these variables. The MISSING keyword stands for both user- and system-missing values.

4.16 SPECIFYING CONDITIONAL TRANSFORMATIONS

The IF command allows you to make transformations in your data contingent on logical conditions. IF consists of a *logical expression* in parentheses followed by a *target variable* and an *assignment expression*. For example, the command

```
IF (AGE GE 18) VOTER=1.
```

uses the logical expression "(AGE GE 18)" and the assignment expression "VOTER=1" to assign the value 1 to the target variable VOTER for all cases with AGE values greater than or equal to 18. You construct the assignment expression in the same way as the expression in a COMPUTE statement (see Section 4.6). Thus, you can specify conditional operations like

```
IF (INCOME GE 25000) TAXES=INCOME * .33.
```

SPSS/PC+ evaluates logical expressions as true or false and executes the specified assignment only when the expression is true; otherwise, the system-missing value is assigned.

You can use string variables in IF transformations, as in the command,

```
IF (AGE GE 18) AGEGRP='ADULTS'
```

The same conditions that apply to using string variables with COMPUTE (see Sections 4.5 and 4.7) also apply when using them with IF.

4.17 Specifying Conditions for Transformations

The logical expression on the IF command can be a complex statement involving variables, constants, functions, nested parentheses, and so on. You must include either a relational operator (such as EQ or GE) or a missing-value function (MISSING or SYSMIS) in the logical expression. Other operations and functions are optional. (See Sections 4.8 through 4.12 for a description of other available operators and functions.)

4.18 Comparing Values in a Logical Expression

A *relation* is a logical expression that compares two values using a *relational operator.* For example, the command

```
IF (COST EQ .43) NEWCOST=2.
```

compares the equivalence of the variable COST and the value .43. The following relational operators are available:

EQ *Equal to.* The logical expression is true if the expression on the left is equal to the expression on the right.

NE *Not equal to.* The logical expression is true if the left and right expressions are not equal.

LT *Less than.* The logical expression is true if the expression on the left is less than the expression on the right.

LE *Less than or equal to.* The logical expression is true if the expression on the left is less than or equal to the expression on the right.

GT *Greater than.* The logical expression is true if the expression on the left is greater than the expression on the right.

GE *Greater than or equal to.* The logical expression is true if the expression on the left is greater than or equal to the expression on the right.

The expressions in a relation can be variables, constants, arithmetic expressions, or functions, as in the commands

```
IF (SCORE1+SCORE2 GT TESTA) NEWSCORE=1.
```

and

```
IF (LOGINCOM GE 5) CLASS=1.
```

Use blanks to separate the relational operator from the expressions. You can use parentheses and extra blanks to make a command more readable, as in the command

```
IF ( (SCORE1 + SCORE2)  GE 90) GRADE=4.
```

Parentheses are required around the logical expression.

4.19 Joining Relations

You can join two or more relations by using the logical operators AND and OR. For example, the command

```
IF (HOMEWORK GE 85 AND MIDTERM GE 90 AND FINAL GE 90) GRADE=4.
```

assigns GRADE the value 4 only when HOMEWORK is at least 85 and MIDTERM and FINAL are at least 90. When AND is used, the logical expression is true only when *all* relations joined by AND are true. When OR is used, the logical expression is true when *any* of the relations joined by OR are true. Thus, the command

```
IF ((A EQ 4 AND B EQ 3) OR (A EQ 3 AND B EQ 4)) C=1
```

assigns the value 1 to *C* for any case that has a value of 4 for *A* and 3 for *B* *or* to any case with the value 3 for *A* and 4 for *B*.

You must specify operators and expressions explicitly; the specification (X EQ 1 OR 2) in place of (X EQ 1 OR X EQ 2) is invalid.

4.20 Reversing the Logic of an Expression

The NOT logical operator reverses the true or false status of the expression that immediately follows it. For example, the command

```
IF (NOT RACE EQ 1 AND SEX EQ 0) GROUP=1.
```

assigns value 1 to GROUP for cases where RACE does not equal 1 and SEX *is* equal to 0. This is not equivalent to the command

```
IF (NOT(RACE EQ 1 AND SEX EQ 0)) GROUP=1.
```

which assigns value 1 to GROUP for cases where RACE does not equal 1 and SEX does *not* equal 0.

4.21 The Order of Evaluation

IF evaluates arithmetic operators and functions in a logical expression in the same order as does the COMPUTE command (see Section 4.8). Functions and arithmetic operators are evaluated first, then relational operators, then NOT, then AND, and then OR. In the expression (NOT SCORESUM/5 EQ 10), the value of SCORESUM is divided by 5, the result compared to 10, and the true-false status of this comparison reversed by NOT.

If you specify both AND and OR, AND is executed before OR. For example, in the command

```
IF (HOMEWORK GE 90 AND MIDTERM GE 90 OR FINAL GE 95) GRADE=4.
```

the logical expression is true for a case with HOMEWORK and MIDTERM values of at least 90, or for a FINAL value of at least 95. You can use parentheses to clarify or change the order of evaluation. Thus, the command

```
IF ((HOMEWORK GE 90 AND MIDTERM GE 90) OR FINAL GE 95) GRADE=4.
```

is equivalent to the previous command. The command

```
IF (HOMEWORK GE 90 AND (MIDTERM GE 90 OR FINAL GE 95)) GRADE=4.
```

is not equivalent. In this statement, a case must have a score of at least 90 for HOMEWORK, as well as a value of at least 90 for MIDTERM *or* 95 for FINAL, to be assigned the grade 4.

4.22 Missing Values

If the truth of a logical expression cannot be determined because of missing values, the command is not executed. In a relation with only one relational operator, the logical expression is indeterminate if the expression on either side of the operator has a missing value. For example, if you specify

```
IF (FINAL GT MIDTERM) TEST=1.
```

and either FINAL or MIDTERM is missing for a case, SPSS/PC+ cannot tell whether one variable is greater than the other. In such a case, SPSS/PC+ leaves the target variable unchanged if it is an existing variable. If it is a new variable, it retains its initialized sytem-missing value.

When several relations are joined by AND or OR, SPSS/PC+ automatically returns the missing value if *any* of the relations in the expression have missing values.

4.23 Missing-Value Logical Functions

You can use the functions MISSING and SYSMIS to specify missing values as criteria for performing or not performing transformations.

MISSING *Return 1 if the value is missing and 0 otherwise.*

SYSMIS *Return 1 if the value is system-missing and 0 otherwise.*

For example, the command

```
IF (SYSMIS(SCORE1)) GRADE=0.
```

determines if SCORE1 is system-missing. If it is, GRADE is assigned the value 0.

The command

```
IF (NOT(MISSING(GRADE))) GRAD=1.
```

evaluates whether the value of GRADE is not equal to the user- or system-missing values. Each case that has a valid value for GRADE is assigned the value 1 for GRAD.

You can also use the VALUE function on an IF command to ignore the user-missing status of values. For example, the commands

```
RECODE AGE (5 THRU 20 = 1) (20 THRU 65 = 2) (65 THRU HI = 3).
MISSING VALUE AGE (3).
IF (VALUE(AGE) GT 0 ) GRPAGE=1.
```

collapse the values of AGE into three values and designate the value 3 as user-missing. The IF command specifies that any case with a value greater than 0 on AGE be given the value 1 on GRPAGE. Because the VALUE keyword on the IF command tells SPSS/PC+ to ignore user-missing values, cases with value 3 on AGE are given the value 1 on GRPAGE.

The VALUE function should always be used in conjunction with a relational operator in a logical expression on the IF command.

4.24 EXECUTING DATA TRANSFORMATIONS

When the data are read, transformation commands are evaluated and executed in the order in which they appear. Thus, the order in which you specify your commands can be important. For the commands

```
RECODE POLACT1 POLACT2 (1 THRU 2=1)(3 THRU 4=2)(ELSE=SYSMIS).
COUNT POLACT=POLACT1,POLACT2(1).
```

the order of execution is critical, since the COUNT command assumes that the RECODE command has already been executed.

Transformations are not carried out until the data are read. You must include a procedure command (or another command that causes SPSS/PC+ to read the data) for the transformations to be executed. Additionally, unless a system file is saved or the data are written out in some way, the transformations are in effect only for a single SPSS/PC+ session.

4.25 USING DATA DEFINITIONS WITH TRANSFORMATIONS

You can use the data-definition commands VARIABLE LABELS, VALUE LABELS, MISSING VALUE, and so on, to describe any variable created or altered by transformations, as in the commands

```
RECODE SEX RACE SURGERY (1=0)(0=1)(ELSE=SYSMIS).
COMPUTE LOGAGE=LN(AGE).
VARIABLE LABELS LOGAGE 'NATURAL LOG OF AGE'.
VALUE LABELS SEX 1 'FEMALE' 0 'MALE'
             RACE 1 'BLACK' 0 'WHITE'
             SURGERY 1 'SURGERY PERFORMED' 0'NO SURGERY'.
```

Since a variable must already exist in your active file before it can be defined, the data-definition commands must follow the transformation commands that create the variable.

Contents

Chapter 5 Selecting, Weighting, and Ordering Cases

Suppose you are interested in plotting the relationship between cost and alcohol content in the beer data, but you only want the plot to include domestic beers. Or you wish to examine the distribution of calories only for the beers that have been assigned ratings of "very good." These are two examples of situations where you want to select a subset of cases from a file, based on some particular criterion. You can use the SELECT IF command to select a subset of cases for an entire SPSS/PC+ session. The PROCESS IF command selects a subset of cases only for the following instruction.

When there are many cases in a file, you may want to select a random sample of them for processing. This decreases the time needed for analysis, and may provide you with useful preliminary results. For example, if you had information on 10,000 cases, you might want to obtain plots or histograms for a random subset of them. These plots should reflect the overall trends present in the data. The SAMPLE command selects a random sample of cases from a file.

To restrict analysis to the first *n* cases in a data set, use the N command. This command is particularly useful if you want to get an idea of the output produced by a procedure without having to wait for all cases to be processed. For example, if you are preparing a report and want to make sure that you have included all the necessary information in an appropriate format, running it on a small number of "test" cases is an efficient strategy.

The WEIGHT command allows you to assign different "weights" to cases for an analysis. For example, if your sample does not reflect the true proportion of cases with particular attributes in the population, you can assign appropriate weights to adjust for this when estimating certain statistics. However, the sample sizes and significance levels associated with such weighted analyses cannot be interpreted in the usual fashion.

On occasion, you may wish to sort the observations in a file based on values of certain variables. For example, you might want to sort the beers in ascending order on the basis of alcohol content. Use the SORT CASES command to do this. You can then list the cases in this order, or prepare reports using the sorted file (see Chapter 18).

5.1 SELECTING CASES PERMANENTLY

The SELECT IF command selects cases for analysis if they meet criteria you specify. You specify these criteria in a logical expression that SPSS/PC+ can evaluate as true or false. For example, the command

```
SELECT IF (SEX EQ 1).
```

selects cases with the value 1 for SEX for analysis.

You can include the SELECT IF command anywhere in an SPSS/PC+ session, except between the BEGIN DATA and END DATA commands. Once you use SELECT IF, the selection specified is in effect for all subsequent procedures. If you use another SELECT IF command, it selects a subset of cases from the first selected subset rather than from the original data set.

Multiple SELECT IF commands should be used with caution, as you can end up selecting no cases. For example, if the commands

```
SELECT IF (SEX EQ 1).
FREQUENCIES VARIABLES=TEMP FIRSTEKG SECNDEKG.
SELECT IF (SEX EQ 2).
FREQUENCIES VARIABLES=TEMP FIRSTEKG SECNDEKG.
```

are used, there will be no cases for the second FREQUENCIES procedure to analyze, since SEX cannot equal both 1 and 2. If you want to temporarily select cases for one procedure, use the PROCESS IF command (see Section 5.3).

5.2 Specifying the Logical Expression

The logical expression on SELECT IF is specified in the same way as the logical expression on the IF command (see Chapter 4). The logical expression must be enclosed in parentheses and include either a relational operator or a missing-value function. You can construct complex selection criteria by using the logical operators AND, OR, and NOT in the logical expression. If the logical expression cannot be determined for a case because of missing values, that case is not selected.

To select cases with missing values for analysis, use the SYSMIS and MISSING functions. For example, the command

```
SELECT IF (MISSING(AGE)).
```

selects all cases with missing values for AGE.

You can use the VALUE function to include cases with user-missing values in a selection of cases. A SELECT IF logical expression containing the VALUE function should also contain a relational operator. Thus, the command

```
SELECT IF (VALUE(SCORE) LE 40).
```

selects all cases with SCORE values less than or equal to 40, including values that have been defined as missing. (See Chapter 4 for more information on missing values in logical expressions.)

5.3 SELECTING CASES TEMPORARILY

The PROCESS IF command temporarily selects cases for analysis by the immediately following procedure. Like SELECT IF, its specification consists of a logical expression, and it can use any of the relational operators (EQ, NE, GT, GE, LT, or LE). However, the logical operators AND, OR, and NOT cannot be used with PROCESS IF. Thus, a specification like (SALES GE 10000 AND DIVISION EQ 3) is not valid on this command.

If you specify several PROCESS IF commands immediately before a procedure command, only the last one takes effect. For example, if the commands

```
PROCESS IF (SEX EQ 1).
PROCESS IF (SEX EQ 2).
FREQUENCIES VARIABLES=TEMP FIRSTEKG SECNDEKG.
```

are used, the second PROCESS IF command overrides the first, and the FREQUENCIES procedure will analyze all cases for which SEX equals 2.

5.4 DRAWING A TEMPORARY SAMPLE

The SAMPLE command draws a random subsample of cases for analysis in the immediately following procedure. You can include SAMPLE anywhere in an SPSS/PC+ session except between the BEGIN DATA and END DATA commands.

To sample a proportion of cases, specify the proportion on the SAMPLE command, as in

```
SAMPLE .25.
```

This command samples approximately one-fourth of the cases.

If you know the total number of cases, you can specify the number of cases to be sampled, as in the command

```
SAMPLE 50 FROM 200.
```

This command draws a random sample of 50 cases only if there are exactly 200 total cases. If there are fewer than 200 cases, proportionately fewer cases are sampled. If there are more than 200 cases, the subsample is taken only from the first 200 cases.

If SAMPLE follows a SELECT IF or PROCESS IF command, the sample is drawn from the selected subset of cases. Conversely, if SAMPLE precedes SELECT IF or PROCESS IF, the specified subset of cases is selected from the sample. If you specify more than one SAMPLE command prior to a procedure, only the last SAMPLE command is executed.

5.5 SELECTING THE FIRST n CASES

The N command is used to select the first *n* cases in a file. For example, if your file has 1000 cases, but you want to analyze only the first 100 cases, specify

```
N 100.
```

You can enter the N command at any point in an SPSS/PC+ session. Once specified, it limits the number of cases analyzed by all subsequent procedures. More than one N command can be used in a single session, but once you execute a procedure using a given *n*, you cannot increase the number of cases for subsequent procedures; you can only decrease the number of cases in the working data set.

The commands PROCESS IF, SELECT IF, and SAMPLE are executed before N if the commands occur together (even if N is specified first). For example, if you specify

```
N 100.
SAMPLE .5.
```

approximately half of the total cases are sampled, and then the first 100 of these are selected for analysis.

5.6 WEIGHTING CASES

You can adjust the weighting of cases for analysis by using the WEIGHT command. For example, if you have a sample in which males have been oversampled (i.e., there is a much higher proportion of males in the sample than in the population), you may want to give the data for males less weight in your computations. The WEIGHT command is also useful for entering data that have been crosstabulated (see Chapter 9).

The only specification on WEIGHT is the name of the variable to be used for weighting, as in the command

```
WEIGHT BY WGHTVAR.
```

Only one weighting variable can be specified on the command, and it must be numeric. The values of the weighting variable need not be integers, but missing or negative values are treated as zeros. Most SPSS/PC+ procedures can handle noninteger weights, with the exception of PLOT and NPAR TESTS. When weighting is used, files that are saved retain the weighting in the system-variable $WEIGHT.

The weighting variable can be an already existing variable or a variable created through transformation statements. For example, suppose men have been oversampled by a factor of 2. To compensate for this, you can weight male cases by one half, as in the commands

```
COMPUTE WT=1.
IF (SEX EQ 2) WT=.5.
WEIGHT BY WT.
```

If you create a weighting variable with an IF command, it is important to first initialize its weight with a COMPUTE command. Otherwise, cases not covered by the IF command will have missing values for the weighting variable. A case weighted by 1 is unaffected when WEIGHT is executed, but a case weighted by 0 or by a missing value is eliminated.

A WEIGHT command stays in effect for the entire job unless followed by another WEIGHT command or turned off with the command

```
WEIGHT OFF.
```

Weighting is *not* cumulative. That is, a new WEIGHT command reweights the sample rather than altering previously weighted values. For example, if the commands

```
WEIGHT BY WT1.
DESCRIPTIVES ALL.
WEIGHT BY WT2.
DESCRIPTIVES ALL.
```

are used, the first DESCRIPTIVE procedure computes summary statistics based on cases weighted by WT1, and the second DESCRIPTIVES procedure computes summary statistics based on cases weighted by WT2.

When weighting is in effect, significance tests are usually based on the weighted sample size. If the weighted number of cases exceeds the sample size, the *p*-values given for these tests will be too small. If the weighted number of cases is smaller than the actual sample size, the *p*-values calculated will be too large. You can avoid these problems by using weight factors that add up to the sample size.

5.7 SORTING DATA

You can use the SORT CASES command to reorder your data according to the values of a specified variable or variables. Specify SORT CASES with the BY keyword, followed by the name(s) of the variable(s) to be used for sorting. The variables specified can be numeric or string. String variables are sorted by alphabetical order.

Cases can be sorted in ascending or descending order. Ascending order is the default. To sort cases in descending order (in which the values for the sorting variable are ordered from highest to lowest), you specify (D) after the variable name, as in the command

```
SORT CASES BY SALES(D).
```

(D) is also used to reverse the alphabetical order of string variables.

You can specify (A) after a variable name to explicitly request ascending order. When (D) or (A) appears after a list of otherwise unspecified variables, all are sorted in the order specified. For example, the command

```
SORT CASES BY PRODUCT DEPT SALES(D).
```

requests that PRODUCT, DEPT, and SALES all be sorted in descending order.

When several sorting variables are listed, cases are first sorted according to the first variable named. Cases with the same value for the first sorting variable are then sorted according to the second sorting variable, and so on. For example, the command

```
SORT CASES BY PRODUCT(D) DEPT(A) SALES(D).
```

produces the following sorted values for PRODUCT, DEPT, and SALES:

```
4 1  $9,750
4 2 $18,083
4 2 $15,608
4 2 $15,132
4 2 $12,438
4 2 $11,240
4 2 $10,050
3 1 $17,051
3 2 $39,000
3 2 $19,682
3 2 $13,650
3 2  $9,777
3 2  $9,507
3 2  $8,872
3 2  $8,239
1 1 $35,750
1 1 $17,111
1 1 $13,910
```

Contents

Chapter 6 Restructuring Files: Procedures JOIN and AGGREGATE

SPSS/PC+ provides two commands for restructuring files. Procedure JOIN combines the contents of system files and creates a new active file. AGGREGATE creates a system file with cases containing summary measures for groups of cases. In this chapter, we will use school enrollment records to illustrate the use of JOIN and AGGREGATE to update and summarize the information in your SPSS/PC+ files.

6.1 OVERVIEW

Consider a simple class roster compiled at pre-registration time. As more students join the class during the first week, their records need to be added to the existing roster. A simple addition of cases, as illustrated in Figure 6.1a, is all that's needed. In SPSS/PC+, this simple addition of cases from one file to another file is accomplished with the JOIN command specifying keyword ADD. Sections 6.2 through 6.3 discuss adding files together.

Figure 6.1a Adding cases to a file

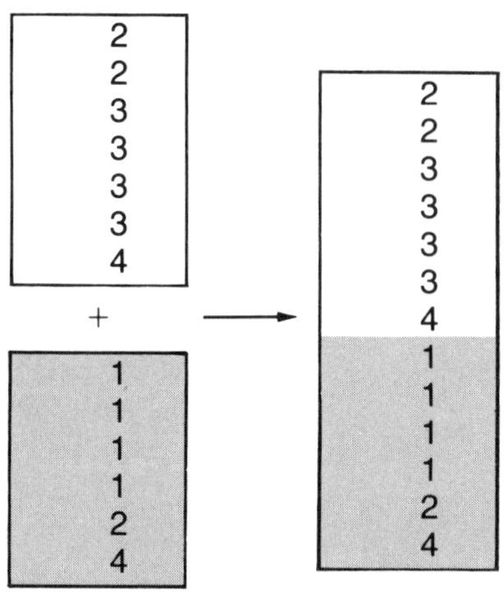

Suppose the instructor wants the new cases added according to a particular order of some variable, such as class year. He could either perform a simple addition and then sort the file or specify that the addition of cases should use the values of a key variable for interleaving, as shown in Figure 6.1b. Interleaving cases from one file with cases in another file is accomplished with JOIN ADD specifying a key variable on the BY subcommand. Section 6.4 discusses the BY subcommand and how to use it.

Figure 6.1b Interleaving cases with a key

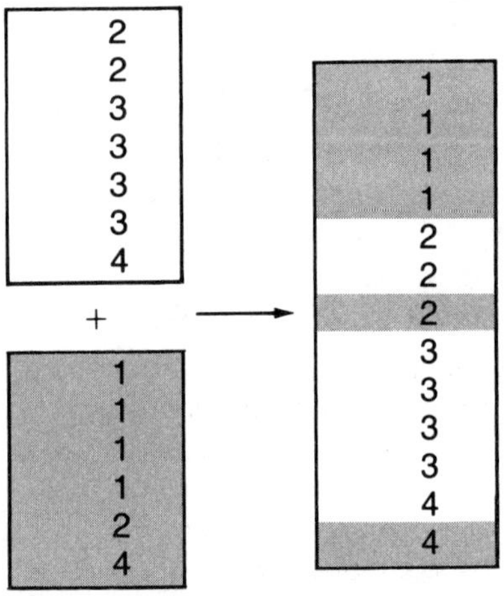

Once the instructor has a class roster, he might decide to add data related to the course to each student record. If the instructor has files that contain exactly the same cases sorted in the same order, he can perform a simple parallel match of files, shown in Figure 6.1c. Procedure JOIN specifying keyword MATCH combines the contents of the first record from each input file, then the contents of the second record from each input file, and so on. Section 6.10 describes how JOIN MATCH combines variables from parallel files.

Figure 6.1c Matching parallel data files

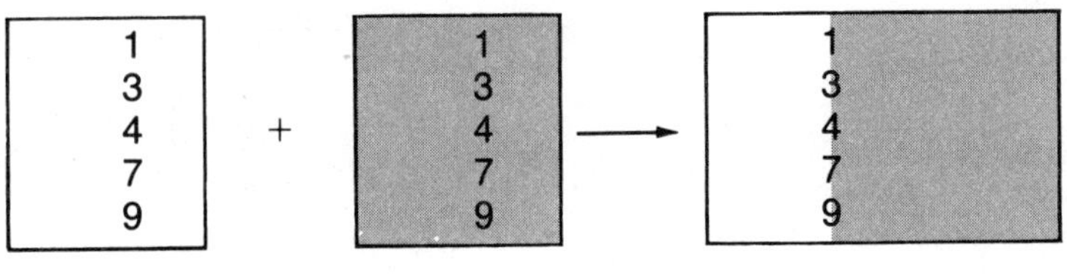

As the quarter progresses, an instructor can use the class roster as a grade book, appending test scores, grades on term papers, and so forth. By using a key variable such as student identification number, he can ensure that the correct test score is matched to the student who took the test. Figure 6.1d shows matching files with a key variable.

By using the key variable, the instructor doesn't have to worry about having test scores for each student in the roster file. For example, students who have dropped the course won't have a grade and are absent from the grade file. These students are assigned a missing value in the resulting file. In addition, each unique value of the key variable is retained in the resulting file. Thus, if a student was not on the class roster but took the exam, the grade is retained in the resulting file. Section 6.11 discusses matching nonparallel files with key variables.

Figure 6.1d Matching nonparallel data files with a key

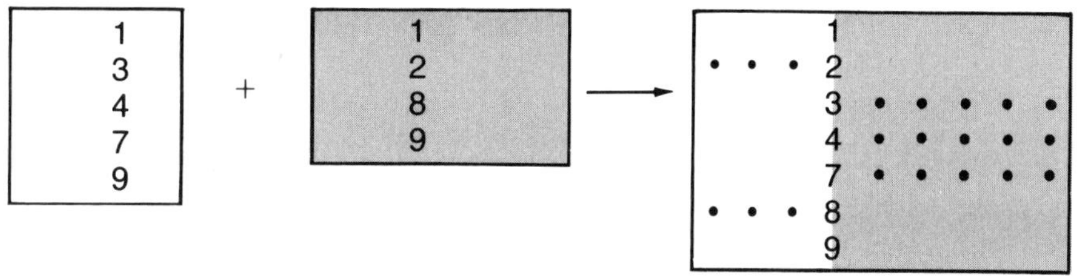

Files containing different types of data can be combined by using JOIN MATCH with a key and designating one file a TABLE file. For example, the instructor might build one file combining the grade-book files for each class he taught that year. He might also build a table file containing one case for each class, with summary information such as the number of exams, the final exam average, high and low grades, etc., recorded for each class. By using the class variable as a key and designating the file with the summary information a table, he can append the summary variables to individual student records. Figure 6.1e shows cases from one file combined with variables from a table. A table match adds variables from the table file to cases only when there is a match on a key variable. Section 6.13 discusses the use of file and table matches.

Figure 6.1e Matching table data to a file with a key

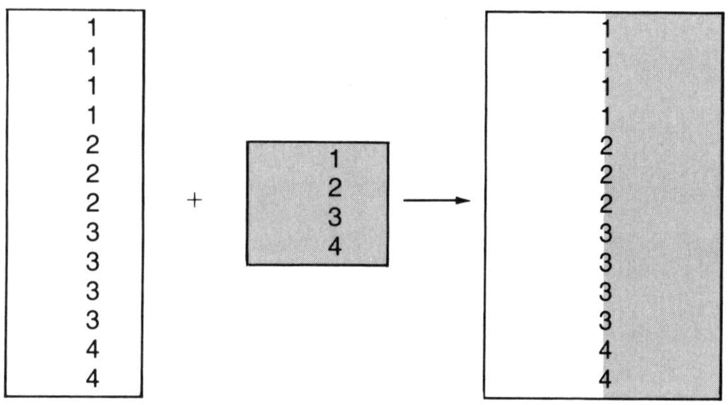

To create the table file of summary information described in the preceding example, the instructor could use his hand calculator to compute class averages and scan the file for the highest and lowest scores for each class. This process would have to be repeated for each class. Or the instructor could have SPSS/PC+ build the table file using procedure AGGREGATE. AGGREGATE, discussed in

Sections 6.14 through 6.23, creates a file containing variables representing attributes (means, standard deviations, minimum value, etc.) of different groups, with one case for each group. Figure 6.1f illustrates how AGGREGATE creates cases based on common values of a grouping variable.

Figure 6.1f Aggregating cases

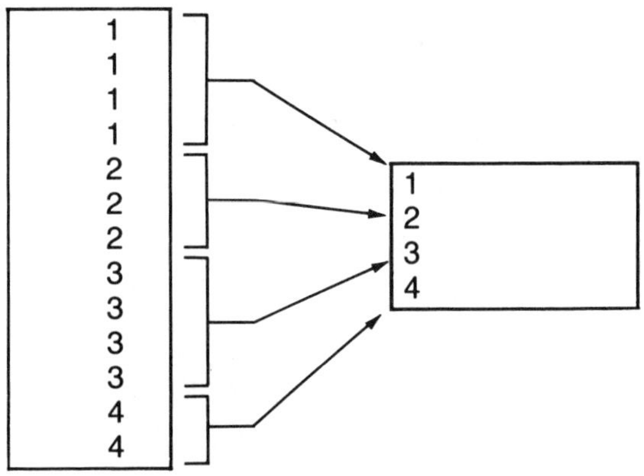

The ADD and MATCH keywords on procedure JOIN are mutually exclusive and must be specified before any other keywords or subcommands. JOIN ADD combines cases from two to five input files. JOIN MATCH combines variables from two to five input files. JOIN MATCH and JOIN ADD combine variables and cases from SPSS/PC+ *system files,* one of which can be the most recently defined active file. JOIN always creates a new active file for use by other commands. Procedure AGGREGATE groups cases and creates variables from an active file and either replaces the active file or writes an SPSS/PC+ system file.

When you use JOIN to interleave cases or to match variables based on values of keys, each input file must be sorted in ascending order of each key variable (see Section 6.12). AGGREGATE does not require that the active file be sorted.

6.2 ADDING FILES

You can use JOIN with keyword ADD to combine the cases from two to five system files into a rectangular file. Instead of using only previously saved SPSS/PC+ system files, one of the files can be the current active file.

6.3 Adding Cases to the Active File

Students who show up at the first day of class can be added to a previously defined system file containing students who have pre-registered for a course. The file SOC100.LAT (Figure 6.3a) contains data on students who showed up for the first day of class and are not included in the pre-registration file. This file contains the same variables as the pre-registration system file but on different students. The commands

```
DATA LIST FILE='SOC100A.LAT'/COURSE 1-3 SECTION 5 (A) MEETTIME 7-10
     DAYS 12-14(A) STUNUM 16-20 LASTNAME 22-33(A) FSTNAME 35-44(A) CLASSYR
46.
VARIABLE LABELS CLASSYR 'Class'.
VALUE LABELS CLASSYR 1 'Freshman' 2 'Sophmore' 3 'Junior' 4 'Senior'.

TITLE 'Combined Soc100 Section A Roster'.
```

provide definitions for an active file with just the late-registering students.

Figure 6.3a Data to add to pre-registration file

```
100 a 0900 mwf 10009 Atmore       Samuel     1
100 a 0900 mwf 24365 McDowell     Denise     1
100 a 0900 mwf 27111 Hutchinson   Christine  1
100 a 0900 mwf 55287 Swift        Jennifer   1
100 a 0900 mwf 62432 Jamieson     Bailey     1
100 a 0900 mwf 89765 Paulsen      Patrick    2
100 a 0900 mwf 54338 Darfler      Marleena   4
```

To add cases for students who pre-registered for the course, issue the command

```
JOIN ADD FILE=*
   /FILE='S100A.PSF'.

LIST.
SAVE OUTFILE='SOC100A.NSF'.
```

The JOIN ADD command adds cases from the system file S100A.PSF to the active file of late-registering students (the asterisk refers to the active file). Note that the order of cases in the resulting active file is determined by the order in which files are named on the FILE subcommands.

After the JOIN command is executed, the active file contains the combined set of cases. The LIST command produces a listing of the contents of this file, shown in Figure 6.3b. The SAVE command creates an SPSS/PC+ system file containing all the cases. The title "Combined Soc100 Section A Roster" specified on the TITLE command is also included in the system file.

Figure 6.3b Listing of combined files

```
COURSE SECTION MEETTIME DAYS STUNUM LASTNAME     FSTNAME    CLASSYR

   100 a             900 mwf   10009 Atmore      Samuel           1
   100 a             900 mwf   24365 McDowell    Denise           1
   100 a             900 mwf   27111 Hutchinson  Christine        1
   100 a             900 mwf   55287 Swift       Jennifer         1
   100 a             900 mwf   62432 Jamieson    Bailey           1
   100 a             900 mwf   89765 Paulsen     Patrick          2
   100 a             900 mwf   54338 Darfler     Marleena         4
   100 a             900 mwf   43289 Sweeney     John             2
   100 a             900 mwf   23763 Baker       Catherine        2
   100 a             900 mwf   22304 Jones       Barbara          3
   100 a             900 mwf   27001 Jacobsen    Richard          3
   100 a             900 mwf   35760 Roberts     Elizabeth        3
   100 a             900 mwf   64352 Atkinson    Thomas           3
   100 a             900 mwf   79885 Klein       Janet            4

Number of cases read =       14    Number of cases listed =      14
```

6.4 Interleaving Cases

You can use JOIN ADD to interleave cases in ascending order of one or more variables by naming one or more key variables on the BY subcommand. You can specify up to ten variables as keys for interleaving cases on BY. Cases must be sorted in ascending order of the key variable before they can be interleaved. Thus, you should use SORT prior to saving system files or sort the current active file (see Section 6.12).

For example, you can create a combined file that contains students from the two files described in Section 6.3 grouped according to class year. Both the system file SOC100A.PSF and the raw data set SOC100A.LAT must be sorted in ascending order of variable CLASSYR. To interleave cases on the basis of CLASSYR, use the following command:

```
JOIN ADD FILE=*
        /FILE='S100A.PSF'
        /BY=CLASSYR.

LIST.
SAVE OUTFILE='SOC100A.SF'.
```

Figure 6.4 shows the results of the LIST command, with all the cases in order of variable CLASSYR.

Figure 6.4 Interleaved cases on the active file

```
COURSE SECTION MEETTIME DAYS STUNUM LASTNAME     FSTNAME    CLASSYR

   100 a             900 mwf   10009 Atmore      Samuel           1
   100 a             900 mwf   24365 McDowell    Denise           1
   100 a             900 mwf   27111 Hutchinson  Christine        1
   100 a             900 mwf   55287 Swift       Jennifer         1
   100 a             900 mwf   62432 Jamieson    Bailey           1
   100 a             900 mwf   89765 Paulsen     Patrick          2
   100 a             900 mwf   43289 Sweeney     John             2
   100 a             900 mwf   23763 Baker       Catherine        2
   100 a             900 mwf   22304 Jones       Barbara          3
   100 a             900 mwf   27001 Jacobsen    Richard          3
   100 a             900 mwf   35760 Roberts     Elizabeth        3
   100 a             900 mwf   64352 Atkinson    Thomas           3
   100 a             900 mwf   54338 Darfler     Marleena         4
   100 a             900 mwf   79885 Klein       Janet            4

Number of cases read =      14    Number of cases listed =        14
```

When you combine cases from two or more files and name a BY variable for interleaving, you can expect a warning message whenever there are duplicate values of the key variable across files. The example above produces such a message because the active file and the system file, S100A.SF, both contain cases with values 2 and 4 for the variable CLASSYR. If your input files do not contain duplicate values, no warning message is issued.

6.5 Managing Variables

Sometimes you will want to include only a subset of variables from your input files in the file you create with JOIN. Or you may have a situation where the same variable has different names on different files that you want to combine.

With the DROP and KEEP subcommands on JOIN, you can select a subset of variables for inclusion in the resulting file. With the RENAME subcommand, you can supply new names for any variables in the resulting file. In addition, you can use the MAP subcommand to produce a table of the variables in all input files and in the resulting file after any subsetting or renaming.

Sections 6.6 through 6.8 show you how to use MAP to see what's in the files you create with JOIN and how to manage the variables within them using DROP, KEEP, and RENAME.

6.6 Viewing the Active File Dictionary

The MAP subcommand produces a listing of the current status of the active file dictionary. Included in the listing are the names of each input file and of the variables from each file.

Consider the following JOIN command, which combines cases from five previously saved system files:

```
JOIN ADD FILE='SOC100A.SF'
    /FILE='SOC100B.SF'
    /FILE='SOC222A.SF'
    /FILE='SOC310A.SF'
    /FILE='HDS444A.SF'
    /MAP.

LIST.
SAVE OUTFILE='ALLSTUD.S85'.
```

This example creates an active file that contains all students registered in every class and all variables that exist in any of the input files. Figure 6.6a shows the map produced by the MAP subcommand. Variables DEPT, CLASSTYP, and PASS-FAIL from the input file HDS444A.SF are present in the resulting active file. If a case comes from an input file that does not contain a particular variable, SPSS/PC+ assigns the system-missing value (numeric variables) or a blank (string variables).

Figure 6.6a Map of concatenated system files

```
RESULT         SOC100A.SF     SOC100B.SF     SOC222A.SF     SOC310A.SF     HDS444A.SF
-------------  -------------  -------------  -------------  -------------  -------------
COURSE         COURSE         COURSE         COURSE         COURSE         COURSE
SECTION        SECTION        SECTION        SECTION        SECTION        SECTION
MEETTIME       MEETTIME       MEETTIME       MEETTIME       MEETTIME       MEETTIME
DAYS           DAYS           DAYS           DAYS           DAYS           DAYS
STUNUM         STUNUM         STUNUM         STUNUM         STUNUM         STUNUM
LASTNAME       LASTNAME       LASTNAME       LASTNAME       LASTNAME       LASTNAME
FSTNAME        FSTNAME        FSTNAME        FSTNAME        FSTNAME        FSTNAME
CLASSYR        CLASSYR        CLASSYR        CLASSYR        CLASSYR        CLASSYR
DEPT                                                                       DEPT
CLASSTYP                                                                   CLASSTYP
PASSFAIL                                                                   PASSFAIL
```

Figure 6.6b shows a listing of the cases in the file. Cases from SOC100A.SF come first, followed by cases from the next named file, and so on. Each case contains a value for variables DEPT, CLASSTYP, and PASSFAIL, even though these variables occur in only one of the five original files.

Figure 6.6b Listing of concatenated system files

```
      S
    C E                 S
    O C                 T
    U T          D      U                                         D
    R I          A      N                                         E
    S O          Y      U                                         P
    E N MEETTIME S      M LASTNAME     FSTNAME    CLASSYR T  CLASSTYP PASSFAIL

  100 a      900 mwf 10009 Atmore      Samuel           1                    .
  100 a      900 mwf 24365 McDowell    Denise           1                    .
  100 a      900 mwf 27111 Hutchinson  Christine        1                    .
  100 a      900 mwf 55287 Swift       Jennifer         1                    .
  100 a      900 mwf 62432 Jamieson    Bailey           1                    .
  100 a      900 mwf 89765 Paulsen     Patrick          2                    .
  100 a      900 mwf 43289 Sweeney     John             2                    .
  100 a      900 mwf 23763 Baker       Catherine        2                    .
  100 a      900 mwf 22304 Jones       Barbara          3                    .
  100 a      900 mwf 27001 Jacobsen    Richard          3                    .
  100 a      900 mwf 35760 Roberts     Elizabeth        3                    .
  100 a      900 mwf 64352 Atkinson    Thomas           3                    .
  100 a      900 mwf 54338 Darfler     Marleena         4                    .
  100 a      900 mwf 79885 Klein       Janet            4                    .
  100 b     1100 mwf 10229 Smithe      Peter            1                    .
  100 b     1100 mwf 34365 Farroro     Dennis           1                    .
  100 b     1100 mwf 25111 Huber       Joan             1                    .
  100 b     1100 mwf 55587 Westerman   David            1                    .
  100 b     1100 mwf 62442 Mount       Sigrid           1                    .
  100 b     1100 mwf 89763 Harris      Emily            4                    .
  100 b     1100 mwf 70395 Jones       James            2                    .
  100 b     1100 mwf 53763 Halley      Beth             2                    .
  100 b     1100 mwf 82224 Manova      Svetlana         2                    .
  100 b     1100 mwf 97291 Sullivan    Linda            3                    .
  100 b     1100 mwf 50762 Dexter      Richard          3                    .
  100 b     1100 mwf 62752 Colby       Jeffrey          3                    .
  100 b     1100 mwf 54358 Fiorello    Francis          3                    .
  222 a     1200 tth 12239 Smith       Patricia         1                    .
  222 a     1200 tth 35467 Freiberg    Gerhardt         1                    .
  222 a     1200 tth 65141 Mills       Charles          2                    .
  222 a     1200 tth 53287 West        James            2                    .
  222 a     1200 tth 39742 Maus        Mickey           2                    .
  222 a     1200 tth  9763 Herrmann    Monty            2                    .
  222 a     1200 tth 48364 Daveport    Joyce            2                    .
  222 a     1200 tth 81043 Hill        Robert           2                    .
  222 a     1200 tth 42524 Mason       Perry            2                    .
  222 a     1200 tth 95591 Evans       Linda            3                    .
  222 a     1200 tth 61932 Bates       Oliver           3                    .
  222 a     1200 tth 93132 Hughes      Olivia           3                    .
  222 a     1200 tth 44359 Bates       Lucy             4                    .
  310 a     1400 mwf 42305 O'Kane      Stanley          1                    .
  310 a     1400 mwf 60611 Hallet      Robin            2                    .
  310 a     1400 mwf 14420 Robinson    Thomas           2                    .
  310 a     1400 mwf  4901 Nace        Laura            3                    .
  310 a     1400 mwf 99999 Liebmann    Geoffrey         3                    .
  310 a     1400 mwf 43995 Countryman  Katharyn         3                    .
  310 a     1400 mwf 60618 Adams       William          3                    .
  310 a     1400 mwf 14520 Stephenson  Lynn             4                    .
  310 a     1400 mwf 22112 Jensen      Lars             4                    .
  444 a     1430 tth 25851 Smith       Jack             3 hds sem            1
  444 a     1430 tth 87034 Cobbleigh   Nicholas         4 hds sem            1
  444 a     1430 tth 13792 Heisenhuer  Jane             4 hds sem            2
  444 a     1430 tth 89763 Harris      Emily            4 hds sem            1
  444 a     1430 tth 91913 Chambers    Dianne           4 hds sem            2

Number of cases read =      54    Number of cases listed =      54
```

6.7 Selecting a Subset of Variables

You can create an active file that contains all the variables from each input file or select a subset of variables from each file. The subcommands DROP and KEEP indicate which variables from an input file should be omitted or retained on the resulting active file. DROP and KEEP apply only to the immediately preceding named file. For example,

```
JOIN ADD FILE='SOC100A.SF'/DROP=SECTION, MEETTIME, DAYS
    /FILE='SOC100B.SF'/DROP=SECTION, MEETTIME, DAYS
    /FILE='SOC222A.SF'/DROP=SECTION, MEETTIME, DAYS
    /FILE='SOC310A.SF'/DROP=SECTION, MEETTIME, DAYS
    /FILE='HDS444A.SF'/KEEP=COURSE STUNUM LASTNAME FSTNAME CLASSYR
    /BY CLASSYR
    /MAP.
```

joins five system files, interleaving cases on the basis of values of CLASSYR. All variables except SECTION, MEETTIME, and DAYS are taken from the first four system files. Only variables COURSE, STUNUM, LASTNAME, FSTNAME, and CLASSYR are included from the fifth system file. The output from the MAP subcommand is shown in Figure 6.7a. It is placed last so that it will show the status at the end of the JOIN ADD operation. The listing of the contents of the active file from the LIST command is shown in Figure 6.7b.

Figure 6.7a Map of combined system files with specific variables

RESULT	SOC100A.SF	SOC100B.SF	SOC222A.SF	SOC310A.SF	HDS444A.SF
COURSE	COURSE	COURSE	COURSE	COURSE	COURSE
STUNUM	STUNUM	STUNUM	STUNUM	STUNUM	STUNUM
LASTNAME	LASTNAME	LASTNAME	LASTNAME	LASTNAME	LASTNAME
FSTNAME	FSTNAME	FSTNAME	FSTNAME	FSTNAME	FSTNAME
CLASSYR	CLASSYR	CLASSYR	CLASSYR	CLASSYR	CLASSYR

Figure 6.7b Listing of interleaved cases after variables are dropped

COURSE	STUNUM	LASTNAME	FSTNAME	CLASSYR
100	10009	Atmore	Samuel	1
100	24365	McDowell	Denise	1
100	27111	Hutchinson	Christine	1
100	55287	Swift	Jennifer	1
100	62432	Jamieson	Bailey	1
100	10229	Smithe	Peter	1
100	34365	Farroro	Dennis	1
100	25111	Huber	Joan	1
100	55587	Westerman	David	1
100	62442	Mount	Sigrid	1
222	12239	Smith	Patricia	1
222	35467	Freiberg	Gerhardt	1
310	42305	O'Kane	Stanley	1
100	89765	Paulsen	Patrick	2
100	43289	Sweeney	John	2
100	23763	Baker	Catherine	2
100	70395	Jones	James	2
100	53763	Halley	Beth	2
100	82224	Manova	Svetlana	2
222	65141	Mills	Charles	2
222	53287	West	James	2
222	39742	Maus	Mickey	2
222	9763	Herrmann	Monty	2
222	48364	Daveport	Joyce	2
222	81043	Hill	Robert	2
222	42524	Mason	Perry	2
310	60611	Hallet	Robin	2
310	14420	Robinson	Thomas	2
100	22304	Jones	Barbara	3
100	27001	Jacobsen	Richard	3
100	35760	Roberts	Elizabeth	3
100	64352	Atkinson	Thomas	3
100	97291	Sullivan	Linda	3
100	50762	Dexter	Richard	3
100	62752	Colby	Jeffrey	3
100	54358	Fiorello	Francis	3
222	95591	Evans	Linda	3
222	61932	Bates	Oliver	3
222	93132	Hughes	Olivia	3
310	4901	Nace	Laura	3
310	99999	Liebmann	Geoffrey	3
310	43995	Countryman	Katharyn	3
310	60618	Adams	William	3
444	25851	Smith	Jack	3
100	54338	Darfler	Marleena	4
100	79885	Klein	Janet	4
100	89763	Harris	Emily	4
222	44359	Bates	Lucy	4
310	14520	Stephenson	Lynn	4
310	22112	Jensen	Lars	4
444	87034	Cobbleigh	Nicholas	4
444	13792	Heisenhuer	Jane	4
444	89763	Harris	Emily	4
444	91913	Chambers	Dianne	4

Number of cases read = 54 Number of cases listed = 54

6.8 Renaming Variables

Sometimes, variables containing the same information have different variable names in different files. For example, the student's identification number might be called STUDNUM in one file and ID in another. You can reconcile such differences during a JOIN ADD with the RENAME subcommand, as in:

```
GET FILE='SOC100.ASF'.

JOIN ADD FILE='SOC100.BSF'/RENAME (STUDNUM=ID)
     /FILE=* /KEEP=LASTNAME FSTNAME ID CLASSYR
     /MAP.
```

The GET command retrieves the system file SOC100.ASF and makes it the active file. JOIN ADD specifies SOC100.BSF as the first input file. The RENAME subcommand changes the name of variable STUDNUM in SOC100.BSF to ID in the resulting active file. The values of LASTNAME, FSTNAME, ID and CLASSYR for cases in the active file (*) are added to SOC100.BSF. The MAP display is shown in Figure 6.8.

The RENAME subcommand ensures that values are contained under the same variable name. If RENAME had not been specified in the example above, the combined file would contain both variables STUDNUM and ID. Cases from the first input file SOC100.BSF would have the system-missing value for ID, while cases from the file SOC100.ASF would be system-missing for STUDNUM.

Figure 6.8 Map of case additions with renamed variable

```
RESULT        SOC100.BSF    *
------------  ------------  ------------
ID            ID            ID
LASTNAME      LASTNAME      LASTNAME
FSTNAME       FSTNAME       FSTNAME
CLASSYR       CLASSYR       CLASSYR
MAJOR         MAJOR
```

The RENAME subcommand can also be used to distinguish variables that record different attributes but have the same name on different files. For example, two files might each contain a variable called CLASS. In one file CLASS refers to course name, and in the second file it refers to the student's year in school. To avoid confusing one variable with the other, you could rename CLASS in the second file to SCHLYEAR. Cases in the second file would then have missing values for CLASS.

6.9 MATCHING FILES

Once a class roster has been developed, you can use JOIN MATCH to append variables with test grades or other information to each student's record. If each input file contains corresponding cases in the same order but with different variables, the files are *parallel* (see Figure 6.1c). A parallel or sequential match combines the variables from the first case from each input file, then the second case from each input file, and so on, without regard to any identifying values that may be present.

Nonparallel files have more or less overlapping sets of cases (see Figure 6.1d). Often cases in one input file are missing from another, or cases may be duplicated in one or the other files. You can append variables to particular cases by specifying that cases be combined according to a common value on one or more key variables (such as an identification number) present in each input file. The BY subcommand names the key variables.

Parallel and nonparallel files contain cases with the same kinds of observations or units of analysis. For example, one file can contain a student roster while another file contains student grades. Each file has individual students as the unit of analysis. Matching the two files results in a student gradebook file.

You can also use JOIN MATCH to combine two files containing different types of data. For example, one file contains test and final grades for students enrolled in five different classes. Another file contains the number of students enrolled in each course and the average final grade for each course offered in the college. By designating the course file a *table* and using the course as a key variable, SPSS/PC+ "looks-up" the data in the table file and matches variables to cases in the other input file. A table file contributes variables only from cases that contain corresponding values on key variables. The table file can contain records that don't necessarily correspond to cases in the other input files.

You can use the DROP, KEEP, RENAME, and MAP subcommands described in Sections 6.6 through 6.8 with JOIN MATCH. Up to five input files can be matched, one of which may be the active file. Sections 6.10 through 6.13 describe applications using parallel, nonparallel, and table look-up matches.

6.10 Parallel Files—Same Cases, Different Variables

After assembling a student roster, an instructor receives additional information on each student, such as the student's major and whether the course is required for graduation. The instructor can define an active file containing these variables and add them to an existing system file containing the roster. If each input file contains corresponding cases in the same order, a parallel match can be specified.

Figure 6.10a shows a raw data file containing information on students' majors and required course status. These variables can be added to the class roster saved in Section 6.4 using the following commands:

```
TITLE 'Grade Book'.
DATA LIST FILE='MAJINFO.DAT'/STUNUM 7-10 LASTNAME 13-24 (A)
         FSTNAME 26-35 (A) REQUIRED 39 MAJOR 41-43 (A).
VALUE LABELS REQUIRED 1 'Must Pass'.

JOIN MATCH FILE='SOC100A.SF'/FILE=*.

LIST.

SAVE OUTFILE='SOC100A.GRA'.
```

The DATA LIST command defines an active file with variables STUNUM, LASTNAME, FSTNAME, REQUIRED, and MAJOR. The JOIN MATCH command joins cases from the system file SOC100A.SF with the variables from the active file, represented by an asterisk.

Figure 6.10a Raw data to be added to class roster

```
100 a 10009 Atmore       Samuel     1
100 a 24365 McDowell     Denise     1
100 a 27111 Hutchinson   Christine  1
100 a 55287 Swift        Jennifer   1
100 a 62432 Jamieson     Bailey     1
100 a 89765 Paulsen      Patrick    2 1 psy
100 a 43289 Sweeney      John       2 1 soc
100 a 23763 Baker        Catherine  2   adm
100 a 22304 Jones        Barbara    3 1 adm
100 a 27001 Jacobsen     Richard    3 1 pol
100 a 35760 Roberts      Elizabeth  3 1 his
100 a 64352 Atkinson     Thomas     3   csi
100 a 79885 Klein        Janet      4 1 che
100 a 54338 Darfler      Marleena   4 1 che
```

The order in which files are named determines the order of variables in the resulting active file. Variables from the first-named file are first, followed by variables from the next-named file, and so on. When two or more files have a variable with the same name, values in the resulting file are taken from the first file named on the JOIN command. Figure 6.10b shows the listing of the active file after the parallel match has been performed.

Figure 6.10b Listing of active file after parallel match

```
                               D                                                     M
                               A                                                     A
                               Y                                                     J
                               O                                                     O
COURSE SECTION MEETTIME S   STUNUM LASTNAME      FSTNAME      CLASSYR REQUIRED R

   100 a                900 mwf  10009 Atmore       Samuel            1        .
   100 a                900 mwf  24365 McDowell     Denise            1        .
   100 a                900 mwf  27111 Hutchinson   Christine         1        .
   100 a                900 mwf  55287 Swift        Jennifer          1        .
   100 a                900 mwf  62432 Jamieson     Bailey            1        .
   100 a                900 mwf  89765 Paulsen      Patrick           2        1 psy
   100 a                900 mwf  43289 Sweeney      John              2        1 soc
   100 a                900 mwf  23763 Baker        Catherine         2        . adm
   100 a                900 mwf  22304 Jones        Barbara           3        1 adm
   100 a                900 mwf  27001 Jacobsen     Richard           3        1 pol
   100 a                900 mwf  35760 Roberts      Elizabeth         3        1 his
   100 a                900 mwf  64352 Atkinson     Thomas            3        . csi
   100 a                900 mwf  54338 Darfler      Marleena          4        1 che
   100 a                900 mwf  79885 Klein        Janet             4        1 che

Number of cases read =      14    Number of cases listed =      14
```

6.11 Nonparallel Files—Different Cases, Different Variables

In many circumstances input files do not contain corresponding cases. Some cases may be present in one file and not in another. For example, students are not always present in class on the day of an examination: some may have overslept, others may have dropped the class. The instructor might have one file with records for all students who registered for the class and another file with records for students who took the exam. By using a key variable such as student number, the instructor can append test scores to the appropriate student record.

Up to ten variables can be named as keys. The key variables must be present in each input file, and each file must be sorted in ascending order on the key variables. The key variables are named on the BY subcommand, which follows all other input file specifications.

For example, the raw data file containing the student identification number (STUNUM) and exam grades is shown in Figure 6.11a. To combine this file with the student roster created in Section 6.10, you must first sort each file in order of the key variable:

```
GET FILE='SOC100A.GRA'/DROP=MEETTIME DAYS.
SORT CASES BY STUNUM.

SAVE OUTFILE='SORTS100.ASF'.

DATA LIST FILE='TEST1.DAT'/STUNUM 1-5 SCORE1 7-8.
SORT CASES BY STUNUM.

TITLE 'GRADES THROUGH EXAM1'.
JOIN MATCH FILE='SORTS100.ASF'/FILE= * /
           BY STUNUM/
           MAP.
LIST.
SAVE OUTFILE='GRAS1AOS.SYF'.
```

The GET subcommand retrieves the system file containing the roster, omitting variables MEETTIME and DAYS. The SORT command sorts cases in ascending order of STUNUM. The SAVE command saves the sorted version of the file for use with JOIN MATCH. The DATA LIST command defines the raw data file, TEST1.DAT, which contains two variables, STUNUM and SCORE. The second SORT command ensures that cases are in the same order as the roster file.

The FILE subcommands on the JOIN command specify system file SORTS-100.ASF as the first file and the active file as the second file. The BY subcommand indicates that cases from the two input files are to be matched only when they have the same value on key variable STUNUM. The MAP subcommand requests a map of the variables contained on each file (Figure 6.11b), and the LIST command shows the data saved into the system file GRAS1A0S.SYF (Figure 6.11c). Note that students who did not take the exam have the system-missing value for variable SCORE1.

Figure 6.11a Student number and exam grade file

```
10009 76
24365 86
27111 68
62432 91
43289 86
23763 55
22304 76
35760 84
64352 85
79885 78
54338 80
```

Figure 6.11b Map from nonparallel match

```
RESULT          SORTS100.ASF *
-------------   ------------- -------------
COURSE          COURSE
SECTION         SECTION
STUNUM          STUNUM        STUNUM
LASTNAME        LASTNAME
FSTNAME         FSTNAME
CLASSYR         CLASSYR
REQUIRED        REQUIRED
MAJOR           MAJOR
SCORE1                        SCORE1
```

Figure 6.11c Active file listing from nonparallel match

```
Page  10   GRADES THROUGH EXAM1                                           6/28/85

COURSE SECTION STUNUM LASTNAME     FSTNAME    CLASSYR REQUIRED MAJOR SCORE1

   100 a          10009 Atmore      Samuel           1        .            76
   100 a          22304 Jones       Barbara          3        1 adm        76
   100 a          23763 Baker       Catherine        2        . adm        55
   100 a          24365 McDowell    Denise           1        .            86
   100 a          27001 Jacobsen    Richard          3        1 pol         .
   100 a          27111 Hutchinson  Christine        1        .            68
   100 a          35760 Roberts     Elizabeth        3        1 his        84
   100 a          43289 Sweeney     John             2        1 soc        86
   100 a          54338 Darfler     Marleena         4        1 che        80
   100 a          55287 Swift       Jennifer         1        .             .
   100 a          62432 Jamieson    Bailey           1        .            91
   100 a          64352 Atkinson    Thomas           3        . csi        85
   100 a          79885 Klein       Janet            4        1 che        78
   100 a          89765 Paulsen     Patrick          2        1 psy         .

Number of cases read =       14    Number of cases listed =       14
```

When you match files with the BY subcommand, the resulting active file contains values for every variable on each input file. If a variable is not a key variable but is contained on more than one input file, the value from the first-named file is used in the resulting active file. If two variables have the same name but represent different information, you can use the RENAME subcommand to ensure that no information is lost (see Section 6.8).

If an input file contains cases with the same value or combination of values for key variables, variables from the other input files are matched only to the first case. Each subsequent duplicate case is assigned system-missing values for the remaining variables from the other input files. If values should be appended to all matching cases, use a TABLE match (see Section 6.13).

6.12 Sorting Key Variables

When a BY subcommand is used on JOIN, all input files must be sorted in ascending order of the key variables. The best way to ensure that files are sorted in the correct order is to sort each input file with the SORT command prior to issuing the JOIN command. For all input files except the active file, you will need to resave the file as a system file after specifying SORT, as shown in the example in Section 6.11.

When multiple keys are specified, an easy rule of thumb is to specify the variables on the SORT command in the same order as they are named on the BY subcommand. For example,

```
GET FILE='PREREG.SYF'.
SORT BY DEPT COURSE.
SAVE OUTFILE='PREREGS.SYF'.

GET FILE='LATEREG.SYF'.
SORT BY DEPT COURSE.

JOIN MATCH FILE='PREREGS.SYF'/ FILE=* /BY DEPT COURSE
```

sorts both files in ascending order of COURSE within categories of DEPT. Table 6.12 illustrates values of the variables sorted in ascending order.

Table 6.12 Sorting order for DEPT COURSE

DEPT	COURSE
hds	100
hds	101
hds	444
mat	100
mat	222
soc	100
soc	222
soc	310

6.13 Table and Case Files—Group and Case Data

A *table lookup* match joins variables from one file designated as a table to groups of corresponding cases based on common values of one or more key variables. The table file should contain only one case for each combination of key variables.

For example, an instructor might create a system file containing summary information on each class. Figure 6.13a shows a listing from this system file. Each case contains a unique combination of values for department, course number, and section. The instructor also has a system file containing all grades for all students he taught that year. A listing of this file is shown in Figure 6.13b. Each line contains individual student grades.

Figure 6.13a Listing of ALLCOURS.SYF file

```
DEPT COURSE  SECTION    EXAMS  CLASSAV

hds    272   a             4    92.02
hds    444   a             2    87.00
soc    100   a             4    86.82
soc    100   b             4    85.62
soc    222   a             6    84.92
soc    310   a             4    87.75
soc    444   a             3    94.35
```

Figure 6.13b Listing of GRADES84.SYF file

DEPT	COURSE	SECTION	STUNUM	LASTNAME	TEST1	TEST2	TEST3	TEST4	TEST5	TEST6	GRADE
hds	444	a	25851	Smith	81	93	.	.	.	.	87
hds	444	a	87034	Cobbleigh	91	96	.	.	.	.	94
hds	444	a	13792	Heisenhuer	93	94	.	.	.	.	94
hds	444	a	89763	Harris	66	85	.	.	.	.	76
hds	444	a	91913	Chambers	93	75	.	.	.	.	84
soc	100	a	10009	Atmore	97	95	95	92	.	.	95
soc	100	a	24365	McDowell	96	87	76	99	.	.	90
soc	100	a	27111	Hutchinson	73	81	91	75	.	.	80
soc	100	a	55287	Swift	.	.	.	.	.	.	.
soc	100	a	62432	Jamieson	89	84	81	87	.	.	85
soc	100	a	89765	Paulsen	.	.	.	.	.	.	.
soc	100	a	43289	Sweeney	95	75	76	82	.	.	82
soc	100	a	23763	Baker	86	79	90	99	.	.	89
soc	100	a	22304	Jones	67	95	98	78	.	.	85
soc	100	a	27001	Jacobsen	.	.	.	.	.	.	.
soc	100	a	35760	Roberts	94	88	87	95	.	.	91
soc	100	a	64352	Atkinson	68	82	90	85	.	.	81
soc	100	a	54338	Darfler	86	90	92	85	.	.	88
soc	100	a	79885	Klein	96	79	85	96	.	.	89
soc	100	b	10229	Smithe	81	71	85	86	.	.	81
soc	100	b	34365	Farroro	88	71	94	96	.	.	87
soc	100	b	25111	Huber	71	83	81	78	.	.	78
soc	100	b	55587	Westerman	76	96	91	98	.	.	90
soc	100	b	62442	Mount	87	84	77	98	.	.	87
soc	100	b	89763	Harris	95	98	84	100	.	.	94
soc	100	b	70395	Jones	68	95	98	90	.	.	88
soc	100	b	53763	Halley	91	91	92	81	.	.	89
soc	100	b	82224	Manova	92	82	97	81	.	.	88
soc	100	b	97291	Sullivan	67	80	89	80	.	.	79
soc	100	b	50762	Dexter	80	90	85	98	.	.	88
soc	100	b	62752	Colby	88	72	79	77	.	.	79
soc	100	b	54358	Fiorello	65	84	95	96	.	.	85
soc	222	a	12239	Smith	84	100	80	96	83	93	89
soc	222	a	35467	Freiberg	75	98	80	95	86	95	88
soc	222	a	65141	Mills	74	78	94	76	81	90	82
soc	222	a	53287	West	73	99	77	75	80	91	83
soc	222	a	39742	Maus	70	95	82	83	79	80	82
soc	222	a	9763	Herrmann	87	82	78	.	.	.	.
soc	222	a	48364	Daveport	84	76	94	93	97	77	87
soc	222	a	81043	Hill	71	87	87	98	91	80	86
soc	222	a	42524	Mason	69	78	84	85	82	93	82
soc	222	a	95591	Evans	80	76	92	83	93	77	84
soc	222	a	61932	Bates	93	75	86	97	82	85	86
soc	222	a	93132	Hughes	80	88	84	92	99	81	87
soc	222	a	44359	Bates	70	89	84	96	85	74	83
soc	310	a	42305	O'Kane	75	87	84	87	.	.	83
soc	310	a	60611	Hallet	75	85	91	88	.	.	85
soc	310	a	14420	Robinson	69	87	89	97	.	.	86
soc	310	a	4901	Nace	91	96	97	82	.	.	92
soc	310	a	99999	Liebmann	92	92	81	82	.	.	87
soc	310	a	43995	Countryman	80	80	85	93	.	.	85
soc	310	a	60618	Adams	72	.	.	.	.	.	.
soc	310	a	14520	Stephenson	94	90	96	87	.	.	92
soc	310	a	22112	Jensen	80	91	98	100	.	.	92

Number of cases read = 54 Number of cases listed = 54

With these two files, the instructor can use JOIN to match variables from the course file to individual student records and compute a variable showing how each student did in relation to the class averages:

```
JOIN MATCH FILE='GRADES84.SYF'/TABLE='ALLCOURS.SYF'
          /BY DEPT COURSE SECTION
          /MAP.

COMPUTE GRADDIFF=GRADE-CLASSAV.

LIST VAR=DEPT COURSE SECTION STUNUM LASTNAME GRADE GRADDIFF.
```

The FILE subcommand on JOIN names the student file, and the TABLE subcommand designates the course file a table. The BY subcommand names the three key variables, and the MAP subcommand requests a map of the variables in each file (see Figure 6.13c). The COMPUTE command creates a new variable based on the difference between GRADE, which comes from the GRADES84.SYF file, and CLASSAV, which comes from the ALLCOURS.SYF file.

To perform the match, JOIN reads a case in the student file and "looks up" a match of key variables in the table file. When a match is found, the values from the table file are appended to the case. Nonmatching cases from the table file are ignored. Nonmatching cases from other input files are assigned system-missing values for variables named in the TABLE file. When JOIN finds more than one case in the student file for a particular combination of DEPT, COURSE, and SECTION, a warning message indicating there are duplicate cases is issued. Usually this message can be ignored.

The results of the LIST command, shown in Figure 6.13d show some of the variables in the resulting active file. Note that no cases appear for courses in the table file that did not have matches in the gradebook file.

Figure 6.13c Map from file and table match

RESULT	GRADES84.SYF	ALLCOURS.SYF
COURSE	COURSE	COURSE
SECTION	SECTION	SECTION
STUNUM	STUNUM	
LASTNAME	LASTNAME	
DEPT	DEPT	DEPT
TEST1	TEST1	
TEST2	TEST2	
TEST3	TEST3	
TEST4	TEST4	
TEST5	TEST5	
TEST6	TEST6	
GRADE	GRADE	
EXAMS		EXAMS
CLASSAV		CLASSAV

Figure 6.13d Listing of cases from table look-up match

DEPT	COURSE	SECTION	STUNUM	LASTNAME	GRADE	GRADDIFF
hds	444	a	25851	Smith	87	0.0
hds	444	a	87034	Cobbleigh	94	7.00
hds	444	a	13792	Heisenhuer	94	7.00
hds	444	a	89763	Harris	76	-11.00
hds	444	a	91913	Chambers	84	-3.00
soc	100	a	10009	Atmore	95	8.18
soc	100	a	24365	McDowell	90	3.18
soc	100	a	27111	Hutchinson	80	-6.82
soc	100	a	55287	Swift	.	.
soc	100	a	62432	Jamieson	85	-1.82
soc	100	a	89765	Paulsen	.	.
soc	100	a	43289	Sweeney	82	-4.82
soc	100	a	23763	Baker	89	2.18
soc	100	a	22304	Jones	85	-1.82
soc	100	a	27001	Jacobsen	.	.
soc	100	a	35760	Roberts	91	4.18
soc	100	a	64352	Atkinson	81	-5.82
soc	100	a	54338	Darfler	88	1.18
soc	100	a	79885	Klein	89	2.18
soc	100	b	10229	Smithe	81	-4.62
soc	100	b	34365	Farroro	87	1.38
soc	100	b	25111	Huber	78	-7.62
soc	100	b	55587	Westerman	90	4.38
soc	100	b	62442	Mount	87	1.38
soc	100	b	89763	Harris	94	8.38
soc	100	b	70395	Jones	88	2.38
soc	100	b	53763	Halley	89	3.38
soc	100	b	82224	Manova	88	2.38
soc	100	b	97291	Sullivan	79	-6.62
soc	100	b	50762	Dexter	88	2.38
soc	100	b	62752	Colby	79	-6.62
soc	100	b	54358	Fiorello	85	-.62
soc	222	a	12239	Smith	89	4.08
soc	222	a	35467	Freiberg	88	3.08
soc	222	a	65141	Mills	82	-2.92
soc	222	a	53287	West	83	-1.92
soc	222	a	39742	Maus	82	-2.92
soc	222	a	9763	Herrmann	.	.
soc	222	a	48364	Daveport	87	2.08
soc	222	a	81043	Hill	86	1.08
soc	222	a	42524	Mason	82	-2.92
soc	222	a	95591	Evans	84	-.92
soc	222	a	61932	Bates	86	1.08
soc	222	a	93132	Hughes	87	2.08
soc	222	a	44359	Bates	83	-1.92
soc	310	a	42305	O'Kane	83	-4.75
soc	310	a	60611	Hallet	85	-2.75
soc	310	a	14420	Robinson	86	-1.75
soc	310	a	4901	Nace	92	4.25
soc	310	a	99999	Liebmann	87	-.75
soc	310	a	43995	Countryman	85	-2.75
soc	310	a	60618	Adams	.	.
soc	310	a	14520	Stephenson	92	4.25
soc	310	a	22112	Jensen	92	4.25

Number of cases read = 54 Number of cases listed = 54

6.14 OBTAINING GROUP DATA: PROCEDURE AGGREGATE

Procedure AGGREGATE computes summary measures such as the sum and mean across groups of cases and produces either an SPSS/PC+ system file or a new active file containing one case per group. The variables on the resulting aggregated file are summary measures.

For example, consider a file of grades for each student enrolled in each class taught by an instructor. AGGREGATE can be used to create a class file containing such items as mean grade for each class, number of students completing the class, and percentage of students failing a class. In the new file, each case is a class and the values are aggregated information on students.

AGGREGATE often is used in conjuction with JOIN MATCH. For example, once you have obtained class averages with AGGREGATE, you can add the average grade for the class to each student's record and compare each student's grade to the average. In this operation, the aggregated file is used as a table file in the JOIN command.

To use AGGREGATE, you must specify three sets of information: the aggregated file, the variables that define groups, and the functions that create the aggregated variables. Optionally, you can specify missing-value treatments and whether the input file has been sorted according to the grouping variables.

6.15 Specifying the File Destination

The file produced by AGGREGATE either replaces the active file or is written as an SPSS/PC+ system file. The OUTFILE subcommand determines the destination of the aggregated file and must be the first specification.

To create and save a system file, specify a DOS filename enclosed in apostrophes on the OUTFILE subcommand. For example, assume you are creating an aggregated file from a class roster file. The sequence of commands is

```
GET FILE='ALLCLASS.SYF'.
AGGREGATE OUTFILE='ENROLLN.SYF'
        /BREAK=DEPT COURSE SECTION
        /NUMSTU=NU.
```

In this example, the new aggregated system file is written to the file ENROLLN.SYF, and the active file remains unchanged. To replace the active file with the aggregated file, specify an asterisk instead of the filename on the OUTFILE subcommand, as in:

```
GET FILE='ALLCLASS.SYF'.
AGGREGATE OUTFILE=*
        /BREAK=DEPT COURSE SECTION
        /NUMSTU=NU.
```

When you specify the active file on OUTFILE, the aggregated file is not permanently saved unless you use the SAVE command after the AGGREGATE procedure.

6.16 Grouping Cases

Procedure AGGREGATE summarizes groups of cases. A *break group* is a set of cases on the input file that have the same values for a variable or set of variables. Each break group defines one case on the new aggregated file.

For example, in a file of students registered for different classes, each case is a student and includes variables for class name, course number, department, student identification number, major, year in school, and other attributes. Each of these variables can be used individually or jointly to group the students. If you were to aggregate by course number, all cases with the same course number would constitute a group. If you were to aggregate by department and course number, all students taking Sociology 100 would be in one group, all students taking Sociology 200 would be in another, all students in Human Development 100 in another, and so forth. Each combination of department and course number is a break group.

The BREAK subcommand defines the break group variables and must follow the OUTFILE subcommand. You can name as many variables as you want on the BREAK subcommand, and you can use keyword TO to refer to a set of adjacent variables on the file. For example, to name variables DEPT, COURSE, and SECTION as the grouping variables, specify:

```
GET FILE='ALLCLASS.SYF'.
AGGREGATE OUTFILE=*
       /BREAK=DEPT COURSE SECTION
       /NUMSTU=NU.

LIST.
```

The GET command names the ALLCLASS.SYF system file, which contains information on all students in all classes. The AGGREGATE command directs the aggregated file to the active file and uses values for variables DEPT and COURSE to group cases. By default, AGGREGATE arranges cases in ascending order on each variable named on the BREAK subcommand. Figure 6.16a shows the contents of the resulting active file from the LIST command. The aggregated file contains five cases, one for each unique combination of DEPT and COURSE. Note that AGGREGATE saves all variables named on the BREAK subcommand on the resulting aggegated file. Each break variable retains all dictionary information from the input file.

Figure 6.16a Aggregated file in default order

```
DEPT COURSE  SECTION  NUMSTU

hds     444  a             5
soc     100  a            14
soc     100  b            13
soc     222  a            13
soc     310  a             9

Number of cases read =        5    Number of cases listed =        5
```

The BREAK subcommand also lets you specify the arrangement of cases on the resulting file. For example, the following AGGREGATE command specifies that the aggregated active file is to be sorted in descending order of variable DEPT, in ascending order of variable COURSE, and descending order of variable SECTION.

```
GET FILE='ALLCLASS.SYF'.
AGGREGATE OUTFILE=*
/BREAK=DEPT (D) COURSE (A) SECTION (D)
       /NUMSTU=NU.

LIST.
```

Figure 6.16b shows the resulting active file sorted in the designated order.

Figure 6.16b Sorted aggregated file

```
DEPT COURSE  SECTION  NUMSTU

soc     100  b            13
soc     100  a            14
soc     222  a            13
soc     310  a             9
hds     444  a             5

Number of cases read =        5    Number of cases listed =         5
```

You can specify sorting order for any variable named on the BREAK subcommand. The sort designation applies to all preceding undesignated variables. In the absence of a designation, the resulting aggregated file is sorted in ascending order on each variable. String values sorted in ascending order follow the ASCII code value sequence: first numbers, then upper-case letters, and finally lower-case letters (see the Appendix in your IBM *BASIC Reference* manual). Sorting strings in descending order reverses the sequence, starting with lower-case letters.

6.17 Using a Presorted Active File

When your active file is already sorted in the order you want your aggregated file, you can use the PRESORTED subcommand. When you specify PRESORTED, each time a *different* value or combination of values is encountered on variables named on the BREAK subcommand a new aggregate case is created. If the input file is not sorted and PRESORTED is specified, AGGREGATE produces multiple cases for combinations of the break variables.

For example, the commands

```
GET FILE='ALLCLASS.SYF'.
AGGREGATE OUTFILE=*
        /PRESORTED
        /BREAK=DEPT COURSE SECTION
        /NUMSTU=NU.

LIST.
```

tell SPSS/PC+ that the ALLCLASS.SYF system file is already sorted by DEPT, COURSE, and SECTION.

When it is used, PRESORTED must be specified between the OUTFILE and BREAK subcommands. You cannot specify sort order for the resulting aggregated file by designating (A) or (D) for variables named on the BREAK subcommand (see Section 6.16). If the PRESORTED subcommand is used and sort order is specified on the BREAK subcommand, the aggregated file is sorted in the order of the input file and a warning is issued stating that the sort directions specified on BREAK are ignored.

6.18 Creating Aggregated Variables

Each variable on an aggregated file is created by applying an aggregate function to a variable on the active file. The simplest specification is a *target variable list* followed by an equals sign, the function keyword, and list of *source variables* in parentheses. The aggregate functions available are listed in Section 6.19.

For example, using the data shown in Figure 6.18a, the commands

```
DATA LIST FILE='ALLGRADE.DAT'/ COURSE 2-4 SECTION 6 (A)
      STUNUM 8-12 LASTNAME 14-25 (A) DEPT 27-29 (A) EXAMS 31
      TEST1 TO TEST6 33-56.
FORMATS TEST1 TO TEST6 (F2.0).
COMPUTE FINGRADE=RND((TEST1+TEST2+TEST3+TEST4+TEST5+TEST6)/EXAMS).
FORMATS FINGRADE (F3).

AGGREGATE OUTFILE='CLASSAV1.TOT'
         /BREAK=DEPT COURSE SECTION
         /CLASSAV=MEAN(FINGRADE).
```

define an active file and then use it to create a new variable CLASSAV as the mean of variable FINGRADE for each section of a course within in each department. CLASSAV is the target variable for mean values of FINGRADE for each break group.

Figure 6.18a ALLGRADE.DAT data set

```
100 a 10009 Atmore       soc 4  97  95  95  92   0   0
100 a 24365 McDowell     soc 4  96  87  76  99   0   0
100 a 27111 Hutchinson   soc 4  73  81  91  75   0   0
100 a 55287 Swift        soc 4                   0   0
100 a 62432 Jamieson     soc 4  89  84  81  87   0   0
100 a 89765 Paulsen      soc 4                   0   0
100 a 43289 Sweeney      soc 4  95  75  76  82   0   0
100 a 23763 Baker        soc 4  86  79  90  99   0   0
100 a 22304 Jones        soc 4  67  95  98  78   0   0
100 a 27001 Jacobsen     soc 4                   0   0
100 a 35760 Roberts      soc 4  94  88  87  95   0   0
100 a 64352 Atkinson     soc 4  68  82  90  85   0   0
100 a 54338 Darfler      soc 4  86  90  92  85   0   0
100 a 79885 Klein        soc 4  96  79  85  96   0   0
100 b 10229 Smithe       soc 4  81  71  85  86   0   0
100 b 34365 Farroro      soc 4  88  71  94  96   0   0
100 b 25111 Huber        soc 4  71  83  81  78   0   0
100 b 55587 Westerman    soc 4  76  96  91  98   0   0
100 b 62442 Mount        soc 4  87  84  77  98   0   0
100 b 89763 Harris       soc 4  95  98  84 100   0   0
100 b 70395 Jones        soc 4  68  95  98  90   0   0
100 b 53763 Halley       soc 4  91  91  92  81   0   0
100 b 82224 Manova       soc 4  92  82  97  81   0   0
100 b 97291 Sullivan     soc 4  67  80  89  80   0   0
100 b 50762 Dexter       soc 4  80  90  85  98   0   0
100 b 62752 Colby        soc 4  88  72  79  77   0   0
100 b 54358 Fiorello     soc 4  65  84  95  96   0   0
222 a 12239 Smith        soc 6  84 100  80  96  83  93
222 a 35467 Freiberg     soc 6  75  98  80  95  86  95
222 a 65141 Mills        soc 6  74  78  94  76  81  90
222 a 53287 West         soc 6  73  99  77  75  80  91
222 a 39742 Maus         soc 6  70  95  82  83  79  80
222 a  9763 Herrmann     soc 6  87  82  78       0   0
222 a 48364 Daveport     soc 6  84  76  94  93  97  77
222 a 81043 Hill         soc 6  71  87  87  98  91  80
222 a 42524 Mason        soc 6  69  78  84  85  82  93
222 a 95591 Evans        soc 6  80  76  92  83  93  77
222 a 61932 Bates        soc 6  93  75  86  97  82  85
222 a 93132 Hughes       soc 6  80  88  84  92  99  81
222 a 44359 Bates        soc 6  70  89  84  96  85  74
310 a 42305 O'Kane       soc 4  75  87  84  87   0   0
310 a 60611 Hallet       soc 4  75  85  91  88   0   0
310 a 14420 Robinson     soc 4  69  87  89  97   0   0
310 a  4901 Nace         soc 4  91  96  97  82   0   0
310 a 99999 Liebmann     soc 4  92  92  81  82   0   0
310 a 43995 Countryman   soc 4  80  80  85  93   0   0
310 a 60618 Adams        soc 4  72               0   0
310 a 14520 Stephenson   soc 4  94  90  96  87   0   0
310 a 22112 Jensen       soc 4  80  91  98 100   0   0
444 a 25851 Smith        hds 2  81  93   0   0   0   0
444 a 87034 Cobbleigh    hds 2  91  96   0   0   0   0
444 a 13792 Heisenhuer   hds 2  93  94   0   0   0   0
444 a 89763 Harris       hds 2  66  85   0   0   0   0
444 a 91913 Chambers     hds 2  93  75   0   0   0   0
```

The target and source variable lists must be of equal length, as in:

```
AGGREGATE OUTFILE='CLASSAV2.TOT'
         /BREAK=DEPT COURSE SECTION
         /CLASSAV TEST1AVE =MEAN(FINGRADE TEST1).
```

This specification creates two aggregated variables: CLASSAV is the mean of FINGRADE, and TEST1AVE is the mean of TEST1.

You can use keyword TO in both the target and source variable lists. For example, to create averages for each test, specify:

```
AGGREGATE OUTFILE='CLASSAV3.TOT'
          /BREAK=DEPT COURSE SECTION
          /TESTAVE1 TO TESTAVE6=MEAN(TEST1 TO TEST6).
```

Any number of functions can be used to create variables. You can define up to 200 variables, including the break variables. Separate each function specification with a slash, as in:

```
AGGREGATE OUTFILE='CLASSAVE.TOT'
          /BREAK=DEPT COURSE SECTION
          /TESTAVE1 AVEFINAL=MEAN(TEST1 FINGRADE)
          /TESTSD1 SDFINAL=SD(TEST1 FINGRADE).

GET FILE='CLASSAVE.TOT'.
LIST.
```

Figure 6.18b shows the results from the LIST command.

Figure 6.18b Listing of aggregated variables and values

```
DEPT COURSE SECTION TESTAVE1 AVEFINAL  TESTSD1  SDFINAL

hds     444 a          84.80    87.00    11.63     7.55
soc     100 a          86.09    86.82    11.53     4.64
soc     100 b          80.69    85.62    10.40     4.91
soc     222 a          77.69    84.92     7.60     2.54
soc     310 a          80.89    87.75     9.28     3.69

Number of cases read =        5    Number of cases listed =        5
```

6.19 Functions

The following functions are available in procedure AGGREGATE:

SUM(varlist)	*Sum across cases.* Dictionary formats are F8.2.
MEAN(varlist)	*Mean across cases.* Dictionary formats are F8.2.
SD(varlist)	*Standard deviation across cases.* Dictionary formats are F8.2.
MAX(varlist)	*Maximum value across cases.* Complete dictionary information is copied from the source variables to the target variables.
MIN(varlist)	*Minimum value across cases.* Complete dictionary information is copied from the source variables to the target variables.
PGT(varlist,value)	*Percentage of cases greater than value.* Dictionary formats are F5.1.
PLT(varlist,value)	*Percentage of cases less than value.* Dictionary formats are F5.1.
PIN(varlist,value1,value2)	*Percentage of cases between value1 and value2 inclusive.* Dictionary formats are F5.1.
POUT(varlist,value1,value2)	*Percentage of cases not between value1 and value2.* Cases where the source variable equals value1 or value2 are not counted. Dictionary formats are F5.1.
FGT(varlist,value)	*Fraction of cases greater than value.* Dictionary formats are F5.3.
FLT(varlist,value)	*Fraction of cases less than value.* Dictionary formats are F5.3.

FIN(varlist,value1,value2) *Fraction of cases between value1 and value2 inclusive.* Dictionary formats are F5.3.

FOUT(varlist,value1,value2) *Fraction of cases not between value1 and value2.* Cases where the source variable equals value1 or value2 are not counted. Dictionary formats are F5.3.

N(varlist) *Weighted number of cases in break group.* Dictionary formats are F7.0 for unweighted files and F8.2 for weighted files.

NU(varlist) *Unweighted number of cases in break group.* Dictionary formats are F7.0.

NMISS(varlist) *Weighted number of missing cases.* Dictionary formats are F7.0 for unweighted files and F8.2 for weighted files.

NUMISS(varlist) *Unweighted number of missing cases.* Dictionary formats are F7.0.

FIRST(varlist) *First nonmissing observed value in break group.* Complete dictionary information is copied from the source variables to the target variables.

LAST(varlist) *Last nonmissing observed value in break group.* Complete dictionary information is copied from the source variables to the target variables.

The percentage functions (PGT, PLT, PIN, and POUT) return values between 0 and 100 inclusive. The fraction functions (FGT, FLT, FIN, and FOUT) return values between 0 and 1 inclusive.

6.20 Function Arguments

Only numeric variables can used with SUM, MEAN, and SD. Both long and short string variables can be used with all other functions. For example, to obtain the percentage of females when SEX is coded M and F, specify either

```
PCTFEM=PLT(SEX,'M')
```

or

```
PCTFEM=PIN(SEX,'F','F')
```

Blanks and commas can be used interchangeably to separate arguments to functions.

Functions PGT, PLT, PIN, POUT, FGT, FLT, FIN, and FOUT take values as arguments. PGT, PLT, FGT, and FLT have one argument that is a value. PIN, POUT, FIN, and FOUT have two value arguments. For example, the specification

```
LOQUART1 LOFINAL=PLT(TEST1 FINGRADE,75)
```

assigns the percentage of cases with values less than 75 for TEST1 to LOQUART1 and for FINGRADE to LOFINAL. The specification

```
CGRADE=FIN(FINGRADE,70,79)
```

assigns the fraction of cases having final grade scores of 70 to 79 to CGRADE. For PIN, POUT, FIN, and FOUT, the first argument should be lower than the second argument. If the first argument is higher, AGGREGATE automatically reverses them and prints a warning message.

The N and NU functions do not require arguments. Without arguments they return the number of weighted and unweighted cases in a break group. If you supply a variable list, they return the weighted and unweighted number of nonmissing cases for the variables specified.

6.21 Labels and Formats

With the exception of the functions MAX, MIN, FIRST, and LAST, which copy complete dictionary information from the source variable, new variables created by AGGREGATE have the default dictionary formats described in Section 6.19 and no labels. To label a new variable, specify a label in apostrophes immediately following the variable name, as in:

```
AGGREGATE OUTFILE='CLASSAV2.TOT'
 /BREAK=DEPT COURSE
 /CLASSAV 'Final Average' TEST1AVE '1st Test Average'=MEAN(FINGRADE
TEST1).
```

The label applies only to the immediately preceding variable.

If you are specifying the aggregated file as the new active file, you can also use the VARIABLE LABELS command to add labels, as in:

```
AGGREGATE OUTFILE=*
          /BREAK=DEPT COURSE
          /CLASSAV TEST1AVE=MEAN(FINGRADE TEST1).
VARIABLE LABELS CLASSAV 'Final Average' / TEST1AVE '1st Test Average'.
```

Use the FORMATS command to change dictionary formats for an active file created from AGGREGATE, as in:

```
AGGREGATE OUTFILE=*
          /BREAK=DEPT COURSE
          /CLASSAV TEST1AVE=MEAN(FINGRADE TEST1).
VARIABLE LABELS CLASSAV 'Final Average' / TEST1AVE '1st Test Average'.
FORMATS CLASSAV TEST1AVE (F4.1).
```

The formats for CLASSAV and TEST1AVE are changed from the default width of eight with two decimal places (F8.2) to a width of four columns with one decimal digit (F4.1).

If the aggregate file is saved as a system file, the file must be retrieved with a GET command before the FORMATS and VARIABLE LABELS commands can be issued.

6.22 Missing Data

By default, all nonmissing cases are used in the computation of aggregate variables. To force target variables to system-missing if any of the cases in the group are missing on the source variable, use the MISSING subcommand. The MISSING subcommand has one keyword specification, COLUMNWISE.

The MISSING subcommand follows the OUTFILE subcommand, as in:

```
AGGREGATE OUTFILE='CLASSREP.SYF'
    /MISSING=COLUMNWISE
    /BREAK=DEPT COURSE
    /TESTAVE1 AVEFINAL=MEAN(TEST1 FINGRADE).
    /TESTSD1 SDFINAL=SD(TEST1 FINGRADE).

GET FILE='CLASSREP.SYF'.
LIST.
```

The results of the LIST command are shown in Figure 6.22. You can compare these values with those shown in Figure 6.18b.

Figure 6.22 Listing of aggregated values with columnwise treatment

```
DEPT COURSE SECTION TESTAVE1 AVEFINAL  TESTSD1  SDFINAL

hds     444 a          84.80    87.00    11.63     7.55
soc     100 a            .        .        .         .
soc     100 b          80.69    85.62    10.40     4.91
soc     222 a          77.69      .       7.60       .
soc     310 a          80.89      .       9.28       .

Number of cases read =        5    Number of cases listed =        5
```

The MISSING subcommand has no effect on the N, NU, NMISS, or NUMISS functions. For example, N(TEST1) returns the same result for the default and for columnwise deletion.

6.23 Including Missing Values

To force a function to treat user-missing values as valid, follow the function name with a period, as in:

```
LOFINAL=PLT.(FINGRADE,75)
```

LOFINAL will equal the percentage of cases within the group with values less than 75 for FINGRADE even if some of the values are defined as missing.

To obtain the first value of AGE in a break group whether it is missing or not, specify:

```
FIRSTAGE = FIRST.(AGE)
```

If the first case in a break group has a user-missing value on AGE, FIRSTAGE is set to that value. Since variables created with FIRST have the same dictionary information as their source variables, the value for FIRSTAGE is still treated as user-missing on the aggregated file.

The period is ignored when used with N, NU, NMISS, and NUMISS if these functions have no argument. On the other hand, NMISS.(AGE) gives the number of cases on which AGE has the system-missing value. The effect of specifying the period on N, NU, NMISS, and NUMISS is illustrated by the following:

N = N. = N(AGE)+NMISS(AGE) = N.(AGE)+NMISS.(AGE)

NU = NU. = NU(AGE)+NUMISS(AGE) = NU.(AGE)+NUMISS.(AGE)

That is, the function N (the same as N. with no argument) is equal to the sum of cases with valid and with missing values for AGE, which is also equal to the sum of cases with either valid or user-missing values and with system-missing values for AGE. The same holds for the NU, NMISS, and NUMISS functions.

Table 6.23 demonstrates the effect of the MISSING subcommand and of including user-missing values. Each entry in the table is the number of cases used to compute the specified function for a particular break group of variable EDUC. The BREAK group has 10 nonmissing cases, 5 user-missing cases, and 2 system-missing cases. With the exception of the MEAN function, columnwise treatment produces the same results as the default for every function.

Table 6.23 Alternative missing-value treatments

Function	Default	Columnwise
N	17	17
N.	17	17
N(EDUC)	10	10
N.(EDUC)	15	15
MEAN(EDUC)	10	0
MEAN.(EDUC)	15	0
NMISS(EDUC)	7	7
NMISS.(EDUC)	2	2

6.24 ANNOTATED EXAMPLE FOR JOIN AND AGGREGATE

At the end of the semester, an instructor puts together a report outlining the individual performance of each student that took his classes. The commands below use JOIN and AGGREGATE to combine class grade books into a single file, develop summary variables of overall class performance, combine class and student variables, and then produce a listing of individual performance compared to class averages.

```
JOIN ADD FILE='SOC100A.FSF'
        /FILE='SOC100B.FSF'
        /FILE='SOC222A.FSF'
        /FILE='SOC310A.FSF'
        /FILE='HDS444A.FSF'
        /MAP.

SORT BY DEPT COURSE SECTION.

AGGREGATE OUTFILE='AVERAGE.FSF'
          /PRESORTED
          /BREAK=DEPT COURSE SECTION
          /AVEGRADE=MEAN(FINGRADE)
          /SDGRADE=SD(FINGRADE)
          /PCTFAIL=PLT(FINGRADE,65).

JOIN MATCH FILE=* /TABLE='AVERAGE.FSF'
          /BY DEPT COURSE SECTION.

COMPUTE GRADEDEV=FINGRADE-AVEGRADE.

LIST VARS=DEPT COURSE SECTION LASTNAME STUNUM FINGRADE GRADEDEV.

SAVE OUTFILE='YEAREND.FSF'.
```

- The JOIN ADD command combines cases from each of five system files. Variables from the files and in the resulting active file are listed with the MAP subcommand.
- The SORT command orders cases in ascending order of SECTION within categories of COURSE within categories of DEPT.
- The OUTFILE subcommand on AGGREGATE indicates that the aggregated file should be directed to system file AVERAGE.FSF. The PRESORTED specification states that the file is already sorted.
- The BREAK subcommand creates one case for each distinct combination of DEPT, COURSE, and SECTION. Each case contains values of the break variables, as well as three variables produced with three different aggregate functions, MEAN, SD, and PLT.
- The JOIN command matches cases in the active file with cases in the aggregate file, which is designated a table file. Cases are matched based on the key variables DEPT, COURSE, and SECTION. A new active file is available for other SPSS/PC+ commands.
- The COMPUTE command creates a variable based on the difference between the student's grade and the class average.
- The LIST command produces a simple listing of the named variables from the current active file.
- The SAVE command saves the active file for use in subsequent sessions.

Figure 6.24a Map of variables in input and resulting files

```
RESULT         SOC100A.FSF   SOC100B.FSF   SOC222A.FSF   SOC310A.FSF   HDS444A.FSF
-------------  ------------  ------------  ------------  ------------  ------------
COURSE         COURSE        COURSE        COURSE        COURSE        COURSE
SECTION        SECTION       SECTION       SECTION       SECTION       SECTION
STUNUM         STUNUM        STUNUM        STUNUM        STUNUM        STUNUM
LASTNAME       LASTNAME      LASTNAME      LASTNAME      LASTNAME      LASTNAME
DEPT           DEPT          DEPT          DEPT          DEPT          DEPT
EXAMS          EXAMS         EXAMS         EXAMS         EXAMS         EXAMS
TEST1          TEST1         TEST1         TEST1         TEST1         TEST1
TEST2          TEST2         TEST2         TEST2         TEST2         TEST2
TEST3          TEST3         TEST3         TEST3         TEST3         TEST3
TEST4          TEST4         TEST4         TEST4         TEST4         TEST4
TEST5          TEST5         TEST5         TEST5         TEST5         TEST5
TEST6          TEST6         TEST6         TEST6         TEST6         TEST6
FINGRADE       FINGRADE      FINGRADE      FINGRADE      FINGRADE      FINGRADE
```

Figure 6.24b LISTING of cases after table lookup JOIN MATCH

DEPT	COURSE	SECTION	LASTNAME	STUNUM	FINGRADE	GRADEDEV
hds	444	a	Smith	25851	87	0.0
hds	444	a	Cobbleigh	87034	94	7.00
hds	444	a	Heisenhuer	13792	94	7.00
hds	444	a	Harris	89763	76	-11.00
hds	444	a	Chambers	91913	84	-3.00
soc	100	a	Atmore	10009	95	8.18
soc	100	a	McDowell	24365	90	3.18
soc	100	a	Hutchinson	27111	80	-6.82
soc	100	a	Swift	55287	.	.
soc	100	a	Jamieson	62432	85	-1.82
soc	100	a	Paulsen	89765	.	.
soc	100	a	Sweeney	43289	82	-4.82
soc	100	a	Baker	23763	89	2.18
soc	100	a	Jones	22304	85	-1.82
soc	100	a	Jacobsen	27001	.	.
soc	100	a	Roberts	35760	91	4.18
soc	100	a	Atkinson	64352	81	-5.82
soc	100	a	Darfler	54338	88	1.18
soc	100	a	Klein	79885	89	2.18
soc	100	b	Smithe	10229	81	-4.62
soc	100	b	Farroro	34365	87	1.38
soc	100	b	Huber	25111	78	-7.62
soc	100	b	Westerman	55587	90	4.38
soc	100	b	Mount	62442	87	1.38
soc	100	b	Harris	89763	94	8.38
soc	100	b	Jones	70395	88	2.38
soc	100	b	Halley	53763	89	3.38
soc	100	b	Manova	82224	88	2.38
soc	100	b	Sullivan	97291	79	-6.62
soc	100	b	Dexter	50762	88	2.38
soc	100	b	Colby	62752	79	-6.62
soc	100	b	Fiorello	54358	85	-.62
soc	222	a	Smith	12239	89	4.08
soc	222	a	Freiberg	35467	88	3.08
soc	222	a	Mills	65141	82	-2.92
soc	222	a	West	53287	83	-1.92
soc	222	a	Maus	39742	82	-2.92
soc	222	a	Herrmann	9763	.	.
soc	222	a	Daveport	48364	87	2.08
soc	222	a	Hill	81043	86	1.08
soc	222	a	Mason	42524	82	-2.92
soc	222	a	Evans	95591	84	-.92
soc	222	a	Bates	61932	86	1.08
soc	222	a	Hughes	93132	87	2.08
soc	222	a	Bates	44359	83	-1.92
soc	310	a	O'Kane	42305	83	-4.75
soc	310	a	Hallet	60611	85	-2.75
soc	310	a	Robinson	14420	86	-1.75
soc	310	a	Nace	4901	92	4.25
soc	310	a	Liebmann	99999	87	-.75
soc	310	a	Countryman	43995	85	-2.75
soc	310	a	Adams	60618	.	.
soc	310	a	Stephenson	14520	92	4.25
soc	310	a	Jensen	22112	92	4.25

Number of cases read = 54 Number of cases listed = 54

Contents

Chapter 7 Data Tabulation: Procedure FREQUENCIES

Few people would dispute the effects of "rainy days and Mondays" on the body and spirit. It has long been known that more suicides occur on Mondays than other days of the week. Recently an excess of cardiac deaths on Mondays has also been noted (Rabkin et al., 1980). This chapter looks at data from the Western Electric study, in which the incidence of coronary heart disease in 2,017 men was monitored for 20 years. Here, the day of the week on which deaths occurred is examined to see if an excess of deaths occurred on Mondays.

7.1 A FREQUENCY TABLE

A first step in analyzing data on day of death might be to count the number of deaths occurring on each day of the week. Figure 7.1a contains this information.

Figure 7.1a Frequency of death by day of week

DAYOFWK DAY OF DEATH

Value Label	Value	Frequency	Percent	Valid Percent	Cum Percent
SUNDAY	1	19	7.9	17.3	17.3
MONDAY	2	11	4.6	10.0	27.3
TUESDAY	3	19	7.9	17.3	44.5
WEDNESDAY	4	17	7.1	15.5	60.0
THURSDAY	5	15	6.3	13.6	73.6
FRIDAY	6	13	5.4	11.8	85.5
SATURDAY	7	16	6.7	14.5	100.0
MISSING	9	130	54.2	MISSING	
	TOTAL	240	100.0	100.0	

Valid Cases 110 Missing Cases 130

Each row of the frequency table describes a particular day of the week. The last row represents cases for which the day of death is not known or that have not died. For the table in Figure 7.1a, there are 110 cases for which day of death is known. The first column *(value label)* gives the name of the day, while the second column contains the *value,* which is the symbol given to the computer to represent the day.

The number of people dying on each day is in the third column *(frequency).* Monday is the least-frequent death day with 11 deaths. These 11 deaths are 4.6% (11/240) of all cases. This *percentage* is in the fourth column. However, of the 240 people, 130 had no day of death. The 11 deaths on Monday are 10.0% of the total deaths for which death days are known (11/110). This *valid percentage* is in the fifth column.

The last column of the table contains the *cumulative percentage.* For a particular day, this percentage is the sum of the valid percentages of that day and all other days that precede it in the table. For example, the cumulative percentage for Tuesday is 44.5, which is the sum of the percentage of deaths that occurred on Sunday, Monday, and Tuesday. It is calculated as

$$\frac{19}{110} + \frac{11}{110} + \frac{19}{110} = \frac{49}{110} = 44.5\% \qquad \textbf{Equation 7.1}$$

Figure 7.1b is a frequency table of day of death for cases who experienced sudden coronary death. This is a particularly interesting category since it is thought that sudden death may be related to stressful events such as return to the work environment. In Figure 7.1b there does not appear to be a clustering of deaths on any particular day. Sunday has 22.2% of the deaths, while Thursday has 8.3%. Since the number of sudden deaths in the table is small, the magnitude of the observed fluctuations is not very impressive.

Figure 7.1b Frequency of sudden cardiac death by day of the week

DAYOFWK DAY OF DEATH

Value Label	Value	Frequency	Percent	Valid Percent	Cum Percent
SUNDAY	1	8	22.2	22.2	22.2
MONDAY	2	4	11.1	11.1	33.3
TUESDAY	3	4	11.1	11.1	44.4
WEDNESDAY	4	7	19.4	19.4	63.9
THURSDAY	5	3	8.3	8.3	72.2
FRIDAY	6	6	16.7	16.7	88.9
SATURDAY	7	4	11.1	11.1	100.0
	TOTAL	36	100.0	100.0	

Valid Cases 36 Missing Cases 0

7.2 Visual Displays

While the numbers in the frequency table can be studied and compared, it is often useful to present results in a visually interpretable form. Figure 7.2a is a pie chart of the data displayed in Figure 7.1a. Each slice represents a day of the week. The size of the slice depends on the frequency of death for that day. Monday is represented by 10.0% of the pie chart since 10.0% of the deaths for which the day is known occurred on Monday.

Figure 7.2a Frequency of death by day of week (From SPSS Graphics, a mainframe product)

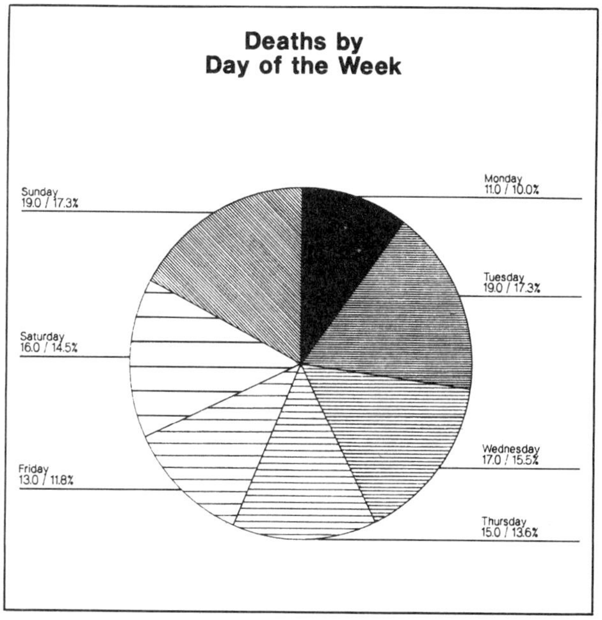

Another way to represent the data is with a bar chart, as shown in Figure 7.2b. There is a bar for each day, and the length of the bar is proportional to the number of deaths observed on that day. Inside each bar is the number of cases occurring on that day.

Figure 7.2b Frequency of death by day of the week

```
DAYOFWK   DAY OF DEATH

         SUNDAY ################################################## 19
         MONDAY ############################## 11
        TUESDAY ################################################## 19
      WEDNESDAY ############################################## 17
       THURSDAY ######################################## 15
         FRIDAY ################################### 13
       SATURDAY ########################################## 16

Valid Cases      110       Missing Cases    130
```

Only values that actually occur in the data are represented in the bar chart from procedure FREQUENCIES. For example, if no deaths took place on Thursday, no space would be left for Thursday and the bar for Wednesday would be followed by the one for Friday. If you chart the number of cars per family, the bar describing 6 cars may be next to the one for 25 cars if no family has 7 to 24 cars. Therefore, you should pay attention to where categories with no cases may occur.

Although the basic information presented by frequency tables, pie charts, and bar charts is the same, the visual displays enliven the data. Differences among the days of the week are apparent at a glance, eliminating the need to pore over columns of numbers.

7.3 What Day?

Although the number of sudden cardiac deaths is small in this study, the data in Figure 7.1b indicate that the number of deaths on Mondays is not particularly large. In fact, Sunday has the most deaths, slightly over 22%. A recent study of over a thousand sudden cardiac deaths in Rochester, Minnesota, also found a slightly increased incidence of death on weekends for men (Beard et al., 1982). The authors speculate that for men, this might mean "the home environment is more stressful than the work environment." But one should be wary of explanations that are not directly supported by data. It is only too easy to find a clever explanation for any statistical finding. (See Chapter 16 for further analysis of these data.)

7.4 Histograms

A frequency table or bar chart of all values for a variable is a convenient way of summarizing a variable that has a relatively small number of distinct values. Variables such as sex, country, and astrological sign are necessarily limited in the number of values they can have. For variables that can take on many different values, such as income to the penny or weight in ounces, a tally of the cases with each observed value may not be very informative. In the worst situation, when all cases have different values, a frequency table is little more than an ordered list of those values.

Variables that have many values can be summarized by grouping the values of the variables into intervals and counting the number of cases with values within each interval. For example, income can be grouped into $5,000 intervals such as 0–4999, 5000–9999, 10000–14999, and so forth, and the number of observations in each group can be tabulated. Such grouping should be done using SPSS/PC+ during the actual analysis of the data. As indicated in Chapter 2, the values for variables should be entered into the data file in their original, ungrouped form.

A histogram is a convenient way to display the distribution of such grouped values. Consider Figure 7.4, which is a histogram for body weight in pounds of the sample of 240 men from the Western Electric Study. The first column indicates the number of cases with values within the interval, while the second column gives the midpoint, or middle value, for the interval. Each row of symbols represents the number of cases with values in the interval. For example, the second row of the histogram has 10 symbols, which represent 10 men who weighed between 130 and 140 pounds in 1958. The number of cases represented by each symbol depends on the size of the sample and the maximum number of cases falling into an interval. Intervals that have no observations are included in the histogram but no symbols are displayed. This differs from a bar chart, which does not leave space for the empty categories.

A histogram can be used in any situation in which it is reasonable to group adjacent values. Histograms should not be used to display variables in which there is no underlying order to the values. For example, if 100 different religions are arbitrarily assigned codes of 1 to 100, grouping values into intervals is meaningless. Either a bar chart or a histogram in which each interval corresponds to a single value should be used to display such data.

Figure 7.4 A histogram of body weight

```
WT58        BODY WEIGHT, 1958 -- LBS

      Count   Midpoint

          5     125.00  #####
         10     135.00  ##########
         20     145.00  ####################
         36     155.00  ####################################
         43     165.00  ###########################################
         39     175.00  #######################################
         31     185.00  ###############################
         19     195.00  ###################
         15     205.00  ###############
          9     215.00  #########
         10     225.00  ##########
          1     235.00  #
          0     245.00
          1     255.00  #
          0     265.00
          1     275.00  #
                        I....+....I....+....I....+....I....+....I....+....I
                        0         10        20        30        40        50
                                  Histogram Frequency

Valid Cases      240        Missing Cases      0
```

7.5 Screening Data

Frequency tables, bar charts, and histograms can serve purposes other than summarizing data. Unexpected codes in the tables may indicate errors in data entry or coding. Cases with death days coded as 0 or 8 are in error if the numbers 1 through 7 represent the days of the week and 9 stands for unknown. Since errors in the data should be eliminated as soon as possible, it is a good idea to run frequency tables as the first step in analyzing data.

Frequency tables and visual displays can also help you identify cases with values that are unusual but possibly correct. For example, a tally of the number of cars in families may show a family with 25 cars. Although such a value is possible, especially if the survey did not specify cars in working condition, it raises suspicion and should be examined to ensure that it is really correct.

Incorrect data values distort the results of statistical analyses, and correct but unusual values may require special treatment. In either case, early identification is valuable.

7.6 RUNNING PROCEDURE FREQUENCIES

The FREQUENCIES procedure produces frequency tables, bar charts (for discrete variables), histograms (for continuous variables), and various descriptive statistics (means, standard deviations, percentiles, and so forth).

FREQUENCIES operates via subcommands. Only one, the VARIABLES subcommand, is required.

7.7 Specifying the Variables

The VARIABLES subcommand names the variables to be analyzed. Simply specify the names of the variables you want to analyze, as in the command

```
FREQUENCIES VARIABLES=RAISE82 AGE DEPT.
```

No other specification or subcommand is needed when only frequency tables are desired. You can use the keyword ALL to name all variables in the file, and the keyword TO to refer to consecutive variables in the file.

7.8 Formatting Options

Several formatting options are available via the FORMAT subcommand. You can control the formatting of tables and the sorting of categories within a table. The FORMAT subcommand affects all of the variables listed on the VARIABLES subcommand. You can use only one FORMAT subcommand per FREQUENCIES command, but you can request several formatting options per FORMAT subcommand.

7.9 Table Formats

The following FORMAT keywords are used to control the formatting of tables:

NOLABELS *Do not display value labels.* By default, FREQUENCIES displays the value labels defined by the VALUE LABELS command (see Chapter 4).

DOUBLE *Double-space frequency tables.*

NEWPAGE *Begin each table on a new page.* By default, FREQUENCIES displays as many tables on a page as it can.

CONDENSE *Use condensed format.* Frequency counts are displayed in three columns. Value labels and percentages are not displayed, and valid and cumulative percentages are rounded off to integers.

ONEPAGE *Use conditional condensed format.* Requests condensed format for tables that would require more than one page with the default format. All other tables are displayed in the default format. If CONDENSE and ONEPAGE are both specified, all tables are displayed in condensed format.

For example, the command

```
FREQUENCIES VARIABLES=RAISE82 AGE DEPT
/FORMAT=NEWPAGE.
```

requests frequency tables for variables RAISE82, AGE, and DEPT. Each table begins on a new page.

7.10 The Order of Values

By default, numeric values in a frequency table are listed in ascending order and string values in alphabetical order. Three other methods of sorting values can be requested by using one of the following keywords on the FORMAT subcommand:

DVALUE *Sort values in descending order.*

AFREQ *Sort values in ascending order of frequency.*

DFREQ *Sort value in descending order of frequency.*

If more than one sorting method is requested, the last one specified is used.

7.11 Suppressing Tables

If you have a lot of variables with many values, or if you want only descriptive statistics (or histograms or bar charts), you may want to suppress the display of frequency tables. Use the following FORMAT keywords to do this:

LIMIT(n) *Do not display tables for variables with more categories than the specified value.*

NOTABLE *Suppress all frequency tables.*

If LIMIT and NOTABLE are both specified, no tables are displayed. When tables are suppressed, the number of cases with missing values and the number of cases with valid values are still displayed.

7.12 Requesting Bar Charts and Histograms

Both bar charts and histograms can be requested with one FREQUENCIES command. Use the BARCHART subcommand to obtain bar charts for all variables listed on the VARIABLES subcommand, and the HISTOGRAM subcommand to obtain histograms for all numeric variables. If you want only those bar charts that will fit on one page and histograms for all other numeric variables, use the HBAR subcommand.

7.13 The BARCHART Subcommand

To obtain bar charts, use subcommand BARCHART, as in the command

```
FREQUENCIES VARIABLES=DAYOFWK/BARCHART.
```

which was used to produce Figure 7.2b. No further specifications are required.

By default, all tabulated values are plotted. The scale for the horizontal axis is in terms of frequencies and is determined by the largest frequency in the data. With optional BARCHART specifications you can specify minimum and maximum bounds for plotting and request a horizontal scale based on percentages. You can also specify the maximum frequency to be used for the horizontal scale.

MIN(n) *Use the lower bound n.* Values below this minimum are not plotted.

MAX(n) *Use the upper bound n.* Values above this maximum are not plotted.

PERCENT(n) *Scale the horizontal axis in percentages.* n specifies the maximum percentage for any value and is not required. If n is too small or not specified, SPSS/PC+ uses 5, 10, 25, 50, or 100, depending on the largest percentage in the data.

FREQ(n) *Scale the horizontal axis in frequencies, with n as the maximum frequency.* If no n is specified, or if n is too small, SPSS/PC+ uses 10, 20, 50, 100, 200, 500, 1000, 2000, and so on, depending on the largest frequency in the data.

These optional specifications can be entered in any order. For example, the command

```
FREQUENCIES VARIABLES=DAYOFWK/BARCHART MAX(5) MIN(2).
```

could be used to request a barchart for the variable DAYOFWK with values from 2 through 5.

7.14 The HISTOGRAM Subcommand

Histograms are obtained by specifying the HISTOGRAM subcommand. No further specifications are required, although several formatting options are available. In the default format, all tabulated values are included, the horizontal axis is scaled by frequencies, and the scale is determined by the largest frequency in the data. The default number of intervals is 21 (or fewer if the range of values is less than 21).

All of the BARCHART formatting specifications described in Section 7.13 can be used with HISTOGRAM to alter the histogram format. In addition, you can specify the interval width and have a normal curve superimposed on the histogram by using the following optional specifications:

INCREMENT(n) *Use an interval width equal to n.*

NORMAL *Superimpose a normal curve.* A normal curve with the same mean and variance as the plotted variable is superimposed on the histogram. All valid values, including those excluded by MIN and MAX, are used in calculating the mean and variance.

The HISTOGRAM specifications can be entered in any order. For example, the command

```
FREQUENCIES  VARIABLES=WT58/ FORMAT=NOTABLE
/HISTOGRAM MIN(120) MAX(280) INCREMENT(10).
```

was used to produce the output in Figure 7.4.

7.15 The HBAR Subcommand

When the HBAR subcommand is used, bar charts are displayed for numeric variables if the chart will fit on one page; otherwise HBAR produces a histogram. HBAR produces bar charts for short string variables and for the short-string portion of long string variables, regardless of the number of values.

All of the HISTOGRAM formatting options can be used with HBAR.

7.16 Requesting Percentiles and Ntiles

Use the PERCENTILES and NTILES subcommands to obtain percentiles for all variables specified on the VARIABLES subcommand. If more than one PERCENTILES or NTILES subcommand is specified, one table with the values for all requested percentiles is displayed.

Percentiles, the values below which given percentages of cases fall, are obtained by specifying PERCENTILES, followed by an optional equals sign and a list of percentages. For example, the command

```
FREQUENCIES VARIABLES=VARZ/PERCENTILES=10 25 33.3 66.7 75.
```

requests the values for percentiles 10, 25, 33.3, 66.7, and 75 for variable VARZ. When a requested percentile cannot be calculated, a period is displayed.

Ntiles, the values that divide the sample into groups with equal numbers of cases, are obtained by specifying NTILES, followed by an optional equals sign and an integer indicating the number of subgroups. For example, the command

```
FREQUENCIES VARIABLES=VARZ/NTILES=4.
```

requests quartiles (percentiles 25, 50, and 75) for variable VARZ.

7.17 Optional Statistics

The STATISTICS subcommand is used to request various statistics for all variables listed on the VARIABLES subcommand. Use the keywords shown below to obtain these statistics.

MEAN *Mean.*

SEMEAN *Standard error of the mean.*

MEDIAN *Median.* The median is not available if AFREQ or DFREQ is specified in the FORMAT subcommand.

MODE *Mode.*

STDDEV *Standard deviation.*

VARIANCE *Variance.*

SKEWNESS *Skewness.*

SESKEW *Standard error of the skewness statistic.*

KURTOSIS *Kurtosis.*

SEKURT *Standard error of the kurtosis statistic.*

RANGE *Range.*

MINIMUM *Minimum.*

MAXIMUM *Maximum.*

SUM *Sum.*

DEFAULT *Mean, standard deviation, minimum, and maximum.*

ALL *All available statistics.*

NONE *No statistics.*

You can specify as many keywords as you wish on the STATISTICS subcommand. For example, the command

```
FREQUENCIES VARIABLES=RAISE82 AGE
/STATISTICS=MEAN SKEWNESS RANGE SUM.
```

requests the mean, skewness, range, and sum for variables RAISE82 and AGE.

If STATISTICS is specified without any keywords, the default statistics are displayed.

7.18 Missing Values

Both user-missing and system-missing values are included in frequency tables. They are labeled as missing and are not included in the valid or cumulative percentages. They are also not used in calculating descriptive statistics and do not appear in bar charts or histograms.

One optional missing-value treatment is available. This option is requested with the keyword INCLUDE on the MISSING subcommand.

INCLUDE *Include cases with user-missing values.* Cases with user-missing values are included in the percentages, statistics, and plots.

7.19 Annotated Example

The following commands produced the output in Figure 7.1a.

```
DATA LIST / DAYOFWK 1 WT58 2-4.
VARIABLE LABELS DAYOFWK 'DAY OF DEATH'
              /WT58 'BODY WEIGHT, 1958 -- LBS'.
VALUE LABELS DAYOFWK 1 'SUNDAY' 2 'MONDAY' 3 'TUESDAY' 4 'WEDNESDAY'
                     5 'THURSDAY' 6 'FRIDAY' 7 'SATURDAY' 9 'MISSING'.
MISSING VALUE DAYOFWK (9).
BEGIN DATA.
lines of data
END DATA.
FREQUENCIES  VARIABLES=DAYOFWK/BARCHART.
FINISH.
```

- The DATA LIST command defines the variable names and column locations for the variables used in the analysis.
- The VARIABLE LABELS and VALUE LABELS commands assign descriptive labels for the variables.
- The MISSING VALUE command defines the value 9 as missing for the variable DAYOFWK.
- The FREQUENCIES command requests a frequency table and a barchart for the variable DAYOFWK.

Contents

Chapter 8 Descriptive Statistics: Procedures FREQUENCIES and DESCRIPTIVES

Survey data that rely on voluntary information are subject to many sources of error. People deliberately distort the truth, inadvertently fail to recall events correctly, or refuse to participate. Refusals influence survey results by failing to provide information about a particular type of person—one who refuses to answer surveys at all or avoids certain types of questions. For example, if college graduates tend to be unwilling to answer polls, results of surveys will be biased.

One possible way to examine the veracity of responses is to compare them to official records. Systematic differences between the two sources jeopardize the usefulness of the survey. Unfortunately, for many sensitive questions such as illicit drug use, abortion history, or even income, official records are usually unavailable.

Wyner (1980) examined the differences between the true and self-reported numbers of arrests obtained from 79 former heroin addicts enrolled in the Vera Institute of Justice Supported Employment Experiment. As part of their regular quarterly interviews, participants were asked about their arrest histories in New York City. The self-reported value was compared to arrest record data coded from New York City Police Department arrest sheets. The goal of the study was not only to quantify the extent of error but also to identify factors related to inaccurate responses.

8.1 EXAMINING THE DATA

Figure 8.1a shows histograms for the three variables—true number of arrests, reported arrests, and the discrepancy between the two. From a histogram it is possible to see the *shape* of the distribution, that is, how likely the different values are, how much spread or *variability* there is among the values, and where typical values are concentrated. Such characteristics are important because of the direct insight they provide into the data and because many statistical procedures are based on assumptions about the underlying distributions of variables.

The distributions of the reported and true number of arrests have a somewhat similar shape. Neither distribution has an obvious central value, although the self-reported values have the tallest peak at 4 to 5 arrests, while the actual number of arrests has its peak at 2 to 3 arrests. The distribution of self-reported arrests also has a peak at 20 to 21 arrests. The peaks corresponding to intervals which contain 5, 15, and 20 arrests arouse the suspicion that people may be more likely to report their arrest records as round numbers. Examination of the true number of arrests shows no corresponding peaks at multiples of five.

Figure 8.1a Reported and true arrests (Histograms from SPSS/PC+ FREQUENCIES)

```
ACTUAL     ACTUAL NUMBER OF ARRESTS

   Count   Midpoint

       3       1.00  #########
      15       3.00  ###############################################
       9       5.00  ############################
       8       7.00  #########################
      10       9.00  ################################
       9      11.00  ############################
       7      13.00  ######################
       6      15.00  ###################
       3      17.00  #########
       3      19.00  #########
       3      21.00  #########
       1      23.00  ###
       0      25.00
       1      27.00  ###
       1      29.00  ###
                     I....+....I....+....I....+....I....+....I....+....I
                     0         4         8        12        16        20
                                        Percent

Valid Cases      79      Missing Cases     0

- - - - - - - - - - - - - - - - - - - - - - - - - - - - - - - - - - - -

SELF       SELF-REPORTED ARRESTS

   Count   Midpoint

       7       1.00  ######################
       9       3.00  ############################
      13       5.00  #########################################
      11       7.00  ###################################
      11       9.00  ###################################
       6      11.00  ###################
       2      13.00  ######
       5      15.00  ################
       2      17.00  ######
       1      19.00  ###
      10      21.00  ################################
       1      23.00  ###
       1      25.00  ###
                     I....+....I....+....I....+....I....+....I....+....I
                     0         4         8        12        16        20
                                        Percent

Valid Cases      79      Missing Cases     0

- - - - - - - - - - - - - - - - - - - - - - - - - - - - - - - - - - - -

ERRORS     REPORTED ARRESTS MINUS ACTUAL ARRESTS

   Count   Midpoint

       2     -13.00  #####
       0     -11.00
       2      -9.00  #####
       5      -7.00  #############
       6      -5.00  ################
       7      -3.00  ###################
      16      -1.00  ###########################################
      15       1.00  ########################################
      12       3.00  ################################
       7       5.00  ###################
       1       7.00  ###
       2       9.00  #####
       2      11.00  #####
       1      13.00  ###
       1      15.00  ###
                     I....+....I....+....I....+....I....+....I....+....I
                     0         5        10        15        20        25
                                        Percent

Valid Cases      79      Missing Cases     0
```

The distribution of the differences between reported and true number of arrests is not as irregularly shaped as the two distributions from which it is derived. It has two adjacent peaks with midpoint values of −1 and +1. Most cases cluster around the peak values, and cases far from these values are infrequent. Figure 8.1b is a condensed frequency table for the response errors (the adjusted and cumulative percentages are rounded to the nearest integer). Almost 47% of the sample (37 cases) reported their arrest record to within two arrests of the true value. Only 22% (17 cases) misrepresented their records by more than 5 arrests.

Underreporting is somewhat more likely than exaggeration, with 39% of the cases overestimating and 48% of the cases underestimating.

Figure 8.1b Error in reported arrests (Condensed frequency table from SPSS/PC+ FREQUENCIES)

ERRORS REPORTED ARRESTS MINUS ACTUAL ARRESTS

VALUE	FREQ	PCT	CUM PCT	VALUE	FREQ	PCT	CUM PCT	VALUE	FREQ	PCT	CUM PCT
-14	2	3	3	-2	6	8	35	7	1	1	92
-9	2	3	5	-1	10	13	48	8	1	1	94
-8	3	4	9	0	10	13	61	9	1	1	95
-7	2	3	11	1	5	6	67	10	1	1	96
-6	1	1	13	2	6	8	75	11	1	1	97
-5	5	6	19	3	6	8	82	12	1	1	99
-4	3	4	23	4	4	5	87	15	1	1	100
-3	4	5	28	5	3	4	91				

Valid Cases 79 Missing Cases 0

8.2 Percentile Values

Percentiles are values above and below which certain percentages of the cases fall. For example, 95% of the cases have values less than or equal to the 95th percentile. From the cumulative percentage column in the frequency table in Figure 8.1b, the value for the 95th percentile is 9.

Figure 8.2 contains some commonly used percentiles for the distributions in Figure 8.1a. The three percentiles (25%, 50%, and 75%) divide the observed distributions into approximately four equal parts. The actual and self-reported number of arrests have the same 25th percentile, the value 4. This means that about 75% of the values are greater than or equal to 4, and 25% less than 4.

Figure 8.2 Percentiles for reported and actual arrests and errors

ACTUAL ACTUAL NUMBER OF ARRESTS

Percentile	Value	Percentile	Value	Percentile	Value
25.00	4.000	50.00	8.000	75.00	13.000

Valid Cases 79 Missing Cases 0

SELF SELF-REPORTED ARRESTS

Percentile	Value	Percentile	Value	Percentile	Value
25.00	4.000	50.00	7.000	75.00	14.000

Valid Cases 79 Missing Cases 0

ERRORS REPORTED ARRESTS MINUS ACTUAL ARRESTS

Percentile	Value	Percentile	Value	Percentile	Value
25.00	-3.000	50.00	0.0	75.00	3.000

Valid Cases 79 Missing Cases 0

8.3 SUMMARIZING THE DATA

Although frequency tables and bar charts are useful for summarizing and displaying data (see Chapter 7), further condensation and description is often desirable. A variety of summary measures that convey information about the data in single numbers can be computed. The choice of summary measure, or *statistic*, as it is often called, depends upon characteristics of the data as well as of the statistic. One important characteristic of the data that must be considered is the *level of measurement* of each variable being studied.

8.4 Levels of Measurement

Measurement is the assignment of numbers or codes to observations. Levels of measurement are distinguished by ordering and distance properties. A computer does not know what measurement underlies the values it is given. You must determine the level of measurement of your data and apply appropriate statistical techniques.

The traditional classification of levels of measurement into nominal, ordinal, interval, and ratio scales was developed by S. S. Stevens (1946). This remains the basic typology and is the one used throughout this manual. Variations exist, however, and issues concerning the statistical effect of ignoring levels of measurement have been debated (see, for example, Borgatta & Bohrnstedt, 1980).

8.5 Nominal Measurement

The nominal level of measurement is the "lowest" in the typology because no assumptions are made about relations between values. Each value defines a distinct category and serves merely as a label or name (hence, "nominal" level) for the category. For instance, the birthplace of an individual is a nominal variable. For most purposes, there is no inherent ordering among cities or towns. Although cities can be ordered according to size, density, or air pollution, a city thought of as "place of birth" is a concept that is normally not tied to any order. When numeric values are attached to nominal categories, they are merely identifiers. None of the properties of numbers such as relative size, addition, or multiplication, can be applied to these numerically coded categories. Therefore, statistics that assume ordering or meaningful numerical distances between the values do not ordinarily give useful information about nominal variables.

8.6 Ordinal Measurement

When it is possible to rank or order all categories according to some criterion, the ordinal level of measurement is achieved. For instance, classifying employees into clerical, supervisory, and managerial categories is an ordering according to responsibilities or skills. Each category has a position lower or higher than another category. Furthermore, knowing that supervisory is higher than clerical and that managerial is higher than supervisory automatically means that managerial is higher than clerical. However, nothing is known about how much higher; no distance is measured. Ordering is the sole mathematical property applicable to ordinal measurements, and the use of numeric values does not imply that any other property of numbers is applicable.

8.7 Interval Measurement

In addition to order, interval measurements have the property of meaningful distance between values. A thermometer, for example, measures temperature in degrees which are the same size at any point on the scale. The difference between 20°C and 21°C is the same as the difference between 5°C and 6°C. However, an interval scale does not have an inherently determined zero point. In the familiar Celsius and Fahrenheit systems, 0° is determined by an agreed-upon definition, not by the absence of heat. Consequently, interval-level measurement allows us to study differences between items but not their proportionate magnitudes. For example, it is incorrect to say that 80°F is twice as hot as 40°F.

8.8 Ratio Measurement

Ratio measurements have all the ordering and distance properties of an interval scale. In addition, a zero point can be meaningfully designated. In measuring physical distances between objects using feet or meters, a zero distance is naturally defined as the absence of any distance. The existence of a zero point means that ratio comparisons can be made. For example, it is quite meaningful to say that a 6-foot-tall adult is twice as tall as a 3-foot-tall child or that a 500-meter race is five times as long as a 100-meter race.

Because ratio measurements satisfy all the properties of the real number system, any mathematical manipulations appropriate for real numbers can be applied to ratio measures. However, the existence of a zero point is seldom critical for statistical analyses.

8.9 Summary Statistics

Figure 8.9 contains a variety of summary statistics that are useful in describing the distributions of reported arrests, true number of arrests, and the discrepancy between the two. The statistics can be grouped into three categories according to what they quantify: central tendency, dispersion, and shape.

Figure 8.9 Statistics describing arrest data

```
ACTUAL     ACTUAL NUMBER OF ARRESTS

Mean          9.253     Std Err        .703     Median        8.000
Mode          3.000     Std Dev       6.248     Variance     39.038
Kurtosis       .597     S E Kurt       .535     Skewness       .908
S E Skew       .271     Range        28.000     Minimum       1.000
Maximum      29.000     Sum         731.000

--------------------------------------------------------------------------

SELF       SELF-REPORTED ARRESTS

Mean          8.962     Std Err        .727     Median        7.000
Mode          5.000     Std Dev       6.458     Variance     41.704
Kurtosis      -.485     S E Kurt       .535     Skewness       .750
S E Skew       .271     Range        25.000     Minimum        0.0
Maximum      25.000     Sum         708.000

Valid Cases       79    Missing Cases     0

--------------------------------------------------------------------------

ERRORS     REPORTED ARRESTS MINUS ACTUAL ARRESTS

Mean          -.291     Std Err        .587     Median         0.0
Mode         -1.000     Std Dev       5.216     Variance     27.209
Kurtosis      1.102     S E Kurt       .535     Skewness       .125
S E Skew       .271     Range        29.000     Minimum     -14.000
Maximum      15.000     Sum         -23.000

Valid Cases       79    Missing Cases     0
```

8.10 Measures of Central Tendency

The mean, median, and mode are frequently used to describe the location of a distribution. The *mode* is the most frequently occurring value (or values). For the true number of arrests, the mode is 3 (see Figure 8.9); for the self-reported values, it is 5. The distribution of the difference between the true and self-reported values is multimodal. That is, it has more than one mode since the values −1 and 0 occur with equal frequency. SPSS/PC+ however, displays only one mode, as shown in Figure 8.9. The mode can be used for data measured at any level. It is usually not the preferred measure for interval and ordinal data since it ignores much of the available information.

The *median* is the value above and below which one half of the observations fall. For example, if there are 79 observations the median is the 40th largest observation. When there is an even number of observations, no unique center value exists, so the mean of the two middle observations is usually taken as the median value. For the arrest data, the median is 0 for the differences, 8 for the true arrests, and 7 for reported arrests. For ordinal data the median is usually a good measure of central tendency since it uses the ranking information. The median should not be used for nominal data since ranking of the observations is not possible.

The *mean*, also called the arithmetic average, is the sum of the values of all observations divided by the number of observations. Thus

$$\bar{X} = \sum_{i=1}^{N} \frac{X_i}{N}$$ **Equation 8.10**

where N is the number of cases and X_i is the value of the variable for the ith case. Since the mean utilizes the distance between observations, the measurements should be interval or ratio. Mean race, religion, and auto color are meaningless. For dichotomous variables coded as 0 and 1, the mean has a special interpretation: it is the proportion of cases coded 1 in the data.

The three measures of central tendency need not be the same. For example, the mean number of true arrests is 9.25, the median is 8, and the mode is 3 (see Figure 8.9). The arithmetic mean is greatly influenced by outlying observations, while the median is not. Adding a single case with 400 arrests would increase the mean from 9.25 to 14.1, but it would not affect the median. Therefore, if there are values far removed from the rest of the observations, the median may be a better measure of central tendency than the mean.

For symmetric distributions, the observed mean, median, and mode are usually close in value. For example, the mean of the differences between reported and true arrest values is -0.291, the median is 0, and the modes are -1 and 0. All three measures give similar estimates of central tendency in this case.

8.11 Measures of Dispersion

Two distributions can have the same values for measures of central tendency and yet be very dissimilar in other respects. For example, if the true number of arrests for five cases in two methadone clinics is

CLINIC A: 0, 1, 10, 14, 20
CLINIC B: 8, 8, 9, 10, 10

the mean number of arrests (9) is the same in both. However, even a cursory examination of the data indicates that the two clinics are different. In the second clinic, all cases have fairly comparable arrest records while in the first the records are quite disparate. A quick and useful index of dissimilarity, or dispersion, is the *range*. It is the difference between the *maximum* and *minimum* observed values. For clinic B the range is 2, while for clinic A it is 20. Since the range is computed only from the minimum and maximum values, it is sensitive to extremes.

Although the range is a useful index of dispersion, especially for ordinal data, it does not take into account the distribution of observations between the maximum and minimum. A commonly used measure of variation that is based on all observations is the *variance*. For a sample, the variance is computed by summing the squared differences from the mean for all observations and then dividing by one less than the number of observations. In mathematical notation this is

$$S^2 = \sum_{i=1}^{N} \frac{(X_i - \bar{X})^2}{N - 1}$$ **Equation 8.11**

If all observations are identical—that is, if there is no variation—the variance is 0. The more spread out they are, the greater the variance. For the methadone clinic example above, the sample variance for Clinic A is 73, while for Clinic B it is 1.

The square root of the variance is termed the *standard deviation*. While the variance is in units squared, the standard deviation is expressed in the same units of measurement as the observations. This is an appealing property since it is much clearer to think of variability in terms of the number of arrests instead of the number of arrests squared.

8.12 The Normal Distribution

For many variables, most observations are concentrated near the middle of the distribution. As distance from the central concentration increases, the frequency of observation decreases. Such distributions are often described as "bell-shaped." An example is the *normal* distribution (see Figure 8.12a). A broad range of observed phenomena in nature and in society are approximately normally distributed. For example, the distributions of variables such as height, weight, and blood pressure are approximately normal. The normal distribution is by far the most important theoretical distribution in statistics and serves as a reference point for describing the form of many distributions of sample data.

The normal distribution is symmetric: when it is folded in the center, the two sides are identical. The three measures of central tendency—the mean, median, and mode—coincide exactly (see Section 8.10). As shown in Figure 8.12a, 95% of all observations in the normal distribution fall within two standard deviations (σ) of the mean (μ), and 68% within one standard deviation. The exact theoretical proportion of cases falling into various regions of the normal curve can be found in tables given in most introductory statistics textbooks.

In SPSS/PC+, you can superimpose a normal distribution on a histogram. For example, in Figure 8.12b, which is a histogram of differences in arrest

Figure 8.12a A normal curve

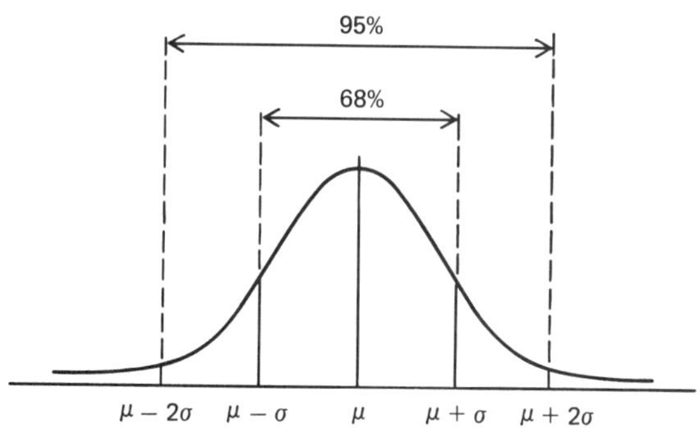

Figure 8.12b Histogram of errors with the normal curve superimposed

```
ERRORS    REPORTED ARRESTS MINUS ACTUAL ARRESTS

   Count   Midpoint

     2      -14.5  :#####
     0      -13.0  .
     0      -11.5    .
     0      -10.0       .
     5       -8.5  #######:########
     2       -7.0  ######       .
     6       -5.5  ################:##
     3       -4.0  #########              .
    10       -2.5  #########################:######
    10       -1.0  ############################:####
    15         .5  ############################:##################
     6        2.0  ###################       .
    10        3.5  #####################:##########
     3        5.0  #########        .
     1        6.5  ###        .
     1        8.0  ###    .
     2        9.5  ####:#
     1       11.0  ##:
     1       12.5  :##
     0       14.0  .
     1       15.5  ###
                   I....+....I....+....I....+....I....+....I....+....I
                   0         4         8         12        16        20
                                       Percent

Valid Cases      79      Missing Cases     0
```

records, the colons and periods indicate what the distribution of cases would be if the variable had a normal distribution with the same mean and variance.

8.13 Measures of Shape

A distribution that is not symmetric but has more cases, or more of a "tail," toward one end of the distribution than the other is called *skewed.* If the tail is toward larger values, the distribution is positively skewed or skewed to the right. If the tail is toward smaller values, the distribution is negatively skewed or skewed to the left.

Another characteristic of the form of a distribution is called *kurtosis*, the extent to which, for a given standard deviation, observations cluster around a central point. If cases within a distribution cluster more than those in the normal distribution (that is, the distribution is more peaked), the distribution is called *leptokurtic.* A leptokurtic distribution also tends to have more observations straggling into the extreme tails than does a normal distribution. If cases cluster less than in the normal distribution (that is, it is flatter), the distribution is termed *platykurtic.*

Although examination of a histogram provides some indication of possible skewness and kurtosis, it is often desirable to compute formal indexes that measure these properties. Values for skewness and kurtosis are 0 if the observed distribution is exactly normal. Positive values for skewness indicate a positive skew, while positive values for kurtosis indicate a distribution that is more peaked than normal. For samples from a normal distribution, measures of skewness and kurtosis typically will not be exactly zero but will fluctuate about zero because of sampling variation.

8.14 Standard Scores

It is often desirable to describe the relative position of an observation within a distribution. Knowing that a person achieved a score of 80 in a competitive examination conveys little information about performance. Judgment of performance would depend on whether 80 is the lowest, the median, or the highest score.

One way of describing the location of a case in a distribution is to calculate its *standard score.* This score, sometimes called the Z score, indicates how many standard deviations above or below the mean an observation falls. It is calculated by finding the difference between the value of a particular observation X_i and the mean of the distribution, and then dividing this difference by the standard deviation:

$$Z_i = \frac{X_i - \overline{X}}{S}$$

Equation 8.14

The mean of Z scores is 0, and the standard deviation is 1.

For example, a participant with 5 actual arrests would have a Z score of (5−9.25)/6.25, or −0.68. Since the score is negative, the case had fewer arrests than the average for the individuals studied.

Standardization permits comparison of scores from different distributions. For example, an individual with Z scores of −0.68 for actual arrests and 1.01 for the difference between reported and actual arrests had fewer arrests than the average but exaggerated more than the average.

When the distribution of a variable is approximately normal and the mean and variance are known or are estimated from large samples, the Z score of an observation provides more specific information about its location. For example, if actual arrests and response error were normally distributed, 75% of cases would have more arrests than the example individual but only 16% would have exaggerated as much (75% of a standard normal curve lies above a Z score of −0.68, and 16% lies above a score of 1.01).

8.15 Who Lies?

The distribution of the difference between reported and actual arrests indicates that response error exists. Although observing a mean close to zero is comforting, misrepresentation is obvious. What then are the characteristics that influence willingness to be truthful?

Wyner identifies three factors that are related to inaccuracies: the number of arrests before 1960, the number of multiple-charge arrests, and the perceived desirability of being arrested. The first factor is related to a frequently encountered difficulty—the more distant an event in time, the less likely it is to be correctly recalled. The second factor, underreporting of multiple-charge arrests, is probably caused by the general social undesirability of serious arrests. Finally, persons who view arrest records as laudatory are likely to inflate their accomplishments.

8.16 STATISTICS AVAILABLE WITH PROCEDURE FREQUENCIES

In addition to frequency tables, bar charts, and histograms, procedure FREQUENCIES calculates univariate statistics for all variables named on the VARIABLES subcommand. To request statistics, use the STATISTICS subcommand followed by an equals sign and the keywords that correspond to the statistics you want. For example, the command

```
FREQUENCIES  VARIABLES=ACTUAL SELF ERRORS/FORMAT=NOTABLE
/STATISTICS=ALL.
```

produces the output in Figure 8.9. See Chapter 7 for a complete list of the statistics available with FREQUENCIES.

8.17 Percentiles

You can use FREQUENCIES to request percentiles for all variables specified on the VARIABLES subcommand. Include the PERCENTILES subcommand followed by an equals sign and a list of percentiles between 0 and 100. For example, the command

```
FREQUENCIES  VARIABLES=ACTUAL SELF ERRORS/FORMAT=NOTABLE
/PERCENTILES=25 50 75.
```

produces Figure 8.2.

In SPSS/PC+, percentiles are calculated by sorting the values from the smallest to the largest and finding the values below and above which the requisite number of cases fall. Therefore, it is possible for several percentiles to have the same value. For example, if the values are

0 1 1 1 1

all percentiles greater than the 20th are 1.

8.18 RUNNING PROCEDURE DESCRIPTIVES

Procedure DESCRIPTIVES calculates all of the statistics provided by procedure FREQUENCIES, except the median and the mode, and provides a compact table of statistics. Because it does not sort values into a frequency table and displays summaries of several variables on a page, it is an efficient procedure for computing descriptive statistics.

Procedure DESCRIPTIVES requires a list of variables for which statistics are to be computed. The optional STATISTICS subcommand indicates the statistics to be computed, and the OPTIONS subcommand specifies treatment of missing values and formatting options.

Descriptive statistics can be computed only for numeric variables. If a string variable is specified in the variable list, a warning is issued and no statistics are displayed for that variable.

8.19 Specifying the Variables

The VARIABLES subcommand names the variables for which statistics are to be calculated. For example, to calculate the default statistics (mean, standard deviation, minimum, and maximum) for variables TRUE, SELF, and ERRORS, specify

```
DESCRIPTIVES VARIABLES=TRUE SELF ERRORS.
```

The actual keyword VARIABLES can be omitted, as in the command

```
DESCRIPTIVES TRUE SELF ERRORS.
```

You can also use the TO keyword to refer to a set of consecutive variables on the active file, and keyword ALL to refer to all user-defined variables.

8.20 Optional Statistics

By default the DESCRIPTIVES procedure calculates the mean, standard deviation, minimum, and maximum. Additional statistics can be requested with the STATISTICS subcommand:

Statistic 1 *Mean.*

Statistic 2 *Standard error of mean.*

Statistic 5 *Standard deviation.*

Statistic 6 *Variance.*

Statistic 7 *Kurtosis.*

Statistic 8 *Skewness.*

Statistic 9 *Range.*

Statistic 10 *Minimum.*

Statistic 11 *Maximum.*

Statistic 12 *Sum.*

Statistic 13 *Mean, standard deviation, minimum, and maximum.* This is the same as the default.

ALL *All available statistics.*

If the STATISTICS subcommand is used, only the statistics requested are displayed. Thus, if you are using the STATISTICS subcommand, you must specify Statistic 13 to get the default statistics.

8.21 Missing Values

By default, DESCRIPTIVES includes only cases with valid values for a variable in the calculation of statistics for that variable. Use one of the following options on the OPTIONS subcommand for alternative treatments of missing values.

Option 1 *Include user-missing values.* Cases that have user-missing values are included in the calculation of statistics for all variables named on the command.

Option 5 *Exclude missing values listwise.* A case with missing values for any of the variables is excluded from computations for all of the variables.

8.22 Formatting Options

By default, DESCRIPTIVES displays the statistics and a 40-character variable label for each variable on one line. If the statistics requested do not fit within the available width, DESCRIPTIVES will first truncate the variable label and then use serial format. (You can use the SET command to change the width of the display; see Command Reference: SET). Serial format provides larger field widths and permits more decimal places for very large or very small numbers than does the default format.

Optionally, you can request any of the following on the OPTIONS subcommand:

Option 2 *Suppress variable labels.*

Option 6 *Serial format.* The requested statistics are displayed below each variable name.

Option 7 *Narrow format.* Use narrow format, regardless of the width defined on SET.

Option 8 *Suppress variable names.* The variable name will be displayed only if there is no variable label.

8.23 Annotated Example

Figure 8.23 shows the default statistics produced by the DESCRIPTIVES command in the following SPSS/PC+ command file:

```
DATA LIST  FIXED / ACTUAL 1-2 SELF 3-4 ERRORS 5-7.
VARIABLE LABELS  ACTUAL 'ACTUAL NUMBER OF ARRESTS'
                 /SELF 'SELF-REPORTED ARRESTS'
                 /ERRORS 'REPORTED ARRESTS MINUS ACTUAL ARRESTS'.
BEGIN DATA.
lines of data
END DATA.
DESCRIPTIVES  ACTUAL SELF ERRORS.
FINISH.
```

- The DATA LIST command reads one record per case with three variables. The variable named ACTUAL is recorded in columns 1 and 2, the variable named SELF is in columns 3 and 4, and the variable named ERRORS is in columns 5 through 7.
- The VARIABLE LABELS command assigns labels to all three variables.
- The DESCRIPTIVES command requests descriptive statistics for the three variables.

Figure 8.23 Default statistics available with DESCRIPTIVES

```
Number of Valid Observations (Listwise) =        79.00

Variable        Mean   Std Dev   Minimum   Maximum      N  Label

ACTUAL          9.25      6.25      1.00     29.00     79  ACTUAL NUMBER OF ARRESTS
SELF            8.96      6.46      0.0      25.00     79  SELF-REPORTED ARRESTS
ERRORS          -.29      5.22    -14.00     15.00     79  REPORTED ARRESTS MINUS ACTUAL ARRESTS
```

Contents

Chapter 9 Crosstabulation and Measures of Association: Procedure CROSSTABS

Newspapers headline murders in subway stations, robberies on crowded main streets, suicides cheered by onlookers. All are indications of the social irresponsibility and apathy said to characterize city residents. Since overcrowding, decreased sense of community, and other urban problems are usually blamed, one might ask whether small town residents are more responsible and less apathetic than their urban counterparts.

Hansson and Slade (1977) used the "lost letter technique" to test the hypothesis that altruism is higher in small towns than in cities, unless the person needing assistance is a social deviant. In this technique, stamped and addressed letters are "lost," and the rate at which they are returned is examined. A total of 216 letters were lost in Hansson and Slade's experiment. Half were dropped within the city limits of Tulsa, Oklahoma, the others in 51 small towns within a 50-mile radius of Tulsa. The letters were addressed to three fictitious people at a post-office box in Tulsa: M. J. Davis; Dandee Davis, c/o Pink Panther Lounge; and M. J. Davis, c/o Friends of the Communist Party. The first person is considered a normal "control," the second a person whose occupation is questionable, and the third a subversive or political deviant.

9.1 CROSSTABULATION

To see whether the return rate is similar for the three addresses, the letters found and mailed and those not mailed must be tallied separately for each address. Figure 9.1 is a *crosstabulation* of address type and response. The number of cases (letters) for each combination of values of the two variables is displayed in a *cell* in the table, together with various percentages. These cell entries provide information about relationships between the variables.

Figure 9.1 Crosstabulation of status of letter by address

```
Crosstabulation:      RETURNED  FOUND AND MAILED
                   By ADDRESS   ADDRESS ON LETTER

              Count I
            Row Pct ICONTROL IDANDEE  ICOMMUNISI
ADDRESS->   Col Pct I        I        IT       I  Row
            Tot Pct I       1I       2I       3I Total
RETURNED    --------+--------+--------+--------+
                  1 I    35  I    32  I    10  I    77
  YES               I  45.5  I  41.6  I  13.0  I  35.6
                    I  48.6  I  44.4  I  13.9  I
                    I  16.2  I  14.8  I   4.6  I
                    +--------+--------+--------+
                  2 I    37  I    40  I    62  I   139
  NO                I  26.6  I  28.8  I  44.6  I  64.4
                    I  51.4  I  55.6  I  86.1  I
                    I  17.1  I  18.5  I  28.7  I
                    +--------+--------+--------+
             Column      72       72       72      216
              Total    33.3     33.3     33.3    100.0

 Chi-Square     D.F.      Significance        Min E.F.     Cells with E.F.< 5
 ----------     ----      ------------        --------     ------------------

   22.56265       2          0.0000             25.667          None

Number of Missing Observations =          0
```

In Figure 9.1, the address is called the *column* variable since each address is displayed in a column of the table. Similarly, the status of the letter—whether it was returned or not—is called the *row* variable. With three categories of the column variable and two of the row, there are six cells in the table.

9.2 Cell Contents and Marginals

The first entry in the table is the number of cases, or *frequency*, in that cell. It is labeled as COUNT in the key printed in the upper-left corner of the table. For example, 35 letters addressed to the control were returned, and 62 letters addressed to the Communist were not returned. The second entry in the table is the *row percentage* (ROW PCT). It is the percentage of all cases in a row that fall into a particular cell. Of the 77 letters returned, 45.5% were addressed to the control, 41.6% to Dandee, and 13.0% to the Communist.

The *column percentage* (COL PCT), the third item in each cell, is the percentage of all cases in a column that occur in a cell. For example, 48.6% of the letters addressed to the control were returned and 51.4% were not. The return rate for Dandee is similar (44.4%), while that for the Communist is markedly lower (13.9%).

The last entry in the table is the *table percentage* (TOT PCT). The number of cases in the cell is expressed as a percentage of the total number of cases in the table. For example, the 35 letters returned to the control represent 16.2% of the 216 letters in the experiment.

The numbers to the right and below the table are known as *marginals*. They are the counts and percentages for the row and column variables taken separately. In Figure 9.1, the row marginals show that 77 (35.6%) of the letters were returned, while 139 (64.4%) were not.

9.3 Choosing Percentages

Row, column, and table percentages convey different types of information, so it is important to choose carefully among them.

In this example, the row percentage indicates the distribution of address types for returned and "lost" letters. It conveys no direct information about the return rate. For example, if twice as many letters were addressed to the control, an identical return rate for all letters would give row percentages of 50%, 25%, and 25%. However, this does not indicate that the return rate is higher for the control. In addition, if each category had the same number of returned letters, the row percentages would have been 33.3%, 33.3%, and 33.3%, regardless of whether one or all letters were returned.

The column percentage is the percentage of letters returned and not returned for each address. By looking at column percentages across rows, one can compare return rates for the address types. Interpretation of this comparison would not be affected if unequal numbers of letters had been addressed to each category.

Since it is always possible to interchange the rows and columns of any table, general rules about when to use row and column percentages cannot be given. The percentages you use depend on the nature of the two variables. If one of the two variables is under experimental control, it is termed an *independent variable*. This variable is hypothesized to affect the response, or *dependent variable*. If variables can be classified as dependent and independent, the following guideline may be helpful: If the independent variable is the row variable, select row percentages; if the independent variable is the column variable, select column percentages. In this example the dependent variable is the status of the letter, whether it was mailed or not. The type of address is the independent variable. Since the independent variable is the column variable in Figure 9.1, column percentages should be used for comparisons of return rates.

9.4 Adding a Control Variable

Since Figure 9.1 combines results from both the city and the towns, differences between the locations are obscured. Two separate tables, one for the city and one for the towns, are required. Figure 9.4 shows crosstabulations of response and address for each of the locations. SPSS/PC+ produces a separate table for each value of the location (control) variable.

Figure 9.4 Crosstabulations of status of letter by address controlled for location

```
Crosstabulation:      RETURNED  FOUND AND MAILED
                   By ADDRESS   ADDRESS ON LETTER
    Controlling for LOCATION    LOCATION LOST
                                          = 1  CITY

              Count  ICONTROL IDANDEE  ICOMMUNISI
ADDRESS->   Col Pct  I        I         IT       I  Row
                     I      1 I       2 I      3 I  Total
RETURNED    --------+--------+---------+--------+
                   1 I    16  I    14   I     9  I    39
  YES                I  44.4  I  38.9   I  25.0  I  36.1
                     +--------+---------+--------+
                   2 I    20  I    22   I    27  I    69
  NO                 I  55.6  I  61.1   I  75.0  I  63.9
                     +--------+---------+--------+
              Column     36        36        36      108
               Total   33.3      33.3      33.3    100.0

 Chi-Square     D.F.        Significance        Min E.F.       Cells with E.F.< 5
 ----------     ----        ------------        --------       ------------------

   3.13043        2             0.2090            13.000             None

Crosstabulation:      RETURNED  FOUND AND MAILED
                   By ADDRESS   ADDRESS ON LETTER
    Controlling for LOCATION    LOCATION LOST
                                          = 2  TOWN

              Count  ICONTROL IDANDEE  ICOMMUNISI
ADDRESS->   Col Pct  I        I         IT       I  Row
                     I      1 I       2 I      3 I  Total
RETURNED    --------+--------+---------+--------+
                   1 I    19  I    18   I     1  I    38
  YES                I  52.8  I  50.0   I   2.8  I  35.2
                     +--------+---------+--------+
                   2 I    17  I    18   I    35  I    70
  NO                 I  47.2  I  50.0   I  97.2  I  64.8
                     +--------+---------+--------+
              Column     36        36        36      108
               Total   33.3      33.3      33.3    100.0

 Chi-Square     D.F.        Significance        Min E.F.       Cells with E.F.< 5
 ----------     ----        ------------        --------       ------------------

  24.92932        2             0.0000            12.667             None

Number of Missing Observations =        0
```

These tables show interesting differences between cities and towns. Although the overall return rates are close, 36.1% for the city and 35.2% for the towns, there are striking differences between the addresses. Only 2.8% of the Communist letters were returned in towns, while 25.0% of them were returned in Tulsa. (At least two of the Communist letters were forwarded by small-town residents to the FBI for punitive action!) The return rates for both the control (52.8%) and Dandee (50.0%) are higher in towns.

The results support the hypothesis that, in small towns, suspected social deviance influences the response more than in big cities, although it is surprising that Dandee and the Pink Panther Lounge were deemed worthy of as much assistance as they received. If the Communist letter is excluded, inhabitants of small towns are somewhat more helpful than city residents, returning 51% of the other letters, in comparison to the city's 42%.

9.5 GRAPHICAL REPRESENTATION OF CROSSTABULATIONS

As with frequency tables, visual representation of a crosstabulation often simplifies the search for associations. Figure 9.5 is a bar chart of letters returned from the crosstabulations shown in Figure 9.4. In a bar chart, the length of each bar represents the frequencies or percentages for each category of a variable. In Figure 9.5, the percentages plotted are the column percentages shown in Figure 9.4 for the returned letters only. This chart clearly shows that the return rates for the control and Dandee are high compared to the return rate for the Communist. Also, it demonstrates more vividly than the crosstabulation that the town residents' return rates for the control and Dandee are higher than city residents' return rates but that the reverse is true for the Communist.

Figure 9.5 Status of letter by address by location (From SPSS Graphics, a mainframe product)

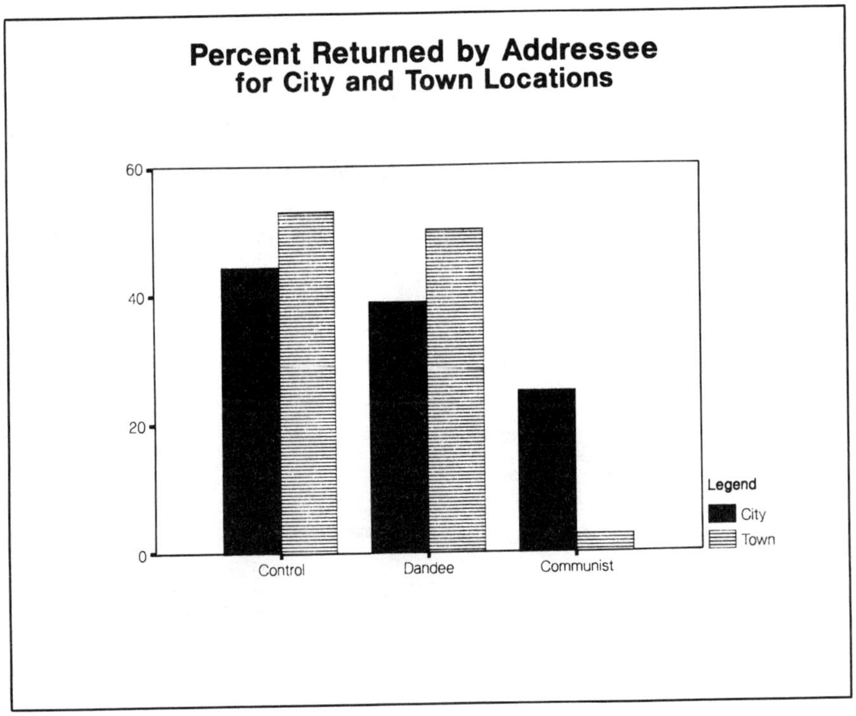

9.6 USING CROSSTABULATION FOR DATA SCREENING

Errors and unusual values in data entry that cannot be spotted with FREQUENCIES can sometimes be identified using crosstabulation. For example, a case coded as a male with a history of three pregnancies would not be identified as suspicious in FREQUENCIES tables of sex and number of pregnancies. When considered separately, the code for male is acceptable for variable sex and the value 3 is acceptable for number of pregnancies. Jointly, however, the combination is unexpected.

Whenever possible, crosstabulations of related variables should be obtained so that anomalies can be identified and corrected before further statistical analysis of the data.

9.7 CROSSTABULATION STATISTICS

Although examination of the various row and column percentages in a crosstabulation is a useful first step in studying the relationship between two variables, row and column percentages do not allow for quantification or testing of that relationship. For these purposes, it is useful to consider various indexes that measure the extent of association as well as statistical tests of the hypothesis that there is no association.

9.8 The Chi-Square Test of Independence

The hypothesis that two variables of a crosstabulation are *independent* of each other is often of interest to researchers. Two variables are by definition independent if the probability that a case falls into a given cell is simply the product of the marginal probabilities of the two categories defining the cell.

For example, in Figure 9.1 if returns of the letter and address type are independent, the probability of a letter being returned to a Communist is the product of the probability of a letter being returned and the probability of a letter being addressed to a Communist. From the table, 35.6% of the letters were returned and 33.3% of the letters were addressed to a friend of the Communist party. Thus, if address type and status of the letter are independent, the probability of a letter being returned to the Communist is estimated to be

$$P(\text{return})\ P(\text{Communist}) = 0.356 \times 0.333 = 0.119$$ **Equation 9.8a**

The *expected* number of cases in that cell is 25.7, which is 11.9% of the 216 cases in the sample. From the table, the *observed* number of letters returned to the Communist is 10 (4.6%), nearly 16 fewer than expected if the two variables are independent.

To construct a statistical test of the independence hypothesis, repeat the above calculations for each cell in the table. The probability under independence of an observation falling into cell (ij) is estimated by

$$P(\text{row} = i \text{ and column} = j) = \left(\frac{\text{count in row } i}{N}\right)\left(\frac{\text{count in column } j}{N}\right)$$ **Equation 9.8b**

To obtain the expected number of observations in cell (ij), the probability is multiplied by the total sample size

$$E_{ij} = N\left(\frac{\text{count in row } i}{N}\right)\left(\frac{\text{count in column } j}{N}\right) = \frac{(\text{count in row } i)\ (\text{count in column } j)}{N}$$ **Equation 9.8c**

Figure 9.8 contains the observed and expected frequencies and the *residuals*, which are the observed minus the expected frequencies for the data in Figure 9.1.

Figure 9.8 Observed, expected, and residual values

```
Crosstabulation:      RETURNED  FOUND AND MAILED
                   By ADDRESS   ADDRESS ON LETTER

             Count  ICONTROL IDANDEE  ICOMMUNISI
ADDRESS-> Exp Val I         I         IT        I  Row
          ResidualI       1 I       2 I       3 I Total
RETURNED  --------+---------+---------+---------+
                1 I    35   I    32   I    10   I    77
   YES            I  25.7   I  25.7   I  25.7   I  35.6%
                  I   9.3   I   6.3   I -15.7   I
                  +---------+---------+---------+
                2 I    37   I    40   I    62   I   139
   NO             I  46.3   I  46.3   I  46.3   I  64.4%
                  I  -9.3   I  -6.3   I  15.7   I
                  +---------+---------+---------+
            Column     72        72        72       216
             Total   33.3%     33.3%     33.3%    100.0%

Number of Missing Observations =         0
```

A statistic often used to test the hypothesis that the row and column variables are independent is the *Pearson chi-square*. It is calculated by summing over all cells the squared residuals divided by the expected frequencies.

$$\chi^2 = \sum_i \sum_j \frac{(O_{ij} - E_{ij})^2}{E_{ij}}$$

Equation 9.8d

The calculated chi-square is compared to the critical points of the theoretical chi-square distribution to produce an estimate of how likely (or unlikely) this calculated value is if the two variables are in fact independent. Since the value of the chi-square depends on the number of rows and columns in the table being examined, one must know the *degrees of freedom* for the table. The degrees of freedom can be viewed as the number of cells of a table that can be arbitrarily filled when the row and column totals (marginals) are fixed. For an $R \times C$ table, the degrees of freedom are $(R-1) \times (C-1)$, since once $(R-1)$ rows and $(C-1)$ columns are filled, frequencies in the remaining row and column cells must be chosen so that marginal totals are maintained.

In this example, there are two degrees of freedom (1×2), and the chi-square value is 22.56 (see Figure 9.1). If type of address and return rate are independent, the probability that a random sample would result in a chi-square value of at least that magnitude is less than 0.00005. On the SPSS/PC+ output, the probability is rounded to four decimal places which in this case is 0.0000. This probability is also known as the *observed significance level* of the test. If the probability is small enough (usually less than 0.05 or 0.01), the hypothesis that the two variables are independent is rejected.

Since the observed significance level in Figure 9.1 is very small (based on the combined city and town data), the hypothesis that address type and return rate are independent is rejected. When the chi-square test is calculated for the city and town data separately (Figure 9.4), different results are obtained. The observed significance level of the city data is 0.209, so the independence hypothesis is not rejected. For the towns, the observed significance level is less than 0.00005, and the hypothesis that address and return rate are independent is rejected. These results support the theory that city and town residents respond differently.

The chi-square test is a test of independence; it provides little information about the strength or form of the association between two variables. The magnitude of the observed chi-square depends not only on the goodness of fit of the independence model, but also on the sample size. If the sample size for a particular table increases n-fold, so does the chi-square value. Thus, large chi-square values can arise in applications where residuals are small relative to expected frequencies but where the sample size is large.

Certain conditions must be met for the chi-square distribution to be a good approximation of the distribution of the statistic in the equation given above. The data must be random samples from multinomial distributions and the expected values must not be too small. While it has been recommended that all expected frequencies be at least 5, recent studies indicate that this is probably too stringent and can be relaxed (Everitt, 1977). If there are cells with expected values less than 5, SPSS/PC+ prints the number of such cells and the minimum expected value.

In hope of improving the approximation in the case of a 2 × 2 table, *Yates' correction for continuity* is sometimes applied. Yates' correction for continuity involves subtracting 0.5 from positive differences between observed and expected frequencies (the residuals) and adding 0.5 to negative differences before squaring. For a discussion of some of the controversy regarding the merits of this correction, see Conover (1974) and Mantel (1974).

An alternative test for the 2 × 2 table is based on the hypergeometric distribution. Exact probabilities of obtaining the observed results if the two variables are independent and the marginals fixed are calculated. This is called *Fisher's exact test*. It is most useful when the total sample size and the expected values are small. SPSS/PC+ calculates Fisher's exact test when the sample size in a 2 × 2 table is 20 or less.

9.9 Measures of Association

In many research situations, the strength and nature of the dependence of variables is of central concern. Indexes that attempt to quantify the relationship between variables in a cross-classification are called *measures of association*. No single measure adequately summarizes all possible types of association. Measures vary in their interpretation and in the way they define perfect and intermediate association. These measures also differ in the way they are affected by various factors such as marginals. For example, many measures are "margin sensitive" in that they are influenced by the marginal distributions of the rows and columns. Such measures reflect information about the marginals along with information about association.

A particular measure may have a low value for a given table, not because the two variables are not related but because they are not related in the way to which the measure is sensitive. No single measure is best for all situations. The type of data, the hypothesis of interest, as well as the properties of the various measures must all be considered when selecting an index of association for a given table. It is not, however, reasonable to compute a large number of measures and then to report the most impressive as if it were the only one examined.

The measures of association available in SPSS/PC+ CROSSTABS are computed only from bivariate tables (see Figure 9.4). For example, if three dichotomous variables are specified in the table, two sets of measures are computed, one for each subtable produced by the values of the controlling variable. In general, if relationships among more than two variables are to be studied, examination of bivariate tables is only a first step. For an extensive discussion of various more sophisticated multivariate procedures for the analysis of qualitative data, see Fienberg (1977), Everitt (1977), and Haberman (1978).

9.10 Nominal Measures

Consider measures that assume only that both variables in the table are nominally measured. As such, these measures can only provide some indication of the strength of association between variables; they cannot indicate direction or anything about the nature of the relationship. The measures provided are of two types: those based on the chi-square statistic and those that follow the logic of proportional reduction in error, denoted PRE.

9.11 Chi-Square-Based Measures

As explained above, the chi-square statistic itself is not a good measure of the degree of association between two variables. But its widespread use in tests of independence has encouraged the use of measures of association based upon it. Each of these measures based on the chi-square attempts to modify the chi-square statistic to minimize the influence of sample size and degrees of freedom as well as to restrict the range of values of the measure to those between 0 and 1. Without such adjustments, comparison of chi-square values from tables with varying dimensions and sample sizes is meaningless.

The *phi coefficient* modifies the chi-square by dividing it by the sample size and taking the square root of the result:

$$\phi = \sqrt{\frac{\chi^2}{N}}$$ **Equation 9.11a**

For tables in which one dimension is greater than 2, phi need not lie between 0 and 1 since the chi-square value can be greater than the sample size. To obtain a measure that must lie between 0 and 1, Pearson suggested the use of

$$C = \sqrt{\frac{\chi^2}{\chi^2 + N}}$$ **Equation 9.11b**

which is called the *coefficient of contingency*. Although the value of this measure is always between 0 and 1, it cannot generally attain the upper limit of 1. The maximum value possible depends upon the number of rows and columns. For example, in a 4 × 4 table, the maximum value of C is 0.87.

Cramér introduced the following variant:

$$V = \sqrt{\frac{\chi^2}{N(k-1)}}$$ **Equation 9.11c**

where k is the smaller of the number of rows and columns. This statistic, known as *Cramér's V*, can attain the maximum of 1 for tables of any dimension. If one of the table dimensions is 2, V and phi are identical.

The chi-square-based measures are hard to interpret. Although when properly standardized they can be used to compare strength of association in several tables, the "strength of association" being compared is not easily related to an intuitive concept of association.

9.12 Proportional Reduction in Error

Common alternatives to chi-square-based measurements are those based on the idea of *proportional reduction in error* (PRE), introduced by Goodman and Kruskal (1954). With PRE measures, the meaning of association is clearer. These measures are all essentially ratios of a measure of error in predicting the values of one variable based on knowledge of that variable alone and the same measure of error applied to predictions based on knowledge of an additional variable.

For example, Figure 9.12 is a crosstabulation of depth of hypnosis and success in treatment of migraine headaches by suggestion (Cedercreutz, 1978). The best guess of the results of treatment when no other information is available is the outcome category with the largest proportion of observations (the modal category). In Figure 9.12, "no change" is the largest outcome category, with 45% of the subjects. The estimate of the probability of incorrect classification is 1 minus the probability of the modal category:

$$P(1) = 1 - 0.45 = 0.55$$ **Equation 9.12a**

Figure 9.12 Depth of hypnosis and success of treatment

```
Crosstabulation:      HYPNOSIS  DEPTH OF HYPNOSIS
                   By MIGRAINE  OUTCOME

             Count  ICURED    IBETTER  INO
MIGRAINE-> Tot Pct  I         I        ICHANGE  I  Row
                    I       1 I      2 I      3 I  Total
HYPNOSIS   --------+--------+--------+--------+
                 1  I    13  I     5  I        I    18
  DEEP              I  13.0  I   5.0  I        I  18.0
                    +--------+--------+--------+
                 2  I    10  I    26  I    17  I    53
  MEDIUM            I  10.0  I  26.0  I  17.0  I  53.0
                    +--------+--------+--------+
                 3  I        I     1  I    28  I    29
  LIGHT             I        I   1.0  I  28.0  I  29.0
                    +--------+--------+--------+
            Column      23       32       45      100
             Total    23.0     32.0     45.0    100.0

 Chi-Square     D.F.        Significance        Min E.F.      Cells with E.F.< 5
 ----------     ----        ------------        --------      ------------------

  65.52525        4            0.0000             4.140        1 OF    9 ( 11.1%)

                                                   With HYPNOSIS      With MIGRAINE
        Statistic                     Symmetric    Dependent          Dependent
        ---------                     ---------    -------------      -------------

Lambda                                 0.35294        0.29787            0.40000
Uncertainty Coefficient                0.35514        0.36537            0.34547

        Statistic                       Value      Significance
        ---------                       -----      ------------

Cramer's V                             0.57239
Contingency Coefficient                0.62918

Number of Missing Observations =          0
```

Information about the depth of hypnosis can be used to improve the classification rule. For each hypnosis category, the outcome category that occurs most frequently for that hypnosis level is predicted. Thus, "no change" is predicted for participants achieving a "light" level of hypnosis, "better" for those achieving a "medium" level, and "cured" for those achieving a "deep" level. The probability of error when depth of hypnosis is used to predict outcome is the sum of the probabilities of all the cells that are not row modes:

$$P(2)=0.05+0.10+0.17+0.01=0.33$$

Equation 9.12b

Goodman and Kruskal's *lambda*, with outcome as the predicted (dependent) variable, is calculated as

$$\lambda_{\text{outcome}} = \frac{P(1) - P(2)}{P(1)} = \frac{0.55 - 0.33}{0.55} = 0.40$$

Equation 9.12c

Thus, a 40% reduction in error is obtained when depth of hypnosis is used to predict outcome.

Lambda always ranges between 0 and 1. A value of 0 means the independent variable is of no help in predicting the dependent variable. A value of 1 means that the independent variable perfectly specifies the categories of the dependent variable (perfection can occur only when each row has at most one nonzero cell). When the two variables are independent, lambda is 0; but a lambda of 0 need not imply statistical independence. As with all measures of association, lambda is constructed to measure association in a very specific way. In particular, lambda reflects the reduction in error when values of one variable are used to predict values of the other. If this particular type of association is absent, lambda is 0. Other measures of association may find association of a different kind even when lambda is 0. A measure of association sensitive to every imaginable type of association does not exist.

For a particular table, two different lambdas can be computed, one using the row variable as the predictor and the other using the column variable. The two do not usually have identical values, so care should be taken to specify which is the dependent variable, that is, the variable whose prediction is of primary interest. In some applications, dependent and independent variables are not clearly distinguished. Then, a symmetric version of lambda, which predicts the row variable and column variable with equal frequency, can be computed. When the lambda statistic is requested, SPSS/PC+ prints the symmetric lambda as well as the two asymmetric lambdas.

9.13 Ordinal Measures

Although relationships among ordinal variables can be examined using nominal measures, other measures reflect the additional information available from ranking. Consideration of the kind of relationships that may exist between two ordered variables leads to the notion of direction of relationship and to the concept of *correlation*. Variables are positively correlated if cases with low values for one variable also tend to have low values for the other and cases with high values on one also tend to be high on the other. Negatively correlated variables show the opposite relationship: the higher the first variable, the lower the second tends to be.

Several measures of association for a table of two ordered variables are based on the comparison of the values of both variables for all possible *pairs* of cases or observations. A pair of cases is *concordant* if the values of both variables for one case are higher (or both are lower) than the corresponding values for the other case. The pair is *discordant* if the value of one variable for a case is larger than the corresponding value for the other case, and the direction is reversed for the second variable. When the two cases have identical values on one or on both variables, they are *tied*.

Thus, for any given pair of cases with measurements on variables X and Y, the pair may be concordant or discordant, or tied in one of three ways: they may be tied on X but not on Y, they may be tied on Y but not on X, or they may be tied on both variables. When data are arranged in crosstabulated form, the number of concordant, discordant, and tied pairs can be easily calculated since all possible pairs can be conveniently determined.

If the preponderance of pairs is concordant, the association is said to be positive: as ranks of variable X increase (or decrease), so do ranks of variable Y. If the majority of pairs is discordant, the association is negative: as ranks of one variable increase, those of the other tend to decrease. If concordant and discordant pairs are equally likely, no association is said to exist.

The ordinal measures presented here all have the same numerator: the number of concordant pairs (P) minus the number of discordant pairs (Q) calculated for all distinct pairs of observations. They differ primarily in the way in which $P-Q$ is normalized. The simplest measure involves subtracting Q from P and dividing by the total number of pairs. If there are no pairs with ties, this measure (Kendall's tau-*a*) is in the range from -1 to $+1$. If there are ties, the range of possible values is narrower; the actual range depends on the number of ties. Since all observations within the same row are tied, so also are those in the same column, and the resulting tau-*a* measures are difficult to interpret.

A measure that attempts to normalize $P-Q$ by considering ties on each variable in a pair separately but not ties on both variables in a pair is tau-*b*:

$$\tau_b = \frac{P - Q}{\sqrt{(P + Q + T_X)\,(P + Q + T_Y)}}$$

Equation 9.13a

where T_X is the number of pairs tied on X but not on Y, and T_Y is the number of pairs tied on Y but not on X. If no marginal frequency is 0, tau-*b* can attain +1 or −1 only for a square table.

A measure that can attain, or nearly attain, +1 or −1 for any $R \times C$ table is tau-*c*:

$$\tau_c = \frac{2m(P - Q)}{N^2(m - 1)} \quad \text{Equation 9.13b}$$

where m is the smaller of the number of rows and columns. The coefficients tau-*b* and tau-*c* do not differ much in value if each margin contains approximately equal frequencies.

Goodman and Kruskal's *gamma* is closely related to the tau statistics and is calculated as

$$G = \frac{P - Q}{P + Q} \quad \text{Equation 9.13c}$$

Gamma can be thought of as the probability that a random pair of observations is concordant minus the probability that the pair is discordant, assuming the absence of ties. The absolute value of gamma is the proportional reduction in error between guessing concordant and discordant ranking of each pair depending on which occurs more often and guessing ranking according to the outcome of the toss of a fair coin. Gamma is 1 if all observations are concentrated in the upper-left to lower-right diagonal of the table. In the case of independence, gamma is 0. However, the converse (that a gamma of 0 necessarily implies independence) need not be true except in the 2 × 2 table.

In the computation of gamma, no distinction is made between the independent and dependent variable; the variables are treated symmetrically. Somers (1962) proposed an asymmetric extension of gamma that differs only in the inclusion of the number of pairs not tied on the independent variable (X) in the denominator. Somers' *d* is

$$d_Y = \frac{P - Q}{P + Q + T_Y} \quad \text{Equation 9.13d}$$

The coefficient d_Y indicates the proportionate excess of concordant pairs over discordant pairs among pairs not tied on the independent variable. The symmetric variant of Somers' *d* uses for the denominator the average value of the denominators of the two asymmetric coefficients.

9.14 Measures Involving Interval Data

If the two variables in the table are measured on an interval scale, various coefficients that make use of this additional information can be calculated. A useful symmetric coefficient that measures the strength of the *linear* relationship is the Pearson correlation coefficient, or *r*. It can take on values from −1 to +1, indicating negative or positive linear correlation. Further description, including a PRE interpretation, is found in Chapter 12.

The *eta* coefficient is appropriate for data in which the dependent variable is measured on an interval scale and the independent variable on a nominal or ordinal scale. When squared, eta can be interpreted as the proportion of the total variability in the dependent variable that can be accounted for by knowing the values of the independent variable. The measure is asymmetric and does not assume a linear relationship between the variables.

9.15 RUNNING PROCEDURE CROSSTABS

The CROSSTABS (alias XTABS) procedure produces tables showing the joint distribution of two or more variables that each have a limited number of distinct values. Cell counts, cell percentages, expected values, residuals, and various measures of association can be obtained. You can also specify the treatment of missing values, obtain measures of association without printing tables, and request an index of tables.

CROSSTABS operates via the TABLES subcommand, which defines the row, column, and control variables, and the OPTIONS and STATISTICS subcommands.

9.16 Specifying the Tables

The TABLES subcommand specifies the variables to be analyzed. Row variables are specified before the first BY, with column variables following. If control variables are included, they follow subsequent BYs. For example, in the command

```
CROSSTABS TABLES=RETURNED BY ADDRESS BY LOCATION.
```

RETURNED is the row variable, ADDRESS is the column variable, and LOCATION is a control variable.

The actual keyword TABLES can be omitted from the command. Thus, the commands

```
CROSSTABS TABLES=RETURNED BY ADDRESS.
```

and

```
CROSSTABS RETURNED BY ADDRESS.
```

are equivalent; either one could be used. Both numeric and string variables may be specified, although long strings are truncated to short ones to define categories.

Tables can have a maximum of ten dimensions (nine BY keywords), with one or more variables listed for each dimension. When control variables are used, a bivariate subtable is produced for each combination of values of the control variables (as in Figure 9.4). The value of the first control variable changes most rapidly and the value of the last control variable changes most slowly. For example, suppose the command

```
CROSSTABS HAPPY BY MARITAL BY SEX BY RACE.
```

is used, with SEX having the values 1 and 2, and RACE having the values 1, 2, and 3. The output will contain, first, the bivariate subtable HAPPY by MARITAL that controls for SEX equal to 1 and RACE equal to 1. The next table will be the HAPPY by MARITAL subtable that controls for SEX equal to 2 and RACE equal to 1. The third table will be the HAPPY by MARITAL subtable that controls for SEX equal to 1 and RACE equal to 2, and so forth. On the other hand, if the command

```
CROSSTABS HAPPY BY MARITAL BY RACE BY SEX.
```

is used, the first table will be the HAPPY by MARITAL subtable that controls for RACE equal to 1 and SEX equal to 1, the second table will be the HAPPY by MARITAL subtable that controls for RACE equal to 2 and SEX equal to 1, the third table will be the HAPPY by MARITAL subtable that controls for RACE equal to 3 and SEX equal to 1, and so forth. Thus, the same subtables are produced, only the order in which they are presented is different.

If more than one variable is specified in the row or column or control variable list, several tables are produced. For example, if the command

```
CROSSTABS TABLES=INCOME MARITAL BY RACE.
```

is used, two bivariate tables, INCOME by RACE and MARITAL by RACE, are produced. This command is *not* equivalent to the command

```
CROSSTABS TABLES=INCOME BY MARITAL BY RACE.
```

which results in three bivariate subtables, INCOME by MARITAL controlling for RACE equal to 1, RACE equal to 2, and RACE equal to 3.

The keyword TO can be used to indicate a set of adjacent variables in the active file, as in the command

```
CROSSTABS HAPPY BY MARITAL TO SEX.
```

This command results in tables for HAPPY by MARITAL, HAPPY by SEX, and HAPPY by all of the variables between MARITAL and SEX in the file.

A slash must be used to separate table lists. For example, the command

```
CROSSTABS TABLES=HAPPY BY SEX/MARITAL BY RACE.
```

specifies two bivariate tables, HAPPY by SEX and MARITAL by RACE.

9.17 Specifying Cell Contents

By default, only the number of cases in each cell is displayed. The OPTIONS subcommand is used to obtain other information in the cells:

Option 13 *Suppress cell counts.* You must specify at least one other option pertaining to cell contents when Option 13 is used.

Option 18 *Display all cell information.* Displays cell count; row, column, and total percentages; expected values; residuals; standardized residuals; and adjusted standardized residuals.

If Options 18 and 13 are both specified, only Option 18 is in effect: cell counts will not be suppressed.

Percentages are obtained by using the following options:

Option 3 *Display row percentages.*

Option 4 *Display column percentages.*

Option 5 *Display two-way table total percentages.*

Thus, the command used to obtain the table in Figure 9.1 was

```
CROSSTABS TABLES=RETURNED BY ADDRESS
/OPTIONS=3 4 5
/STATISTICS=1.
```

The options used to request expected frequencies or residuals are

Option 14 *Display expected frequencies.*

Option 15 *Display chi-square residuals.*

Option 16 *Display standardized chi-square residuals.*

Option 17 *Display adjusted standardized chi-square residuals.* (See Haberman, 1978.)

For example, the output in Figure 9.8 was obtained by specifying

```
CROSSTABS TABLES=RETURNED BY ADDRESS
/OPTIONS=14 15.
```

9.18 Obtaining Measures of Association

Measures of association for each two-way table or subtable are obtained by using the STATISTICS subcommand:

Statistic 1 *Chi-square.* Fisher's exact test is computed using the rounded values of the cell entries when there are fewer than 20 cases in a 2 × 2 table that does not result from missing rows or columns in a larger table; Yates' corrected chi-square is computed for all other 2 × 2 tables.

Statistic 2 *Phi for 2 × 2 tables, Cramér's* V *for larger tables.*

Statistic 3 *Contingency coefficient.*

Statistic 4 *Lambda, symmetric and asymmetric.*

Statistic 5 *Uncertainty coefficient, symmetric and asymmetric.*

Statistic 6 *Kendall's tau*-b. CROSSTABS does not calculate this statistic for tables containing string variables.

Statistic 7 *Kendall's tau*-c. CROSSTABS does not calculate this statistic for tables containing string variables.

Statistic 8 *Gamma.* CROSSTABS prints zero-order gammas for 2-way tables and conditional gammas for the 2-way subtables of 3-way to 10-way tables. It does not calculate gammas for tables containing string variables.

Statistic 9 *Somers'* d, *symmetric and asymmetric.* CROSSTABS does not calculate these statistics for tables containing string variables.

Statistic 10 *Eta.* Available for numeric data only.

Statistic 11 *Pearson's* r. Available for numeric data only.

ALL *All available statistics.* Specify ALL on the STATISTICS subcommand.

For example, the output in Figure 9.12 was obtained by using the command

```
CROSSTABS HYPNOSIS BY MIGRAINE
 /OPTIONS=5
 /STATISTICS=1 2 3 4 5.
```

If the STATISTICS subcommand is omitted, no statistics are printed.

9.19 Missing Values

By default, cases with missing values are excluded on a table-by-table basis. A case with missing values for any of the variables specified for a table is not used in the table or in the calculation of statistics. The number of missing cases is always printed after the end of the table, following the last subtable and any requested statistics.

If you want to include user-missing values in a table, you must specify Option 1 on the OPTIONS subcommand.

Option 1 *Include user-missing values.* Cases with user-missing values are treated as valid and are included in the table and in the calculation of percentages and statistics.

9.20 Formatting Options

By default, tables and subtables are displayed with variable labels and value labels (when defined). Although value labels may be up to 20 characters long, CROSSTABS uses only the first 16 characters. The value labels for the column variable are printed on two lines, with eight characters per line. If you don't want these labels broken up at an awkward point, you can redefine them using the VALUE LABELS command (see Chapter 3), as in the command

```
VALUE LABELS HYPNOSIS 1 'DEEP' 2 'MEDIUM' 3 'LIGHT'/
             MIGRAINE 1 'CURED' 2 'BETTER' 3 'NO      CHANGE'.
```

The following options can be used to modify the display.

Option 2 *Suppress variable and value labels.*

Option 6 *Suppress value labels, but display variable labels.*

Option 8 *Order rows by descending value.* The default is ascending.

Option 12 *Suppress tables.*

Option 19 *Suppress values.* Values are displayed only if no value labels have been defined.

If statistics are requested with the STATISTICS subcommand and Option 12 is specified, only the statistics are printed. If statistics are not requested and Option 12 is specified, no output is displayed.

9.21 Entering Crosstabulated Data

Frequently, you already have a crosstabulation that you want to present in a different way or for which you want to produce additional statistics. You can enter the crosstabulated data rather than the original observations into SPSS/PC+ and proceed with your analysis. Each cell of the table is considered a case. For each case (cell of the table), enter the cell counts along with the values of the row, column, and control variables. Define this file as you would any other data file. Then use the WEIGHT command (see Command Reference) to specify that each case should be counted as many times as specified by the cell frequency (see Section 9.22).

9.22 Annotated Example

The table in Figure 9.1 was obtained by using the following SPSS/PC+ commands:

```
DATA LIST  FIXED / FREQ 1-5 RETURNED 7 ADDRESS 9.
WEIGHT BY FREQ.
VARIABLE LABELS
  RETURNED 'FOUND AND MAILED'/
  ADDRESS 'ADDRESS ON LETTER'.
VALUE LABELS
  RETURNED  1 'YES' 2 'NO'/
  ADDRESS  1 'CONTROL' 2 'DANDEE' 3 'COMMUNIST'.
BEGIN DATA.
   35 1 1
   37 2 1
   32 1 2
   40 2 2
   10 1 3
   62 2 3
END DATA.
CROSSTABS  TABLES=RETURNED BY ADDRESS
 /OPTIONS=3,4,5
 /STATISTICS=1.
FINISH.
```

- The DATA LIST command defines the variables for the cells, which are treated like cases. The cell frequency is entered in columns 1 through 5 and is defined as the variable FREQ. The value of RETURNED for each cell is entered in column 7 and the value of ADDRESS in column 9. The first line of data, therefore, represents the first cell of the table, where RETURNED has value 1 and ADDRESS has value 1. The second line of data is the second cell in the first column of the table, and so forth.
- The WEIGHT command uses FREQ, the cell frequencies, as the weighting variable. The value of this variable for each case is used to weight the cases: SPSS/PC+ reads the single case as if it occurred as many times as the number specified by FREQ. For example, the first case is counted as 35 cases. Thus, the WEIGHT command enables you to easily reproduce crosstabulated data.
- The VARIABLE LABELS and VALUE LABELS commands assign descriptive labels to the variables and their values.
- The CROSSTABS command specifies the desired crosstabulation.

If you want to rearrange the variables in the crosstabulated input data or drop certain variables from the table, you can do so by specifying only the variables that you want, in the order that you want. For example, the command

```
CROSSTABS TABLES=ADDRESS BY RETURNED
```

defines ADDRESS as the row variable and RETURNED as the column variable.

Contents

Chapter 10 Describing Subpopulations: Procedure MEANS

The 1964 Civil Rights Act prohibits discrimination in the workplace based on sex or race. Employers who violate the act by unfair hiring or advancement practices can be prosecuted. Numerous lawsuits have been filed on behalf of women, blacks, and other groups offered equal protection under the law.

The courts have ruled that statistics can be used as *prima facie* evidence of discrimination, and many lawsuits depend heavily on complex statistical analyses, which attempt to demonstrate that similarly qualified individuals are not treated equally. Identifying and measuring all variables that legitimately influence promotion and hiring is difficult, if not impossible, especially for nonroutine jobs. Years of schooling and prior work experience can be quantified, but what about the more intangible attributes such as enthusiasm and creativity? How are they to be objectively measured so as not to become convenient smoke screens for concealing discrimination?

10.1 SEARCHING FOR DISCRIMINATION

In this chapter, employee records for 474 individuals hired between 1969 and 1971 by a bank engaged in Equal Employment Opportunity (EEO) litigation are analyzed. Two types of unfair employment practices are of particular interest: shunting (placing some employees in lower job categories than others with similar qualifications) and salary and promotion inequities.

Although extensive and intricate statistical analyses are usually involved in studies of this kind (see, for example, Roberts, 1980), the discussion here is necessarily limited. The SPSS/PC+ MEANS procedure is used to calculate average salaries for groups of employees based on race and sex. Additional grouping variables are introduced to help "explain" some of the observed variability in salary.

10.2 Who Does What?

Figure 10.2 is a crosstabulation of job category at the time of hiring with sex and race characteristics. The first three job classifications contain 64% of white males (adding column percents), 94% of both nonwhite males and white females, and

100% of nonwhite females. Among white males, 17% are in the college trainee program, compared to 4% of white females.

Figure 10.2 Crosstabulation of job category by sex-race

```
Crosstabulation:        JOBCAT      EMPLOYMENT CATEGORY
                     By SEXRACE

               Count  IWHITE    IMINORITYIWHITE    IMINORITYI
SEXRACE->    Col Pct  IMALES    IMALES   IFEMALES  IFEMALES I   Row
             Tot Pct  I    1.00I    2.00I    3.00I    4.00I  Total
JOBCAT        --------+--------+--------+--------+--------+
                    1 I    75  I    35  I    85  I    32  I    227
  CLERICAL            I  38.7  I  54.7  I  48.3  I  80.0  I   47.9
                      I  15.8  I   7.4  I  17.9  I   6.8  I
                      +--------+--------+--------+--------+
                    2 I    35  I    12  I    81  I     8  I    136
  OFFICE TRAINEE      I  18.0  I  18.8  I  46.0  I  20.0  I   28.7
                      I   7.4  I   2.5  I  17.1  I   1.7  I
                      +--------+--------+--------+--------+
                    3 I    14  I    13  I        I        I     27
  SECURITY OFFICER    I   7.2  I  20.3  I        I        I    5.7
                      I   3.0  I   2.7  I        I        I
                      +--------+--------+--------+--------+
                    4 I    33  I     1  I     7  I        I     41
  COLLEGE TRAINEE     I  17.0  I   1.6  I   4.0  I        I    8.6
                      I   7.0  I    .2  I   1.5  I        I
                      +--------+--------+--------+--------+
                    5 I    28  I     2  I     2  I        I     32
  EXEMPT EMPLOYEE     I  14.4  I   3.1  I   1.1  I        I    6.8
                      I   5.9  I    .4  I    .4  I        I
                      +--------+--------+--------+--------+
                    6 I     3  I     1  I     1  I        I      5
  MBA TRAINEE         I   1.5  I   1.6  I    .6  I        I    1.1
                      I    .6  I    .2  I    .2  I        I
                      +--------+--------+--------+--------+
                    7 I     6  I        I        I        I      6
  TECHNICAL           I   3.1  I        I        I        I    1.3
                      I   1.3  I        I        I        I
                      +--------+--------+--------+--------+
               Column     194       64      176       40      474
                Total    40.9     13.5     37.1      8.4    100.0

Number of Missing Observations =          0
```

Although these observations are interesting, they do not imply discriminatory placement into beginning job categories because the qualifications of the various groups are not necessarily similar. If women and nonwhites are more qualified than white males in the same beginning job categories, discrimination may be suspected.

10.3 Level of Education

One easily measured employment qualification is years of education. Figure 10.3a shows the average years of education for the entire sample (labeled For Entire Population) and then for each of the two sexes (labeled SEX and MALES or FEMALES) and then for each of the two race categories within each sex category (labeled MINORITY and WHITE or NONWHITE).

Figure 10.3a Education broken down by race within sex

```
Summaries of   EDLEVEL     EDUCATIONAL LEVEL
By levels of   SEX         SEX OF EMPLOYEE
               MINORITY    MINORITY CLASSIFICATIONS

Variable          Value  Label                      Sum        Mean     Std Dev    Variance     Cases

For Entire Population                            6395.0000    13.4916     2.8848     8.3223       474

SEX                   0  MALES                   3723.0000    14.4302     2.9793     8.8764       258
  MINORITY            0  WHITE                   2895.0000    14.9227     2.8484     8.1132       194
  MINORITY            1  NONWHITE                 828.0000    12.9375     2.8888     8.3452        64

SEX                   1  FEMALES                 2672.0000    12.3704     2.3192     5.3785       216
  MINORITY            0  WHITE                   2172.0000    12.3409     2.4066     5.7917       176
  MINORITY            1  NONWHITE                 500.0000    12.5000     1.9081     3.6410        40

  Total Cases =        474
```

The entire sample has an average of 13.49 years of education. Males have more years of education than females—an average of 14.43 years compared to 12.37. White males have the highest average level of education, almost 15 years, which is 2 years more than nonwhite males and approximately 2.5 years more than either group of females.

Figure 10.3b Education by sex-race and job category

```
Summaries of   EDLEVEL     EDUCATIONAL LEVEL
By levels of   JOBCAT      EMPLOYMENT CATEGORY
               SEXRACE

Variable          Value  Label                          Mean     Std Dev    Cases

For Entire Population                                 13.4916     2.8848      474

JOBCAT                1  CLERICAL                     12.7753     2.5621      227
  SEXRACE          1.00  WHITE    MALES               13.8667     2.3035       75
  SEXRACE          2.00  MINORITYMALES                13.7714     2.3147       35
  SEXRACE          3.00  WHITE    FEMALES             11.4588     2.4327       85
  SEXRACE          4.00  MINORITYFEMALES              12.6250     2.1213       32

JOBCAT                2  OFFICE TRAINEE               13.0221     1.8875      136
  SEXRACE          1.00  WHITE    MALES               13.8857     1.4095       35
  SEXRACE          2.00  MINORITYMALES                12.5833     2.6097       12
  SEXRACE          3.00  WHITE    FEMALES             12.8148     1.9307       81
  SEXRACE          4.00  MINORITYFEMALES              12.0000     0.0           8

JOBCAT                3  SECURITY OFFICER             10.1852     2.2194       27
  SEXRACE          1.00  WHITE    MALES               10.2857     2.0542       14
  SEXRACE          2.00  MINORITYMALES                10.0769     2.4651       13

JOBCAT                4  COLLEGE TRAINEE              17.0000     1.2845       41
  SEXRACE          1.00  WHITE    MALES               17.2121     1.3407       33
  SEXRACE          2.00  MINORITYMALES                17.0000     0.0           1
  SEXRACE          3.00  WHITE    FEMALES             16.0000     0.0           7

JOBCAT                5  EXEMPT EMPLOYEE              17.2813     1.9713       32
  SEXRACE          1.00  WHITE    MALES               17.6071     1.7709       28
  SEXRACE          2.00  MINORITYMALES                14.0000     2.8284        2
  SEXRACE          3.00  WHITE    FEMALES             16.0000     0.0           2

JOBCAT                6  MBA TRAINEE                  18.0000     1.4142        5
  SEXRACE          1.00  WHITE    MALES               18.3333     1.1547        3
  SEXRACE          2.00  MINORITYMALES                19.0000     0.0           1
  SEXRACE          3.00  WHITE    FEMALES             16.0000     0.0           1

JOBCAT                7  TECHNICAL                    18.1667     1.4720        6
  SEXRACE          1.00  WHITE    MALES               18.1667     1.4720        6

  Total Cases =    474
```

In Figure 10.3b, the cases are further subdivided by their combined sex-race characteristics and by their initial job category. For each cell in the table, the average years of education, the standard deviation, and number of cases are displayed. White males have the highest average years of education in all job categories except MBA trainees, where the only nonwhite male MBA trainee has nineteen years of education. From this table, it does not appear that females and nonwhites are overeducated when compared to white males in similar job categories. However, it is important to note that group means provide information about a particular class of employees. While discrimination may not exist for a class as a whole, some individuals within that class may be victims (or beneficiaries) of discrimination.

10.4 Beginning Salaries

The average beginning salary for the 474 persons hired between 1969 and 1971 is $6,806. The distribution by the four sex-race categories is shown in Figure 10.4a.

Figure 10.4a Beginning salary by sex-race

```
Summaries of   SALBEG      BEGINNING SALARY
By levels of   SEXRACE

Variable        Value  Label                     Sum        Mean    Std Dev   Variance    Cases

For Entire Population                     3226250.00   6806.4346  3148.2553 9911511.19     474

SEXRACE          1.00  WHITE    MALES     1675680.00   8637.5258  3871.1017 14985428.4     194
SEXRACE          2.00  MINORITYMALES       419424.000  6553.5000  2228.1436 4964624.00      64
SEXRACE          3.00  WHITE    FEMALES    939926.000  5340.4886  1225.9605 1502979.07     176
SEXRACE          4.00  MINORITYFEMALES     191220.000  4780.5000   771.4188 595086.923      40

  Total Cases =     474
```

White males have the highest beginning salaries—an average of $8,638—followed by nonwhite males. Since males are in higher job categories than females, this difference is not surprising.

Figure 10.4b Beginning salary by sex-race and job category

```
Summaries of   SALBEG       BEGINNING SALARY
By levels of   JOBCAT       EMPLOYMENT CATEGORY
               SEXRACE

Variable          Value  Label                       Mean

For Entire Population                           6806.4346

JOBCAT                1  CLERICAL               5733.9471
  SEXRACE          1.00  WHITE    MALES         6553.4400
  SEXRACE          2.00  MINORITYMALES          6230.7429
  SEXRACE          3.00  WHITE    FEMALES       5147.3176
  SEXRACE          4.00  MINORITYFEMALES        4828.1250

JOBCAT                2  OFFICE TRAINEE         5478.9706
  SEXRACE          1.00  WHITE    MALES         6262.2857
  SEXRACE          2.00  MINORITYMALES          5610.0000
  SEXRACE          3.00  WHITE    FEMALES       5208.8889
  SEXRACE          4.00  MINORITYFEMALES        4590.0000

JOBCAT                3  SECURITY OFFICER       6031.1111
  SEXRACE          1.00  WHITE    MALES         6102.8571
  SEXRACE          2.00  MINORITYMALES          5953.8462

JOBCAT                4  COLLEGE TRAINEE        9956.4878
  SEXRACE          1.00  WHITE    MALES        10467.6364
  SEXRACE          2.00  MINORITYMALES         11496.0000
  SEXRACE          3.00  WHITE    FEMALES       7326.8571

JOBCAT                5  EXEMPT EMPLOYEE       13258.8750
  SEXRACE          1.00  WHITE    MALES        13255.2857
  SEXRACE          2.00  MINORITYMALES         15570.0000
  SEXRACE          3.00  WHITE    FEMALES      10998.0000

JOBCAT                6  MBA TRAINEE           12837.6000
  SEXRACE          1.00  WHITE    MALES        14332.0000
  SEXRACE          2.00  MINORITYMALES         13992.0000
  SEXRACE          3.00  WHITE    FEMALES       7200.0000

JOBCAT                7  TECHNICAL             19996.0000
  SEXRACE          1.00  WHITE    MALES        19996.0000

  Total Cases =     474
```

Figure 10.4b shows beginning salaries subdivided by race, sex, and job category. For most of the job categories, white males have higher beginning salaries than the other groups. There is a $1,400 salary difference between white males and white females in the clerical jobs and a $1,000 difference in the general office trainee classification. In the college trainee program, white males averaged over $3,000 more than white females. However, Figure 10.3b shows that white females in the college trainee program had only an undergraduate degree, while white males had an average of 17.2 years of schooling.

10.5 Introducing More Variables

The differences in mean beginning salaries between males and females are somewhat suspect. It is, however, unwise to conclude that salary discrimination exists since several important variables, such as years of prior experience, have not been considered. It is necessary to control (or to adjust statistically) for other relevant variables. Using procedure MEANS to crossclassify cases by the variables of interest and compare salaries across the subgroups is one way of achieving control. However, as the number of variables increases, the number of cases in each cell rapidly diminishes, making statistically meaningful comparisons difficult. To circumvent these problems, regression methods, which achieve control by specifying certain statistical relations that may describe what is happening, are used. Regression methods are described in Chapter 17.

10.6 RUNNING PROCEDURE MEANS

MEANS calculates the means, sums, standard deviations, and variances of a variable for subgroups defined by other variables (the independent variables). Optional specifications on MEANS provide one-way analysis of variance and a test of linearity. MEANS operates via the TABLES, OPTIONS, and STATISTICS subcommands.

10.7 Specifying the Tables

The minimum specification required by the MEANS procedure is the TABLES subcommand with a single tables list. The tables list must specify at least one dependent variable, the keyword BY, and at least one independent variable, as in the command

```
MEANS TABLES=SALBEG BY SEXRACE.
```

used to produce the output in Figure 10.4a.

The actual keyword TABLES can be omitted. For example, the output in Figure 10.4a could also have been obtained with the command

```
MEANS SALBEG BY SEXRACE.
```

Although the dependent variable must be numeric, independent variables can be numeric or string. Long strings are truncated to short strings to define categories. More than one dependent variable and independent variable can be specified, and the TO keyword can be used to name a set of adjacent variables in the active file, as in the command

```
MEANS RAISE80 TO RAISE83 BY DEPT TO AGE.
```

This command produces MEANS tables for all variables between (and including) RAISE80 and RAISE83 in the file, broken down first by DEPT, then by each of the variables between DEPT and AGE in the file, and finally by AGE.

When several BY keywords are used in the tables list, the variable after the last BY changes most quickly. (This is the opposite of the CROSSTABS format, in which the last variable changes most slowly.) For example, the command

```
MEANS TABLES=SALBEG BY JOBCAT BY SEXRACE.
```

produces the table shown in Figure 10.4b, which shows beginning salary (SALBEG) for value 1 of JOBCAT broken down by each of the four values of SEXRACE. Then, beginning salary for value 2 of JOBCAT is shown, again broken down by the four values of SEXRACE, and so forth.

When you specify multiple variables in a dimension of a table, they will be processed from left to right. For example, the command

```
MEANS RAISE82 RAISE83 BY DEPT AGE BY SEX RACE.
```

produces eight tables. The first is RAISE82 by DEPT by SEX, the second is RAISE82 by DEPT by RACE, the third is RAISE82 by AGE by SEX, and so on. The last table is RAISE83 by AGE by RACE.

More than one tables list can be specified on a single MEANS command if they are separated by a slash. For example, the command

```
MEANS TABLES=RAISE81 BY SEX/EDLEVEL BY RACE.
```

requests two tables, RAISE81 by SEX and EDLEVEL by RACE.

10.8 Optional Statistics

By default, procedure MEANS displays means, standard deviations, and the number of cases for subgroups. You can modify the display by specifying the following options on the OPTIONS subcommand:

Option 5 *Suppress group counts.* The number of cases in each group is not displayed.

Option 6 *Display group sums.*

Option 7 *Suppress group standard deviations.*

Option 11 *Suppress group means.*

Option 12 *Display group variances.*

You can also obtain a one-way analysis of variance and a test of linearity by specifying the following on the STATISTICS subcommand:

Statistic 1 *One-way analysis of variance including eta and eta^2.*

Statistic 2 *Test of linearity.* Includes the sums of squares, mean squares, and degrees of freedom associated with the linear and nonlinear components, as well as the F ratio, Pearson's r, and Pearson's r^2. The linearity test is not calculated if the grouping variable is a string variable.

ALL *Display all statistics.* Produces the same display as Statistic 2.

If a two-way or higher-order breakdown is specified, the second and subsequent grouping variables are ignored in the analysis of variance table. For example, the command

```
MEANS SALBEG BY JOBCAT BY SEXRACE
/STATISTICS=1.
```

produces a breakdown of SALBEG by JOBCAT within SEXRACE, but an analysis of variance is calculated only for the SALBEG by JOBCAT table. Two-way and higher-order analyses of variance can be obtained by using the ANOVA procedure (see Chapter 15), and a more complete one-way analysis of variance can be obtained by using the ONEWAY procedure (see Chapter 14).

10.9 Missing Values

By default, MEANS deletes cases with missing values on a table-by-table basis: a case with missing values for any of the variables specified for a table is not used in building that table.

You can specify two alternative missing-value treatments on the OPTIONS subcommand:

Option 1 *Include cases with user-missing values.* Cases with user-missing values are included in all tables and statistics.

Option 2 *Exclude cases with user-missing dependent values.* Any case with missing values for the independent variables but with a valid value for the dependent variable is included in the analysis.

10.10 Formatting Options

By default, MEANS displays variable names and variable labels at the beginning of each table. Within the table, groups defined by the independent variables are identified by variable name, values, and value labels. Use the OPTIONS subcommand to change these defaults.

Option 3 *Suppress all labels.* No variable or value labels are displayed.

Option 8 *Suppress value labels.* No value labels are displayed for independent variables.

Option 9 *Suppress independent variable names.*

Option 10 *Suppress independent variable values.*

10.11 Annotated Example

The following SPSS/PC+ commands were used to obtain the output in Figure 10.4a:

```
DATA LIST / MINORITY 8 SEX 16 JOBCAT 24 EDLEVEL 31-32 SALBEG 36-40.
VARIABLE LABELS MINORITY 'MINORITY CLASSIFICATIONS'
                /SEX 'SEX OF EMPLOYEE'
                /JOBCAT 'EMPLOYMENT CATEGORY'
                /EDLEVEL 'EDUCATIONAL LEVEL'
                /SALBEG 'BEGINNING SALARY'.
VALUE LABELS MINORITY 0 'WHITE' 1 'NONWHITE'
             /SEX 0 'MALES' 1 'FEMALES'
             /JOBCAT 1 'CLERICAL' 2 'OFFICE TRAINEE' 3 'SECURITY OFFICER'
                    4 'COLLEGE TRAINEE' 5 'EXEMPT EMPLOYEE'
                    6 'MBA TRAINEE' 7 'TECHNICAL'.

COMPUTE SEXRACE=1.
IF              (MINORITY EQ 1 AND SEX EQ 0) SEXRACE=2.
IF              (MINORITY EQ 0 AND SEX EQ 1) SEXRACE=3.
IF              (MINORITY EQ 1 AND SEX EQ 1) SEXRACE=4.
VALUE LABELS    SEXRACE 1 'WHITE   MALES' 2 'MINORITYMALES'
                3 'WHITE   FEMALES' 4 'MINORITYFEMALES'.
BEGIN DATA.
lines of data
END DATA.
MEANS TABLES=SALBEG BY SEXRACE
      /OPTIONS=6,12.
FINISH.
```

- The DATA LIST command gives the variable names and column locations for the variables used in the analysis.
- The VARIABLE LABELS and VALUE LABELS commands give descriptive labels to the variables and the values.
- The COMPUTE command and the three IF commands create a new variable, SEXRACE, that combines the sex and race variables already in the file. COMPUTE sets the initial value of SEXRACE to 1; this will be the white-male category. The IF commands then specify the value 2 for nonwhite males, the value 3 for white females, and the value 4 for nonwhite females.
- The VALUE LABELS command defines labels for the values of the new SEXRACE variable. These labels are formatted to avoid awkward breaks when displayed in the CROSSTABS table in Figure 10.2 (see Chapter 9).
- The MEANS command requests a table of SALBEG broken down by SEXRACE. Options 6 and 12 request the printing of group sums and group variances.

Contents

Chapter 11 Testing Differences Between Two Means: Procedure T-TEST

Would you buy a disposable raincoat, vegetables in pop-top cans, or investment counseling via closed-circuit television? These products and 17 others were described in questionnaires administered to 100 married couples (Davis & Ragsdale, 1983). Respondents were asked to rate on a scale of 1 (definitely want to buy) to 7 (definitely do not want to buy) their likelihood of buying each product. Of the 100 couples, 50 received questionnaires with pictures of the products and 50 received questionnaires without pictures. In this chapter we will examine whether pictures affect consumer preferences and whether husbands' and wives' responses differ.

11.1 TESTING HYPOTHESES

The first part of the table in Figure 11.1 contains basic descriptive statistics for the buying scores of couples receiving questionnaires with and without pictures. A couple's buying score is simply the sum of all ratings assigned to products by the husband and wife individually. Low scores indicate buyers while high scores indicate reluctance to buy. The 50 couples who received questionnaires without pictures (Group 1) had a mean score of 168 while the 48 couples who received forms with pictures had an average score of 159. (Two couples did not complete the questionnaire and are not included in the analysis.) The standard deviations show that scores for the second group were somewhat more variable than those for the first.

Figure 11.1 Family buying scores by questionnaire type

```
Independent samples of  VISUAL

Group 1:  VISUAL  EQ 0            Group 2:  VISUAL  EQ 1

t-test for:  FAMSCORE

                     Number                     Standard   Standard
                    of Cases      Mean         Deviation     Error

     Group 1           50       168.0000         21.787      3.081
     Group 2           48       159.0833         27.564      3.979

                   I Pooled Variance Estimate I Separate Variance Estimate
                   I                          I
     F    2-Tail   I    t    Degrees of 2-Tail I    t    Degrees of   2-Tail
   Value   Prob.   I  Value    Freedom   Prob. I  Value    Freedom     Prob.
                   I                          I
    1.60   0.106   I   1.78      96      0.078 I   1.77     89.43      0.080
```

If one is willing to restrict the conclusions to the 98 couples included in the study, it is safe to say that couples who received forms with pictures indicated a greater willingness to purchase the products than couples who received forms without pictures. However, this statement is not very satisfying. What is needed is some type of statement about the effect of the two questionnaire types for all couples—or at least some larger group of couples—not just those actually studied.

11.2 Samples and Populations

The totality of all cases about which conclusions are desired is called the *population*, while the observations actually included in the study are the *sample*. The couples in this experiment can be considered a sample from the population of couples in the United States.

The field of statistics helps us draw inferences about populations based on observations obtained from *random samples,* or samples in which the characteristics and relationships of interest are independent of the probabilities of being included in the sample. The necessity of a good research design cannot be overemphasized. Unless precautions are taken to ensure that the sample is from the population of interest and that the cases are chosen and observed without bias, the results obtained from statistical analyses may be misleading. For example, if the sample for this study contained only affluent suburban couples, conclusions about all couples might be unwarranted.

If measurements are obtained from an entire population, the population can be characterized by the various measures of central tendency, dispersion, and shape described in Chapter 8. The results describe the population exactly. Usually, however, one obtains information from a random sample, and the results serve as *estimates* of the unknown population values. Special notation is used to identify population values, termed *parameters*, and to distinguish them from sample values, termed *statistics*. The mean of a population is denoted by μ, and the variance by σ^2. The symbols $\bar{X}$ and S^2 are reserved for the mean and variance of samples.

11.3 Sampling Distributions

The sample that is actually included in a study is just one of many random samples that could have been selected from a population. For example, if the population consists of married couples in the United States, the number of different samples that could be chosen for inclusion in a study is mind-boggling. The estimated value of a population parameter depends on the particular sample chosen. Different samples usually produce different estimates.

Figure 11.3 is a histogram of 400 means. Each mean is calculated from a random sample of 25 observations, from a population which has a normal distribution with a mean value of zero and a standard deviation of 1. The estimated means are not all the same. Instead, they have a distribution. Most sample means are fairly close to zero, the population mean. The mean of the 400 means is 0.010 and the standard deviation of these means is 0.205. In fact, the distribution of the means appears approximately normal.

Although Figure 11.3 gives some idea of the appearance of the distribution of sample means of size 25 from a standard normal population, it is only an approximation since all possible samples of size 25 have not been taken. If the number of samples taken is increased to 1000, an even better picture of the distribution could be obtained. As the number of samples of a fixed size increases, the observed (or empirical) distribution of the means approaches the underlying, or theoretical, distribution.

Figure 11.3 Means of 400 samples of size 25 from a normal distribution

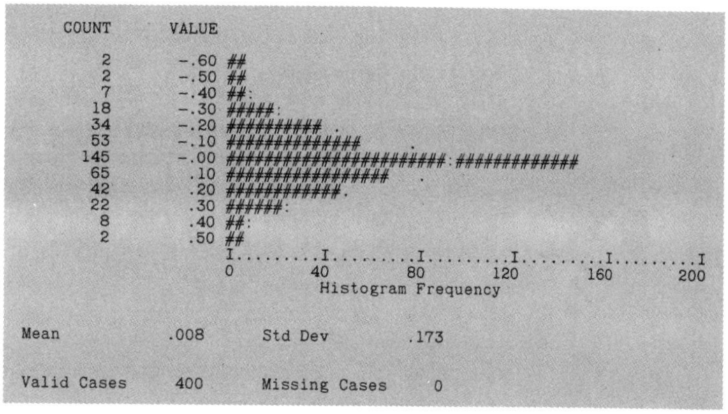

The theoretical distribution of all possible values of a statistic obtained from a population is called the *sampling distribution* of the statistic. The mean of the sampling distribution is called the *expected value* of the statistic. The standard deviation is termed the *standard error*. The sampling distributions of most commonly used statistics calculated from random samples are tabulated and readily accessible. Knowing the sampling distribution of a statistic is very important for hypothesis testing, since from it one can calculate the probability of obtaining an observed sample value if a particular hypothesis is true. For example, from Figure 11.3, it appears quite unlikely that a sample mean based on a sample of size 25 from a standard normal distribution would be greater than 0.5 if the population mean were zero.

11.4 Sampling Distribution of the Mean

Since hypotheses about population means are often of interest, the sampling distribution of the mean is particularly important. If samples are taken from a normal population, the sampling distribution of the sample mean is also normal. As expected, the observed distribution of the 400 means in Figure 11.3 is approximately normal. The theoretical distribution of the sample mean, based on all possible samples of size 25, is exactly normal.

Even when samples are taken from a nonnormal population, the distribution of the sample means will be approximately normal for sufficiently large samples. This is one reason for the importance of the normal distribution in statistical inference. Consider Figure 11.4a, which shows a sample from a uniform distribution. In a uniform distribution all values of a variable are equally likely, and hence the proportion of cases in each bin of the histogram is roughly the same.

Figure 11.4a Values from a uniform distribution

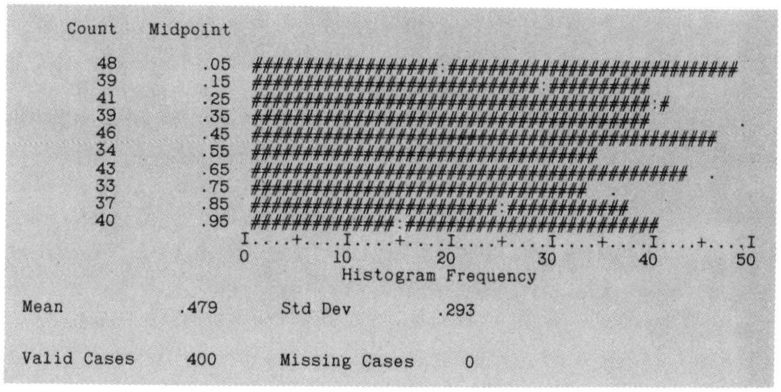

Figure 11.4b is a histogram of 400 means calculated from samples of size 25 from a uniform distribution. Note that the observed distribution is approximately normal even though the distribution from which the samples were taken is markedly nonnormal.

Figure 11.4b Distribution of 400 means calculated from samples of size 25 from a uniform distribution

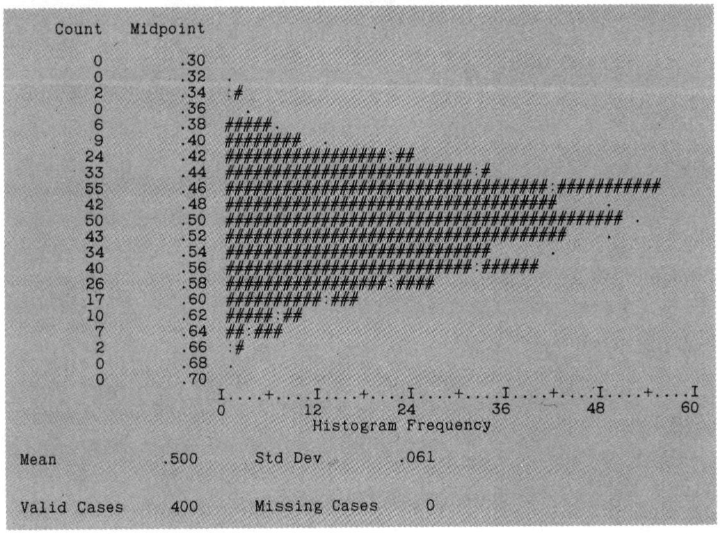

Both the size of a sample and the shape of the distribution from which samples are taken affect the shape of the sampling distribution of the mean. If samples are small and come from distributions that are far from normal, the distribution of the means will not be even approximately normal. As the size of the sample increases, the sampling distribution of the mean will approach normality.

The mean of the theoretical sampling distribution of the means of samples of size n is μ, the population mean. The standard error, which is another name for the standard deviation of the sampling distribution of the mean, is

$$\sigma_{\bar{X}} = \frac{\sigma}{\sqrt{N}}$$ **Equation 11.4a**

where σ is the standard deviation of the population, and N is the sample size.

The standard deviation of the observed sampling distribution of means in Figure 11.3 is 0.205. This is close to the value of the standard error for the theoretical distribution which, from Equation 11.4a, is 1/5, or 0.20.

Usually the value of the standard error is unknown and is estimated from a single sample using

$$S_{\bar{X}} = \frac{S}{\sqrt{N}}$$ **Equation 11.4b**

where S is the *sample* standard deviation. The estimated standard error is displayed in the FREQUENCIES procedure and is also part of the output shown in Figure 11.1. For example, for Group 1 the estimated standard error of the mean is

$$\frac{21.787}{\sqrt{50}} = 3.081 \qquad \textbf{Equation 11.4c}$$

This value is displayed in the column labeled STANDARD ERROR in Figure 11.1.

The standard error of the mean depends on both the sample standard deviation and the sample size. As the size of a sample increases, the standard error decreases. This is intuitively clear, since the more data are gathered, the more confident you can be that the sample mean is not too far from the population mean. Also, as the standard deviation of the observations decreases, the standard error decreases as well. Small standard deviations occur when observations are fairly homogeneous. In this case, means based on different samples should also not vary much.

11.5 THE TWO-SAMPLE T TEST

Consider again whether there is evidence that the type of questionnaire administered influences couples' buying decisions. The question is not whether the two sample means are equal, but whether the two population means are equal.

To test the hypothesis that, in the population, buying scores for the two questionnaire types are the same, the following statistic can be calculated:

$$t = \frac{\overline{X}_1 - \overline{X}_2}{\sqrt{S_1^2/N_1 + S_2^2/N_2}} \qquad \textbf{Equation 11.5a}$$

$\overline{X}_1$ is the sample mean of Group 1, S_1^2 is the variance, and N_1 is the sample size.

Based on the sampling distribution of the above statistic, one can calculate the probability that a difference at least as large as the one observed would occur if the two population means (μ_1 and μ_2) are equal. This probability is called the *observed significance level.* If the observed significance level is small enough, usually less than 0.05, or 0.01, the hypothesis that the population means are equal is rejected.

The t value and its associated probability are given in Figure 11.1 in the section labeled SEPARATE VARIANCE ESTIMATE. The t value is

$$t = \frac{168.0 - 159.08}{\sqrt{\dfrac{21.787^2}{50} + \dfrac{27.564^2}{48}}} = 1.77 \qquad \textbf{Equation 11.5b}$$

If $\mu_1 = \mu_2$, the probability of observing a difference at least as large as the one in the sample is estimated to be about 0.08. Since this probability is greater than 0.05, the hypothesis that mean buying scores in the population are equal for the two types of forms is not rejected. The entry under DEGREES OF FREEDOM in Figure 11.1 is a function of the sample size in the two groups and is used together with the t value in establishing the observed significance level.

Another statistic based on the t distribution can be used to test the equality of means hypothesis. This statistic, known as the *pooled-variance* t *test,* is based on the assumption that the population variances in the two groups are equal and is obtained using a pooled estimate of that common variance. The test statistic is identical to the equation for t given previously except that the individual group variances are replaced by a pooled estimate S_p^2. That is,

$$t = \frac{\bar{X}_1 - \bar{X}_2}{\sqrt{S_p^2/N_1 + S_p^2/N_2}}$$ **Equation 11.5c**

where S_p^2, the pooled variance, is a weighted average of the individual variances and is calculated as

$$S_p^2 = \frac{(N_1 - 1)S_1^2 + (N_2 - 1)S_2^2}{N_1 + N_2 - 2}$$ **Equation 11.5d**

From the output in Figure 11.1, the pooled t test value for the study is 1.78. The degrees of freedom for the pooled t test are 96, the sum of the sample sizes in both groups minus 2. If the pooled-variance t test is used when the population variances are not equal, the probability level associated with the statistic may be in error. The amount of error depends on the inequality of the sample sizes and of the variances. However, using the separate-variance t value when the population variances are equal will usually result in an observed significance level somewhat larger than it should be. For large samples, the discrepancy between the two methods is small. In general, it is a good idea to use the separate-variance t test whenever you suspect that the variances are unequal.

The statistic used to test the hypothesis that the two population variances are equal is the F value, which is the ratio of the larger sample variance to the smaller. In Figure 11.1, this value is ($27.6^2/21.8^2=1.6$). If the observed significance level for the F test is small, the hypothesis that the population variances are equal is rejected, and the separate-variance t test for means should be used. In this example, the significance level for the F test is large, and thus the pooled-variance t test is appropriate.

11.6 Significance Levels

The commonsense interpretation of a small observed significance level is straightforward: it appears unlikely that the two population means are equal. Of course, there is a possibility that the means are equal and the observed difference is due to chance. The *observed significance level* is the probability that a difference at least as large as the one observed would have arisen if the means were really equal.

When the observed significance level is too large to reject the equality hypothesis, the two population means may indeed be equal, or the means may be unequal but the difference cannot be detected. Failure to detect can be due to a true difference that is very small. For example, if a new cancer drug prolongs survival time by only one day when compared to the standard treatment, it is unlikely that such a difference will be detected, especially if survival times vary substantially and the additional day represents a small increment.

There are other reasons why true differences may not be found. If the sample sizes in the two groups are small or the variability large, even substantial differences may not be detected. Significant t values are obtained when the numerator of the t statistic is large when compared to the denominator. The numerator is the difference between the sample means, and the denominator depends on the standard deviations and sample sizes of the two groups. For a given standard deviation, the larger the sample size, the smaller the denominator. Thus, a difference of a given magnitude may be significant if obtained with a sample size of 100, but not significant with a sample size of 25.

11.7 One-Tailed vs. Two-Tailed Tests

A two-tailed test is used to detect a difference in means between two populations regardless of the direction of the difference. For example, in the study of buying scores presented in this chapter, we are interested in whether buying scores without pictures are larger *or* smaller than buying scores with pictures. In applications where one is interested in detecting a difference in one direction—such as whether a new drug is better than the current treatment—a so-called one-tailed test can be performed. The procedure is the same as for the two-tailed test, but the resulting probability value is divided by 2, adjusting for the fact that the equality hypothesis is rejected only when the difference between the two means is sufficiently large and in the direction of interest. In a two-tailed test, the equality hypothesis is rejected for large positive or negative values of the statistic.

11.8 What's the Difference?

It appears that the questionnaire type has no significant effect on couples' willingness to purchase products. Overall buying scores for the two conditions are similar. Pictures of the products do not appear to enhance their perceived desirability. In fact, the pictures actually appear to make several products somewhat less desirable. However, since the purpose of the questionnaires is to ascertain buying intent, including a picture of the actual product may help gauge true product response. Although the concept of disposable raincoats may be attractive, if they make the owner look like a walking trash bag their appeal may diminish considerably.

11.9 USING PROCEDURE CROSSTABS TO TEST HYPOTHESES

The T-TEST procedure is used to test hypotheses about the equality of two means for variables measured on an interval or ratio scale. Procedure CROSSTABS and the chi-square statistic can be used to test hypotheses about a dichotomous variable, such as purchase of a particular product.

Figure 11.9 is a crosstabulation showing the number of husbands who would definitely want to buy vegetables in pop-top cans when shown a picture and when not shown a picture of the product. The vegetables in pop-top cans were chosen by 6.0% of the husbands who were tempted with pictures and 16.0% of the husbands who were not shown pictures. The chi-square statistic provides a test of the hypothesis that the proportion of husbands selecting the vegetables in pop-top cans is the same for the picture and no-picture forms.

Figure 11.9 Husbands' preference for vegetables in pop-top cans

```
Crosstabulation:      H2S
                   By VISUAL

               Count   |WITHOUT |WITH    |
VISUAL->     Col Pct   |PICTURES|PICTURES|   Row
                       |     0  |     1  |  Total
H2S            --------+--------+--------+
                   1   |     8  |     3  |    11
  DEFINITELY BUY       |  16.0  |   6.0  |  11.0
                       +--------+--------+
                   2   |    42  |    47  |    89
  NOT DEFINITE         |  84.0  |  94.0  |  89.0
                       +--------+--------+
              Column        50       50      100
               Total      50.0     50.0    100.0

 Chi-Square     D.F.      Significance        Min E.F.      Cells with E.F.| 5
 ----------     ----      ------------        --------      ----------------

   1.63432        1         0.2011             5.500          None
   2.55362        1         0.1100          ( Before Yates Correction )

Number of Missing Observations =         0
```

The probability of 0.2011 associated with the chi-square statistic in Figure 11.9 is the probability that a difference at least as large as the one observed would occur in the sample if in the population there were no difference in the selection of the product between the two formats. Since the probability is large, the hypothesis of no difference between the two formats is not rejected.

11.10 INDEPENDENT VS. PAIRED SAMPLES

Several factors may contribute to the observed differences in response between two groups. Some of the observed difference between the scores of the picture and no-picture groups may be attributable to questionnaire type. Some of the difference may also be due to differences between individuals. Not all couples have the same buying desires, so even if the type of form does not affect buying, differences between the two groups will probably be observed due to differences between the couples within the two groups.

One method of minimizing the influence of individual variation is to choose the two groups so that the couples within them are comparable on characteristics that can influence buying behavior, such as income, education, family size, and so forth.

It is sometimes possible to obtain pairs of subjects, such as twins, and assign one member of each pair to each of the two treatments. Another frequently used experimental design is to expose the same individual to both types of conditions. (In this design, care must be taken to ensure that the sequential administration of treatments does not influence response by providing practice, decreasing attention span, or affecting the second treatment in other ways.) In both designs, subject-to-subject variability has substantially less effect. These designs are called *paired-samples designs,* since for each subject there is a corresponding, or paired, subject in the other group. In the second design, a person is paired with himself or herself. In an *independent-samples design,* there is no pairing of cases; all observations are independent.

11.11 Analysis of Paired Data

Interpreting the significance of results from paired experiments follows the same basic steps that were discussed previously for independent samples, except that different computations are used. For each pair of cases, the difference in the responses is calculated. The statistic used to test the hypothesis that the mean difference in the population is zero is

$$t = \frac{\overline{D}}{S_D/\sqrt{N}}$$

Equation 11.11

where $\overline{D}$ is the observed difference between the two means and S_D is the standard deviation of the differences of the paired observations. The sampling distribution of t, if the differences are normally distributed with a mean of zero, is Student's t with $N-1$ degrees of freedom, where N is the number of pairs. If the pairing is effective, the standard error of the difference will be smaller than the standard error obtained if two independent samples with N subjects each were chosen. However, if the variables chosen for pairing do not affect the responses under study, pairing may result in a test that is less powerful since true differences can be detected less frequently.

For example, to test the hypothesis that there is no difference between husbands' and wives' buying scores, a paired *t* test should be calculated. A paired test is appropriate since husbands and wives constitute matched observations. Hopefully, including both members of a couple controls for some nuisance effects like socioeconomic status, age, and so forth. The observed differences are more likely to be attributable to differences in sex.

Figure 11.11 contains output from the paired *t* test. The entry under number of cases is the number of pairs of observations. The mean difference is the difference between the mean scores for males and females. The *t* value is the mean difference divided by the standard error of the difference (0.55/1.73=0.32). The two-tailed probability for this test is 0.75, so there is insufficient evidence to reject the null hypothesis that married males and females have similar mean buying scores.

Figure 11.11 Husbands' versus wives' buying scores

```
Paired samples t-test:  HSSCALE   HUSBANDS BUYING SCORE
                        WSSCALE   WIVES BUYING SCORE

Variable     Number                 Standard   Standard
             of Cases     Mean     Deviation     Error

HSSCALE          98     82.0918      14.352      1.450
WSSCALE          98     81.5408      15.942      1.610

(Difference) Standard    Standard  I          2-Tail I     t     Degrees of   2-Tail
    Mean     Deviation     Error   I   Corr.  Prob.  I   Value     Freedom     Prob.
                                   I                 I
   0.5510      17.095      1.727   I   0.367  0.000  I   0.32        97        0.750
```

The correlation coefficient between husbands' and wives' scores is 0.367. A positive correlation indicates that pairing has been effective in decreasing the variability of the mean difference. The larger the correlation coefficient, the greater the benefit of pairing. See Chapter 13 for further discussion of correlation.

11.12 HYPOTHESIS TESTING: A REVIEW

The purpose of hypothesis testing is to help draw conclusions about population parameters based on results observed in a random sample. The procedure remains virtually the same for tests of most hypotheses.

- A hypothesis of no difference (called a *null hypothesis*) and its alternative are formulated.
- A test statistic is chosen to evaluate the null hypothesis.
- For the sample, the test statistic is calculated.
- The probability, if the null hypothesis is true, of obtaining a test value at least as extreme as the one observed is determined.
- If the observed significance level is judged small enough, the null hypothesis is rejected.

11.13 The Importance of Assumptions

In order to perform a statistical test of any hypothesis, it is necessary to make certain assumptions about the data. The particular assumptions depend on the statistical test being used. Some procedures require stricter assumptions than others. For *parametric tests,* some knowledge about the distribution from which samples are selected is required.

The assumptions are necessary to define the sampling distribution of the test statistic. Unless the distribution is defined, correct significance levels cannot be calculated. For the pooled *t* test, the assumption is that the observations are random samples from normal distributions with the same variance.

For many procedures, not all assumptions are equally important. Moderate violation of some assumptions may not always be serious. Therefore, it is important to know for each procedure not only what assumptions are needed but also how severely their violation may influence results. For example the *F* test for equality of variances is quite sensitive to departures from normality, while the *t* test for equality of means is less so.

The responsibility for detecting violations of assumptions rests with the researcher. Unfortunately, unlike the experimenter in chemistry, no explosions or disintegrating terminals threaten the investigator who does not comply with good statistical practice. However, from a research viewpoint, the consequences can be just as severe.

Wherever possible, tests of assumptions—often called diagnostic checks of the model—should be incorporated as part of the hypothesis-testing procedures. Throughout SPSS/PC+, attempts have been made to provide facilities for examining assumptions. For example, in the FREQUENCIES procedure, histograms and measures of skewness and kurtosis provide a convenient check for the normality assumption. Discussion of other such diagnostics is included with the individual procedures.

11.14 RUNNING PROCEDURE T-TEST

The T-TEST procedure computes the Student's *t* statistic for testing the significance of a difference between means of independent or paired samples. When independent samples are used, T-TEST calculates the separate-variance and pooled-variance *t* statistics and the *F* test for homogeneity of variances. The two-tailed observed significance level is also displayed for each *t* statistic.

11.15 Requesting Independent-Samples Tests

The variable and criterion to be used to divide the sample into two independent groups are specified by the GROUPS subcommand. It is followed by the VARIABLES subcommand, which names the variable or variables to be tested.

11.16 Defining the Groups

Any of the following three methods can be used to define two groups on the GROUPS subcommand:

- If a single number in parentheses follows the grouping variable, all cases whose value for the grouping variable is greater than or equal to this number go into one group; the remaining cases go into the other group. For example, the output in Figure 11.1 could be produced by the command:

  ```
  T-TEST GROUPS=VISUAL(1)/VARIABLES=FAMSCORE.
  ```

- If two values are specified in parentheses after the grouping variable, one group consists of all cases whose value for the grouping variable is the first number, while the second group consists of all cases whose value for the grouping variable is the second number. Thus, the output in Figure 11.1 could also have been produced by the command:

```
T-TEST GROUPS=VISUAL(0,1)/VARIABLES=FAMSCORE.
```

- If the grouping variable has only two values, 1 and 2, no value list is necessary; only the name of the grouping variable is required. If VISUAL had been coded as 1 or 2 rather than 0 or 1, the shortest command that would produce the output in Figure 11.1 would be:

```
T-TEST GROUPS=VISUAL/VARIABLES=FAMSCORE.
```

Only one variable (numeric or string) may be named in the GROUPS subcommand. When a string variable is used, long strings are truncated to short strings to define the categories. Only one GROUPS subcommand may be used per T-TEST command.

11.17 Specifying the Variables

The VARIABLES subcommand follows the GROUPS subcommand and lists the variables to be analyzed. Up to 50 variables may be specified, but they must all be numeric. The specifications for the variable list follow the usual SPSS/PC+ conventions. The TO keyword may be used to refer to adjacent variables in the data file. Only one VARIABLES subcommand may be used per T-TEST command.

11.18 Requesting Paired-Samples Tests

To obtain a paired *t* test, you must have two separate variables that indicate the values for the two members of the pairs. The only required subcommand is PAIRS, which specifies the variables being compared. String variables may not be used. For example, the output in Figure 11.11 was obtained by specifying

```
T-TEST PAIRS=HSSCALE WSSCALE.
```

If three or more variables are listed in PAIRS, each variable is compared with every other variable. For example, the command

```
T-TEST PAIRS=SURVEY1 SURVEY2 SURVEY3.
```

produces three paired *t* tests, one comparing SURVEY1 and SURVEY2, one comparing SURVEY1 and SURVEY3, and one comparing SURVEY2 and SURVEY3.

The keyword WITH can be used to request paired *t* tests comparing each variable to the left of WITH to every variable to the right of WITH. Thus, the command

```
T-TEST PAIRS=SURVEY1 WITH SURVEY2 SURVEY3.
```

produces two paired *t* tests, one comparing SURVEY1 with SURVEY2 and the other comparing SURVEY1 with SURVEY3.

You can use Option 5 to specify special pairing of variables using the keyword WITH.

Option 5 *Special pairing for paired-samples test.* Must be used with the keyword WITH. The first variable before WITH is compared to the first variable after WITH and so forth. (For more information, see T-TEST in Command Reference.)

Thus, the command

```
T-TEST SURVEY1 TO SURVEY3 WITH SURVEY4 TO SURVEY6
/OPTIONS = 5.
```

compares SURVEY1 and SURVEY4, SURVEY2 and SURVEY5, and SURVEY3 and SURVEY6.

You can specify multiple analysis lists by separating them with a slash, as in

```
T-TEST PAIRS=SURVEY1 SURVEY2 SURVEY3/PRETEST WITH POSTTST1 POSTTST2.
```

11.19 Requesting One-Sample Tests

Although T-TEST is designed to be a two-sample procedure, it can be used to calculate one-sample *t* tests if a COMPUTE command is also used. To test the null hypothesis that the mean of a population is some specified value *m*, you can use the commands

```
COMPUTE MEAN=m.
T-TEST PAIRS=varname MEAN.
```

where varname is the name of the variable you want to test. You can give computed variable MEAN any name you wish (as long as it conforms to the SPSS/PC+ variable-naming conventions).

11.20 Missing Values

By default, T-TEST deletes only cases with missing values for those variables necessary for a particular *t* test. For independent-samples *t* tests, T-TEST excludes cases with missing values for the grouping variable or the variable to be tested. For paired-samples *t* tests, T-TEST excludes cases with missing values for either of the variables in a given pair. Two other missing-value treatments are available and are specified in the OPTIONS subcommand:

Option 1 *Include user-missing values.* User-defined missing values are included in the analysis.

Option 2 *Exclude missing values listwise.* Cases with missing values for any variables listed in T-TEST are excluded from the analysis.

11.21 Formatting Options

By default, T-TEST displays variable labels next to variable names. You can use Option 3 to suppress these labels.

Option 3 *Suppress variable labels.*

11.22 Annotated Example

The complete set of SPSS/PC+ commands needed to obtain the analyses in this chapter is the following:

```
DATA LIST / VISUAL 1 FAMSCORE 4-6 HSSCALE 11-13 WSSCALE 16-18.
VAR LABELS FAMSCORE 'FAMILY BUYING SCORE'/
           WSSCALE 'WIVES BUYING SCORE'/
           HSSCALE 'HUSBANDS BUYING SCORE'.
BEGIN DATA.
0  169    102   67
0  206    109   97
0  156     73   83
:
:
1  170     96   74
1  178     86   92
1  179     92   87
END DATA.
T-TEST GROUPS=VISUAL(1)/VARIABLES=FAMSCORE/
       PAIRS=HSSCALE WSSCALE.
FINISH.
```

- The DATA LIST command gives the variable names and column locations.
- The VAR LABELS command assigns new labels to the variables FAMSCORE, WSSCALE, and HSSCALE.
- The T-TEST command requests an independent-samples test (with VISUAL as the grouping variable) and a paired *t* test (comparing WSSCALE and HSSCALE).

The output for this job is shown in Figures 11.1 and 11.11.

Contents

Chapter 12 Plotting Data: Procedure PLOT

Today the quest for the Fountain of Youth has been replaced by the Search for Slimness. It's almost acceptable to grow old, as long as one remains trim and fit. Programs for weight loss are assuming ever increasing attention, and behavioral psychologists are studying the effectiveness of many different weight-loss strategies. Black and Sherba (1983) studied the effects of two different types of behavior programs on weight loss. One group of subjects was taught behavioral weight-loss techniques, while the second was taught weight-loss techniques and problem-solving behavior. Their data set is examined in this chapter.

12.1 DESCRIBING WEIGHT LOSS

As discussed in Chapter 7, a histogram is a convenient method for displaying the distribution of a variable that can have many values. Figure 12.1a is the percent of excess weight actually lost during the treatment for each of the twelve cases in the study. From this figure we can see that about one third of the participants lost 20% or more of the required weight during treatment. To see if weight loss is maintained, consider Figure 12.1b which is the percent of weight loss one year after treatment. It appears that subjects did not gain back the weight but maintained weight loss.

Figure 12.1a Histogram of weight loss during treatment

```
TREATRED  REDUCTION QUOTIENT AT TREATMENT

     Count   Midpoint

         0     -27.50
         4     -22.50  ########################################
         1     -17.50  ##########
         4     -12.50  ########################################
         1      -7.50  ##########
         1      -2.50  ##########
         1       2.50  ##########
         0       7.50
                       I....+....I....+....I....+....I....+....I....+....I
                       0         1         2         3         4         5
                                 Histogram Frequency

Valid Cases       12       Missing Cases      0
```

Figure 12.1b Histogram of weight loss after one year

```
TWELVRED  REDUCTION QUOTIENT AT TWELVE MONTHS

     Count   Midpoint

         3     -45.00  ##############################
         1     -35.00  ##########
         1     -25.00  ##########
         3     -15.00  ##############################
         2      -5.00  ####################
         2       5.00  ####################
                       I....+....I....+....I....+....I....+....I....+....I
                       0         1         2         3         4         5
                                 Histogram Frequency

Valid Cases       12       Missing Cases      0
```

Although the histograms provide information about the weight loss during treatment and weight loss after twelve months, they reveal nothing about the relationship between the two variables since they each describe single variables. To determine whether lost weight during treatment is maintained or replaced at twelve months, the two variables must be studied together.

Figure 12.1c is a scatterplot of the percent of weight loss during treatment and at one year for the twelve cases. Each symbol 1 on the plot represents one case, showing the values for that case on two variables—loss during treatment and loss at one year. For example, the circled point represents a case with a treatment loss of 25% and a twelve month value of −18%.

Figure 12.1c Scatterplot for weight loss during treatment and after one year

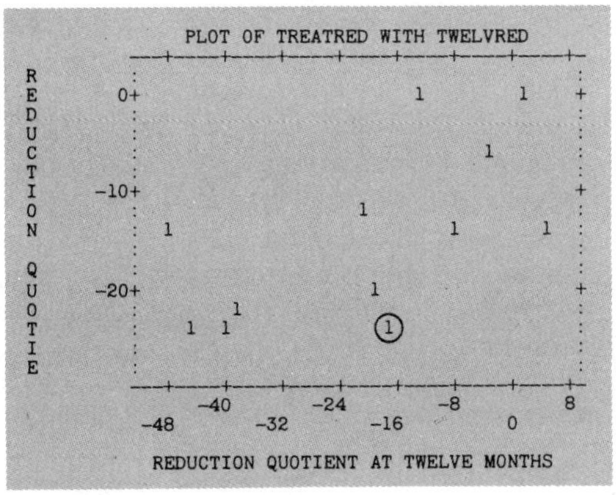

Since plots generated for terminals and printers have a limited number of positions in which to display points, it may not be possible to distinguish cases with similar values for the two variables. When two or more cases with similar values fall on the same point on the scatterplot, a number is displayed indicating how many cases overlap at that point. The scale of the plot depends on the minimum and maximum values for the two variables plotted. If the values for a few cases are far removed from the others, the majority of cases may appear bunched together in order to permit the outlying cases to appear on the same plot.

Figure 12.1d contains the symbols used to represent multiple cases at each point. For example, the symbol D is used when there are 13 coincident points.

Figure 12.1d Scatterplot symbols for multiple cases

```
Frequencies and symbols used (not applicable for control or overlay plots)

      1 - 1       11 - B      21 - L      31 - V
      2 - 2       12 - C      22 - M      32 - W
      3 - 3       13 - D      23 - N      33 - X
      4 - 4       14 - E      24 - O      34 - Y
      5 - 5       15 - F      25 - P      35 - Z
      6 - 6       16 - G      26 - Q      36 - *
      7 - 7       17 - H      27 - R
      8 - 8       18 - I      28 - S
      9 - 9       19 - J      29 - T
     10 - A       20 - K      30 - U
```

12.2 Controlled Scatterplots

Often it is informative to identify each point on a scatterplot by its value on a third variable. For example, cases may be designated as males or females, or as originating from the West, Midwest, or East. Figure 12.2 is the same plot as Figure 12.1c except each case is identified as being a participant in the behavior program (the value 1) or the problem-solving program (the value 2). A dollar sign is displayed if cases from different groups coincide.

Figure 12.2 Scatterplot identifying the two programs

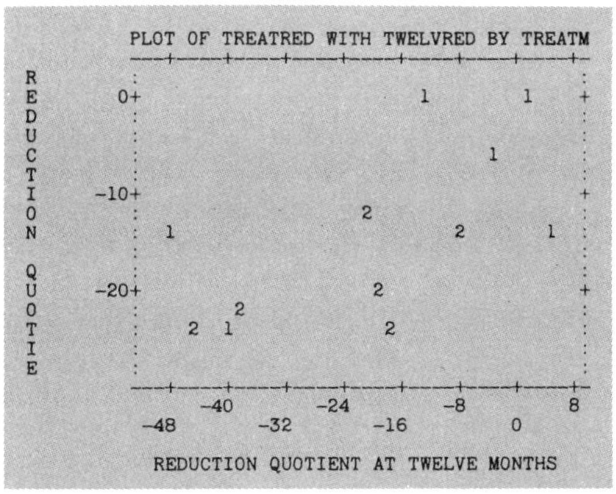

By examining Figure 12.2 one can see if the weight-loss-maintenance relationships are similar for the two groups.

12.3 Plotting Multiple Variables

Weight-loss maintenance may be associated with many variables, including age. Figure 12.3a is a plot of age with weight loss during treatment while Figure 12.3b is a plot of weight loss at twelve months with age. There appears to be a somewhat negative relationship between age and weight loss. Older people appear to have lost a greater percentage of weight than younger ones.

Figure 12.3a Scatterplot of age with weight loss during treatment

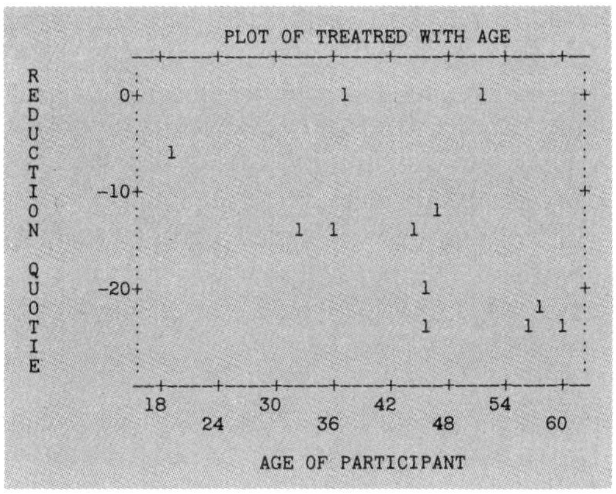

Figure 12.3b Scatterplot of age with weight loss at twelve months

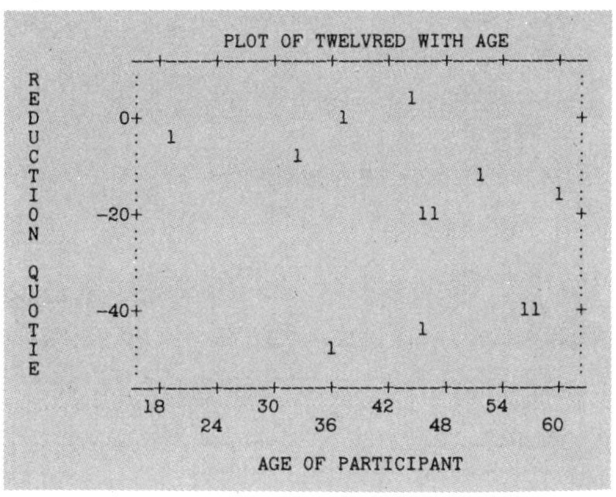

Figures 12.3a and 12.3b can be combined into a single plot as shown in Figure 12.3c. Each case appears twice on Figure 12.3c, once with treatment weight loss (denoted as 1) and once with twelve month loss (denoted as 2). When there are several cases with similar ages one cannot tell which are the matching points. For example, at age 36, there are four points since there are two cases with similar ages (one is 36, one is 35). The $ which is displayed represents multiple occurrences at a given location. However, we cannot tell if these are the two values for the same case, or one value from one case, and one from another.

Figure 12.3c Overlay plot of weight loss during treatment and at twelve months

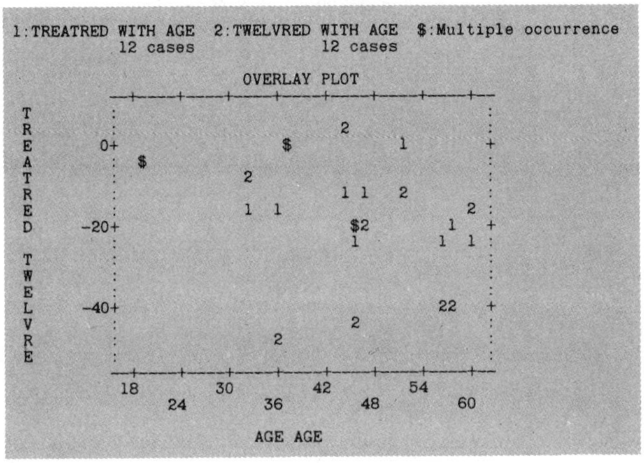

12.4 RUNNING PROCEDURE PLOT

You can use the PLOT procedure to obtain bivariate scatterplots or regression plots (with or without control variables), contour plots, overlay plots, and some regression statistics. (For complete regression analysis, use procedure REGRESSION, described in Chapter 17.) Formatting options enable you to control axis size and scale, the plotting symbols used, and the frequency they represent. You can also label the plot and axes, request reference lines, and plot standardized variables.

12.5 Specifying the Variables

Use the PLOT subcommand to specify the variables to be plotted. Variables to be plotted on the vertical (*Y*) axis are specified first, followed by the WITH keyword, followed by the variables to be plotted on the horizontal (*X*) axis.

By default, PLOT produces bivariate scatterplots. For example, the following command produces the output in Figure 12.1c:

```
PLOT PLOT=TREATRED WITH TWELVRED.
```

You can produce multiple plots with one PLOT subcommand. For example, the command

```
PLOT PLOT=IQ GRE WITH GPA SAT.
```

produces four plots: IQ with GPA, IQ with SAT, GRE with GPA, and GRE with SAT.

You can also specify plots of individual pairs of variables on one PLOT subcommand by using the PAIR keyword. For example, the command

```
PLOT PLOT=IQ GRE WITH GPA SAT (PAIR).
```

produces two plots: IQ with GPA and GRE with SAT.

Multiple plot lists can be specified on the PLOT subcommand if they are separated by semicolons, as in the command

```
PLOT PLOT=IQ WITH GPA EDUC;GRE WITH SAT.
```

This produces three plots: IQ with GPA, IQ with EDUC, and GRE with SAT. You can also specify multiple PLOT subcommands on one PLOT command.

A control variable or contour variable (see Section 12.6) can be specified on the PLOT subcommand by naming it after the BY keyword following the list of horizontal-axis variables. For example, the command

```
PLOT PLOT=TREATRED WITH TWELVRED BY TREATMNT.
```

was used to obtain the plot in Figure 12.2. Only one control or contour variable can be specified on a plot list. PLOT uses the first character of a control variable's value label as a plotting symbol. For example, if SEX is the control variable, with value labels FEMALE and MALE, the observations for females are represented by F and those for males by M. If a variable has no value labels, the first character of the actual value is used as the plotting symbol. When cases with different values for the control variable fall in the same position on the plot, they are represented by a single $.

12.6 Choosing the Type of Plot

Use the FORMAT subcommand to specify the type of plot you want to produce. Four types of plots are available: scatterplots, regression plots, contour plots, and overlay plots. If FORMAT is not used, or is used without further specification, scatterplots are displayed. To specify plot type, use the following keywords on the FORMAT subcommand.

DEFAULT *Bivariate scatterplot.* When there are no control variables each symbol represents the case count at that plot position. When a control variable is specified, each symbol represents the first character of the value label of the control variable.

REGRESSION *Scatterplot plus regression statistics.* The vertical-axis variable is regressed on the horizontal-axis variable, and the regression line intercepts on each axis are indicated with the letter R. In a control plot, regression statistics are pooled over all categories.

CONTOUR(n) *Contour plot with* n *levels.* Contour plots use a continuous variable as the control variable. The control variable is specified after BY on the PLOT subcommand. The contour variable is recoded into *n* intervals of equal width. Up to 35 contour levels can be specified. If *n* is omitted, the default is 10 levels.

OVERLAY *Overlay plots.* All plots specified on the next PLOT subcommand are displayed in one plot frame. A unique plotting symbol is used for each overlaid plot. An additional symbol indicates multiple plot points at the same position. Control plots cannot be overlaid.

For more information on these keywords, refer to Command Reference: PLOT.

Specify the FORMAT subcommand before the PLOT subcommand to which it refers. One FORMAT subcommand can be specified before each PLOT subcommand.

For example, the command

```
PLOT FORMAT=OVERLAY
/PLOT=TREATRED TWELVRED WITH AGE.
```

produces the overlay plot in Figure 12.3c.

Overlay plots are useful when several variables represent the same type of measurement or when the same variable is measured at different times. For example, the command

```
PLOT SYMBOLS='MD'
/VSIZE=30 /HSIZE=70
/FORMAT=OVERLAY
/TITLE 'MARRIAGE AND DIVORCE RATES  1900-1981'
/VERTICAL='RATES PER 1000 POPULATION'
/HORIZONTAL='YEAR' REFERENCE (1918,1945) MIN (1900) MAX (1983)
/PLOT=MARRATE DIVRATE WITH YEAR.
```

produces the overlay plot of marriage and divorce rates over time shown in Figure 12.6a (data taken from the *Information Please Almanac,* 1983).

Figure 12.6a An overlay plot

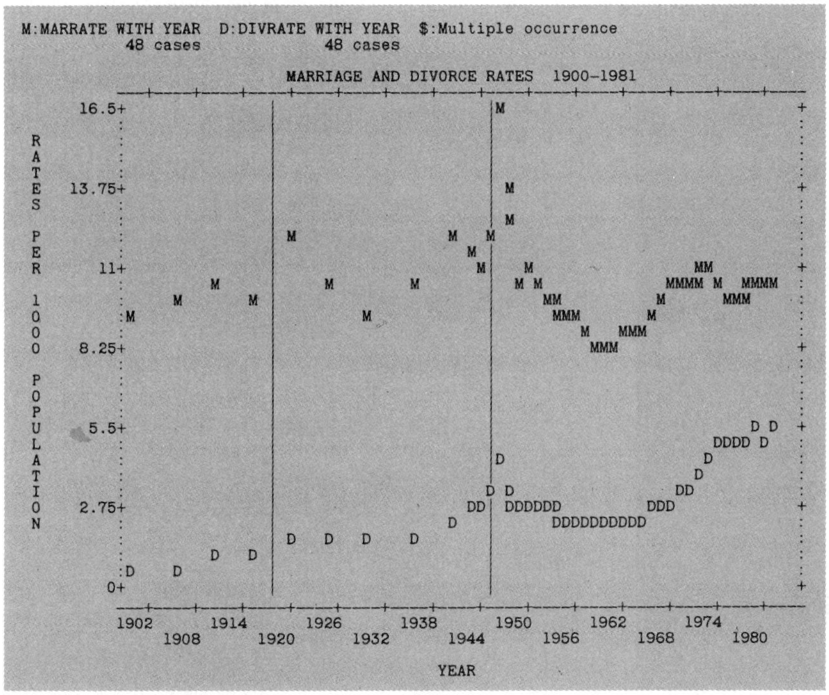

Contour plots evaluate the effect of a continuous variable as a control variable. If you use symbols with different degrees of density, you can produce a visual representation of the density of your contour variable. For example, the command

```
PLOT FORMAT=CONTOUR (10)
/HSIZE=100/VSIZE=60
/SYMBOLS='.-=*+OXOXM','       -OW'
/TITLE='SOLUBILITY OF AMMONIA IN WATER'
/HORIZONTAL='ATMOSPHERIC PRESSURE'
/VERTICAL='TEMPERATURE'
/PLOT=TEMP WITH PRESSURE BY CONCENT.
```

produces the output in Figure 12.6b, representing the concentration of ammonia in water under varying conditions of temperature and atmospheric pressure.

Figure 12.6b A contour plot

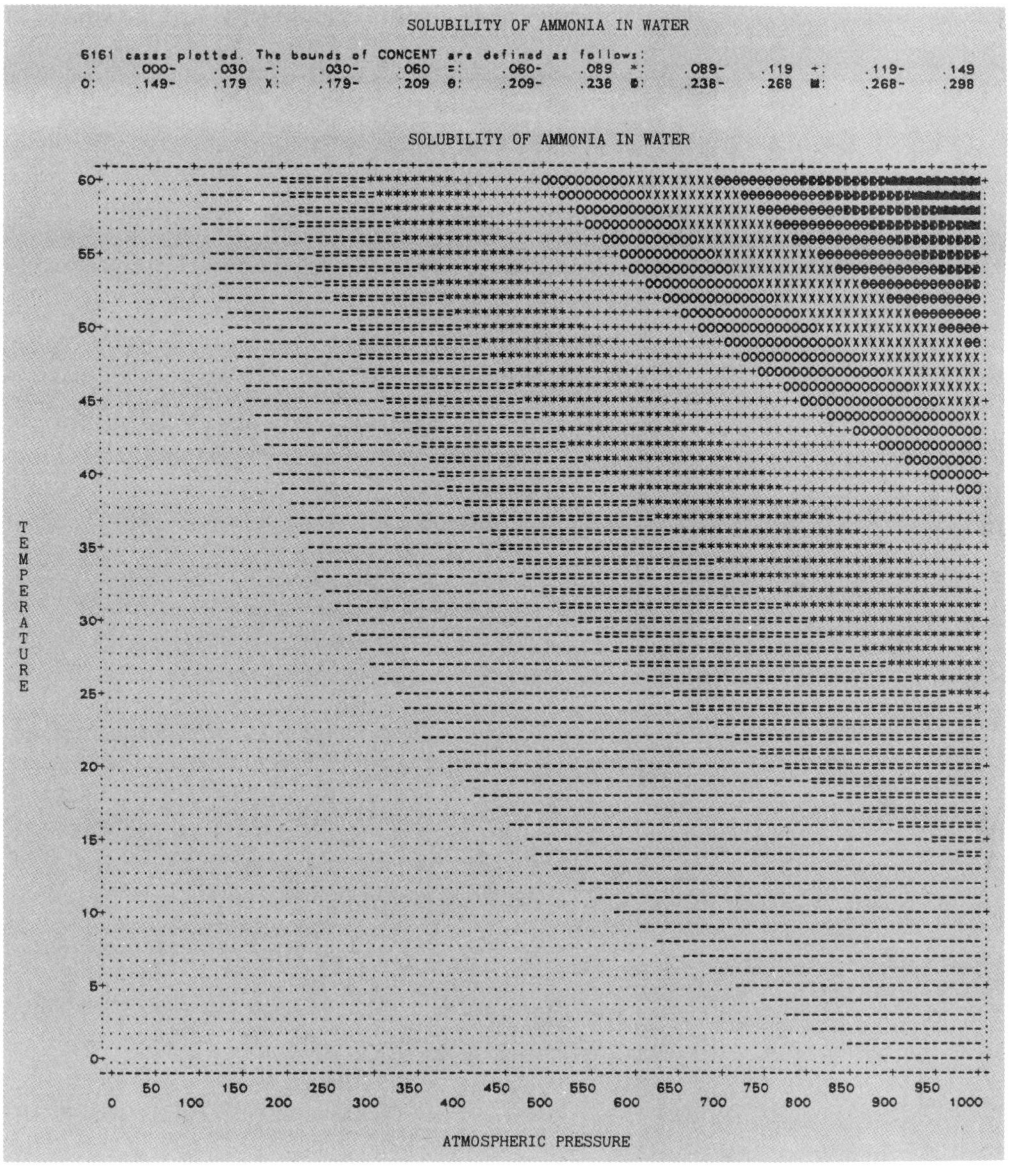

Plots with regression statistics are described and shown in Chapter 13.

12.7 Setting Plot Symbols

A wide range of alphabetical, numeric, and special ASCII graphic characters are available as plot symbols. Use the CUTPOINT and SYMBOLS subcommands to control the display of plot symbols.

The CUTPOINT subcommand specifies the number of cases represented by plotting symbols. For example, you can have the symbol 1 represent one case at a position, the symbol 2 represent two cases at a position, and the symbol 3 represent three or more cases at a position. CUTPOINT can be used only once on a PLOT command and applies to all plots requested. The CUTPOINT subcommand cannot be used for control plots, overlay, or contour plots.

The following specifications are available with CUTPOINT:

EVERY(n) *Frequency intervals of width* n. Assign the first symbol when the number of cases at a position is between 1 and *n*, the second symbol when the number of cases at a position is between $n+1$ and $2n$, and so on. The default symbols used are 1, 2, 3, ... X, Y, Z, with a default *n* of 1. An asterisk (*) represents 36 or more cases.

(value list) *Each value defines a cutpoint.* Assign the first symbol when the number of cases at a position is less than or equal to the first value, the second symbol when the number of cases at a position is greater than the first value and less than or equal to the second value, and so on. Specify values separated by blanks or commas. Up to 35 cutpoints can be specified.

For example, if the command

```
PLOT CUTPOINT=EVERY(4)
/PLOT=SCORE WITH ANXIETY.
```

is used, one to four cases in the same position are represented by a 1, five to eight cases by a 2, and so on. If the command

```
PLOT CUTPOINT=(5,10,25)
/PLOT=SCORE WITH ANXIETY.
```

is used, one to five cases in the same position are represented by a 1, six to ten cases by a 2, 11 to 25 cases by a 3, and more than 25 cases by a 4.

The SYMBOLS subcommand allows you to choose other plotting symbols to represent a plot position. For scatterplots and regression plots, each symbol represents the number of cases at a plot position. For overlay plots, each symbol represents one of the overlaid plots. For contour plots, each symbol represents one level of the contour variable. SYMBOLS cannot be used with control plots. Use the VALUE LABELS command to define appropriate labels when control plots are requested. The SYMBOLS subcommand can be used only once on a PLOT command and applies to all plots requested. You can request one of the following options with the SYMBOLS subcommand:

ALPHANUMERIC *Alphanumeric plotting symbols.* The characters 1 through 9, A through Z, and * are used, in that order. Thus, in a scatterplot, 1 indicates one case at a position, and * represents 36 or more cases at a position. This is the default if the SYMBOLS subcommand is not specified.

NUMERIC *Numeric plotting symbols.* The characters 1 through 9 and * are used, with * indicating 10 or more cases at a position in a scatterplot.

'symbols'['ovprnt'] *List of plot symbols.* In the list of symbol values, the characters specified are not separated by blanks or commas. The list of overprinting symbols is enclosed in a separate set of apostrophes or quotation marks and is separated from the first symbols list with a blank or comma. When overprint symbols are displayed on the screen, only the second overprinting symbol list will be displayed. You can select any special ASCII graphic characters available on your PC as symbols. Look under "ASCII" in the index of your DOS manual for further reference.

If the SYMBOLS subcommand is specified, a table of symbols and their equivalents like the one shown in Figure 12.1d is displayed. If the SYMBOLS subcommand is omitted, the default alphanumeric symbol set is used.

For example, the command

```
PLOT CUTPOINTS=(1,2,3,4)
/SYMBOLS='.:x*X'
/PLOT=INCOME WITH ASTRSIGN.
```

requests a scatterplot with a period (.) representing one case at a position, a colon (:) representing two cases, x representing three cases, an asterisk (*) representing four cases, and X representing five or more cases.

12.8 Specifying Plot Titles

Use the TITLE subcommand to specify a plot title. The title can contain up to 60 characters and must be enclosed in single quotation marks, as in the command

```
PLOT TITLE='CORPORATE TAKEOVERS 1975-1983'
/PLOT=TAKEOVER WITH YEAR.
```

TITLE can be specified once before each PLOT subcommand and applies only to the following PLOT subcommand. Titles longer than the horizontal axis are truncated. If the TITLE subcommand is not used, a default title is displayed, consisting of the names of the variables plotted for scatterplots or the type of the plot requested on FORMAT.

12.9 Scaling and Labeling Plot Axes

The VERTICAL and HORIZONTAL subcommands allow you to control the scaling and labeling of the vertical and horizontal axes, obtain reference lines at specified positions, specify minimum and maximum values, and obtain plots of standardized variables. Resetting minimum or maximum values is especially useful when you want to focus on a subset of a larger plot. Standardized plots are appropriate when you want to overlay plots of variables with very different scales.

The VERTICAL and HORIZONTAL subcommands can be used once before each PLOT subcommand and apply to all plots specified in the following PLOT subcommand. You can request the following specifications on these subcommands:

'label' *Use the axis label specified (up to 40 characters).* The default label is the variable label or, if there is no variable label, the variable name. Labels longer than the axis are truncated.

MIN(n) *Use the minimum value* n *on the axis.* The default is the minimum observed value for the plotted variable.

MAX(n) *Use the maximum value* n *on the axis.* The default is the maximum observed value for the plotted variable or, sometimes, a slightly larger number (to obtain equal-width integer scaling).

UNIFORM *Use the same scaling for all plots.* This keyword is unnecessary when MIN and MAX are specified. If UNIFORM is specified instead of the MIN and MAX keywords, PLOT determines the minimum and maximum observed values across all plotted variables on the PLOT subcommands.

REFERENCE(values) *Draw reference lines at the values specified.* Specify values separated by blanks or commas. The default is no reference lines.

STANDARDIZE *Plot standardized variables.* The default is to plot observed values.

12.10 Setting the Plot Size

The VSIZE and HSIZE subcommands control the height and width, respectively, of the plot. The default size of the PLOT depends on the current page size (see Command Reference: PLOT for defaults). These subcommands override the page size set on the SET command. For example, the command

```
PLOT VSIZE=30/HSIZE=45
/PLOT=SALES WITH REP DISTRICT.
```

requests a height of 30 lines and a width of 45 positions for the plots of SALES with REP and SALES with DISTRICT.

The VSIZE and HSIZE subcommands can be used only once: all plots requested are then drawn to the specified size.

12.11 Missing Values

The MISSING subcommand controls the treatment of missing values. The subcommand can be used only once on each PLOT command. You can request the following options on the MISSING subcommand:

PLOTWISE *Exclude cases with missing values plotwise.* Cases that are missing for any variable within a single plot are not included. In an overlay plot, plotwise deletion applies to each plot that is overlaid. This is the default if no missing-value treatment is specified.

LISTWISE *Exclude cases with missing values listwise.* Cases with missing values for any variable named on the PLOT subcommand are excluded from all plots.

INCLUDE *Include cases with user-defined missing values.* INCLUDE can be used with either PLOTWISE or LISTWISE to include cases with user-missing values while deleting cases with system-missing values according to one or the other treatment.

12.12 Annotated Example

The following commands were used to obtain the output in Figure 12.3c:

```
DATA LIST FREE
 /TREATMNT SEX AGE INITWT PEROVRWT TREATLOS THREELOS SIXLOSS TWELVLOS
  TREATRED THREERED SIXRED TWELVRED.
VARIABLE LABELS
  TREATMNT 'TREATMENT GROUP'/SEX 'SEX OF PARTICIPANT'/
  AGE 'AGE OF PARTICIPANT'/INITWT 'INITAL WEIGHT IN POUNDS'/
  PEROVRWT 'INITIAL PERCENT OVERWEIGHT'/
  TREATLOS 'WEIGHT LOSS AT TREATMENT'/
  THREELOS 'WEIGHT LOSS AT THREE MONTHS'/
  SIXLOSS 'WEIGHT LOSS AT SIX MONTHS'/
  TWELVLOS 'WEIGHT LOSS AT TWELVE MONTHS'/
  TREATRED 'REDUCTION QUOTIENT AT TREATMENT'/
  THREERED 'REDUCTION QUOTIENT AT THREE MONTHS'/
  SIXRED 'REDUCTION QUOTIENT AT SIX MONTHS'/
  TWELVRED 'REDUCTION QUOTIENT AT TWELVE MONTHS'.
VALUE LABELS  SEX 1 'FEMALE' 2 'MALE'.
BEGIN DATA.
lines of data
END DATA.
PLOT FORMAT=OVERLAY/
     PLOT=TREATRED TWELVRED WITH AGE.
FINISH.
```

- The DATA LIST command tells SPSS/PC+ that the variables are to be read in freefield format. It also gives the variable names.
- The VARIABLE LABELS command assigns descriptive labels to the variables.
- The VALUE LABELS command assigns labels to the values of the variable SEX.
- The PLOT command asks for an overlay plot of TREATRED and TWELVRED with AGE.

Contents

Chapter 13 Measuring Linear Association: Procedure CORRELATION

Youthful lemonade-stand entrepreneurs as well as balding executives of billion-dollar corporations share a common concern—increasing sales. Hand-lettered signs affixed to neighborhood trees, television campaigns, siblings and friends canvassing local playgrounds, and international sales forces are known to be effective tactics. However, the impact of various intertwined factors on sales can be difficult to isolate, and much effort in the business world is expended on determining exactly what makes a product sell.

Churchill (1979) describes a study undertaken by the manufacturer of Click ball-point pens on the effectiveness of the firm's marketing efforts. A random sample of forty sales territories is selected, and sales, amount of advertising, and number of sales representatives are recorded. This chapter looks at the relationship between sales and these variables.

13.1 EXAMINING RELATIONSHIPS

Figure 13.1a is a scatterplot of the amount of sales and the number of television spots in each of forty territories. A scatterplot can reveal various types of associations between two variables. Figure 13.1b contains some commonly encountered patterns. In the first panel there appears to be no discernible relationship between the two variables. The variables are related exponentially in the second panel. That is, *Y* increases very rapidly for increasing values of *X*. In the third panel, the relationship between the two variables is U-shaped. Small and large values of the *X* variable are associated with large values of the *Y* variable.

Figure 13.1a Scatterplot showing a linear relationship

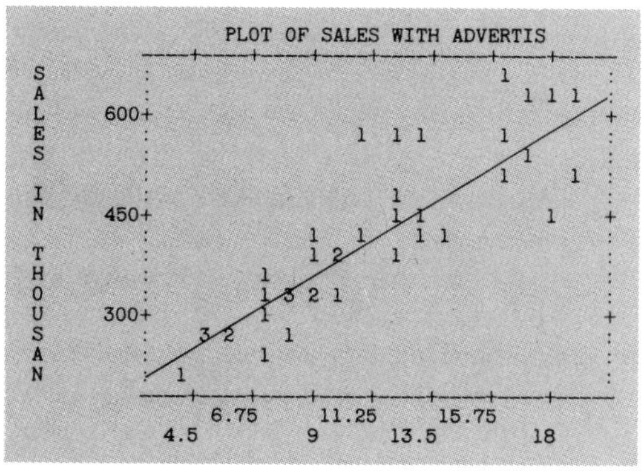

Figure 13.1b Some common relationships

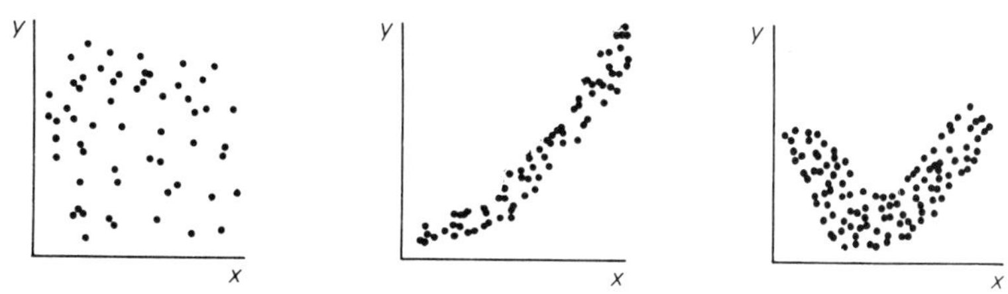

From Figure 13.1a there appears to be a positive association between sales and advertising. That is, as the amount of advertising increases, so does the number of sales. The relationship between sales and advertising may also be termed *linear,* since the observed points cluster more or less around a straight line.

13.2 THE CORRELATION COEFFICIENT

Although a scatterplot is an essential first step in studying the association between two variables, it is often useful to quantify the strength of the association by calculating a summary index. One commonly used measure is the Pearson correlation coefficient, denoted by r. It is defined as

$$r = \frac{\sum_{i=1}^{N} (X_i - \bar{X})(Y_i - \bar{Y})}{(N - 1)S_X S_Y}$$

Equation 13.2

where N is the number of cases and S_x and S_y are the standard deviations of the two variables. The absolute value of r indicates the strength of the linear relationship. The largest possible absolute value is 1, which occurs when all points fall exactly on the line. When the line has a positive slope, the value of r is positive, and when the slope of the line is negative, the value of r is negative (see Figure 13.2a).

Figure 13.2a Scatterplots with correlation coefficients of +1 and −1

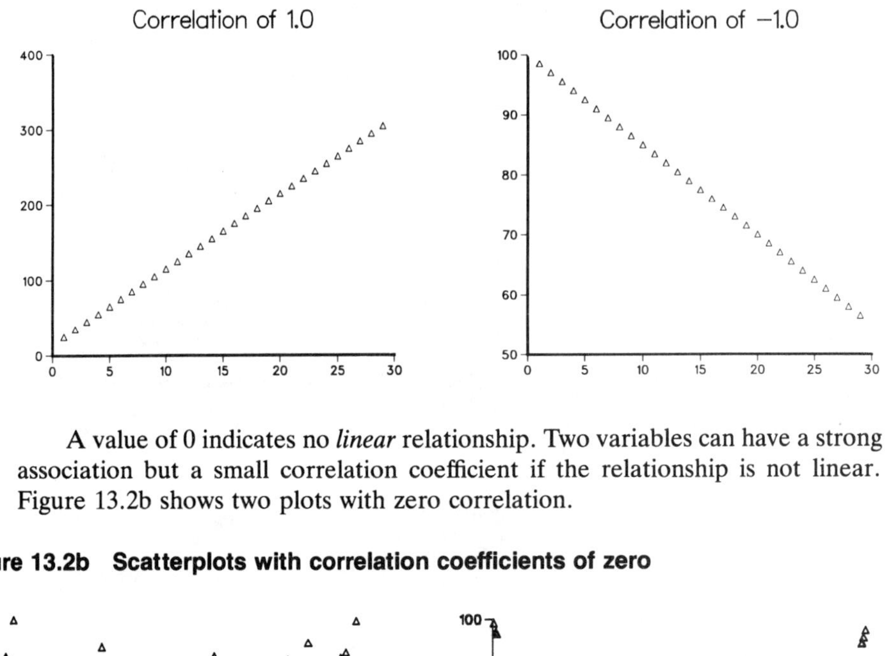

A value of 0 indicates no *linear* relationship. Two variables can have a strong association but a small correlation coefficient if the relationship is not linear. Figure 13.2b shows two plots with zero correlation.

Figure 13.2b Scatterplots with correlation coefficients of zero

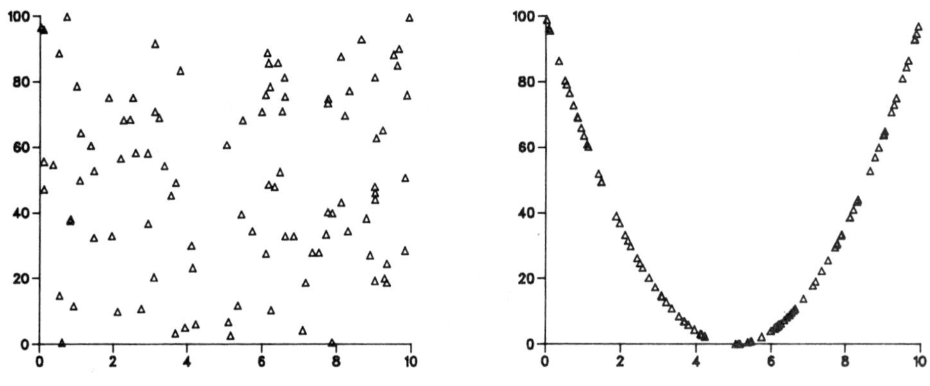

It is important to examine correlation coefficients together with scatterplots since the same coefficient can result from very different underlying relationships. The variables plotted in Figure 13.2c have a correlation coefficient greater than 0.8, as do the variables plotted in Figure 13.1a. But note how different the relationships are between the two sets of variables. In Figure 13.2c there is a strong positive linear association only for part of the graph. The relationship between the two variables is basically nonlinear. The scatterplot in Figure 13.1a is very different. The points cluster more or less around a line. Thus, the correlation coefficient should be used only to summarize the strength of linear association.

Figure 13.2c Scatterplot of percentage no facial hair with year

```
                  PLOT OF CLEAN WITH YEAR
         ---+----+----+----+----+----+----+----+-
P     105+                                      +
E        :                                      :
R        :                                      :
C        :                               11343  :
E        :                             23331    :
N      70+                        1  1211       +
T        :                      1  1 221        :
         :                       1  41          :
N        :   2                   333            :
O        :   1 1               11               :
       35+    1               432               +
F        :   122 1  1       11                  :
A        :     131121      233                  :
C        :      1222212 2242                    :
I        :        1  32312                      :
A       0+             11                       +
         ---+----+----+----+----+----+----+----+-
              1860      1900      1940      1980
         1840      1880      1920      1960

                           YEAR
```

13.3 Some Properties of the Correlation Coefficient

A common mistake in interpreting the correlation coefficient is to assume that correlation implies causation. No such conclusion is automatic. While sales are highly correlated with advertising, they are also highly correlated with other variables, such as the number of sales representatives in a territory. Advertising alone does not necessarily result in increased sales. For example, territories with high sales may simply have more money to spend on TV spots, regardless of whether the spots are effective.

The correlation coefficient is a symmetric measure since interchanging the two variables X and Y in the formula does not change the results. The correlation coefficient is not expressed in any units of measure, and it is not affected by linear transformations such as adding or subtracting constants or multiplying or dividing all values of a variable by a constant.

13.4 Calculating Correlation Coefficients

Figure 13.4 is a table of correlation coefficients for the number of television spots, number of sales representatives, and amount of sales. The entry in each cell is the correlation coefficient. For example, the correlation coefficient between advertising and sales is .8802. This value indicates that there is a fairly strong linear association between the two variables, as shown in Figure 13.1a. The table is symmetric since the correlation between X and Y is the same as the correlation

Figure 13.4 Correlation coefficients

Correlations:	ADVERTIS	REPS	SALES
ADVERTIS	1.0000	.7763**	.8802**
REPS	.7763**	1.0000	.8818**
SALES	.8802**	.8818**	1.0000

N of cases: 40 Significance: * - .01 ** - .001

" . " is printed if a coefficient cannot be computed

Statistics Guide B

between Y and X. The values on the diagonal are all 1 since a variable is perfectly related to itself. The sample size and significance levels are displayed after the table. In this example, 40 cases were used in all computations since all territories had values for the three variables.

13.5 Hypothesis Tests about the Correlation Coefficient

Although the correlation coefficient is sometimes used only as a summary index to describe the observed strength of the association, in some situations description and summary are but a first step. The primary goal may be to test hypotheses about the unknown population correlation coefficient—denoted as ρ—based on its estimate, the sample correlation coefficient r. In order to test such hypotheses, certain assumptions must be made about the underlying joint distribution of the two variables. A common assumption is that independent random samples are taken from a distribution in which the two variables are together distributed normally. If this condition is satisfied, the test that the population coefficient is 0 can be based on the statistic

$$t = r\sqrt{\frac{N-2}{1-r^2}}$$ **Equation 13.5**

which, if $\rho=0$, has a Student's t distribution with $N-2$ degrees of freedom. Either one- or two-tailed tests can be calculated. If nothing is known in advance, a two-tailed test is appropriate. That is, the hypothesis that the coefficient is zero is rejected for both extreme positive and extreme negative values of t. If the direction of the association can be specified in advance, the hypothesis is rejected only for t values that are of sufficient magnitude and in the direction specified.

In SPSS/PC+, coefficients with one-tailed observed significance levels less than 0.01 are designated with a single asterisk; those with one-tailed significance levels less than 0.001, with two asterisks. From Figure 13.4, the probability that a correlation coefficient of at least 0.88 is obtained when there is no linear association in the population between sales and advertising is less than 0.001. Care should be exercised when examining the significance levels for large tables. Even if there is no association between the variables, if many coefficients are computed some would be expected to be statistically significant by chance alone.

Special procedures must be employed to test more general hypotheses of the form $\rho=\rho_0$ where ρ_0 is a constant. If the assumptions of bivariate normality appear unreasonable, a variety of *nonparametric* measures, which make limited assumptions about the underlying distributions of the variables, can be calculated. See Chapter 16 for further discussion.

13.6 Correlation Matrices and Missing Data

For a variety of reasons, data files frequently contain incomplete observations. Respondents in surveys scrawl illegible responses or refuse to answer certain questions. Laboratory animals die before experiments are completed. Patients fail to keep scheduled clinic appointments.

Analysis of data with missing values is troublesome. Before even considering possible strategies, you should determine whether there is evidence that the missing-value pattern is not random. That is, are there reasons to believe that missing values for a variable are related to the values of that variable or other variables? For example, people with low incomes may be less willing to report their financial status than more affluent people. This may be even more pronounced for people who are poor but highly educated.

One simple method of exploring such possibilities is to subdivide the data into two groups—those observations with missing data on a variable and those with complete data—and examine the distributions of the other variables in the file across these two groups. The SPSS/PC+ procedures CROSSTABS and T-TEST are particularly useful for this. For a discussion of more sophisticated methods for detecting nonrandomness, see Frane (1976).

If it appears that the data are not missing randomly, use great caution in attempting to analyze the data. It may be that no satisfactory analysis is possible, especially if there are only a few cases.

If you are satisfied that the missing data are random, several strategies are available. First, if the same few variables are missing for most cases, exclude those variables from the analysis. Since this luxury is not usually available, you can alternatively keep all variables but eliminate the cases with missing values on any of them. This is termed *listwise* missing-value treatment since a case is eliminated if it has a missing value on any variable in the list.

If many cases have missing data for some variables, listwise missing-value treatment could eliminate too many cases and leave you with a very small sample. One common technique is to calculate the correlation coefficient between a pair of variables based on all cases with complete information for the two variables regardless of whether the cases have missing data on any other variable. For example, if a case has values only for variables 1, 3, and 5, it is used only in computations involving variable pairs 1 and 3, 1 and 5, and 3 and 5. This is *pairwise* missing-value treatment.

13.7 Choosing Pairwise Missing-Value Treatment

Several problems can arise with pairwise matrices, one of which is inconsistency. There are some relationships between coefficients that are impossible but may occur when different cases are used to estimate different coefficients. For example, if age and weight and age and height have a high positive correlation, it is impossible in the same sample for height and weight to have a high negative correlation. However, if the same cases are not used to estimate all three coefficients, such an anomaly can occur.

There is no single sample size that can be associated with a pairwise matrix since each coefficient can be based on a different number of cases. Significance levels obtained from analyses based on pairwise matrices must be viewed with caution since little is known about hypothesis testing in such situations.

It should be emphasized that missing-value problems should not be treated lightly. You should base your decision on careful examination of the data and not leave the choices up to system defaults.

13.8 THE REGRESSION LINE

If there is a linear relationship between two variables, a straight line can be used to summarize the data. When the correlation coefficient is +1 or −1, little thought is needed to determine the line that best describes the data: the line passes through all of the observations. When the observations are less highly correlated, many different lines can be drawn to represent the data.

One of the most commonly used procedures for fitting a line to the observations is the method of *least squares*. This method results in a line that minimizes the sum of squared vertical distances from the data points to the line.

The equation for the straight line that relates predicted sales to advertising is

$$\text{PREDICTED SALES} = \text{a} + \text{b(ADVERTISING)} \qquad \textbf{Equation 13.8a}$$

The intercept, a, is the predicted sales when there is no advertising. The slope, b, is the change in predicted sales for a unit change in advertising. That is, it is the amount of change in sales per television spot.

The actual values of a and b calculated with the method of least squares are printed as part of the SPSS/PC+ PLOT output (see Figure 13.8). The least-squares equation for the line is

PREDICTED SALES = 135.4 + 25.31(ADVERTISING) **Equation 13.8b**

Figure 13.1a shows this regression line.

Figure 13.8 Intercept and slope from PLOT

```
40 cases plotted. Regression statistics of SALES on ADVERTIS:
Correlation  .88016 R Squared   .77467  S.E. of Est   59.56023 Sig.  .0000
Intercept(S.E.)  135.43360( 25.90649)  Slope(S.E.)    25.30770(  2.21415)
```

For each pair of variables, two different regression lines can be calculated, since the values of the slope and intercept depend on which variable is dependent (the one being predicted) and which is independent (the one used for prediction). In the SPSS/PC+ PLOT output, the variable plotted on the vertical axis is considered the dependent variable in the calculation of statistics.

13.9 Prediction

Based on the regression equation, it is possible to predict sales from advertising. For example, a territory with 10 television spots per month is expected to have sales of about \$388,400 (135.4 + 25.3(10)). Considerable caution is needed when predictions are made for values of the independent variable which are much larger or much smaller than those used to derive the equation. A relationship which is linear for the observed range of values may not be linear everywhere. For example, estimating Y values for the beginning of Figure 13.2c based on a regression line for the latter part of the plot would result in a very poor fit.

The difference between observed sales and sales predicted by the model is called a *residual*. The residual for a territory with 10 television spots and observed sales of 403.6 is 15.2:

$$\text{RESIDUAL} = \text{OBSERVED} - \text{PREDICTED} = 403.6 - 388.4 = 15.2$$ **Equation 13.9**

Residuals can be calculated for each of the sales territories. Figure 13.9 contains the observed value (SALES), the predicted value (*PRED), and the residual for the first 10 territories (*RESID). The residuals provide an idea of how well the calculated regression line actually fits the data.

Figure 13.9 Residuals from the regression line

```
Casewise Plot of Standardized Residual

*: Selected   M: Missing

          -3.0      0.0      3.0
  Case #   0:........:........:0   SALES      *PRED      *RESID
       1    .        *        .    260.3   261.9721     -1.6721
       2    .       *.        .    286.1   312.5875    -26.4875
       3    .        *        .    279.4   287.2798     -7.8798
       4    .        . *      .    410.8   363.2029     47.5971
       5    .        *        .    438.2   439.1260      -.9260
       6    .       *.        .    315.3   337.8952    -22.5952
       7    .        .       *.    565.1   413.8183    151.2817
       8    .        .*       .    570.0   540.3568     29.6432
       9    .      * .        .    426.1   464.4337    -38.3337
      10    .        *        .    315.0   312.5875      2.4125
  Case #   0:........:........:0   SALES      *PRED      *RESID
          -3.0      0.0      3.0
```

13.10 Goodness of Fit

Although the regression line is a useful summary of the relationship between two variables, the values of the slope and intercept alone do little to indicate how well the line actually fits the data. A goodness-of-fit index is needed.

The observed variation in the dependent variable can be subdivided into two components: the variation "explained" by the regression and the residual from the regression, or

$$\text{TOTAL SS} = \text{REGRESSION SS} + \text{RESIDUAL SS} \quad \textbf{Equation 13.10a}$$

The *total sum of squares* is a measure of overall variation and is given by

$$\text{TOTAL SUM OF SQUARES} = \sum_{i=1}^{N} (Y_i - \overline{Y})^2 \quad \textbf{Equation 13.10b}$$

The total sum of squares for sales is 598,253. (It is N-1 times the variance.)

The *regression sum of squares*, or the sum of squares due to regression, is

$$\text{REGRESSION SUM OF SQUARES} = \sum_{i=1}^{N} (\hat{Y}_i - \overline{Y})^2 \quad \textbf{Equation 13.10c}$$

where $\hat{Y}_i$ is the predicted value for the ith case. The regression sum of squares is a measure of how much variability in the dependent variable is attributable to the linear relationship between the two variables. For this example, the regression sum of squares is 463,451.

The *residual sum of squares*, sometimes called the error sum of squares, is obtained by squaring each of the residuals and then summing them.

$$\text{RESIDUAL SUM OF SQUARES} = \sum_{i=1}^{N} (Y_i - \hat{Y}_i)^2 \quad \textbf{Equation 13.10d}$$

The residual sum of squares for sales is 134,802. The standard deviation of the residuals, called the standard error of the estimate, is

$$\text{SEE} = \sqrt{\frac{\text{RESIDUAL SUM OF SQUARES}}{N - 2}} = \sqrt{\frac{134{,}802}{38}} = 59.56 \quad \textbf{Equation 13.10e}$$

The standard error is displayed in Figure 13.8.

The proportion of the variation in the dependent variable that is *explained* by the linear regression is computed by comparing the total sum of squares and the regression sum of squares:

$$r^2 = \frac{\text{REGRESSION SUM OF SQUARES}}{\text{TOTAL SUM OF SQUARES}} = \frac{463{,}451}{598{,}253} = 0.775 \quad \textbf{Equation 13.10f}$$

If there is no linear association in the sample, the value of r^2 is 0 since the predicted values are just the mean of the dependent variable and the regression sum of squares is 0. If Y and X are perfectly linearly related, the residual sum of squares is 0 and r^2 is 1. The square root of r^2 is r, the Pearson correlation coefficient between the two variables.

13.11 Further Topics in Regression

In this chapter, only the most basic concepts in regression analysis are discussed. Chapter 17 contains detailed descriptions of simple two-variable regression as well as multiple regression analysis.

13.12 RUNNING PROCEDURE CORRELATION

The CORRELATION (alias PEARSON CORR) procedure calculates Pearson product-moment correlations for pairs of variables. The display includes the coefficient (r), an indication of significance level, and the number of cases upon which the coefficients are computed. Means, standard deviations, crossproduct deviations, and covariances are available. You can also specify optional formats and methods of handling missing data. In addition, you can write out a square matrix containing correlation coefficients and the number of cases for use in other SPSS/PC+ procedures (see Chapter 3).

13.13 Specifying the Variables

The VARIABLES subcommand lists all variables to be included in the correlation matrix. For example, to produce the correlation matrix shown in Figure 13.4, specify

```
CORRELATION VARIABLES=ADVERTIS REPS SALES.
```

The actual keyword VARIABLES can be omitted.

The order in which you name the variables is the order in which they are displayed. Use the keyword WITH to obtain the correlations of one set of variables with another set. For example, the command

```
CORRELATION VARIABLES=ADVERTIS WITH REPS SALES.
```

produces two correlations, ADVERTIS with REPS and ADVERTIS with SALES. You can specify several analysis lists by separating them with slashes.

Only numeric variables can be named on the VARIABLES subcommand. Long or short string variables on an analysis list will prevent execution of CORRELATION.

13.14 Optional Statistics

By default, the full correlation matrix and the number of valid cases on which the matrix is based are displayed. A correlation that cannot be computed is displayed as a period (.). One-tailed probabilities of less than .01 are indicated by an asterisk (*) and those less than .001 by two asterisks (**). The display uses the width defined on the SET command (see Command Reference: SET).

The following additional options and statistics are available using the OPTIONS and STATISTICS subcommands:

Option 3 *Display two-tailed probabilities.* Two-tailed probabilities less than .01 are indicated by an asterisk (*) and those less than .001 by two asterisks (**). When Option 5 is also specified, exact two-tailed probabilities are given.

Option 5 *Display count and probability.* The number of cases used to compute each coefficient and exact probability are displayed. This option is ignored when Option 6 is specified.

Statistic 1 *Univariate mean, standard deviation, and count.* Missing values are handled on a variable-by-variable basis, regardless of the missing-value option in effect.

Statistic 2 *Cross-product deviations and covariance.*

ALL *Display all statistics.* Includes statistics available with Statistics 1 and 2.

For example, to obtain the the statistics in Figure 13.14, specify

```
CORRELATION VARIABLES=ADVERTIS REPS SALES
/STATISTICS=1.
```

Figure 13.14 Univariate statistics with Statistic 1

Variable	Cases	Mean	Std Dev
ADVERTIS	40	10.9000	4.3074
REPS	40	5.0000	1.6486
SALES	40	411.2875	123.8540

13.15 Missing Values

By default, SPSS/PC+ excludes a case from the calculation of all correlation coefficients if it has a missing value on any variable named on the VARIABLES subcommand. This is listwise missing-value treatment. Specify Options 1 or 2 on the OPTIONS subcommand for alternative missing-value treatments.

Option 1 *Include cases with user-missing values* in the computations.

Option 2 *Exclude cases with missing values pairwise.* All cases with valid values for the pair of variables used to compute a coefficient are included in the computation of that coefficient (see Section 13.7).

13.16 Writing Matrices

By default, matrices are written only to the display file. Use Option 4 on the OPTIONS subcommand to write matrices to the results file named on the SET command for use in other procedures.

Option 4 *Write count and correlation matrix.* The correlation matrix and number of cases used to compute each coefficient for each analysis list are written to the results file named on the SET command.

You cannot use the keyword WITH in the analysis list if you specify Option 4. See Command Reference: CORRELATION for the format of the matrix file.

13.17 Annotated Example

The following commands produce the output in Figures 13.14 and 13.17:

```
DATA LIST   / ADVERTIS 1-2 EFFIC 4 REPS 6 SALES 8-12 (1).
BEGIN DATA.
lines of data
END DATA.
CORRELATION  ADVERTIS REPS SALES/
             STATISTICS=1.
```

- The DATA LIST command defines the variable names and gives their column locations.
- The CORRELATION command asks for correlations between ADVERTIS, REPS, and SALES. Statistic 1 asks for univariate statistics.

Figure 13.17 Correlation matrix

```
Correlations:  ADVERTIS    REPS        SALES

  ADVERTIS    1.0000       .7763       .8802
  REPS         .7763      1.0000       .8818
  SALES        .8802       .8818      1.0000

N of cases:    40          1-tailed Signif:  * - .01  ** - .001

" . " is printed if a coefficient cannot be computed
```

Contents

Chapter 14 One-way Analysis of Variance: Procedure ONEWAY

Rotund Italians washing down carbohydrate-laden feasts with jugs of chianti, somber Jews ritualistically sipping Sabbath wine, melancholy Irish submerging grief and frustration in a bottle—all are common ethnic stereotypes. Is there any evidence to support these notions? In *Ethnic Drinking Subcultures,* Greeley (1981) examines drinking habits in a sample of five ethnic populations within four major American cities.

A total of 1,107 families completed questionnaires detailing their drinking behavior and ancestral origins. Irish, Italian, Jewish, Swedish, and English families were included. Greeley investigates possible differences in drinking habits and a variety of cultural and psychological explanations for them. In this chapter, only differences in total alcohol consumption are considered.

14.1 DESCRIPTIVE STATISTICS AND CONFIDENCE INTERVALS

Figure 14.2 contains basic descriptive statistics for total yearly alcohol consumption in pints for the adult males in the study. The Italians and Irish are the biggest consumers, drinking an average of 24 pints a year. The Jewish males drink the least, an average of slightly more than 9 pints a year.

The sample mean for a group provides the single best guess for the unknown population value μ_i. It is unlikely that the value of the sample mean is exactly equal to the population parameter, but it is probably not too different. Based on the sample mean, it is possible to calculate a range of values that, with a designated likelihood, includes the population value. Such a range is called a *confidence interval.* For example, as shown in Figure 14.2, the 95% confidence interval for μ_{Irish} is the range 19.61 to 28.89 pints. This means that if repeated samples are selected from a population under the same conditions and 95% confidence intervals are calculated, 95% of the intervals will contain the unknown parameter μ_{Irish}. Since the parameter value is unknown, it is not possible to determine whether a particular interval contains it.

14.2 ANALYSIS OF VARIANCE

Looking at the sample means in Figure 14.2, you might wonder whether the observed differences can be reasonably attributed to chance or whether there is reason to suspect true differences between the five groups. One of the statistical procedures commonly used to test the hypothesis that several population means are equal is *analysis of variance* or ANOVA.

Figure 14.2 Total yearly alcohol consumption for adult males (in pints)

Group	Count	Mean	Standard Deviation	Standard Error	95 Pct Conf Int for Mean		
IRISH	119	24.2500	25.5620	2.3433	19.6097	To	28.8903
ITALIAN	84	24.3120	24.1880	2.6391	19.0629	To	29.5611
JEWISH	41	9.2500	21.6250	3.3773	2.4243	To	16.0757
SWEDISH	74	16.5630	26.7500	3.1096	10.3655	To	22.7605
ENGLISH	90	21.8750	21.5630	2.2729	17.3587	To	26.3913
Total	408	20.8373	24.6519	1.2205	18.4381	To	23.2365

Group	Minimum	Maximum
IRISH	0.0	145.0
ITALIAN	0.0	128.0
JEWISH	0.0	87.0
SWEDISH	0.0	112.0
ENGLISH	0.0	117.0
Total	0.0	145.0

Certain assumptions are required for correct application of the ANOVA test. Independent samples from normally distributed populations with the same variance must be selected. In subsequent discussion, it is assumed that the populations sampled constitute the entire set of populations about which conclusions are desired. For example, the five ethnic groups are considered to be the only ones of interest. They are not viewed as a sample from all possible ethnic groups. This is called a *fixed-effects model.*

14.3 Partitioning Variation

In analysis of variance, the observed variability in the sample is subdivided into two components—variability of the observations within a group about the group mean and variability of the group means. If the amount of alcohol consumed doesn't vary much for individuals within the same ethnic group—for example, all the Swedes seem to drink about the same—but the group means differ substantially, there is evidence to suspect that the population means are not all equal.

The *within-groups sum of squares* is a measure of the variability within groups. It is calculated as

$$\text{SSW} = \sum_{i=1}^{k} (N_i - 1)S_i^2$$ **Equation 14.3a**

where S_i^2 is the variance of group i about its mean, and N_i is the number of cases in group i. For the data shown in Figure 14.2, the within-groups sum of squares is

$$\begin{aligned}\text{SSW} &= 25.56^2(118) + 24.19^2(83) + 21.63^2(40) \\ &\quad + 26.75^2(73) + 21.56^2(89) \\ &= 237{,}986\end{aligned}$$ **Equation 14.3b**

Variability of the group means is measured by the *between-groups sum of squares*, which is

$$\text{SSB} = \sum_{i=1}^{k} N_i(\bar{X}_i - \bar{X})^2$$ **Equation 14.3c**

The mean of the ith group is denoted $\bar{X}_i$, and the mean of the entire sample is $\bar{X}$. For the drinking study, the between-groups sum of squares is

$$\begin{aligned} SSB &= (24.25-20.84)^2(119) + (24.31-20.84)^2(84) \\ &\quad + (9.25-20.84)^2(41) + (16.56-20.84)^2(74) \\ &\quad + (21.88-20.84)^2(90) \\ &= 9{,}353 \end{aligned}$$ **Equation 14.3d**

The sums of squares, and other related statistics, are usually displayed in an analysis of variance table, as shown in Figure 14.3.

Figure 14.3 Analysis of variance table

```
   Variable  AMOUNT       AMOUNT OF ALCOHOL CONSUMED IN PINTS
By Variable  ETHNIC       ETHNIC BACKGROUND

                               Analysis of Variance

                                 Sum of          Mean             F        F
       Source           D.F.     Squares         Squares        Ratio    Prob.

Between Groups            4      9353.8671      2338.4668      3.9599    .0036

Within Groups           403    237986.3869       590.5369

Total                   407    247340.2540
```

The mean squares in Figure 14.3 are obtained by dividing the sums of squares by their degrees of freedom. The between-groups degrees of freedom are $k-1$, where k is the number of groups. The within-groups degrees of freedom are $N-k$, where N is the number of cases in the entire sample.

14.4 Testing the Hypothesis

To test the hypothesis that the five ethnic groups under study consume the same average amount of alcohol—that is, that

$$\mu_{\text{Irish}} = \mu_{\text{Italian}} = \mu_{\text{Jewish}} = \mu_{\text{Swedish}} = \mu_{\text{English}}$$ **Equation 14.4a**

the following statistic is calculated (see Figure 14.3):

$$F = \frac{\text{BETWEEN GROUPS MEAN SQUARE}}{\text{WITHIN GROUPS MEAN SQUARE}} = \frac{2338.47}{590.54} = 3.96$$ **Equation 14.4b**

When the assumptions described in Section 14.2 are met, the observed significance level is obtained by comparing the calculated F to values of the F distribution with $k-1$ and $N-k$ degrees of freedom. The observed significance level is the probability of obtaining an F statistic at least as large as the one calculated when all population means are equal. If this probability is small enough, the hypothesis that all population means are equal is rejected. In this example, the observed significance level is approximately 0.0036 (Figure 14.3). Thus, it appears unlikely that men in the five ethnic populations consume the same mean amount of alcohol.

14.5 MULTIPLE COMPARISON PROCEDURES

A significant *F* statistic indicates only that the population means are probably unequal. It does not pinpoint where the differences are. A variety of special techniques, termed *multiple comparison* tests, are available for determining which population means are different from each other.

You may question the need for special techniques—why not just calculate the *t* test described in Chapter 11 for all possible pairs of means? The problem is that when many comparisons are made, some will appear to be significant even when all population means are equal. With five groups, for example, there are ten possible comparisons between pairs of means. When all population means are equal, the probability that at least one of the ten observed significance levels will be less than 0.05 is about 0.29 (Snedecor, 1967).

Multiple comparison tests protect against calling too many differences significant. These tests set up more stringent criteria for declaring differences significant than does the usual *t* test. That is, the difference between two sample means must be larger to be identified as a true difference.

14.6 The Scheffé Test

Many multiple comparison tests are available, and they all provide protection in slightly different ways (for further discussion, see Winer, 1971). Figure 14.6a is output from the *Scheffé* multiple comparison test for the ethnic drinking data. The Scheffé method is conservative for pairwise comparisons of means. It requires larger differences between means for significance than most of the other methods.

Figure 14.6a The Scheffé multiple comparison test

```
      Variable  AMOUNT       AMOUNT OF ALCOHOL CONSUMED IN PINTS
   By Variable  ETHNIC       ETHNIC BACKGROUND

Multiple Range Test

Scheffe Procedure
Ranges for the 0.050 level -

          4.38    4.38    4.38    4.38

The ranges above are table ranges.
The value actually compared with Mean(J)-Mean(I) is..
      17.1834 * Range * Sqrt(1/N(I) + 1/N(J))

  (*) Denotes pairs of groups significantly different at the 0.050 level

                             J S E I I
                             E W N R T
                             W E G I A
                             I D L S L
                             S I I H I
                             H S S   A
                               H H   N
     Mean       Group

     9.2500     JEWISH
    16.5630     SWEDISH
    21.8750     ENGLISH
    24.2500     IRISH        *
    24.3120     ITALIAN      *
```

The means are ordered and displayed from smallest to largest, as shown in Figure 14.6a. Pairs of means that are significantly different at the 0.05 level in this case are indicated with an asterisk in the lower half of the matrix at the bottom of the output. In this example, the asterisks under the vertical column labeled "Jewish" mean that Jews are significantly different from the Irish and the Italians. No other pair is found to be significantly different. If no pairs are significantly different, a message is displayed and the matrix is suppressed.

The formula above the matrix indicates how large an observed difference must be to attain significance using the particular multiple comparison test. The table ranges are the values for the range variable in the formula.

If the sample sizes in all groups are equal, or an average sample size is used in the computations, a somewhat modified table is also displayed (Figure 14.6b). Instead of indicating which groups are significantly different, means that are not different are grouped. Subset 1 shows that Jews, Swedes, and the English are not different. Subset 2 groups Swedes, English, Irish, and Italians. Jews do not appear in the same subset as Irish and Italians since they are significantly different from these two.

Figure 14.6b Homogeneous subsets

```
  Homogeneous Subsets   (Subsets of groups, whose highest and lowest means
                         do not differ by more than the shortest
                         significant range for a subset of that size)

SUBSET  1

Group        JEWISH        SWEDISH        ENGLISH
Mean         9.2500        16.5630        21.8750
- - - - - - - - - - - - - - - - - - - - - - - - - - - - - - - -

SUBSET  2

Group        SWEDISH       ENGLISH        IRISH          ITALIAN
Mean         16.5630       21.8750        24.2500        24.3120
- - - - - - - - - - - - - - - - - - - - - - - - - - - - - - - -
```

14.7 EXPLANATIONS

Both cultural and psychological explanations for differences in drinking habits among ethnic groups have been suggested (Greeley, 1981). In Jewish culture, the religious symbolism associated with drinking, as well as strong cultural norms against drunkenness, seem to discourage alcohol consumption. For the Irish, alcohol is a vehicle for promotion of fun and pleasure, as well as a potent tranquilizer for dissipating grief and tension. Such high expectations of alcohol make it a convenient escape and foster dependency. Italians have accepted drinking as a natural part of daily life. Alcohol is treated almost as a food and not singled out for its special pleasures.

14.8 Tests for Equality of Variance

As previously discussed, one of the assumptions needed for applying analysis of variance properly is that of equality of variances. That is, all of the populations from which random samples are taken must not only be normal but must also have the same variance, σ^2. Several procedures are available for testing this assumption of *homogeneity of variance*. Unfortunately, many of them are not very useful since they are influenced by characteristics of the data other than the variance.

Figure 14.8 Tests for homogeneity of variance

```
Tests for Homogeneity of Variances

     Cochrans C = Max. Variance/Sum(Variances) =   .2479, P =  .248 (Approx.)
     Bartlett-Box F =                             1.349 , P =  .249
     Maximum Variance / Minimum Variance          1.539
```

Figure 14.8 shows the results of the three tests for homogeneity of variance available in SPSS/PC+ for the ethnic drinking study. The significance levels are relatively large, so the hypothesis that the populations have the same variance cannot be rejected. Since the variances do not appear to be unequal, there is no reason to worry. Even if the variances appeared to be different, but the sample sizes in all groups were similar, there would be no cause for alarm since the ANOVA test is not particularly sensitive to violations of equality of variance under such conditions. However, if sample sizes are quite dissimilar and variances are unequal, you should consider transforming the data or using a statistical procedure that requires less stringent assumptions (see Chapter 16).

14.9 RUNNING PROCEDURE ONEWAY

The ONEWAY procedure produces a one-way analysis of variance. Although ANOVA (see Chapter 15) can also produce a one-way analysis of variance, ONEWAY performs several optional tests not available in ANOVA. Both ONEWAY and ANOVA calculate sums of squares, mean squares, degrees of freedom, and *F* tests, but only ONEWAY allows you to test for nonlinear trends, specify contrasts, and use multiple comparison tests. ONEWAY also reads and writes matrix materials.

ONEWAY operates via subcommands, including the OPTIONS and STATISTICS subcommands. You must first specify a dependent variable list and an independent variable with its range of values. The optional subcommands that produce contrasts, tests for trends, and multiple comparisons appear after this specification and can be entered in any order.

14.10 Specifying the Variables

The VARIABLES subcommand is the only required specification for ONEWAY and consists of the name of at least one dependent variable, the keyword BY, and the name of the independent variable followed by its minimum and maximum values enclosed in parentheses and separated by a comma. The actual keyword VARIABLES can be omitted. Thus, the output in Figure 14.3 could be obtained by specifying either

```
ONEWAY VARIABLES=AMOUNT BY ETHNIC(1,5).
```

or

```
ONEWAY AMOUNT BY ETHNIC(1,5).
```

The minimum and maximum values are the lowest and highest values of the independent variable to be used in the analysis. Cases with values for the independent variable outside this range are excluded from the analysis.

Only one independent variable can be used in an analysis list. When more than one dependent variable is specified, a separate one-way analysis of variance is produced for each one.

14.11 Specifying Multiple Comparison Tests

Use the RANGES subcommand to specify any of seven multiple comparison tests available in ONEWAY. You can specify multiple RANGES subcommands, and each one requests one test. The keywords for RANGES are shown below. Each specifies a type of multiple comparison test, and some can be followed by a number in parentheses indicating the significance level. If a significance level is not specified, ONEWAY uses a .05 significance level.

LSD(p) *Least significant difference.* Any significance level between 0 and 1 can be specified.

DUNCAN(p) *Duncan's multiple-range test.* Only the significance levels .01, .05, or .10 can be specified.

SNK *Student-Newman-Keuls test.* Only the .05 significance level can be used.

BTUKEY *Tukey's alternate test.* Only the .05 significance level can be used.

TUKEY *Honestly significant difference.* Only the .05 significance level can be used.

MODLSD(p) *Modified LSD.* Any significance level between 0 and 1 can be specified.

SCHEFFE(p) *Scheffé's test.* Any significance level between 0 and 1 can be specified.

For example, Figures 14.6a and 14.6b were obtained by specifying

```
ONEWAY AMOUNT BY ETHNIC(1,5)
/RANGES=SCHEFFE.
```

ONEWAY produces two types of output, depending on the design and multiple comparison test. Multiple comparisons for all groups are produced whenever RANGES is used. In this type of output, group means are listed in ascending order. Asterisks in the matrix of group names indicate which means are significantly different. For example, the asterisks in Figure 14.6a indicate that Jews are significantly different from the Irish and the Italians.

For balanced designs or when Option 10 (harmonic means) is specified, ONEWAY produces homogeneous subsets of means. Figure 14.6b shows output with homogeneous subsets of means. In this example, two subsets are produced. The first subset includes Jews and Swedes, and the second subset includes Swedes, English, Irish, and Italians. The means of the groups within a subset are *not* significantly different.

You can specify any other type of range for multiple comparisons by listing specific range values. Up to $k-1$ range values can be specified in ascending order, where k is the number of groups and where the range value times the standard error of the combined subset is the critical value. If fewer than $k-1$ values are specified, the last value is used for the remaining values. You can specify repetitions with $n*r$, where n is the number of repetitions and r is the range value. To use a single critical value for all subsets, specify one range value. For example, the command

```
ONEWAY VARIABLES=AMOUNT BY ETHNIC(1,5)
/RANGES=4.38.
```

produces the Scheffé test shown in Figure 14.6a.

By default, the multiple comparison test uses the harmonic mean of the sizes of the two groups being compared. If you want the harmonic mean of *all* group sizes to be used, specify Option 10 on the OPTIONS subcommand. When Option 10 is specified, ONEWAY calculates homogeneous subsets.

Option 10 *Harmonic mean of all group sizes used as the sample size for each group in range tests.* If the harmonic mean is used for unbalanced designs, ONEWAY determines homogeneous subsets for all range tests.

14.12 Partitioning Sums of Squares

The POLYNOMIAL subcommand partitions the between-groups sums of squares into linear, quadratic, cubic, and higher-order trend components. Its specifications consist of a single number, which indicates the degree of the highest-order polynomial to be used. For example, the command

```
ONEWAY AMOUNT BY ETHNIC(1,5)
/POLYNOMIAL=2.
```

specifies a polynomial of order 2 (a quadratic) as the highest-order polynomial. The number specified must be a positive integer less than or equal to 5 and less than the number of groups. The POLYNOMIAL subcommand follows the variable specifications and can be used only once.

When the design is balanced and POLYNOMIAL is used, ONEWAY computes the sum of squares for each order polynomial from weighted polynomial contrasts, using the values of the independent variable as the metric. These contrasts are orthogonal, so that the sums of squares for each order polynomial are statistically independent. If the design is unbalanced, but there is equal spacing between the values of the independent variable, ONEWAY computes sums of squares using the unweighted polynomial contrasts.

14.13 Specifying Contrasts

The CONTRAST subcommand specifies *a priori* contrasts to be tested by the t statistic. Its specification is simply a list of coefficients, with each coefficient corresponding to a value of the independent variable. For example, the command

```
ONEWAY AMOUNT BY ETHNIC(1,5)
/CONTRAST=1 1 -2 0 0.
```

specifies comparison of the combined means of the Irish and Italians with the mean for Jews. The last two groups are not included in the contrast. Fractional coefficients can also be used, as in the command

```
ONEWAY AMOUNT BY ETHNIC(1,5)
/CONTRAST=.5 .5 -1 0 0.
```

For most applications, the coefficients should sum to 0. Sets of coefficients that do not sum to 0 are used, but a warning message is displayed.

The notation $n*c$ can be used to specify the same coefficient for n consecutive means. For example, the command

```
ONEWAY AMOUNT BY ETHNIC(1,5)
/CONTRAST=2*1 -2 2*0.
```

specifies a contrast coefficient of 1 for the Irish and Italians, a coefficient of -2 for Jews, and a coefficient of 0 for Swedes and the English. You must specify a contrast coefficient for every group implied by the range given for the independent variable, even if a group has no cases. Trailing zeros need not be specified, however, so that the command

```
ONEWAY AMOUNT BY ETHNIC(1,5)
/CONTRAST=2*1 -2.
```

is equivalent to the previous command.

Only one set of contrast coefficients can be specified per CONTRAST subcommand. However, you can specify multiple CONTRAST subcommands.

Output for each contrast list includes the value of the contrast, the standard error of the contrast, the *t* statistic, the degrees of freedom for *t*, and the two-tailed observed significance level of *t*. Both pooled-variance and separate-variance estimates are displayed.

14.14 Optional Statistics

You can specify three optional sets of statistics or the keyword ALL on the STATISTICS subcommand:

Statistic 1 *Group descriptive statistics.* For each group, the count, mean, standard deviation, standard error, minimum, maximum, and 95% confidence interval are displayed for each dependent variable.

Statistic 2 *Fixed- and random-effects statistics.* For the fixed-effects model, the standard deviation, standard error, and 95% confidence interval are displayed. For the random-effects model, the standard error, the 95% confidence interval, and the estimate of the between-component variance are displayed.

Statistic 3 *Homogeneity-of-variance-tests.* Cochran's *C*, the Bartlett-Box *F*, and Hartley's *F* Max are displayed.

ALL *Display all statistics.*

For example, the output in Figures 14.2 and 14.8 was obtained by specifying

```
ONEWAY AMOUNT BY ETHNIC(1,5)
/STATISTICS=1 3.
```

14.15 Missing Values

By default, ONEWAY deletes cases with missing values on an analysis-by-analysis basis; that is, a case with missing values for the dependent variable or the independent variable is not used in that analysis. Cases with values for the independent variable outside the range specified are also excluded from the analysis. Two alternative missing-value treatments are available and are specified on the OPTIONS subcommand:

Option 1 *Include user-missing values.* Cases with user-defined missing values are included in the analysis.

Option 2 *Exclude missing values listwise.* Cases with missing values for any variable in the analysis list are excluded from the analysis.

Neither option will cause ONEWAY to include cases with values for the independent value that are outside the specified range.

14.16 Formatting Options

By default, ONEWAY displays variable labels and identifies groups as GRP1, GRP2, and so forth. The display can be modified by specifying the following options on the OPTIONS subcommand:

Option 3 *Suppress variable labels.*

Option 6 *Display value labels for groups.* Use the first eight characters of the value labels defined for the independent variable as group labels.

Option 6 was used to obtain the group identifications in Figure 14.6a.

14.17 Matrix Materials

ONEWAY can read and write matrix materials. It writes out frequencies, means, and standard deviations in a format it can read, and it also reads frequencies, means, and the pooled variance.

To write matrix materials to a file, specify Option 4 on the OPTIONS subcommand.

Option 4 *Write matrix materials to a file.* Writes a vector of category frequencies, a vector of means, and a vector of standard deviations for each dependent variable. Vectors are written 80 characters per line, with each vector beginning on a new line. The format for the frequencies vector is F10.2, and for the means and standard deviations vectors, F10.4. Thus, each line has a maximum of eight values.

If Option 4 is used, the file to which the material is written is controlled by the SET command (see Command Reference).

ONEWAY reads two types of matrix materials, specified with Options 7 and 8 on the OPTIONS subcommand:

Option 7 *Read matrix of counts, means, and standard deviations.* ONEWAY expects a vector of counts for each group, followed by a vector of group means and a vector of group standard deviations like those written by Option 4.

Option 8 *Read matrix of counts, means, pooled variance, and degrees of freedom.* ONEWAY expects a vector of counts for each group, followed by a vector of means for each group, followed by the pooled variance (the within-groups mean square) and the degrees of freedom for the pooled variance. If the degrees of freedom are omitted, ONEWAY takes them to be $n-k$, where n is the number of cases and k is the number of groups. Statistics 1, 2, and 3, and the separate-variance estimate for contrasts cannot be computed.

If Option 7 or 8 is used, the MATRIX keyword must be specified on the DATA LIST command. For either option, each vector begins on a new line and can be entered in fixed or freefield format. Unless matrix materials produced by ONEWAY are to be read, it is easier to use freefield input with the FREE keyword on the DATA LIST command. (For more information, see DATA LIST and ONEWAY in Command Reference.)

14.18 Annotated Example

The SPSS/PC+ commands used to produce the output in this chapter are

```
DATA LIST MATRIX FREE / AMOUNT ETHNIC.
VAR LABELS  AMOUNT 'AMOUNT OF ALCOHOL CONSUMED IN PINTS'/
            ETHNIC 'ETHNIC BACKGROUND'.
VALUE LABELS   ETHNIC 1 'IRISH' 2 'ITALIAN' 3 'JEWISH' 4 'SWEDISH'
                      5 'ENGLISH'.
BEGIN DATA.
119.      84.       41.       74.       90.
 24.25     24.312     9.25     16.563    21.875
 25.562    24.188    21.625    26.75     21.563
END DATA.
ONEWAY AMOUNT BY ETHNIC(1,5)
       /RANGES=SCHEFFE
       /OPTIONS=6,7
       /STATISTICS=1,3.
FINISH.
```

- The DATA LIST command indicates that the variables AMOUNT and ETHNIC are being read in as a freefield matrix.
- The VAR LABELS command assigns descriptive labels to AMOUNT and ETHNIC.
- The VALUE LABELS command assigns labels to values of ETHNIC, indicating the ethnic group represented by each value.
- The BEGIN DATA and END DATA commands enclose the matrix materials.
- The ONEWAY command names AMOUNT as the dependent variable and ETHNIC as the independent variable. The minimum and maximum values for ETHNIC are 1 and 5.
- The RANGES subcommand calculates multiple comparisons between means using the Scheffé test.
- The OPTIONS subcommand requests the display of value labels for each group and specifies matrix input.
- The STATISTICS subcommand requests descriptive statistics and homogeneity-of-variance tests.

Contents

Chapter 15 Analysis of Variance: Procedure ANOVA

Despite constitutional guarantees, any mirror will testify that all citizens are not created equal. The consequences of this inequity are pervasive. Physically attractive persons are perceived as more desirable social partners, more persuasive communicators, and generally more likeable and competent. Even cute children and attractive burglars are disciplined more leniently than their homely counterparts (Sigall & Ostrove, 1975).

Much research on physical attractiveness focuses on its impact on heterosexual relationships and evaluations. Its effect on same-sex evaluations has received less attention. Anderson and Nida (1978) examined the influence of attractiveness on the evaluation of writings by college students. In the study, 144 male and 144 female students were asked to appraise essays purportedly written by college freshmen. A slide of the "author" was projected during the rating as part of "supplemental information." Half of the slides were of authors of the same sex as the rater; the other half were of authors of the opposite sex. The slides had previously been determined to be of high, medium, and low attractiveness. Each participant evaluated one essay for creativity, ideas, and style. The three scales were combined to form a composite measure of performance.

15.1 DESCRIPTIVE STATISTICS

Figure 15.1 contains average composite scores for the essays, subdivided by the three categories of physical attractiveness and the two categories of sex similarity. The table is similar to the summary table shown for the one-way analysis of variance in Chapter 14. The difference here is that there are two independent (or grouping) variables, attractiveness and sex similarity. The first mean displayed (25.11) is for the entire sample. The number of cases (288) is shown in parentheses. Then for each of the independent variables, mean scores are displayed for each of the categories. The attractiveness categories are ordered from low (coded 1) to high (coded 3). Evaluations in which the rater and author are of the same sex are coded as 1, while opposite-sex evaluations are coded as 2. Finally, a table of means is displayed for cases classified by both grouping variables. Attractiveness is the row variable, and sex is the column variable. Each mean is based on the responses of 48 subjects.

Figure 15.1 Table of group means

```
                          * * *  C E L L   M E A N S  * * *

                SCORE
             BY ATTRACT
                SEX

TOTAL POPULATION

     25.11
  (   288)

ATTRACT
         1          2          3

     22.98      25.78      26.59
  (    96)  (    96)  (    96)

SEX
         1          2

     25.52      24.71
  (   144)  (   144)

             SEX
                    1          2
ATTRACT
         1      22.79      23.17
             (    48)  (    48)

         2      28.63      22.92
             (    48)  (    48)

         3      25.13      28.04
             (    48)  (    48)
```

The overall average score is 25.11. Highly attractive individuals received the highest average score (26.59), while those rated low in physical appeal had the lowest score (22.98). There doesn't appear to be much difference between the average scores assigned to same (25.52) and opposite-sex (24.71) individuals. Highly attractive persons received an average rating of 25.13 when evaluated by individuals of the same sex and 28.04 when evaluated by students of the opposite sex.

15.2 ANALYSIS OF VARIANCE

Three hypotheses are of interest in the study: Does attractiveness relate to the composite scores? Does sex similarity relate to the scores? And is there an interaction between the effects of attractiveness and sex? The statistical technique used to evaluate these hypotheses is an extension of the one-way analysis of variance outlined in Chapter 14. The same assumptions as before are needed for correct application: the observations should be independently selected from normal populations with equal variances. Again, discussion here is limited to the situation in which both grouping variables are considered fixed. That is, they constitute the populations of interest.

The total observed variation in the scores is subdivided into four components: the sums of squares due to attractiveness, sex, their interaction, and the residual. This can be expressed as

$$\text{TOTAL SS} = \text{ATTRACTIVENESS SS} + \text{SEX SS} + \text{INTERACTION SS} + \text{RESIDUAL SS} \quad \textbf{Equation 15.2}$$

Figure 15.2 is the analysis of variance table for this study. The first column lists the sources of variation. The sums of squares attributable to each of the components are given in the second column. The sums of squares for each independent variable alone are sometimes termed the "main effect" sums of squares. The "explained" sum of squares is the total sum of squares for the main effect and interaction terms in the model.

The degrees of freedom for sex and attractiveness, listed in the third column, are one fewer than the number of categories. For example, since there are three levels of attractiveness, there are two degrees of freedom. Similarly, sex has one degree of freedom. Two degrees of freedom are associated with the interaction term (the product of the degrees of freedom of each of the individual variables). The degrees of freedom for the residual are $N-1-k$, where k equals the degrees of freedom for the explained sum of squares.

Figure 15.2 Analysis of variance table

* * * A N A L Y S I S   O F   V A R I A N C E * * *

SCORE
BY ATTRACT
SEX

SOURCE OF VARIATION	SUM OF SQUARES	DF	MEAN SQUARE	F	SIGNIF OF F
MAIN EFFECTS	733.700	3	244.567	3.276	0.022
ATTRACT	686.850	2	343.425	4.600	0.011
SEX	46.850	1	46.850	0.628	0.429
2-WAY INTERACTIONS	942.350	2	471.175	6.311	0.002
ATTRACT SEX	942.350	2	471.175	6.311	0.002
EXPLAINED	1676.050	5	335.210	4.490	0.000
RESIDUAL	21053.140	282	74.656		
TOTAL	22729.190	287	79.196		

The mean squares shown in Figure 15.2 are obtained by dividing each sum of squares by its degrees of freedom. Hypothesis tests are based on the ratios of the mean squares of each source of variation to the mean square for the residual. When the assumptions are met and the true means are in fact equal, the distribution of the ratio is an F with the degrees of freedom for the numerator and denominator terms.

15.3 Testing for Interaction

The F value associated with the attractiveness and sex interaction is 6.311. The observed significance level is approximately 0.002. Therefore, it appears that there is an interaction between the two variables. What does this mean?

Figure 15.3a Cell means (From SPSS Graphics, a mainframe product)

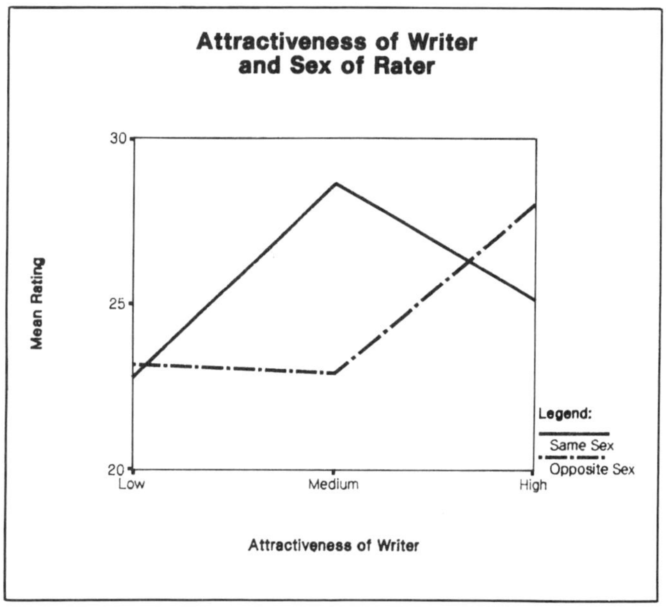

Consider Figure 15.3a, which is a plot of the cell means in Figure 15.1. Notice how the mean scores relate not only to the attractiveness of the individual and to the sex of the rater, but also to the particular combination of the values of the variables. Opposite-sex raters assign the highest scores to highly attractive individuals. Same-sex raters assign the highest scores to individuals of medium attractiveness. Thus, the ratings for each level of attractiveness depend on the sex variable. If there were no interaction between the two variables, a plot like that shown in Figure 15.3b might result, where the difference between the two types of raters is the same for the three levels of attractiveness.

Figure 15.3b Cell means with no interaction (From SPSS Graphics, a mainframe product)

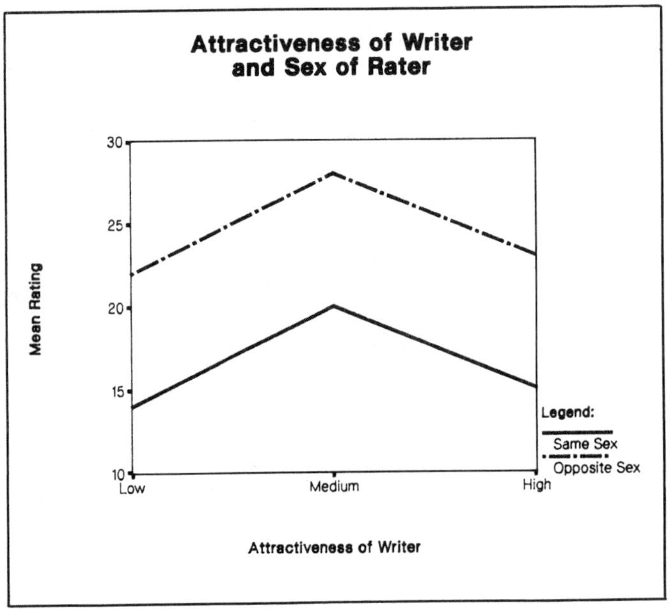

15.4 Tests for Sex and Attractiveness

Once the presence of interaction has been established, it is not particularly useful to continue hypothesis testing since the two variables *jointly* affect the dependent variable. If there is no significant interaction, the grouping variables can be tested individually. The F value associated with attractiveness would provide a test of the hypothesis that attractiveness does not affect the rating. Similarly, the F value associated with sex would test the hypothesis that sex has no main effect on evaluation.

Note that the small F value associated with sex does not indicate that response is unaffected by sex, since sex *is* included in the significant interaction term. Instead, it shows that when response is averaged over attractiveness levels, the two sex category means are not significantly different.

15.5 EXPLANATIONS

Several explanations are consistent with the results of this study. Since most people consider themselves moderately attractive, the highest degree of identification should be with same-sex individuals of moderate attractiveness. The higher empathy may result in the higher scores. An alternative theory is that moderately attractive individuals are generally perceived as more desirable same-sex friends: they have more favorable personality profiles and don't encourage unfavorable comparisons. Their writing scores may benefit from their perceived popularity.

Although we don't want friends who outshine us, handsome (and beautiful) dates provide a favorable reflection and enhance our status. Physical beauty is advantageous for heterosexual relationships, but not same-sex friendships. This prejudice may affect all evaluations of highly attractive members of the opposite sex. Regardless of the explanation, certain practical conclusions are apparent. Students, choose your instructors carefully! Authors, think twice before including your photo on the book jacket!

15.6 EXTENSIONS

Analysis of variance techniques can be used with any number of grouping variables. For example, the data in Table 15.1 originated from a more complicated experiment than described here. There were four factors—essay quality, physical attractiveness, sex of writer, and sex of subject. The original data were analyzed with a 3 × 3 × 2 × 2 ANOVA table. (The numbers indicate how many categories each grouping variable has.) The conclusions from our simplified analysis are the same as those from the more elaborate analysis.

Each of the cells in our experiment had the same number of subjects. This greatly simplifies the analysis and its interpretation. When unequal sample sizes occur in the cells, the total sum of squares cannot be partitioned into nice components that sum to the total. A variety of techniques are available for calculating sums of squares in such "non-orthogonal" designs. The methods differ in the way they adjust the sums of squares to account for other effects in the model. Each method results in different sums of squares and tests different hypotheses. However, when all cell frequencies are equal, the methods yield the same results. For discussion of various procedures for analyzing designs with unequal cell frequencies, see Kleinbaum and Kupper (1978) and Overall and Klett (1972).

15.7 RUNNING PROCEDURE ANOVA

The ANOVA procedure performs an *n*-way analysis of variance. Multiple factors (independent variables) and several dependent variables can be analyzed in one ANOVA procedure (see Command Reference for limitations). Cell means and group sizes can be displayed in addition to the analysis of variance table.

For one-way analysis of variance, the ONEWAY procedure is often preferable (see Chapter 14). Contrasts, multiple comparisons, tests for trends, and homogeneity-of-variance statistics are produced by ONEWAY but not by ANOVA.

ANOVA operates via subcommands, including the optional STATISTICS and OPTIONS subcommands. The default model is the full factorial design, although other models can be analyzed. Covariates can be specified, but ANOVA does not perform a full analysis of covariance. There are three methods available for decomposing sums of squares, and the order of entry for covariates and factor main effects can be specified. Multiple classification analysis tables are also available.

15.8 Specifying the Variables

The only required specification is the VARIABLES subcommand with a list of one or more dependent variables, the keyword BY, and one to five factors (independent variables) followed by their minimum and maximum values enclosed in parentheses and separated by a comma. The actual keyword VARIABLES can be omitted. So, the analysis of variance table in Figure 15.2 could be obtained by specifying either

```
ANOVA SCORE BY ATTRACT(1,3) SEX(1,2).
```

or

```
ANOVA VARIABLES=SCORE BY ATTRACT(1,3) SEX(1,2).
```

The values of the factors are arbitrary from a statistical point of view. They are not arbitrary, however, from a computational point of view, as they define the dimensions of the table of means and variances from which the analysis of variance is obtained. If the factor variables do not have consecutive values, they should be recoded using the RECODE command (see Command Reference) before running ANOVA. Otherwise, SPSS/PC+ may not have enough computer resources to complete the job.

15.9 Specifying Full Factorial Models

The full factorial model is the default. If two or more dependent variables are named, separate analyses of variance are produced for each. They are not analyzed jointly. Interaction terms up to fifth order are analyzed.

If two or more factor variables have the same value range, the range can be listed following the last factor it applies to, as in the command

```
ANOVA CHOL BY OVERWT AGE RACE(1,4).
```

The value-range specification need not correspond exactly to a variable's actual range of values. If the specified range is smaller than the actual range, cases with values outside the specified range are excluded from the analysis. If a range larger than the actual range is specified, however, the memory required to process the ANOVA command is needlessly increased.

Only integer-valued variables should be used as factors. If a noninteger-valued variable is included in the factor list, its values will be truncated to integers. This may result in categories that are not what you intended. The RECODE command should be used before the ANOVA command to transform noninteger factor values into integer values.

More than one design can be specified on an ANOVA command by using multiple analysis lists separated by a slashes, as in the command

```
ANOVA PRESTIGE BY INCOME(1,3) RACE SEX (1,2)
/PRSTCHNG BY INCOME(1,3) RACE SEX (1,2) AGE (1,5).
```

15.10 Requesting Cell Means and Counts

The means and counts table is displayed when Statistic 3 is specified on the STATISTICS subcommand.

Statistic 3 *Display cell means and counts for the dependent variable.* For each dependent variable, a separate table is displayed, showing the means and cell counts for each combination of factor values that define the effect (ignoring all other factors). Means of covariates are not displayed.

For example, the tables in Figure 15.1 were produced by specifying

```
ANOVA SCORE BY ATTRACT(1,3) SEX (1,2)
/STATISTICS=3.
```

Tables for the SEX, ATTRACT, and SEX by ATTRACT interaction effects are given in this output.

If Option 9 is used, this table is not available (see Section 15.13). Since Options 3 through 6 (see Section 15.11) suppress interaction terms, cell means corresponding to the suppressed interaction terms are not displayed when any of these options is used.

15.11 Suppressing Interaction Effects

By default, all interaction effects up to and including fifth-order effects are examined. You can suppress various orders of interaction effects by using the following options on the OPTIONS subcommand:

Option 3 *Suppress all interaction terms.* Only main effects and covariate effects are in the ANOVA table; the interaction sums of squares are pooled into the error (residual) sum of squares.

Option 4 *Delete all three-way and higher-order interaction terms.* The sums of squares for three-way and higher-order interactions are pooled into the error sums of squares.

Option 5 *Delete all four-way and higher-order interaction terms.* The sums of squares for four-way and higher-order interactions are pooled into the error sum of squares.

Option 6 *Delete all five-way interaction terms.* The sums of squares for the five-way and higher-order interactions are pooled into the error sum of squares.

ANOVA will not examine six-way or higher-order interactions, and the sums of squares for such terms are always pooled into the error sum of squares.

15.12 Specifying Covariates

List covariates after the factor list, following the keyword WITH, as in the command

```
ANOVA SCORE BY ATTRACT(1,3) WITH SELFATRT.
```

By default, the covariates are assessed first, with main effects assessed after adjusting for the covariates. You can specify the order in which blocks of covariates and factor main effects are assessed with the following options:

Option 7 *Process covariates concurrently with main effects for factors.*

Option 8 *Process covariates after the main effects of the factors have been included.*

Note that the order of entry is irrelevant if Option 9 (the regression approach) is specified (see Section 15.13).

Unstandardized regression coefficients for the covariates are displayed if Statistic 2 is specified on the STATISTICS subcommand.

Statistic 2 *Display unstandardized regression coefficients for covariates.* These regression coefficients are computed when the covariates are entered into the equation. Their values depend on the design specified by Options 7 through 10 or by default. (See Section 15.13 for a discussion of Options 9 and 10.) These coefficients are displayed immediately below the analysis of variance table.

15.13 Decomposing Sums of Squares

Three methods are available for decomposing sums of squares. The default method is the *classical experimental approach,* in which each type of effect is assessed separately in the following order (unless Option 7 or 8 has been specified):

- Effects of covariates.
- Main effects of factors.
- Two-way interaction effects.
- Three-way interaction effects.
- Four-way interaction effects.
- Five-way interaction effects.

Each effect is adjusted for all other effects of that type and for all previously entered effects.

The second method, requested by specifying Option 9 on the OPTIONS subcommand, is the *regression approach.*

Option 9 *Regression approach.* All effects are assessed simultaneously, with each effect adjusted for all other effects in the model.

The following restrictions apply when this method is used:

- For each independent variable, at least one case must have the lowest value of the independent variable, since the lowest values are used as the reference category. If this does not hold, no ANOVA table is produced and a message identifying the first problematic variable is displayed.
- When the independent variables are crosstabulated, no cell defined by the smallest value of any independent variable can be empty. If this restriction is violated, one or more orders of interaction effects are suppressed, and a warning message is displayed. This constraint does not apply to cells defined by values of independent variables that do not occur in the data. For example, if two independent variables each have the values 1, 2, and 4, the (1,1), (1,2), (2,1), (4,1), and (1,4) cells can not be empty. The (1,3), (2,3), (3,3), (4,3), (3,1), (3,2), and (3,4) cells are automatically empty, however, and this does not violate the restriction. The (2,2), (2,4), (4,2), and (4,4) cells can be empty, although the degrees of freedom will be reduced accordingly if they are.

The lowest nonempty value of each independent variable should therefore be specified in the value range. Specifying a value range of (0,9) for a variable that actually has only the values 1 through 9 results in a violation of both restrictions, and no ANOVA table will be produced.

The *hierarchical approach* is the third method available for decomposing sums of squares. This method differs from the classical experimental approach only in the way covariates and factor main effects are treated.

Option 10 *Hierarchical approach.* Factor main effects and covariate effects are assessed hierarchically. The order in which they are assessed is determined by the order in which they are listed on the ANOVA command. The factor main effects are adjusted only for the factor main effects already assessed, and the covariate effects are adjusted only for the covariates already assessed.

To understand the three approaches, consider the command

```
ANOVA TESTSCOR BY SEX(1,2) MTVN(1,5) ATT(1,3).
```

Table 15.13 summarizes the way each of the three approaches would analyze the design. In the default classical experimental approach, each main effect is assessed with the two other main effects held constant. The two-way interactions are assessed with all main effects and all other two-way interactions held constant, and the three-way interaction is assessed with all main effects and two-way interactions held constant. The regression approach assesses each factor or interaction while holding all other factors and interactions constant. In the hierarchical approach, the order in which the factors are listed on the ANOVA command determines the order in which they are assessed.

Table 15.13 Terms adjusted for in each approach

Effect	Classical approach (default)	Regression approach (Option 9)	Hierarchical approach (Option 10)
SEX	MTVN, ATT	All others	None
MTVN	SEX, ATT	All others	SEX
ATT	SEX, MTVN	All others	SEX, MTVN
SEX*MTVN	SEX, MTVN, ATT, SEX*ATT, MTVN*ATT	All others	SEX, MTVN, ATT, SEX*ATT, MTVN*ATT
SEX*ATT	SEX, MTVN, ATT, SEX*MTVN, MTVN*ATT	All others	SEX, MTVN, ATT, SEX*MTVN, MTVN*ATT
MTVN*ATT	SEX, MTVN, ATT, SEX*MTVN, SEX*ATT	All others	SEX, MTVN, ATT, SEX*MTVN, SEX*ATT
SEX*MTVN*ATT	SEX, MTVN, ATT, SEX*MTVN, SEX*ATT, MTVN*ATT	All others	SEX, MTVN, ATT, SEX*MTVN, SEX*ATT, MTVN*ATT

15.14 Summary of Analysis Methods

Various methods for decomposing sums of squares and specifying the order of entry of covariates have been discussed. Table 15.14 summarizes the types of analyses produced when combinations of these methods are used.

Table 15.14 Combinations of Options 7 through 10

	Assessments between types of effects	Assessments within the same type of effect
Default	**Covariates** *then* **Factors** *then* **Interactions**	**Covariates:** Adjust for all other covariates **Factors:** Adjust for covariates and all other factors **Interactions:** Adjust for covariates, factors, and all other interactions of the same and lower orders
Option 7	**Factors** **Covariates** concurrently *then* **Interactions**	**Covariates:** Adjust for factors and all other covariates **Factors:** Adjust for covariates and all other factors **Interactions:** Adjust for covariates, factors, and all other interactions of the same and lower orders
Option 8	**Factors** *then* **Covariates** *then* **Interactions**	**Factors:** Adjust for all other factors **Covariates:** Adjust for factors and all other covariates **Interactions:** Adjust for covariates, factors, and all other interactions of the same and lower orders
Option 9	**Covariates** **Factors** **Interactions** simultaneously	**Covariates:** Adjust for factors, interactions, and all other covariates **Factors:** Adjust for covariates, interactions, and all other factors **Interactions:** Adjust for covariates, factors, and all other interactions
Option 10	**Covariates** *then* **Factors** *then* **Interactions**	**Covariates:** Adjust for preceding covariates **Factors:** Adjust for covariates and preceding factors **Interactions:** Adjust for covariates, factors, and all other interactions of the same and lower orders
Options 7 and 10	**Factors** **Covariates** concurrently *then* **Interactions**	**Factors:** Adjust only for preceding factors **Covariates:** Adjust for factors and preceding covariates **Interactions:** Adjust for covariates, factors, and all other interactions of the same and lower orders
Options 8 and 10	**Factors** *then* **Covariates** *then* **Interactions**	**Factors:** Adjust only for preceding factors **Covariates:** Adjust for factors and preceding covariates **Interactions:** Adjust for covariates, factors, and all other interactions of the same and lower orders

15.15 Multiple Classification Analysis

Multiple classification output consists of the grand mean of the dependent variables, a table of deviations from the grand mean for each factor level (*treatment effects*), and several measures of association. The deviations indicate the magnitude of the effect of each factor level. To obtain this output, specify Statistic 1 on the STATISTICS subcommand.

Statistic 1 *Display multiple classification analysis (MCA) table.* Three types of deviations are displayed: unadjusted deviations, deviations adjusted for the main effects of other factors if covariates are used, and deviations adjusted for the main effects of other factors and for covariates. The adjusted deviations show the effect of a particular factor level after variation due to other factors (or to other factors and covariates) has been taken into account.

The form of analysis specified by Options 7 through 10 and their defaults affects the results in the MCA table. If covariates are used, a complete table can be obtained only if Option 8, Options 8 and 10, or Options 7 and 10 are used. If a model in which factors are not assessed first is specified, deviations adjusted only for factors are not displayed in the MCA table. If Option 9 is used, the MCA table is not produced. Ordinarily, this table is of interest only when no interaction terms are significant.

Several measures of association are also displayed in the MCA table. For each factor in the table, the eta statistic is calculated. Its squared value indicates the proportion of variance "explained" by a given factor. Standardized regression coefficients, called BETA's are also displayed. Finally, multiple R and R^2 are at the bottom of the table. R^2 indicates the proportion of variance in the dependent variable "accounted for" by all factors, covariates, and interaction terms.

15.16 Missing Values

By default, a case with missing values for any variables in the analysis list is excluded from all analyses specified by that list. With Option 1, cases with user-defined missing data are included in the analysis.

Option 1 *Include user-missing values.*

15.17 Formatting Options

By default, variable and value labels (if defined) are included in the output. The width of the output defaults to the width defined on SET (see Command Reference: SET). Two options are available for controlling the format of the output:

Option 2 *Suppress variable and value labels.*

Option 11 *Use narrow format.*

15.18 Annotated Example

The following SPSS/PC+ commands were used to obtain the analysis in this chapter:

```
DATA LIST  / ATTRACT 1-2 SEX 3 SCORE 4-5.
RECODE ATTRACT (1=1) (5=2) (10=3).
VARIABLE LABELS
  ATTRACT 'ATTRACTIVENESS LEVEL'
  /SEX 'SEX SIMILARITY'
  /SCORE 'COMPOSITE SCORE'.
VALUE LABELS
  ATTRACT 1 'LOW' 2 'MEDIUM' 3 'HIGH'
  /SEX 1 'SAME' 2 'OPPOSITE'.
BEGIN DATA.
lines of data
END DATA.
ANOVA  SCORE BY ATTRACT(1,3) SEX(1,2)
      /STATISTICS=3.
FINISH.
```

- The DATA LIST command indicates that the data for each case is on a single record, with ATTRACT in columns 1 and 2, SEX in column 3, and SCORE in columns 4 and 5.
- The RECODE command changes the original values of ATTRACT (1, 5, and 10) to the consecutive values 1, 2, and 3, so that ANOVA processing will be more efficient.
- The VARIABLE LABELS command assigns variable labels, and the VALUE LABELS command defines labels for the new values of ATTRACT and for the values of SEX.
- The ANOVA command names SCORE as the dependent variable and ATTRACT and SEX as the factors. The minimum and maximum values for ATTRACT are 1 and 3, and for SEX, 1 and 2.
- The STATISTICS subcommand displays group means and counts for SCORE.

Contents

Chapter 16 Nonparametric Tests: Procedure NPAR TESTS

Coffee and carrots have recently joined saccharin, tobacco, Laetrile, and interferon on the ever-expanding list of rumored causes of and cures for cancer. This list is necessarily tentative and complicated. The two major sources of evidence—experiments on animals and examination of the histories of afflicted persons—are fraught with problems. It is difficult to predict, based on large doses of suspect substances given to small animals, the consequences of small amounts consumed by humans over a long time span.

In studies of people, lifestyle components are difficult to isolate, and it is challenging—if not impossible—to unravel the contribution of a single factor. For example, what is the role of caffeine based on a sample of overweight, sedentary, coffee- and alcohol-drinking, cigarette-smoking, urban dwellers?

Nutrition is also thought to be an important component in cancer development and progression. For example, the per capita consumption of dietary fats is positively correlated with the incidence of mammary and colon cancer in humans (Wynder, 1976). In a recent study, King et al. (1979) examined the relationship between diet and tumor development in rats. Three groups of animals of the same age, species, and physical condition were injected with tumor cells. The rats were divided into three groups and fed diets of either low, saturated, or unsaturated fat.

One hypothesis of interest is whether the length of time until a tumor develops in rats fed saturated diets differs from the length of time in rats fed unsaturated diets. If it is tenable to assume that tumor-free time is normally distributed, the two-sample *t* test described in Chapter 11 can be used to test the hypothesis that the population means are equal. However, if the distribution of times does not appear to be normal, and especially if the sample sizes are small, statistical procedures that do not require assumptions about the shapes of the underlying distributions should be considered.

16.1 THE MANN-WHITNEY TEST

The *Mann-Whitney test*, also known as the Wilcoxon test, does not require assumptions about the shape of the underlying distributions. It tests the hypothesis that two independent samples come from populations having the same distribution. The form of the distribution need not be specified. The test does not require that the variable be measured on an interval scale; an ordinal scale is sufficient.

16.2 Ranking the Data

To compute the test, the observations from both samples are first combined and ranked from smallest to largest. Consider Table 16.2, which shows a sample of the King data reported by Lee (1979). Case 4 has the shortest elapsed time to development of a tumor, 68 days. It is assigned a rank of 1. The next shortest time is for Case 3, so it is assigned a rank of 2. Cases 5 and 6 both exhibited tumors on the same day. They are both assigned a rank of 3.5, the average of the ranks (3 and 4) for which they are tied. Case 2, the next largest, is given a rank of 5, and Case 1 is given a rank of 6.

Table 16.2 Ranking the data

Saturated			Unsaturated		
Case	Time	Rank	Case	Time	Rank
1	199	6	4	68	1
2	126	5	5	112	3.5
3	81	2	6	112	3.5

16.3 Calculating the Test

The statistic for testing the hypothesis that the two distributions are equal is the sum of the ranks for each of the two groups. If the groups have the same distribution, their sample distributions of ranks should be similar. If one of the groups has more than its share of small or large ranks, there is reason to suspect that the two underlying distributions are different.

Figure 16.3 shows the output from the Mann-Whitney test for the complete King data. For each group, the mean rank and number of cases is given. (The mean rank is the sum of the ranks divided by the number of cases.) Note that the unsaturated-diet group has only 29 cases since one rat died of causes unrelated to the experiment. The entry displayed under *W* is the sum of the ranks for the group with the smaller number of observations. If both groups have the same number of observations, *W* is the rank sum for the group named first on the NPAR TESTS command. For this example, *W* is 963, the sum of the ranks for the saturated-diet group.

Figure 16.3 Mann-Whitney output

```
- - - - - Mann-Whitney U - Wilcoxon Rank Sum W Test

    TUMOR
 by DIET

     Mean Rank     Cases

        26.90         30  DIET = 0  UNSATURATED
        33.21         29  DIET = 1  SATURATED
                      --
                      59  Total

                                         Corrected for Ties
           U                W               Z      2-tailed P
       342.0            963.0         -1.4112         0.1582
```

The number identified on the output as U is the number of times a value in the unsaturated-diet group precedes a value in the saturated-diet group. To understand what this means, consider the data in Table 16.2 again. All three cases in the unsaturated-diet group have smaller ranks than the first case in the saturated-diet group, so they all precede Case 1 in the rankings. Similarly, all three cases in the unsaturated-diet group precede Case 2. Only one unsaturated-diet case (Case 4) is smaller in value than Case 3. Thus, the number of times the value for an unsaturated-diet case precedes the value for a saturated-diet case is 3+3+1=7. The number of times the value of a saturated-diet case precedes the value of an unsaturated-diet case is 2, since Case 3 has a smaller rank than both Cases 5 and 6. The smaller of these two numbers is displayed on the output as U. If the two distributions are equal, values from one group should not consistently precede values in the other.

The significance levels associated with U and W are the same. They can be obtained by transforming the score to a standard normal deviate (Z). If the total sample size is less than 30, an exact probability level based on the distribution of the score is also displayed. From Figure 16.3, the observed significance level for this example is 0.158. Since the significance level is large, the hypothesis that tumor-free time has the same distribution for the two diet groups is not rejected.

16.4 Which Diet?

You should not conclude from these findings that it doesn't matter—as far as tumors are concerned—what kind of fat you (or rats) eat. King et al. found that rats fed the unsaturated diet had a total of 96 tumors at the end of the experiment, while rats fed the saturated diet had only 55 tumors. They also found that large tumors were more common in the unsaturated-diet group than in the saturated-diet group. Thus, unsaturated fats may be more hazardous than saturated fats.

16.5 Assumptions

The Mann-Whitney test requires only that the observations be a random sample and that values can be ordered. These assumptions, especially randomness, are not to be made lightly, but they are less restrictive than those for the two-sample *t* test for means. The *t* test further requires that the observations be selected from normally distributed populations with equal variances. (An approximate test for the case of unequal variances is presented in Chapter 11).

Since the Mann-Whitney test can always be calculated instead of the *t* test, what determines which should be used? If the assumptions needed for the *t* test are met, the *t* test is more powerful than the Mann-Whitney test. That is, the *t* test will detect true differences between the two populations more often than will the Mann-Whitney test since the *t* test uses more information from the data. Substitution of ranks for the actual values loses potentially useful information. On the other hand, using the *t* test when its assumptions are substantially violated may result in an erroneous observed significance level.

In general, if the assumptions of the *t* test appear reasonable, it should be used. When the data are ordinal—or interval but from a markedly nonnormal distribution—the Mann-Whitney test is the procedure of choice.

16.6 NONPARAMETRIC TESTS

Many statistical procedures, like the Mann-Whitney test, require limited distributional assumptions about the data. Collectively these procedures are termed *distribution-free* or *nonparametric tests.* Like the Mann-Whitney test, distribution-free tests are generally less powerful than their parametric counterparts. They are most useful in situations where parametric procedures are not appropriate: when the data are nominal or ordinal, or when interval data are from markedly nonnormal distributions. Significance levels for certain nonparametric tests can be determined regardless of the shape of the population distribution since they are based on ranks.

In the following sections, various nonparametric tests will be used to reanalyze some of the data described in previous chapters. Since the data were chosen to illustrate the parametric procedures, they satisfy assumptions that are more restrictive than those for nonparametric procedures. However, they provide an opportunity for learning new procedures with familiar data and for comparing results from different types of analyses.

16.7 One-Sample Tests

Various one-sample nonparametric procedures are available for testing hypotheses about the parameters of a population. These include procedures for examining differences in paired samples.

16.8 The Sign Test

In Chapter 11, the paired *t*-test for means is used to test the hypothesis that mean buying scores for husbands and wives are equal. Remember that this test requires the assumption that differences are normally distributed.

The *sign test* is a nonparametric procedure used with two related samples to test the hypothesis that the distributions of two variables are the same. This test makes no assumptions about the shape of these distributions.

To compute the sign test, the difference between the buying scores of husbands and wives is calculated for each case. Next, the numbers of positive and negative differences are obtained. If the distributions of the two variables are the same, the numbers of positive and negative differences should be similar.

Figure 16.8 Sign test output

```
- - - - - Sign Test

      HSSCALE   HUSBAND SELF SCALE
 with WSSCALE   WIFE SELF SCALE


          Cases

            56  - Diffs (WSSCALE Lt HSSCALE)                Z =      1.6416
            39  + Diffs (WSSCALE Gt HSSCALE)
             3    Ties                             2-tailed P =       .1007
            ---
            98    Total
```

The output in Figure 16.8 shows that the number of negative differences is 56, while the number of positive differences is 39. The total number of cases is 98, including three with no differences. The observed significance level is .1007. Since this value is large, the hypothesis that the distributions are the same is not rejected.

16.9 The Wilcoxon Signed-Ranks Test

The sign test uses only the direction of the differences between the pairs and ignores the magnitude. A discrepancy of 15 between husbands' and wives' buying scores is treated in the same way as a discrepancy of 1. The *Wilcoxon signed-ranks test* incorporates information about the magnitude of the differences and is therefore more powerful than the sign test.

To compute the Wilcoxon signed-ranks test, the differences are ranked ignoring the signs. In the case of ties, average ranks are assigned. The sums of the ranks for positive and negative differences are then calculated.

Figure 16.9 Wilcoxon signed-ranks test output

```
- - - - - Wilcoxon Matched-pairs Signed-ranks Test

     HSSCALE   HUSBAND SELF SCALE
with WSSCALE   WIFE SELF SCALE

        Mean Rank     Cases

          45.25         56  - Ranks (WSSCALE Lt HSSCALE)
          51.95         39  + Ranks (WSSCALE Gt HSSCALE)
                         3    Ties  (WSSCALE Eq HSSCALE)
                        ---
                        98    Total

        Z =     -.9428                   2-tailed P =  .3458
```

From Figure 16.9, the average rank of the 56 negative differences is 45.25. The average positive rank is 51.95. The three cases that are tied have the same value for both variables. The observed significance level associated with the test is large (.3458), so, again, the hypothesis of no difference is not rejected.

16.10 The Wald-Wolfowitz Runs Test

The runs test is a test of randomness. That is, given a sequence of observations, the runs test examines whether the value of one observation influences the values taken by later observations. If there is no influence (the observations are independent), the sequence is considered random.

A *run* is any sequence of like observations. For example, if a coin is tossed fifteen times and the outcomes recorded, the following sequence might result:

HHHTHHHHTTTTTTT

There are four runs in this sequence: HHH, T, HHHH, and TTTTTTT. The total number of runs is a measure of randomness, since too many runs, or too few, suggest dependence between observations. The Wald-Wolfowitz runs test converts the total number of runs into a *Z* statistic having approximately a normal distribution. The only requirement for this test is that the variable tested be dichotomous (have only two possible values).

Suppose, for example, that a weather forecaster records whether it snows for twenty days in February and obtains the following sequence (1=snow, 0=no snow):

01111111010111111100

To test the hypothesis that the occurrence or nonoccurrence of snow on one day has no effect on whether it snows on later days, the runs test is performed, resulting in the output in Figure 16.10.

Figure 16.10 Runs test

```
- - - - - Runs Test

     SNOW

          Runs:    7          Test Value = 1

          Cases:   5   Lt 1
                  15   Ge 1              Z =  -.6243
                  --
                  20   Total 2-tailed P =   .5324
```

Since the observed significance level is quite large (.5324), the hypothesis of randomness is not rejected. It does not appear, from these data, that snowy (or nonsnowy) days affect the later occurrence of snow.

16.11 The Binomial Test

With data that are binomially distributed, the hypothesis that the probability *p* of a particular outcome is equal to some number is often of interest. For example, you might want to find out if a tossed coin was unbiased. To check this, you could test to see whether the probability of heads was equal to 1/2. The binomial test compares the observed frequencies in each category of a binomial distribution to the frequencies expected under a binomial distribution with the probability parameter *p*.

For instance, a nickel is tossed twenty times, with the following results (1=heads, 0=tails):

10011111101111011011

The output in Figure 16.11 shows a binomial test of the hypothesis that the probability of heads equals 1/2, for these data.

Figure 16.11 Binomial test

```
- - - - - Binomial Test

   HEADS

    Cases

       5    = 0              Test Prob. =    .5000
      15    = 1
      --                     2-tailed P =    .0207
      20    Total
```

The small (.0207) observed significance level indicates that it is not likely that *p* equals 1/2 and it appears that the coin is biased.

16.12 The Kolmogorov-Smirnov One-Sample Test

The Kolmogorov-Smirnov test is used to determine how well a random sample of data fits a particular distribution (uniform, normal, or Poisson). It is based on comparison of the sample cumulative distribution function to the hypothesized cumulative distribution function.

This test can be used with the beer data (see Chapter 2) to see whether it is reasonable to assume that the ALCOHOL variable is normally distributed. The Kolmogorov-Smirnov output in Figure 16.12 shows an observed significance level of .05, small enough to cast doubt on the assumption of normality.

Figure 16.12 Kolmogorov-Smirnov test

```
- - - - - Kolmogorov - Smirnov Goodness of Fit Test

   ALCOHOL   ALCOHOL BY VOLUME (IN %)

   Test Distribution   -  Normal                    Mean:  4.577
                                     Standard Deviation:    .603

          Cases:  35

          Most Extreme Differences
     Absolute        Positive       Negative          K-S Z        2-tailed P
      0.22941         0.15585       -0.22941          1.357           0.050
```

16.13 The One-Sample Chi-Square Test

In Chapter 7, frequencies of deaths for the days of the week are examined. The FREQUENCIES output suggests that all days of the week are equally hazardous in regard to death. To test this conclusion, the *one-sample chi-square test* can be used. This nonparametric test requires only that the data be a random sample.

To calculate the one-sample chi-square statistic, the data are first classified into mutually exclusive categories of interest—days of the week in this example—and then expected frequencies for these categories are computed. Expected frequencies are the frequencies that would be expected if the null hypothesis is true. For the death data, the hypothesis to be tested is that the probability of death is the same for each day of the week. The day of death is known for 110 subjects. The hypothesis implies that the expected frequency of deaths for each weekday is 110 divided by 7, or 15.71. Once the expected frequencies are obtained, the chi-square statistic is computed as

$$\chi^2 = \sum_{i=1}^{k} (O_i - E_i)^2 / E_i \qquad \textbf{Equation 16.13}$$

where O_i is the observed frequency for the ith category, E_i is the expected frequency for the ith category, and k is the number of categories.

Figure 16.13 One-sample chi-square output

- - - - - Chi-square Test

DAYOFWK DAY OF DEATH

	Category	Cases Observed	Expected	Residual
SUNDAY	1.00	19	15.71	3.29
MONDAY	2.00	11	15.71	-4.71
TUESDAY	3.00	19	15.71	3.29
WEDNSDAY	4.00	17	15.71	1.29
THURSDAY	5.00	15	15.71	-.71
FRIDAY	6.00	13	15.71	-2.71
SATURDAY	7.00	16	15.71	.29
	Total	110		

Chi-Square	D.F.	Significance
3.400	6	0.757

If the null hypothesis is true, the chi-square statistic has approximately a chi-square distribution with $k-1$ degrees of freedom. This statistic will be large if the observed and expected frequencies are substantially different. Figure 16.13 shows the output from the one-sample chi-square test for the death data. The observed chi-square value is 3.4, and the associated significance level is 0.757. Since the observed significance level is large, the hypothesis that deaths are evenly distributed over days of the week is not rejected.

16.14 The Friedman Test

The Friedman test is used to compare two or more related samples. (This is an extension of the problem of paired data.) The k variables to be compared are ranked from 1 to k for each case, and the mean ranks for the variables are calculated and compared, resulting in a test statistic with approximately a chi-square distribution.

The Friedman test can be used to analyze data from a psychology experiment concerned with memory. In this experiment, subjects were asked to memorize, first, a two-digit number, then a three-digit number, and, finally, a four-digit number. After each number was memorized, they were shown a single digit and asked if that digit were present in the number memorized. The times taken to reach a decision for the two-, three-, and four-digit numbers are the three related variables of interest.

Figure 16.14 shows the results of the Friedman test, examining the hypothesis that the number of digits memorized has no effect on the time taken to reach a decision. The observed significance level is extremely small, so it appears that the number of digits does affect decision time.

Figure 16.14 Friedman test

```
- - - - - Friedman Two-way ANOVA

     Mean Rank    Variable

          1.21    P2DIGIT
          2.13    P3DIGIT
          2.67    P4DIGIT

          Cases           Chi-Square           D.F.   Significance
            24              26.0833              2          .0000
```

16.15 Tests for Two or More Independent Samples

A variety of nonparametric tests involve comparisons between two or more independent samples. (The Mann-Whitney test is one such test.) In this respect, these tests resemble the *t*-tests and one-way analyses of variance described in Chapters 11 and 14.

16.16 The Two-Sample Median Test

The two-sample median test is used to determine whether two populations have the same median. The two samples are combined and the median for the total distribution is calculated. The number of observations above this median, as well as the number of observations less than or equal to this median, are counted for each sample. The test statistic is based on these counts.

This test can be used to determine whether median sodium levels are the same for the highest-rated and lowest-rated beers in the beer data. The output in Figure 16.16 shows the largest possible p value, 1. Therefore, there is no reason to suspect different medians.

Figure 16.16 Median test of sodium by rating

```
- - - - - Median Test

    SODIUM     SODIUM PER 12 FLUID OUNCES IN MG
  by RATING

                                 RATING
                                    3          1
                       I----------I----------I
             Gt Median I        4 I        5 I
  SODIUM               I----------I----------I
             Le Median I        6 I        6 I
                       I----------I----------I

             Cases          Median              Exact Probability
               21             15                          1.0000
```

16.17 The Two-Sample Wald-Wolfowitz Runs Test

A runs test can be used to test the hypothesis that two samples come from populations with the same distributions. To perform this test, the two samples are combined and the values sorted. A run in this combined and sorted sample consists of a sequence of values belonging to the first sample or a sequence of values belonging to the second sample. If there are too few runs, it suggests that the two populations have different distributions.

The Wald-Wolfowitz test can be used with the beer data to compare calories for the highest-ranked and lowest-ranked beers. The output in Figure 16.17 shows an observed significance level of .0119. Since this is small, the distribution of calories for the highest-ranked beers appears to differ from the distribution of calories for the lowest-ranked beers.

Figure 16.17 Wald-Wolfowitz runs test

```
- - - - - Wald-Wolfowitz Runs Test

   CALORIES  CALORIES PER 12 FLUID OUNCES
 by RATING

        Cases

          10  RATING = 3  FAIR
          11  RATING = 1  VERY GOOD
          --
          21  Total
                                                        Exact
                         Runs              Z        1-tailed P
    Minimum Possible:      6         -2.2335           .0119
    Maximum Possible:      6         -2.2335           .0119

    WARNING -- There are    1 Inter-group Ties involving    5 cases.
```


16.18 The Two-Sample Kolmogorov-Smirnov Test

The Kolmogorov-Smirnov test for two samples provides another method for testing whether two samples come from populations with the same distributions. It is based on a comparison of the distribution functions for the two samples.

This test can be used with the beer data, to compare the alcohol content of the highest-ranked and lowest-ranked beers. Since the observed significance level in Figure 16.18 is small, the alcohol distributions do not appear to be the same. The approximation used to obtain the observed significance level may be inadequate in this case, however, because of the small sample size.

Figure 16.18 Kolmogorov-Smirnov two-sample test

```
- - - - - Kolmogorov - Smirnov 2-Sample Test

   ALCOHOL   ALCOHOL BY VOLUME (IN %)
 by RATING

        Cases

          10  RATING = 3  FAIR
          11  RATING = 1  VERY GOOD
          --
          21  Total

WARNING - Due to small sample size, probability tables should be consulted.

           Most Extreme Differences
     Absolute      Positive       Negative         K-S Z      2-tailed P
     0.60000       0.0            -0.60000         1.373           0.046
```

16.19 The *k*-Sample Median Test

An extension of the two-sample median test, the *k*-sample median test compares the medians of three or more independent samples. Figure 16.19 shows a *k*-sample median test comparing median prices for the highest-, middle-, and lowest-quality beers. The observed significance level is fairly large (.091), indicating no real difference in the median price of the three types of beer.

Figure 16.19 k-sample median test

```
- - - - - Median Test

   PRICE     PRICE PER 6-PACK
 by RATING

                                  RATING
                           1           2           3
                  I-----------I-----------I-----------I
       Gt Median  I         8 I         5 I         3 I
 PRICE            I-----------I-----------I-----------I
       Le Median  I         3 I         9 I         7 I
                  I-----------I-----------I-----------I

       Cases          Median     Chi-Square     D.F.   Significance
        35             2.65        4.7937         2        .0910
```

16.20 The Kruskal-Wallis Test

An experiment described earlier in this chapter investigates the effects of three diets on tumor development in rats. The Mann-Whitney test was calculated to examine the different effects of saturated and unsaturated diets. To test for differences between all three diets, an extension of the Mann-Whitney test can be used. This test is known as the *Kruskal-Wallis one-way analysis of variance*.

The procedure for computing the Kruskal-Wallis test is similar to that used in the Mann-Whitney test. All cases from the groups are combined and ranked. Average ranks are assigned in the case of ties. For each group, the ranks are summed, and the Kruskal-Wallis H statistic is computed from these sums. The H statistic has approximately a chi-square distribution under the hypothesis that the three groups have the same distribution.

Figure 16.20 Kruskal-Wallis one-way analysis of variance output

```
- - - - - Kruskal-Wallis 1-way ANOVA

   TUMOR
by DIET

   Mean Rank    Cases

      34.12       30   DIET = 0    UNSATURATED
      43.50       29   DIET = 1    SATURATED
      56.24       29   DIET = 2    LOW-FAT
                  --
                  88   Total

                                                 Corrected for Ties
      CASES     Chi-Square  Significance     Chi-Square  Significance
         88       11.1257        0.0038        11.2608        0.0036
```

The output in Figure 16.20 shows that the third group, the low-fat-diet group, has the largest average rank. The value of the Kruskal-Wallis statistic (labeled Chi-Square) is 11.1257. When the statistic is adjusted for the presence of ties, the value changes to 11.2608 (labeled Corrected Chi-Square). The small observed significance level suggests that the development time for a tumor is not the same for all three groups.

16.21 RUNNING PROCEDURE NPAR TESTS

You can perform all of the nonparametric tests discussed in this chapter, and several more, with procedure NPAR TESTS. (See Siegel, 1956, for further information about these tests.) In addition to obtaining test statistics, you can request additional statistics, specify missing-value treatments, and use a random subsample of your data for NPAR TESTS. The tests available are summarized in Table 16.21.

Table 16.21 Nonparametric tests available in NPAR TESTS

Data organization	Nominal scale	Ordinal scale
One sample	Chi-square Runs Binomial	Kolmogorov-Smirnov
Two related samples	McNemar	Sign Wilcoxon
k related samples	Cochran's Q	Friedman Kendall
Two independent samples		Median Mann-Whitney Kolmogorov-Smirnov Wald-Wolfowitz Moses
k independent samples		Median Kruskal-Wallis

NPAR TESTS operates via subcommands, including the OPTIONS and STATISTICS subcommands. The general format for an NPAR TESTS command is

```
NPAR TESTS testname[(parameters)]=varlist
```

Each subcommand requests a specific test and lists the variables to be tested. The equals sign is optional. You can use the TO keyword to refer to adjacent variables in the file. The form of the variable list depends on the type of test specified. More than one test may be requested on one NPAR TESTS command. (See Command Reference for limitations.)

16.22 One-Sample Tests

The one-sample tests available in NPAR TESTS are the one-sample chi-square tests, the Kolmogorov-Smirnov test, the runs test, and the binomial test.

16.23 The One-Sample Chi-Square Test

Use the CHISQUARE subcommand to obtain a one-sample chi-square test. The format is:

```
NPAR TESTS CHISQUARE=varlist[(lo,hi)]
```

The range following the variable list is optional. If it is not specified, each distinct value of the variable named is treated as a separate category. If the range is specified, noninteger values are truncated to integers, resulting in categories with only integer values. Cases with values outside the specified range are excluded from the analysis.

By default, the expected frequencies are assumed to be equal. You can specify other frequencies, however, by using the EXPECTED subcommand, which has the format

```
EXPECTED=f1,f2,...,fn
```

where *f1* through *fn* are the expected frequencies to be used. You must specify a frequency greater than zero for each category of the variable. The values listed in EXPECTED are treated as proportions rather than actual numbers of cases expected. That is, the values are summed, and each value is then divided by this sum to calculate the proportion of cases expected in the corresponding category. For example, the command

```
NPAR TESTS CHISQUARE=FLOWERS
/EXPECTED=1,2,2,2,1.
```

specifies 1/8, 2/8, 2/8, 2/8, and 1/8 as the expected proportions for the values of FLOWERS.

The EXPECTED subcommand applies to all variables listed in the preceding CHISQUARE subcommand. If you want to specify different expected frequencies for each variable, use separate CHISQUARE and EXPECTED subcommands for each variable. Several CHISQUARE and EXPECTED subcommands can also be used to test different expected frequencies for the same variable.

The output in Figure 16.13 was produced with the command:

```
NPAR TESTS  CHI-SQUARE=DAYOFWK.
```

For each chi-square test, the output consists of the observed and expected numbers of cases in each category, the residual (observed minus expected) for each category, and the chi-square statistic with its degrees of freedom and observed significance level.

16.24 The Kolmogorov-Smirnov One-Sample Test

Use the K-S subcommand to obtain the Kolmogorov-Smirnov test for one sample. The distributions you can test against are the uniform, normal, and Poisson distributions.

The K-S format is

```
NPAR TESTS K-S(dis[parameters])=varlist
```

where *dis* is one of the distribution keywords UNIFORM, NORMAL, or POISSON. Each of these keywords has optional parameters which are separated from the keyword and each other by commas:

UNIFORM *Uniform distribution.* The optional parameters are the minimum and maximum values (in that order). If these are not specified, K-S uses the sample minimum and maximum values.

NORMAL *Normal distribution.* The optional parameters are the mean and standard deviation (in that order). If these are not specified, K-S uses the sample mean and standard deviation.

POISSON *Poisson distribution.* The one optional parameter is the mean. If this is not specified, K-S uses the sample mean. A word of caution about testing against a Poisson distribution: if the mean of the test distribution is large, evaluating the probabilities is a very time consuming process. If a mean of 100,000 or larger is used, K-S uses a normal approximation to the Poisson distribution.

For example, the command

```
NPAR TESTS K-S(POISSON,5)=HORSKICK.
```

compares the sample distribution of the HORSKICK variable with the Poisson distribution having a mean of 5. The output in Figure 16.12 was produced with the command:

```
NPAR TESTS K-S(NORMAL) = ALCOHOL.
```

The output for the Kolmogorov-Smirnov test includes the distribution used for the test, the most extreme positive, negative, and absolute differences, and the Kolmogorov-Smirnov *Z* with its observed significance level.

The Kolmogorov-Smirnov test assumes that the parameters of the test distribution have been specified in advance (not calculated from the data). When the parameters are estimated from the sample, the distribution of the test statistic changes. SPSS/PC+ makes no corrections for this.

16.25 The Runs Test

Use subcommand RUNS to obtain the runs test. This subcommand has the general format

```
NPAR TESTS RUNS(cutpoint)=varlist
```

The cutting point dichotomizes the variables in the variable list. Even if a variable is dichotomous to begin with, a cutting point must be specified. For example, a variable that takes only the values 0 and 1 would have 1 for the cutting point, as in the command

```
NPAR TESTS RUNS(1)=SNOW.
```

which produces the output in Figure 16.10. You can specify the observed mean, median, or mode or a value for *cutpoint*.

MEAN *Mean.* All values below the observed mean make up one category; all values greater than or equal to the mean make up the other category.

MEDIAN *Median.* All values below the observed median make up one category; all values greater than or equal to the median make up the other category.

MODE *Mode.* All values below the observed mode make up one category; all values greater than or equal to the mode make up the other category.

value *Specified value.* All values below the specified value make up one category; all values greater than or equal to the specified value make up the other category.

The RUNS output shows the cutting point, the number of runs, the number of cases below the cutting point, the number of cases greater than or equal to the cutting point, and the test statistic Z with its observed significance level.

16.26 The Binomial Test

Use the BINOMIAL subcommand to perform the binomial test. Its format is

```
NPAR TESTS BINOMIAL[(p)]=varlist(value or value1,value2)
```

where p is the proportion of cases expected in the *first* category. If p is not specified, a p of .5 is used. A two-tailed test is performed only when p is .5. When p is any other value, a one-tailed test is performed.

If no values are specified after the variable list, each variable named is assumed to have only two values. If one value follows the variable list, it is used as a cutting point: all cases with values less than or equal to the cutting point go in the first category, and all other cases go in the second category. If two values follow the variable list, all cases with *value1* go in the first category and all cases with *value2* go in the second category. For example, the command

```
NPAR TESTS BINOMIAL = HEADS(0,1).
```

produces the output in Figure 16.11.

The BINOMIAL output shows the value of each category, the number of cases in each category, the test proportion of cases, and the probability of the observed population.

16.27 Tests for Two Related Samples

The McNemar test, the sign test, and the Wilcoxon matched-pairs signed-ranks test are available in NPAR TESTS for paired samples. The subcommands for requesting these tests have the general format

```
NPAR TESTS testname=varlist [WITH varlist]
```

where the keyword WITH and the second variable list are optional, signifying that each variable in the first list is to be paired with each variable in the second list. For example, the command

```
NPAR TESTS SIGN=RATING1 WITH RATING2 RATING3.
```

requests two sign tests, one for RATING1 with RATING2 and one for RATING1 with RATING3.

When only one variable list is specified, each variable in the list is paired with every other variable in the list. For example, the command

```
NPAR TESTS SIGN=RATING1 RATING2 RATING3.
```

produces three sign tests: RATING1 with RATING2, RATING1 with RATING3, and RATING2 with RATING3.

You can also pair variables sequentially by specifying Option 3 on the OPTIONS subcommand.

Option 3 *Sequential pairing of variables for two related samples.*

When Option 3 is requested with a single variable list, the first variable in the list is paired with the second, the second variable with the third, and so on. For example, the command

```
NPAR TESTS SIGN=RATING1 RATING2 RATING3
/OPTIONS=3.
```

performs two sign tests: one for RATING1 with RATING2 and one for RATING2 with RATING3.

If Option 3 is specified with the keyword WITH and two variable lists, the first variable in the first list before WITH is paired with the first variable in the second list, and so forth. For example, the command

```
NPAR TESTS SIGN=RATING1 RATING2 WITH RATING3 RATING4
/OPTIONS=3.
```

requests sign tests for RATING1 with RATING3 and RATING2 with RATING4.

16.28 The McNemar Test

The MCNEMAR subcommand requests the McNemar test for two correlated dichotomous variables. Pairs of variables being tested must have the same two values. Variables that are not dichotomous must be recoded before using NPAR TESTS (see Chapter 5). If fewer than ten cases have different values for the two variables, the binomial distribution is used to find the observed significance level. The output displayed includes a 2×2 table and the observed significance level. If a chi-square statistic is calculated, it is also displayed.

The format for MCNEMAR is

```
NPAR TESTS MCNEMAR=varlist [WITH varlist]
```

For example, the command

```
NPAR TESTS MCNEMAR=CONTRCT1 CONTRCT2.
```

requests a McNemar test on the 2×2 table for CONTRCT1 and CONTRCT2.

16.29 The Sign Test

The SIGN subcommand requests sign tests for paired variables. The format of SIGN is

```
NPAR TESTS SIGN=varlist [WITH varlist]
```

For example,the output in Figure 16.8 was produced by the command

```
NPAR TESTS SIGN=HSSCALE WITH WSSCALE.
```

If there are more than 25 cases, the observed significance level of the test statistic is based on a normal approximation. If there are 25 or fewer cases, the binomial distribution is used to compute the exact observed significance level.

16.30 The Wilcoxon Matched-Pairs Signed-Ranks Test

The WILCOXON subcommand requests the Wilcoxon test. Its format is

```
NPAR TESTS WILCOXON=varlist [WITH varlist]
```

For example, the output in Figure 16.9 was obtained with the command

```
NPAR TESTS WILCOXON=HSSCALE WITH WSSCALE.
```

The Wilcoxon test output shows the mean rank for each variable, the number of positive, negative, and tied ranks, and the test statistic Z with its observed significance level.

16.31 Tests for *k* Related Samples

Cochran's Q test, the Friedman test, and Kendall's coefficient of concordance W are tests for k related samples that are available in NPAR TESTS. The general format for requesting these tests is

```
NPAR TESTS testname=varlist
```

16.32 Cochran's Q

The COCHRAN subcommand requests Cochran's Q and has the format

```
NPAR TESTS COCHRAN=varlist
```

The variables to be tested must be dichotomous and coded with the same two values. If they are not, they must be recoded. The output displayed for COCHRAN includes the number of cases in each category for each variable and Cochran's Q with its degrees of freedom and observed significance level.

16.33 The Friedman Test

The Friedman test is requested with the FRIEDMAN subcommand, which has the format

```
NPAR TESTS FRIEDMAN=varlist
```

For example, the command

```
NPAR TESTS FRIEDMAN = P2DIGIT P3DIGIT P4DIGIT.
```

was used to produce the output in Figure 16.14.

The Friedman test output shows the mean rank for each variable and a chi-square statistic with its degrees of freedom and its observed significance level.

16.34 Kendall's Coefficient of Concordance

The KENDALL subcommand requests Kendall's W and has the format

```
NPAR TESTS KENDALL=varlist
```

This test assumes that each case is a judge or rater. If you want to treat your variables as judges, you must transpose your data matrix.

The output produced consists of the number of cases, Kendall's W, and a chi-square statistic with its degrees of freedom and observed significance level.

16.35 Tests for Two Independent Samples

NPAR TESTS performs five tests for two independent samples: the two-sample median test, the Mann-Whitney U test, the Kolmogorov-Smirnov two-sample test, the Wald-Wolfowitz runs test, and the Moses test of extreme reactions. The general format for requesting these tests is

```
NPAR TESTS testname=varlist BY variable(value1,value2)
```

where the variable named after BY is the variable used to group the cases. All cases with *value1* are in the first group and all cases with *value2* are in the second group.

16.36 The Two-Sample Median Test

The MEDIAN subcommand requests the two-sample median test. Its format is

```
NPAR TESTS MEDIAN [(value)]=varlist BY variable(value1,value2)
```

The value in parentheses after MEDIAN is the median to be used for the test. If a value is not specified, the sample median is used. The two values in parentheses following the variable named after BY specify the categories for the grouping variable. A two-sample median test is performed if the first value is greater than the second or if the second value is only one greater than the first. If the second value is more than one greater, a *k*-sample median test is performed (see Section 16.42). For example, the command

```
NPAR TESTS MEDIAN = SODIUM BY RATING(3,1).
```

was used to produce the output in Figure 16.16.

The Median test output shows a 2×2 table of cases greater than and less than or equal to the median, and the observed significance level. When the number of cases is greater than 30, a chi-square statistic is displayed. When the number of cases is 30 or less, Fisher's exact test (one-tailed) is computed.

16.37 The Mann-Whitney *U* Test

Request the Mann-Whitney test with the M-W subcommand, which has the format

```
NPAR TESTS M-W=varlist BY variable(value1,value2)
```

For example, the command

```
NPAR TESTS M-W=TUMOR BY DIET(0,1).
```

was used to produce the output in Figure 16.3.

The output produced by M-W includes the mean rank for each group, the Mann-Whitney *U* statistic, and the Wilcoxon *W*. For samples with less than 30 cases, the exact observed significance level (calculated using the algorithm of Dineen and Blakely, 1973) is displayed. For larger samples, a *Z* statistic with its (approximate) observed significance level is displayed.

16.38 The Kolmogorov-Smirnov Two-Sample Test

The K-S subcommand requests the Kolmogorov-Smirnov two-sample test and has the format

```
NPAR TESTS K-S=varlist BY variable(value1,value2)
```

For example, the command

```
NPAR TESTS K-S=ALCOHOL BY RATING(3,1).
```

was used to produce the output in Figure 16.18.

The output shows the number of cases in each group, the most extreme positive, negative, and absolute differences, and the Kolmogorov-Smirnov *Z* with its observed significance level.

16.39 The Wald-Wolfowitz Test

The Wald-Wolfowitz runs test is obtained with the W-W subcommand, which has the format

```
NPAR TESTS W-W=varlist BY variable(value1,value2)
```

For example, the command

```
NPAR TESTS W-W=CALORIES BY RATING(3,1).
```

was used to produce the output in Figure 16.17.

The output includes the number of cases in each group, the exact number of runs if there are no ties, and the observed significance level. If the sample size is 30 or less, the exact one-tailed observed significance level is calculated. Otherwise, a normal approximation is used to calculate the observed significance level.

16.40 The Moses Test of Extreme Reactions

The MOSES subcommand requests the Moses test of extreme reactions and has the format

```
NPAR TESTS MOSES[(n)]=varlist BY variable(value1,value2)
```

where *n* is the number of cases to be excluded from each end of the (sorted) data. This data trimming is sometimes needed to eliminate distortion due to outliers. If *n* is not specified, 5% of the cases are trimmed from each end of the range of the control group. *Value1* corresponds to the control group.

The output shows the number of cases in each group, the span of the control group with its observed significance level when all cases are included, and the span of the control group with its significance level after outliers have been removed.

16.41 Tests for *k* Independent Samples

Two tests for *k* samples are available in NPAR TESTS, the *k*-sample median test and the Kruskal-Wallis one-way analysis of variance. The general format for requesting these tests is

```
NPAR TESTS testname=varlist BY variable(value1,value2)
```

The variable following the BY keyword is used to split the cases into *k* groups, where *value1* and *value2* specify the minimum and maximum values used to define groups. For example, the command

```
NPAR TESTS MEDIAN=POLRANK BY RACE(1,3).
```

requests that the RACE variable be used to group the data into three categories corresponding to RACE values 1, 2, and 3.

16.42 The *k*-Sample Median Test

The *k*-sample median test is requested with the MEDIAN subcommand, which has the format

NPAR TESTS MEDIAN [(*value*)] BY *variable*(*value1,value2*)

The value in parentheses after MEDIAN is the median to be used for the test. If no value is specified, the sample median is used. For a *k*-sample median test, *value2* must be at least two greater than *value1*. For example, the command

```
NPAR TESTS MEDIAN=SOCATT BY RACE(1,3).
```

requests a three-sample median test for the groups defined by RACE values 1, 2, and 3. But the command

```
NPAR TESTS MEDIAN=SOCATT BY RACE(3,1).
```

requests a two-sample median test for the groups defined by RACE values 1 and 3. The command

```
NPAR TESTS MEDIAN = PRICE BY RATING(1,3).
```

produces the output in Figure 16.19.

The output for the median test includes a $2 \times k$ table and the chi-square test with its degrees of freedom and observed significance level.

16.43 The Kruskal-Wallis One-Way Analysis of Variance

The K-W subcommand requests the Kruskal-Wallis test and has the format

```
NPAR TESTS K-W=varlist BY variable(value1,value2)
```

If the first grouping variable is less than the second, every value in the range *value1* to *value2* defines a group. If the first value is greater than the second, two groups are formed, using the two values. For example, the output in Figure 16.20 was obtained with the command

```
NPAR TESTS K-W=TUMOR BY DIET(0,2).
```

The output displayed consists of the mean rank for each group, the number of cases in each group, the chi-square statistic with its observed significance level uncorrected for ties, and the chi-square statistic with its observed significance level corrected for ties.

16.44 Optional Statistics

You can obtain additional summary statistics for all variables named on NPAR TESTS subcommands by specifying Statistics 1 or 2 on the STATISTICS subcommand. All cases with valid values on a variable are used in calculating the statistics for that variables. The following statistics are available:

Statistic 1 *Univariate statistics.* The mean, maximum, minimum, standard deviation, and count are displayed for each variable named on a subcommand.

Statistic 2 *Quartiles and count.* The values corresponding to the 25th, 50th, and 75th percentiles for each variable named on a subcommand are displayed.

16.45 Missing Values

By default, cases with missing values are excluded on a test-by-test basis. That is, cases with missing values for any of the variables used in a particular test are excluded from calculations for that test only. You may request two alternative missing-value treatments with the OPTIONS subcommand.

Option 1 *Include user-missing values.* Cases with user-missing values are included in all tests requested on the command.

Option 2 *Exclude missing values listwise.* Cases missing on any variable named on any subcommand are excluded from all analyses.

16.46 Subsampling

Since many of the NPAR TESTS procedures are based on ranks, cases must be stored in memory. If you do not have enough computer memory to store all the cases, you will need to use Option 4 on the OPTIONS subcommand to select a random subsample of cases for analysis.

Option 4 *Random sampling* if there is insufficient memory.

Because such sampling would invalidate a runs test, this option is ignored when the RUNS subcommand is used.

16.47 Annotated Example

The following SPSS/PC+ commands were used to obtain the output in Figures 16.3 and 16.20.

```
DATA LIST FREE/DIET   TUMOR.
FORMATS DIET(F1.0).
VALUE LABELS DIET 0 'UNSATURATED' 1 'SATURATED' 2 'LOW-FAT'.
BEGIN DATA.
lines of data
END DATA.
NPAR TESTS M-W TUMOR BY DIET(0,1)/
  K-W TUMOR BY DIET(0,2).
FINISH.
```

- The DATA LIST command tells SPSS/PC+ that the two variables DIET and TUMOR are entered in freefield format.
- The FORMATS command assigns a new format to the variable DIET.
- The VALUE LABELS command assigns descriptive labels to the values of variable DIET.
- NPAR TESTS requests a Mann-Whitney test and a Kruskal-Wallis test.

Contents

Chapter 17 Multiple Regression: Procedure REGRESSION

The 1964 Civil Rights Act prohibits discrimination in the workplace based on sex or race. Employers who violate the act, by unfair hiring or advancement, are liable to prosecution. Numerous lawsuits have been filed on behalf of women, blacks, and other groups on these grounds.

The courts have ruled that statistics can be used as *prima facie* evidence of discrimination. Many lawsuits depend heavily on complex statistical analyses, which attempt to demonstrate that similarly qualified individuals are not treated equally (Roberts, 1980). In this chapter, employee records for 474 individuals hired between 1969 and 1971 by a bank engaged in Equal Employment Opportunity litigation are analyzed. A mathematical model is developed that relates beginning salary and salary progression to various employee characteristics such as seniority, education, and previous work experience. One objective is to determine whether sex and race are important predictors of salary.

The technique used to build the model is linear regression analysis, one of the most versatile data analysis procedures. Regression can be used to summarize data as well as to study relations among variables.

17.1 LINEAR REGRESSION

Before examining a model that relates beginning salary to several other variables, consider the relationship between beginning salary and current (as of March 1977) salary. For employees hired during a similar time period, beginning salary should serve as a reasonably good predictor of salary at a later date. Although superstars and underachievers might progress differently from the group as a whole, salary progression should be similar for the others. The scatterplot of beginning salary and current salary produced by the PLOT procedure and shown in Figure 17.1 supports this hypothesis.

Figure 17.1 Scatterplot of beginning and current salaries

```
                PLOT OF SALNOW WITH SALBEG
         --+----+----+----+----+----+----+----+--
C        :                                        :
U   52500+                                      1 +
R        :                                        :
R        :                    1                   :
E        :          1               1             :
N        :             1 1 1                      :
T   35000+       1   11       1                   +
         :    (1)     2211  1                     :
S        :       1 13 431  11 1 1                 :
A        :     1 21 322311                        :
L        :     343 331221                         :
A   17500+   2755732   1                          +
R        : 15G*C7 1 1                             :
Y        : M***A2                                 :
         : UN91                                   :
         :                                        :
         --+----+----+----+----+----+----+----+--
        4000     12000     20000     28000
             8000      16000     24000     32000

                     BEGINNING SALARY
```

A scatterplot may suggest what type of mathematical functions would be appropriate for summarizing the data. A variety of functions are useful in fitting models to data. Parabolas, hyperbolas, polynomials, trigonometric functions, and many more are potential candidates. For the scatterplot in Figure 17.1, current salaries tend to increase linearly with increases in beginning salary. If the plot indicates that a straight line is not a good summary measure of the relationship, you should consider other possibilities, including attempts to transform the data to achieve linearity (see Section 17.27).

17.2 Outliers

A plot may also indicate the presence of points suspiciously different from the others. Examine such observations, termed *outliers*, carefully to see if they result from errors in gathering, coding, or entering data. The circled point in Figure 17.1 appears to be an outlier. Though neither the value of beginning salary ($6,300) nor the value of current salary ($32,000) is unique, jointly they are unusual.

The treatment of outliers can be difficult. If the point is really incorrect, due to coding or entry problems, you should correct it and rerun the analysis. If there is no apparent explanation for the outlier, consider interactions with other variables as a possible explanation. For example, the outlier may represent an employee who was hired as a low-paid clerical worker while pursuing an MBA degree. After graduation, a rapid rise in position was possible, making education the variable that explains the unusual salary characteristics of the employee.

17.3 Choosing a Regression Line

Since current salary tends to increase linearly with beginning salary, a straight line can be used to summarize the relationship. The equation for the line is

predicted current salary $= B_0 + B_1$(beginning salary) **Equation 17.3a**

The *slope* (B_1) is the dollar change in the fitted current salary for a dollar change in the beginning salary. The *intercept* (B_0) is the theoretical estimate of current salary if there were a beginning salary of 0.

However, the observed data points do not all fall on a straight line but cluster about it. Many lines can be drawn through the data points; the problem is to select among them. The method of *least squares* results in a line that minimizes the sum of squared vertical distances from the observed data points to the line. Any other line has a larger sum. Figure 17.3a shows the least-squares line superimposed on the salary scatterplot. Several vertical distances from points to the line are also shown.

Figure 17.3a Regression line for beginning and current salaries

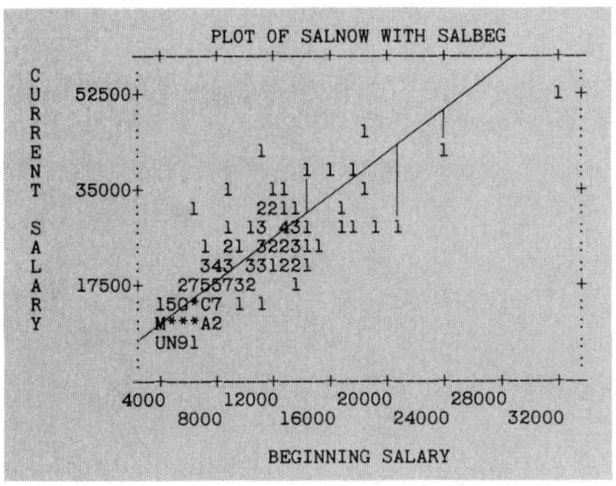

You can use the SPSS/PC+ REGRESSION procedure to calculate the least-squares line. For the data in Figure 17.1, that line is

predicted current salary $= 771.28 + 1.91$(beginning salary) **Equation 17.3b**

The slope and intercept values are shown in the column labeled B in the output shown in Figure 17.3b.

Figure 17.3b Statistics for variables in the equation

```
------------------ Variables in the Equation ------------------

Variable              B        SE B       Beta         T  Sig T

SALBEG          1.90945      .04741     .88012    40.276  .0000
(Constant)    771.28230   355.47194                2.170  .0305
```


17.4 The Standardized Regression Coefficient

The *standardized regression coefficient*, labeled Beta in Figure 17.3b, is defined as

$$BETA = B_1 \frac{S_X}{S_Y}$$ **Equation 17.4**

Multiplying the regression coefficient (B_1) by the ratio of the standard deviation of the independent variable (S_X) to the standard deviation of the dependent variable (S_Y) results in a dimensionless coefficient. In fact, the Beta coefficient is the slope of the least-squares line when both X and Y are expressed as Z scores. The Beta coefficient is further discussed in Section 17.38.

17.5 From Samples to Populations

Generally, more is sought in regression analysis than a description of observed data. One usually wishes to draw inferences about the relationship of the variables in the population from which the sample was taken. How are beginning and current salaries related for all employees, not just those included in the sample? To draw inferences about population values based on sample results, the following assumptions are needed:

Normality and Equality of Variance. For any fixed value of the independent variable X, the distribution of the dependent variable Y is normal, with mean $\mu_{Y/X}$ (the mean of Y for a given X) and a constant variance of σ^2 (see Figure 17.5). This assumption specifies that not all employees with the same beginning salary have the same current salary. Instead, there is a normal distribution of current salaries for each beginning salary. Though the distributions have different means, they have the same variance σ^2.

Figure 17.5 Regression assumptions

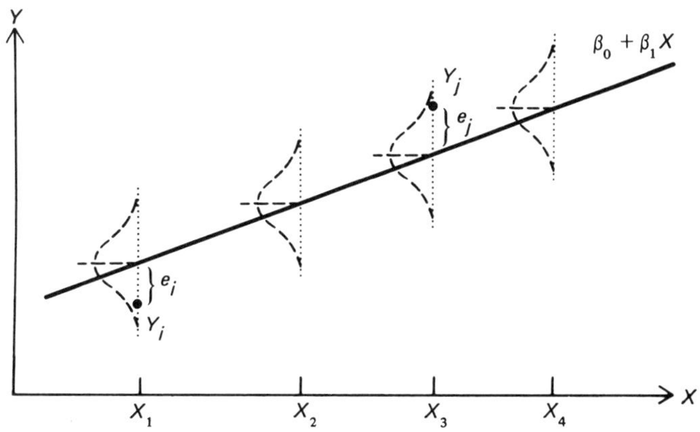

Independence. The Y's are statistically independent of each other. That is, observations are in no way influenced by other observations. For example, observations are *not* independent if they are based on repeated measurements from the same experimental unit. If three observations are taken from each of four families, the twelve observations are not independent.

Linearity. The mean values $\mu_{Y/X}$ all lie on a straight line, which is the population regression line. This line is drawn in Figure 17.5. An alternative way of stating this assumption is that the linear model is correct.

When there is a single independent variable, the model can be summarized by

$$Y_i = \beta_0 + \beta_1 X_i + e_i$$ **Equation 17.5**

The population parameters (values) for the slope and intercept are denoted by β_1 and β_0. The term e_i, often called an error or disturbance, is the difference between the observed value of Y_i and the subpopulation mean at the point X_i. The e_i are assumed to be normally distributed, independent, random variables with a mean of 0 and variance of σ^2 (see Figure 17.5).

17.6 Estimating Population Parameters

Since β_0 and β_1 are unknown population parameters, they must be estimated from the sample. The least-squares coefficients B_0 and B_1, discussed in Section 17.3, are used to estimate the population parameters.

However, the slope and intercept estimated from a single sample typically differ from the population values and vary from sample to sample. To use these estimates for inference about the population values, the sampling distributions of the two statistics are needed. When the assumptions of linear regression are met, the sampling distributions of B_0 and B_1 are normal with means of β_0 and β_1.

The standard error of B_0 is

$$\sigma_{B_0} = \sigma \sqrt{\frac{1}{N} + \frac{\bar{X}^2}{(N-1)S_X{}^2}}$$ **Equation 17.6a**

where $S_X{}^2$ is the sample variance of the independent variable. The standard error of B_1 is

$$\sigma_{B_1} = \frac{\sigma}{\sqrt{(N-1)S_X{}^2}}$$ **Equation 17.6b**

Since the population variance of the errors, σ^2, is not known, it must also be estimated. The usual estimate of σ^2 is

$$S^2 = \frac{\sum_{i=1}^{N} (Y_i - B_0 - B_1 X_i)^2}{N-2}$$ **Equation 17.6c**

The positive square root of σ^2 is termed the *standard error of the estimate,* or the standard deviation of the residuals. (The reason for this name is discussed in Section 17.15.) The estimated standard errors of the slope and intercept are printed in the third column (labeled SE B) in Figure 17.3b.

17.7 Testing Hypotheses

A frequently tested hypothesis is that there is no linear relationship between X and Y—that the slope of the population regression line is 0. The statistic used to test this hypothesis is

$$t = \frac{B_1}{S_{B_1}}$$ **Equation 17.7a**

The distribution of the statistic, when the assumptions are met and the hypothesis of no linear relationship is true, is Student's t distribution with $N-2$ degrees of freedom. The statistic for testing the hypothesis that the intercept is 0 is

$$t = \frac{B_0}{S_{B_0}}$$ **Equation 17.7b**

Its distribution is also Student's t with $N-2$ degrees of freedom.

These t statistics and their two-tailed observed significance levels are displayed in the last two columns of Figure 17.3b. The small observed significance level (less than 0.00005) associated with the slope for the salary data supports the hypothesis that beginning and current salary are linearly related.

17.8 Confidence Intervals

A statistic calculated from a sample provides a point estimate of the unknown parameter. A point estimate can be thought of as the single best guess for the population value. While the estimated value from the sample is typically different from the value of the unknown population parameter, the hope is that it isn't too far away. Based on the sample estimate, it is possible to calculate a range of values that, with a designated likelihood, includes the population value. Such a range is called a *confidence interval.* For example, as shown in Figure 17.8, the 95% confidence interval for β_1, the population slope, is 1.816 to 2.003.

Figure 17.8 Confidence intervals

Variable	95% Confdnce Intrvl B	
SALBEG	1.81629	2.00261
(Constant)	72.77921	1469.78540

Ninety-five percent confidence means that, if repeated samples are drawn from a population under the same conditions and 95% confidence intervals are calculated, 95% of the intervals will contain the unknown parameter β_1. Since the parameter value is unknown, it is not possible to determine whether or not a particular interval contains it.

17.9 Goodness of Fit

An important part of any statistical procedure that builds models from data is establishing how well the model actually fits. This topic encompasses the detection of possible violations of the required assumptions in the data being analyzed. Sections 17.10 through 17.16 are limited to the question of how close to the fitted line the observed points fall. Subsequent sections discuss other assumptions and tests for their violation.

17.10 The R^2 Coefficient

A commonly used measure of the goodness of fit of a linear model is R^2, sometimes called the *coefficient of determination.* It can be thought of in a variety of ways. Besides being the square of the correlation coefficient between variables X and Y, it is the square of the correlation coefficient between Y, the observed value of the dependent variable, and $\hat{Y}$, the predicted value of Y from the fitted line. If for each employee one computes (based on the coefficients in the output in Figure 17.3b) the predicted salary

$$\text{predicted current salary} = 771.28 + 1.91(\text{beginning salary})$$ **Equation 17.10a**

and then calculates the square of the Pearson correlation coefficent between predicted current salary and observed current salary, R^2 is obtained. If all the observations fall on the regression line, R^2 is 1. If there is no linear relationship between the dependent and independent variables, R^2 is 0.

Note that R^2 is a measure of the goodness of fit of a particular model and that an R^2 of 0 does not necessarily mean that there is no association between the variables. Instead, it indicates that there is no *linear relationship*.

In the output in Figure 17.10, R^2 is labeled R Square; its square root is called Multiple R. The sample R^2 tends to be an optimistic estimate of how well the model fits the population. The model usually does not fit the population as well as it fits the sample from which it is derived. The statistic *adjusted* R^2 attempts to correct R^2 to more closely reflect the goodness of fit of the model in the population. Adjusted R^2 is given by

$$R_a{}^2 = R^2 - \frac{p(1 - R^2)}{N - p - 1}$$ **Equation 17.10b**

where p is the number of independent variables in the equation (1 in the salary example).

Figure 17.10 Summary statistics for the equation

```
Multiple R               .88012
R Square                 .77461
Adjusted R Square        .77413
Standard Error       3246.14226
```

17.11 Analysis of Variance

To test the hypothesis of no linear relationship between X and Y, several equivalent statistics can be computed. When there is a single independent variable, the hypothesis that the population R^2 is 0 is identical to the hypothesis that the population slope is 0. The test for $R^2_{pop}=0$ is usually obtained from the *analysis of variance* (ANOVA) table (see Figure 17.11a).

Figure 17.11a Analysis of variance table

```
ANALYSIS OF VARIANCE
                     DF       SUM OF SQUARES        MEAN SQUARE
REGRESSION            1    17092967800.01931  17092967800.0193
RESIDUAL            472     4973671469.79483    10537439.55465

F =      1622.11776          SIGNIF F =   .0000
```

The total observed variability in the dependent variable is subdivided into two components—that which is attributable to the regression (labeled Regression) and that which is not (labeled Residual). Consider Figure 17.11b. For a particular point, the distance from Y_i to $\bar{Y}$ (the mean of the Y's) can be subdivided into two parts.

$$Y_i - \bar{Y} = (Y_i - \hat{Y}_i) + (\hat{Y}_i - \bar{Y})$$ **Equation 17.11a**

Figure 17.11b Components of variability

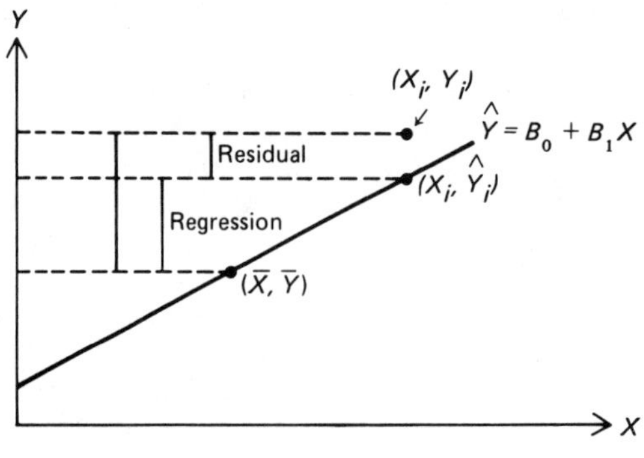

The distance from Y_i, the observed value, to $\hat{Y}_i$, the value predicted by the regression line, or $Y_i - \hat{Y}_i$, is 0 if the regression line passes through the point. It is called the *residual from the regression.* The second component $(\hat{Y}_i - \overline{Y})$ is the distance from the regression line to the mean of the Y's. This distance is "explained" by the regression in that it represents the improvement in the estimate of the dependent variable achieved by the regression. Without the regression, the mean of the dependent variable $(\overline{Y})$ is used as the estimate. It can be shown that

$$\sum_{i=1}^{N} (Y_i - \overline{Y})^2 = \sum_{i=1}^{N} (Y_i - \hat{Y}_i)^2 + \sum_{i=1}^{N} (\hat{Y}_i - \overline{Y})^2$$ **Equation 17.11b**

The first quantity following the equals sign is called the *residual sum of squares* and the second quantity is the *regression sum of squares.* The sum of these is called the *total sum of squares*.

The analysis of variance table displays these two sums of squares under the heading Sum of Squares (Figure 17.11a). The Mean Square for each entry is the Sum of Squares divided by the degrees of freedom (DF). If the regression assumptions are met, the ratio of the mean square regression to the mean square residual is distributed as an F statistic with p and $N-p-1$ degrees of freedom. F serves to test how well the regression model fits the data. If the probability associated with the F statistic is small, the hypothesis that $R^2_{pop}=0$ is rejected. For this example, the F statistic is

$$F = \frac{\text{MEAN SQUARE REGRESSION}}{\text{MEAN SQUARE RESIDUAL}} = 1622$$ **Equation 17.11c**

The observed significance level (SIGNIF F) is less than 0.00005.

The square root of the F value (1622) is 40.28, which is the value of the t statistic for the slope in Figure 17.3b. The square of a t value with k degrees of freedom is an F value with 1 and k degrees of freedom. Therefore, either t or F values can be computed to test that $\beta_i=0$.

Another useful summary statistic is the standard error of the estimate, S, which can also be calculated as the square root of the residual mean square (Section 17.15).

17.12 Another Interpretation of R^2

Partitioning the sum of squares of the dependent variable allows another interpretation of R^2. It is the proportion of the variation in the dependent variable "explained" by the model.

$$R^2 = 1 - \frac{\text{RESIDUAL SUM OF SQUARES}}{\text{TOTAL SUM OF SQUARES}} = 0.775$$ **Equation 17.12a**

Similarly, adjusted R^2 is

$$R_a^2 = 1 - \frac{\text{RESIDUAL SUM OF SQUARES}/(N-p-1)}{\text{TOTAL SUM OF SQUARES}/(N-1)}$$ **Equation 17.12b**

where p is the number of independent variables in the equation (1 in the salary example).

17.13 Predicted Values and Their Standard Errors

By comparing the observed values of the dependent variable to the values predicted by the regression equation, you can learn a good deal about how well a model and the various assumptions fit the data (see the discussion of residuals beginning with Section 17.17). Predicted values are also of interest when the results are used to predict new data. You may wish to predict the mean Y for all cases with a given value of X, denoted X_0, or to predict the value of Y for a single case. For example, you can predict either the mean salary for all employees with a beginning salary of \$10,000 or the salary for a particular employee with a beginning salary of \$10,000. In both situations, the predicted value

$$\hat{Y}_0 = B_0 + B_1X_0 = 771 + 1.91 \times 10{,}000 = 19{,}871 \qquad \textbf{Equation 17.13}$$

is the same. What differs is the standard error.

17.14 Predicting Mean Response

The estimated standard error for the predicted mean Y at X_0 is

$$S_{\hat{Y}} = S\sqrt{\frac{1}{N} + \frac{(X_0 - \overline{X})^2}{(N-1)S_X{}^2}} \qquad \textbf{Equation 17.14a}$$

The equation for the standard error shows that the smallest value occurs when X_0 is equal to $\overline{X}$, the mean of X. The larger the distance from the mean, the greater the standard error. Thus, the mean of Y for a given X is better estimated for central values of the observed X's than for outlying values. Figure 17.14a is a plot from the PLOT procedure of the standard errors of predicted mean salaries for different values of beginning salary.

Figure 17.14a Standard errors for predicted mean responses

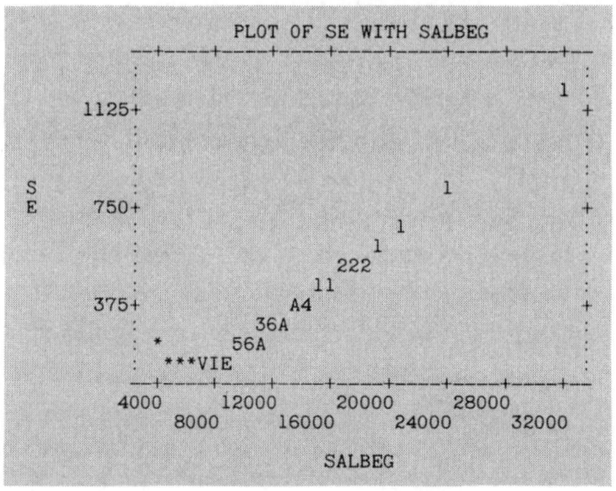

Prediction intervals for the mean predicted salary are calculated in the standard way. The 95% confidence interval at X_0 is

$$\hat{Y} \pm t_{\left(1-\frac{\alpha}{2}, N-2\right)} S_{\hat{Y}} \qquad \textbf{Equation 17.14b}$$

Figure 17.14b shows a typical 95% confidence band for predicted mean responses. It is narrowest at the mean of X and increases as the distance from the mean $(X_0 - \overline{X})$ increases.

Figure 17.14b 95% confidence band for mean prediction

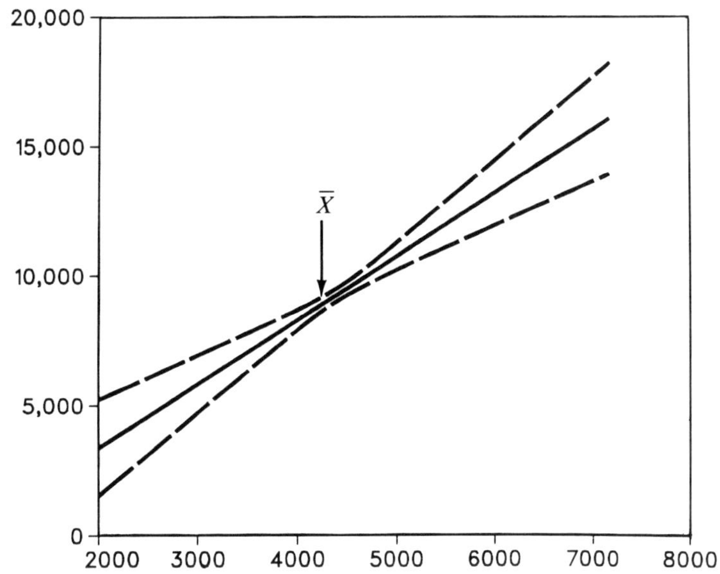

17.15 Predicting a New Value

Although the predicted value for a single new observation at X_0 is the same as the predicted value for the mean at X_0, the standard error is not. The two sources of error when predicting an individual observation are illustrated in Figure 17.15. They are

1 The individual value may differ from the population mean of Y for X_0.

2 The estimate of the population mean at X_0 may differ from the population mean.

Figure 17.15 Sources of error in predicting individual observations

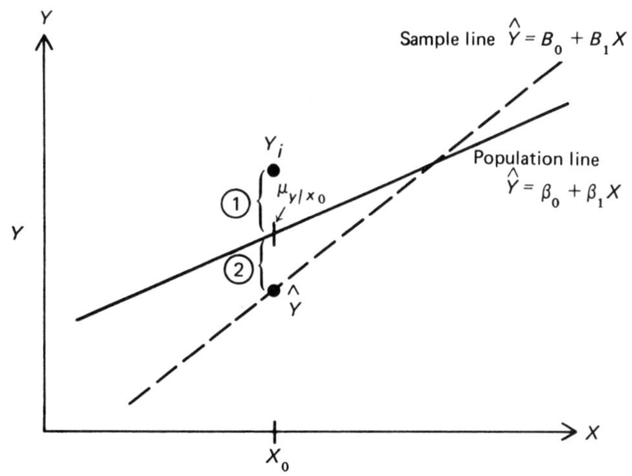

When estimating the mean response, only the second error component is considered. The variance of the individual prediction is the variance of the mean prediction plus the variance of Y_i for a given X. This can be written as

$$S^2_{ind\hat{Y}} = S^2_{\hat{Y}} + S^2 = S^2\left(1 + \frac{1}{N} + \frac{(X_0 - \bar{X})^2}{(N-1)S_X^{\,2}}\right) \qquad \textbf{Equation 17.15a}$$

Prediction intervals for the new observation are obtained by substituting S_{indY} for S_Y in the equation for the confidence intervals for the mean given in Section 17.14. If the sample size is large, the terms $1/N$ and

$$\frac{(X_0 - \bar{X})^2}{(N-1)S_X^{\,2}} \qquad \textbf{Equation 17.15b}$$

are negligible. In that case, the standard error is simply S, which explains the name *standard error of the estimate* for S (see Section 17.6).

17.16 Reading the Casewise Plot

Figure 17.16 shows the output from the beginning and end of a plot of the salary data. The sequence number of the case and an optional labeling variable (SEXRACE) are listed first, followed by the plot of standardized residuals, the observed (SALNOW), predicted (PRED), and residual (RESID) values, and, finally, the standard error of the mean prediction (SEPRED). The variance of an individual prediction can be obtained by adding S^2 to the square of each of the standard error values. You can generate predicted values and the standard errors of the mean responses in the REGRESSION procedure, and you can display both of these values for all cases or for a subset of cases along with a casewise plot.

Figure 17.16 Casewise plot with predicted values and standard errors

```
Casewise Plot of Standardized Residual

*: Selected   M: Missing

                 -3.0            0.0             3.0
Case #  SEXRACE   0:..............:..............:0  SALNOW       *PRED      *RESID     *SEPRED
     1     1.00   .              *.               .   16080  16810.6600   -730.6600    167.1489
     2     1.00   .       *       .               .   41400  46598.0758  -5198.0758    828.6655
     3     1.00   .               .  *            .   21960  20247.6695   1712.3305    219.3531
     4     1.00   .               .  *            .   19200  17383.4949   1816.5051    174.0406
     5     1.00   .      *        .               .   28350  33995.7076  -5645.7076    523.9021
     6     1.00   .               .  *            .   27250  25586.4910   1663.5090    329.1520
     7     1.00   .               .  *            .   16080  13946.4854   2133.5146    149.1662
     8     1.00   .               .    *          .   14100  11082.3108   3017.6892    163.3307
     9     1.00   .               .  *            .   12420  10394.9089   2025.0911    171.0096
    10     1.00   .              *.               .   12300  12800.8156   -500.8156    151.0211
    11     1.00   .               .   *           .   15720  12800.8156   2919.1844    151.0211
    12     1.00   .          *    .               .    8880  12227.9807  -3347.9807    153.9241
   ...
   ...
   ...
   470     4.00   .               *               .    9420   9592.9401   -172.9401    181.5927
   471     4.00   .               .*              .    9780   9134.6721    645.3279    188.3196
   472     4.00   .             * .               .    7680   9249.2391  -1569.2391    186.5956
   473     4.00   .             * .               .    7380   8561.8372  -1181.8372    197.3294
   474     4.00   .           *   .               .    8340  10738.6099  -2398.6099    166.9964
Case #  SEXRACE   0:..............:..............:0  SALNOW       *PRED      *RESID     *SEPRED
                 -3.0            0.0             3.0
```

17.17 Searching for Violations of Assumptions

You usually don't know in advance whether a model such as linear regression is appropriate. Therefore, it is necessary to conduct a search focused on residuals to look for evidence that the necessary assumptions are violated.

17.18 Residuals

In model building, a *residual* is what is left after the model is fit. It is the difference between an observed value and the value predicted by the model.

$$E_i = Y_i - B_0 - B_1X_i = Y_i - \hat{Y}_i \qquad \textbf{Equation 17.18}$$

In regression analysis, the true errors e_i are assumed to be independent normal values with a mean of 0 and a constant variance of σ^2. If the model is appropriate for the data, the observed residuals E_i, which are estimates of the true errors e_i, should have similar characteristics.

If the intercept term is included in the equation, the mean of the residuals is always 0, so it provides no information about the true mean of the errors. Since the sum of the residuals is constrained to be 0, they are *not* strictly independent. However, if the number of residuals is large when compared to the number of independent variables, the dependency among the residuals can be ignored for practical purposes.

The relative magnitudes of residuals are easier to judge when they are divided by estimates of their standard deviations. The resulting standardized residuals are expressed in standard deviation units above or below the mean. For example, the fact that a particular residual is −5198.1 provides little information. If you know that its standardized form is −3.1, you know not only that the observed value is less than the predicted value but also that the residual is larger than most in absolute value.

Residuals are sometimes adjusted in one of two ways. The *standardized residual* for case *i* is the residual divided by the sample standard deviation of the residuals. Standardized residuals have a mean of 0 and a standard deviation of 1. The *Studentized residual* is the residual divided by an estimate of its standard deviation that varies from point to point, depending on the distance of X_i from the mean of X. Usually standardized and Studentized residuals are close in value, but not always. The Studentized residual reflects more precisely differences in the true error variances from point to point.

17.19 Linearity

For the bivariate situation, a scatterplot is a good means for judging how well a straight line fits the data. Another convenient method is to plot the residuals against the predicted values. If the assumptions of linearity and homogeneity of variance are met, there should be no relationship between the predicted and residual values. You should be suspicious of any observable pattern.

For example, fitting a least-squares line to the data in the two left-hand plots in Figure 17.19a yields the residual plots shown on the right. The two residual plots show patterns since straight lines do not fit the data well. Systematic patterns between the predicted values and the residuals suggest possible violations of the linearity assumption. If the assumption was met, the residuals would be randomly distributed in a band about the horizontal straight line through 0, as shown in Figure 17.19b.

Figure 17.19a Standardized residuals scatterplots

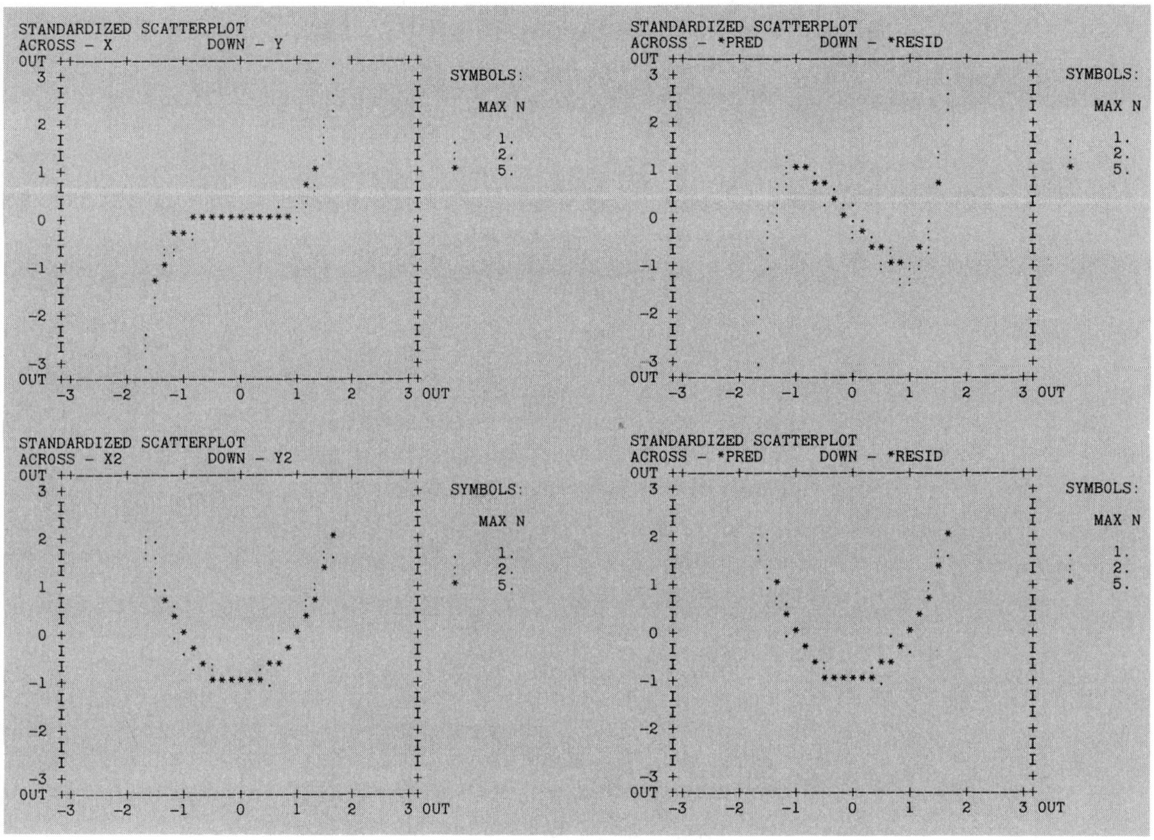

Figure 17.19b Randomly distributed residuals

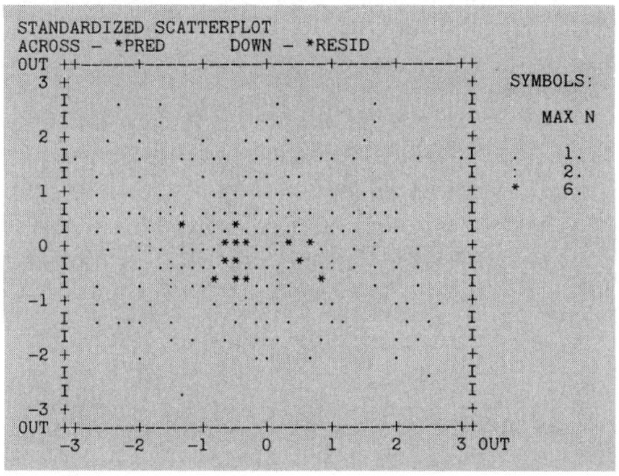

Residuals can also be plotted against individual independent variables. Again, if the assumptions are met, you should see a horizontal band of residuals. Consider plotting the residuals against independent variables not in the equation as well. If the residuals are not randomly distributed, you may want to include the variable in the equation for a multiple regression model (see Sections 17.32 through 17.54).

17.20 Equality of Variance

You can also use the previously described plots to check for violations of the equality of variance assumption. If the spread of the residuals increases or decreases with values of the independent variables or with predicted values, you should question the assumption of constant variance of *Y* for all values of *X*.

Figure 17.20 is a plot of the Studentized residuals against the predicted values for the salary data. The spread of the residuals increases with the magnitude of the predicted values, suggesting that the variability of current salaries increases with salary level. Thus, the equality of variance assumption appears to be violated.

Figure 17.20 Unequal variance

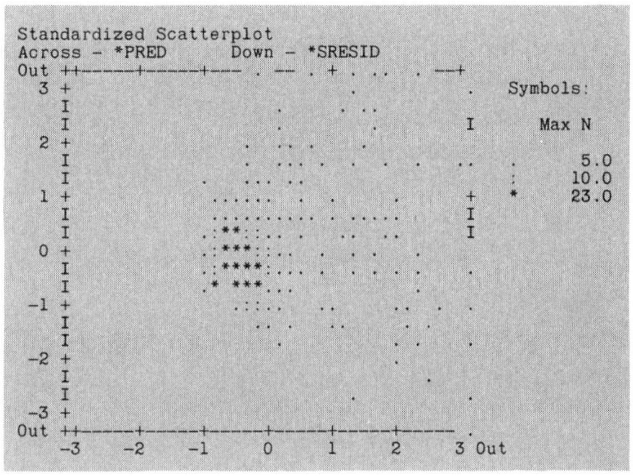

17.21 Independence of Error

Whenever the data are collected and recorded sequentially, you should plot residuals against the sequence variable. Even if time is not considered a variable in the model, it could influence the residuals. For example, suppose you are studying survival time after surgery as a function of complexity of surgery, amount of blood transfused, dosage of medication, and so forth. In addition to these variables, it is also possible that the surgeon's skill increased with each operation and that a patient's survival time is influenced by the number of prior patients treated. The plot of residuals corresponding to the order in which patients received surgery shows a shorter survival time for earlier patients than for later patients (see Figure 17.21). If sequence and the residual are independent, you should not see a discernable pattern.

Figure 17.21 Casewise serial plot

```
CASEWISE PLOT OF STUDENTIZED RESIDUAL

                      -3.0      0.0      3.0        LIFE     *PRED    *RESID  *SRESID
 SEQNUM     TIME       0:.......:........:0
      1    78012        . *      .        .      15.0000   19.5624   -4.5624  -2.2598
      2    78055        . *      .        .      13.5000   17.8974   -4.3974  -2.1856
      3    78122        .  *     .        .       9.9000   13.8390   -3.9390  -1.9871
      4    78134        .    *   .        .      15.5000   18.5218   -3.0218  -1.4997
      5    78233        .   *    .        .      35.0000   38.2933   -3.2933  -1.7466
      6    78298        .     *  .        .      14.7000   16.6487   -1.9487   -.9720
      7    78344        .      * .        .      34.8000   36.0040   -1.2040   -.6258
      8    79002        .       *.        .      20.8000   20.8111    -.0111   -.0055
      9    79008        .        . *      .      15.9000   14.8796    1.0204    .5123
     10    79039        .        *        .      22.0000   21.6436     .3564    .1762
     11    79101        .        .  *     .      13.7000   11.7578    1.9422    .9910
     12    79129        .        .  *     .      14.2000   11.4456    2.7544   1.4082
     13    79178        .        .   *    .      33.2000   30.3847    2.8153   1.4144
     14    79188        .        .   *    .      26.2000   22.4761    3.7239   1.8401
     15    79189        .        .    *   .      37.4000   33.2984    4.1016   2.0920
    ...
```

The *Durbin-Watson* statistic, a test for sequential correlation of adjacent error terms, is defined as

$$D = \frac{\sum_{t=2}^{N} (E_t - E_{t-1})^2}{\sum_{t=1}^{N} E_t^2} \qquad \textbf{Equation 17.21}$$

The differences between successive residuals tend to be small when error terms are positively correlated and large when error terms are negatively correlated. Thus, small values of D indicate positive correlation and large values of D indicate negative correlation. Consult tables of the D statistic for bounds upon which significance tests can be based (Neter & Wasserman, 1974).

17.22 Normality

The distribution of residuals may not appear to be normal for reasons other than actual nonnormality: misspecification of the model, nonconstant variance, a small number of residuals actually available for analysis, etc. Therefore, you should pursue several lines of investigation. One of the simplest is to construct a histogram of the residuals such as the one shown in Figure 17.22a for the salary data.

Figure 17.22a Histogram of Studentized residuals

```
Histogram - Studentized Residual

  N  Exp N       (* = 2 Cases,     . : = Normal Curve)
  7    .37   Out ****
  2    .73  3.00 *
  4   1.85  2.67 :*
  2   4.23  2.33 *.
  6   8.65  2.00 ***.
 12  15.85  1.67 ****** .
  7  26.01  1.33 ****        .
 18  38.23  1.00 *********         .
 35  50.34   .67 ******************      .
 63  59.38   .33 *****************************:**
 87  62.74  0.0  ******************************:*************
114  59.38  -.33 *****************************:***************************
 64  50.34  -.67 ************************:*******
 32  38.23 -1.00 ****************  .
  9  26.01 -1.33 *****       .
  6  15.85 -1.67 ***    .
  1   8.65 -2.00 *  .
  1   4.23 -2.33 *.
  2   1.85 -2.67 :
  0    .73 -3.00
  2    .37   Out *
```

The REGRESSION histogram contains a tally of the observed number of residuals (labeled N) in each interval and the number expected in a normal distribution with the same mean and variance as the residuals (Exp N). The first and last intervals (Out) contain residuals more than 3.16 standard deviations from the mean. Such residuals deserve examination. A histogram of expected *N*'s is superimposed on that of the observed *N*'s. Expected frequencies are indicated by a period. When observed and expected frequencies overlap, a colon is displayed. However, it is unreasonable to expect the observed residuals to be exactly normal—some deviation is expected because of sampling variation. Even if the errors are normally distributed in the population, sample residuals are only approximately normal.

In the histogram in Figure 17.22a, the distribution does not seem normal since there is an exaggerated clustering of residuals toward the center and a straggling tail toward large positive values. Thus, the normality assumption may be violated.

Another way to compare the observed distribution of residuals to that expected under the assumption of normality is to plot the two cumulative distributions against each other for a series of points. If the two distributions are identical, a straight line results. By observing how points scatter about the expected straight line, you can compare the two distributions.

Figure 17.22b is a cumulative probability plot of the salary residuals. Initially, the observed residuals are below the straight line, since there is a smaller number of large negative residuals than expected. Once the greatest concentration of residuals is reached, the observed points are above the line, since the observed cumulative proportion exceeds the expected.

Figure 17.22b A normal probability (P-P) plot

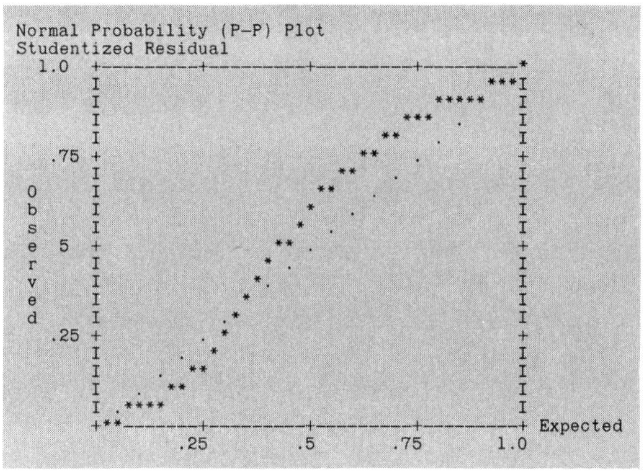

17.23 Locating Outliers

You can spot outliers readily on residual plots since they are cases with very large positive or negative residuals. In the histogram, cases with values greater than +3.16 or less than −3.16 appear in the interval labeled Out. In the scatterplots, they appear on the borders of the plot, again labeled Out. Since you usually want more information about outliers, you can use the casewise plotting facility to display identification numbers and a variety of other statistics for cases having residuals beyond a specified cutoff point.

Figure 17.23 lists information for the nine cases with Studentized residuals greater than 3 in absolute value. Only two of these nine employees have current salaries less than those predicted by the model (Cases 67 and 122), while the others have larger salaries. The second column contains identifier information that indicates that all outliers are white males (SEXRACE=1). They all have large salaries, an average of $33,294, while the average for the sample is only $13,767. Thus, there is some evidence that the model may not fit well for the highly paid cases.

Figure 17.23 Casewise plot of residuals outliers

```
Casewise Plot of Studentized Residual

Outliers = 3.     *: Selected   M: Missing

                   -6.     -3.  3.     6.
  Case #  SEXRACE   0:.......:  :.......:0   SALNOW       *PRED      *RESID
     24     1.00    .        ..*        .    28000  17383.4949  10616.5051
     60     1.00    .        ..        *.    32000  12800.8156  19199.1844
     67     1.00    .       * ..        .    26400  37043.1894 -10643.1894
    114     1.00    .        .. *       .    38800  27511.2163  11288.7837
    122     1.00    .    *   ..         .    26700  40869.7266 -14169.7266
    123     1.00    .        .. *       .    36250  24639.4039  11610.5961
    129     1.00    .        ..     *   .    33500  17383.4949  16116.5051
    149     1.00    .        ..         *    41500  21782.8671  19717.1329
    177     1.00    .        ..   *     .    36500  23295.1513  13204.8487

      9 Outliers found.
```

17.24 Other Unusual Observations: Mahalanobis' Distance

In Section 17.2, an employee was identified as an outlier because the combination of values for beginning and current salaries was atypical. This case, which is Case 60, shows up in Figure 17.23 since it has a large value for the Studentized residual.

Another unusual employee (Case 56) has a beginning salary of \$31,992. Since the average beginning salary for the entire sample is only \$6,806 and the standard deviation is 3148, the case is eight standard deviations above the mean. But since the Studentized residual is not large, this case does not appear in Figure 17.23.

However, cases that have unusual values for the independent variables can have a substantial impact on the results of analysis and should be identified. One measure of the distance of cases from average values of the independent variables is *Mahalanobis' distance*. In the case of a regression equation with a single independent variable, it is the square of the standardized value of X:

$$D_i = \left(\frac{X_i - \overline{X}}{S_X}\right)^2$$ **Equation 17.24**

Thus, for Case 56, the Mahalanobis' distance shown in Figure 17.24 is 64 (8^2). When there is more than one independent variable—where Mahalanobis' distance is most valuable—the computations are more complex.

Figure 17.24 Mahalanobis' distances

Outliers - Mahalanobis' Distance

Case #	SEXRACE	*MAHAL
56	1.00	63.99758
2	1.00	29.82579
122	1.00	20.32559
67	1.00	14.99121
132	1.00	12.64145
55	1.00	12.64145
415	2.00	11.84140
5	1.00	11.32255
172	1.00	10.49188
23	1.00	10.46720

17.25 Influential Cases: Deleted Residuals and Cook's Distance

Certain observations in a set of data can have a large influence on estimates of the parameters. Figure 17.25a shows such a point. The regression line obtained for the data is quite different if the point is omitted. However, the residual for the circled point is not particularly large when the case (Case 8) is included in the computations and does not therefore arouse suspicion (see the plot in Figure 17.25b).

Figure 17.25a Influential observation

```
Standardized Scatterplot
Across - X          Down - Y
Out ++-----+-----+-----+-----+-----+-----++
  3 +                                    +   Symbols:
    I                                    I
    I                                    I      Max N
  2 +                               O    +
    I                                    I    .     1.0
    I                                    I
  1 +                                    +
    I                     .              I
    I            .     .                 I
  0 +               .                    +
    I                                    I
    I             .                      I
 -1 +                .  .                +
    I                                    I
    I                                    I
 -2 +                                    +
    I                                    I
    I                                    I
 -3 +                                    +
Out ++-----+-----+-----+-----+-----+-----++
    -3    -2    -1     0     1     2     3 Out
```

One way to identify an influential case is to compare the residuals for a case when the suspected case is included in the equation and when it is not. The *adjusted predicted value* (ADJPRED) for case i when it is not included in the computation of the regression line is

$$\hat{Y}_i^{(i)} = B_0^{(i)} + B_1^{(i)} X_i$$ **Equation 17.25a**

where the superscript i indicates that the ith case is excluded. The residual calculated for a case when it is not included is called the *deleted residual* (DRESID), computed as

$$Y_i - \hat{Y}_i^{(i)} \qquad \textbf{Equation 17.25b}$$

The deleted residual can be divided by its standard error to produce the *Studentized deleted residual* (SDRESID).

Although the difference between the deleted and ordinary residual for a case is useful as an index of the influence of that case, this measure does not reflect changes in residuals of other observations when the ith case is deleted. *Cook's distance* does consider changes in all residuals when case i is omitted (Cook, 1977). It is defined as

$$C_i = \frac{\sum_{j=1}^{N} (\hat{Y}_j^{(i)} - \hat{Y}_j)^2}{(p + 1)S^2} \qquad \textbf{Equation 17.25c}$$

where p is the number of independent variables.

The casewise plot for the data in Figure 17.25a is shown in Figure 17.25b. The line for Case 8 describes the circled point. It has neither a very large Studentized residual nor a very large Studentized deleted residual. However, the deleted residual is 5.86 ($Y-ADJPRED$=12−6.14), which is somewhat larger than the ordinary residual (1.24). The large Mahalanobis' distance identifies the case as having an X value far from the mean, while the large Cook's D identifies the case as an influential point.

Figure 17.25b Casewise plot to study influential observation

```
Casewise Plot of Standardized Residual

*: Selected   M: Missing

         -3.0           0.0            3.0
 Case #   0:..............:..............:0    Y     *RESID    *SRESID   *SDRESID   *ADJPRED     *MAHAL    *COOK D
      1    .              .     *        .    7     2.9394     1.4819     1.6990     2.9096     1.0947      .4300
      2    .             *.              .    4     -.5758     -.2780     -.2554     4.7349      .6401      .0107
      3    .              . *            .    6      .9091      .4262      .3951     4.9062      .3068      .0184
      4    .        *     .              .    3    -2.6061    -1.2000    -1.2566     6.0252      .0947      .1158
      5    .              . *            .    7      .8788      .4016      .3717     5.9950      .0038      .0116
      6    .      *       .              .    3    -3.6364    -1.6661    -2.0747     7.1791      .0341      .2071
      7    .              . *            .    8      .8485      .3937      .3641     7.0000      .1856      .0138
      8    .              .  *           .   12     1.2425     1.1529     1.1929     6.1426     4.6402     2.4687
 Case #   0:..............:..............:0    Y     *RESID    *SRESID   *SDRESID   *ADJPRED     *MAHAL    *COOK D
         -3.0           0.0            3.0
```

The regression coefficients with and without Case 8 are shown in Figures 17.25c and 17.25d. Both $B_0^{(8)}$ and $B_1^{(8)}$ are far removed from B_0 and B_1, since Case 8 is an influential point.

Figure 17.25c Regression results from all cases

```
--------------------- Variables in the Equation ----------------------

Variable              B        SE B     95% Confdnce Intrvl B      Beta

X                .51514      .21772      -.01759     1.04788     .69476
(Constant)      3.54547     1.41098       .09294     6.99799

----------- in ------------

Variable           T  Sig T

X              2.366  .0558
(Constant)     2.513  .0457
```

Figure 17.25d Regression coefficients without Case 8

```
----------------- Variables in the Equation ------------------

Variable              B        SE B        Beta          T  Sig T

X                .07141      .42738      .07451       .167  .8739
(Constant)      5.14294     1.91132                  2.691  .0433
```

17.26 When Assumptions Appear To Be Violated

When evidence of violation of assumptions appears, you can pursue one of two strategies. You can either formulate an alternative model, such as weighted least squares, or you can transform the variables so that the current model will be more adequate. For example, taking logs, square roots, or reciprocals can stabilize the variance, achieve normality, or linearize a relationship.

17.27 Coaxing a Nonlinear Relationship to Linearity

To try to achieve linearity, you can transform either the dependent or independent variables, or both. If you alter the scale of independent variables, linearity can be achieved without any effect on the distribution of the dependent variable. Thus, if the dependent variable is normally distributed with constant variance for each value of X, it remains so.

When you transform the dependent variable, its distribution is changed. This new distribution must then satisfy the assumptions of the analysis. For example, if logs of the values of the dependent variable are taken, log Y—not the original Y—must be normally distributed with constant variance.

The choice of transformations depends on several considerations. If the form of the true model governing the relationship is known, it should dictate the choice. For instance, if it is known that $\hat{Y}=AC^X$ is an adequate model, taking logs of both sides of the equation results in

$$\log \hat{Y}_i = \underbrace{(\log A)}_{[B_0]} + \underbrace{(\log C)}_{[B_1]} X_i$$

Equation 17.27

Thus log Y is linearly related to X.

Figure 17.27 A transformed relationship

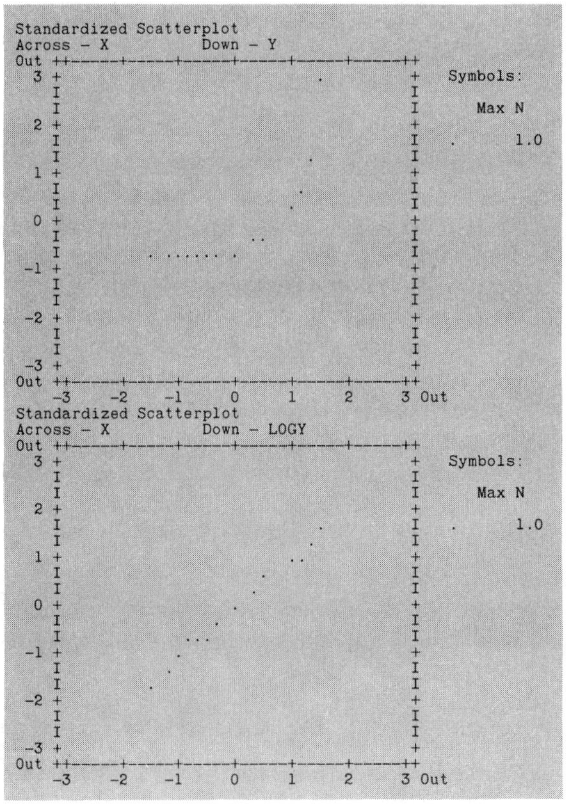

If the true model is not known, you should choose the transformation by examining the plotted data. Frequently, a relationship appears nearly linear for part of the data but is curved for the rest. The first plot in Figure 17.27 is an example. Taking the log of the dependent variable results in the second plot—an improved linear fit.

Other transformations that may diminish curvature are the square root of Y and $-1/Y$. The choice depends, to a certain extent, on the severity of the problem.

17.28 Coping with Skewness

When the distribution of residuals is positively skewed, the log transformation of the dependent variable is often helpful. For negatively skewed distributions, the square transformation is common. It should be noted that the F tests used in regression hypothesis testing are usually quite insensitive to moderate departures from normality.

17.29 Stabilizing the Variance

If the variance of the residuals is not constant, you can try a variety of remedial measures:

- When the variance is proportional to the mean of Y for a given X, use the square root of Y if all Y_i are positive.
- When the standard deviation is proportional to the mean, try the logarithmic transformation.
- When the standard deviation is proportional to the square of the mean, use the reciprocal of Y.
- When Y is a proportion or rate, the arc sine transformation may stabilize the variance.

17.30 Transforming the Salary Data

The assumptions of constant variance and normality appear to be violated with the salary data (see Figures 17.20 and 17.22a). A regression equation using logs of beginning and current salary was developed to obtain a better fit to the assumptions. Figure 17.30a is a scatterplot of Studentized residuals against predicted values when logs of both variables are used in the regression equation.

Figure 17.30a Scatterplot of transformed salary data

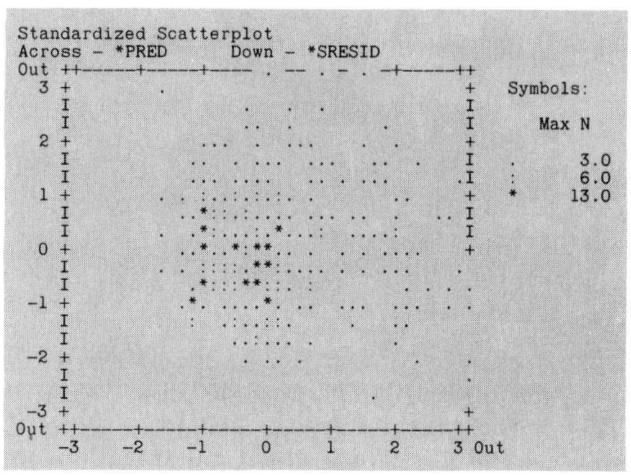

Compare Figures 17.20 and 17.30a and note the improvement in the behavior of the residuals. The spread no longer increases with increasing salary level. Also compare Figures 17.22a and 17.30b and note that the distribution in Figure 17.30b is nearly normal.

Figure 17.30b Histogram of transformed salary data

```
Histogram - Studentized Residual

 N Exp N        (* = 1 Cases,     . : = Normal Curve)
 3   .37   Out ***
 1   .73  3.00 :
 3  1.85  2.67 *:*
 4  4.23  2.33 ***:
10  8.65  2.00 ********:*
14 15.85  1.67 ************** .
21 26.01  1.33 *********************    .
31 38.23  1.00 *******************************      .
48 50.34   .67 ************************************************ .
55 59.38   .33 *******************************************************   .
63 62.74  0.0  **************************************************************:
64 59.38  -.33 **********************************************************:*****
62 50.34  -.67 *************************************************:************
44 38.23 -1.00 *************************************:******
28 26.01 -1.33 *************************:**
14 15.85 -1.67 ************** .
 7  8.65 -2.00 ******* .
 1  4.23 -2.33 *  .
 1  1.85 -2.67 *.
 0   .73 -3.00 .
 0   .37   Out
```

For the transformed data, the multiple R increases slightly to 0.8864, and the outlier plot contains only four cases (compare with Figures 17.10 and 17.23). Thus, the transformation appears to have resulted in a better model.

17.31 A Final Comment on Assumptions

Rarely are assumptions not violated in one way or another in regression analysis and other statistical procedures. However, this is not a justification for ignoring the assumptions. Cranking out regressions with little thought to possible departures from the necessary assumptions can lead to problems in interpreting and applying results. Significance levels, confidence intervals, and other results are sensitive to certain types of violations and cannot be interpreted in the usual fashion if serious departures exist.

By carefully examining residuals and, if need be, using transformations or other methods of analysis, you are in a much better position to pursue analyses that solve the problems you are investigating. Even if everything isn't perfect, you can at least knowledgeably gauge the possible extent of difficulties.

17.32 MULTIPLE REGRESSION MODELS

Beginning salary seems to be a good predictor of current salary, given the evidence shown above. Nearly 80% ($R^2 = 0.77$ from Figure 17.10) of the observed variability in current salaries can be explained by beginning salary levels. But how do variables such as education level, years of experience, race, and sex affect the salary level at which one enters the company?

17.33 Predictors of Beginning Salary

Multiple linear regression extends bivariate regression by incorporating multiple independent variables. The model can be expressed as:

$$Y_i = \beta_0 + \beta_1 X_{1i} + \beta_2 X_{2i} + \ldots + \beta_p X_{pi} + e_i \qquad \textbf{Equation 17.33}$$

The notation X_{pi} indicates the value of the pth independent variable for case i. Again, the β terms are unknown parameters and the e_i terms are independent random variables that are normally distributed with mean 0 and constant variance σ^2. The model assumes that there is a normal distribution of the dependent variable for every combination of the values of the independent variables in the model. For example, if child's height is the dependent variable and age and maternal height are the independent variables, it is assumed that for every combination of age and maternal height there is a normal distribution of children's heights and, though the means of these distributions may differ, all have the same variance.

17.34 The Correlation Matrix

One of the first steps in calculating an equation with several independent variables is to calculate a correlation matrix for all variables, as shown in Figure 17.34. The variables are the log of beginning salary, years of education, sex, years of work experience, race, and age in years. Variables sex and race are represented by *indicator variables,* that is, variables coded as 0 or 1. SEX is coded 1 for female and 0 for male, and MINORITY is coded 1 for nonwhite and 0 for white.

Figure 17.34 The correlation matrix

	LOGBEG	EDLEVEL	SEX	WORK	MINORITY	AGE
LOGBEG	1.000	.686	-.548	.040	-.173	-.048
EDLEVEL	.686	1.000	-.356	-.252	-.133	-.281
SEX	-.548	-.356	1.000	-.165	-.076	.052
WORK	.040	-.252	-.165	1.000	.145	.804
MINORITY	-.173	-.133	-.076	.145	1.000	.111
AGE	-.048	-.281	.052	.804	.111	1.000

The matrix shows the correlations between the dependent variable (LOGBEG) and each independent variable, as well as the correlations between the independent variables. Particularly note any large intercorrelations between the independent variables, since such correlations can substantially affect the results of multiple regression analysis.

17.35 Correlation Matrices and Missing Data

For a variety of reasons, data files frequently contain incomplete observations. Respondents in surveys scrawl illegible responses or refuse to answer certain questions. Laboratory animals die before experiments are completed. Patients fail to keep scheduled clinic appointments. Thus, before computing the correlation matrix, you must usually decide what to do with cases that have missing values for some of the variables.

Before even considering possible strategies, you should determine whether there is evidence that the missing-value pattern is not random. That is, are there reasons to believe that missing values for a variable are related to the values of that variable or other variables? For example, people with low incomes may be less willing to report their financial status than more affluent people. This may be even more pronounced for people who are poor but highly educated.

One simple method of exploring such possibilities is to subdivide the data into two groups—those observations with missing data on a variable and those with complete data—and examine the distributions of the other variables in the file across these two groups. The SPSS/PC+ procedures CROSSTABS and T-TEST are particularly useful for this. For a discussion of more sophisticated methods for detecting nonrandomness, see Frane (1976).

If it appears that the data are not missing randomly, use great caution in attempting to analyze the data. It may be that no satisfactory analysis is possible, especially if there are only a few cases.

If you are satisfied that the missing data are random, several strategies are available. First, if the same few variables are missing for most cases, consider excluding those variables from the analysis. Since this luxury is not usually available, you can alternatively keep all variables but eliminate the cases with missing values on any of them. This is termed *listwise* missing-value treatment since a case is eliminated if it has a missing value on any variable in the list.

If many cases have missing data for some variables, listwise missing-value treatment may eliminate too many cases and leave you with a very small sample. One common technique is to calculate the correlation coefficient between a pair of variables based on all cases with complete information for the two variables, regardless of whether the cases have missing data on any other variable. For example, if a case has values only for variables 1, 3, and 5, it is used only in computations involving variable pairs 1 and 3, 1 and 5, and 3 and 5. This is *pairwise* missing-value treatment.

Several problems can arise with pairwise matrices, one of which is inconsistency. There are some relationships between coefficients that are impossible but may occur when different cases are used to estimate different coefficients. For example, if age and weight and age and height have a high positive correlation, it is impossible in the same sample for height and weight to have a high negative correlation. However, if the same cases are not used to estimate all three coefficients, such an anomaly can occur.

Another problem with pairwise matrices is that no single sample size can be obtained since each coefficient may be based on a different number of cases. In addition, significance levels obtained from analyses based on pairwise matrices must be viewed with caution, since little is known about hypothesis testing in such situations.

Missing-value problems should not be treated lightly. You should always select a missing-value treatment based on careful examination of the data and not leave the choices up to system defaults. In this example, complete information is available for all cases, so missing values are not a problem.

17.36 Partial Regression Coefficients

The summary output when all independent variables are included in the multiple regression equation is shown in Figure 17.36a. The *F* test associated with the analysis of variance table is a test of the null hypothesis that

$$\beta_1=\beta_2=\beta_3=\beta_4=\beta_5=0$$

Equation 17.36a

In other words, it is a test of whether there is a linear relationship between the dependent variable and the entire set of independent variables.

Figure 17.36a Statistics for the equation and analysis of variance table

```
Multiple R            .78420
R Square              .61498
Adjusted R Square     .61086
Standard Error        .09559

Analysis of Variance
                DF      Sum of Squares      Mean Square
Regression       5             6.83039          1.36608
Residual       468             4.27638           .00914

F =      149.50125       Signif F = 0.0
```

The statistics for the independent variables in Figure 17.36b are parallel to those obtained in regression with a single independent variable (see Figure 17.3b). In multiple regression, the coefficients labeled B are called *partial regression coefficients* since the coefficient for a particular variable is adjusted for other independent variables in the equation. The equation that relates the predicted log of beginning salary to the independent variables is

$$\begin{aligned}\text{LOGBEG} = {} & 3.3853 + 0.00102(\text{AGE}) \\ & - 0.10358(\text{SEX}) - 0.05237(\text{MINORITY}) \\ & + 0.03144(\text{EDLEVEL}) + 0.00161(\text{WORK})\end{aligned}$$

Equation 17.36b

Since the dependent variable is in log units, the coefficients can be approximately interpreted in percentage terms. For example, the coefficient of -0.104 for the SEX variable when females are coded as 1 indicates that female salaries are estimated to be about 10% less than male salaries, after statistical adjustment for age, education, work history, and minority status.

Figure 17.36b Statistics for variables in the equation

Variable	B	SE B	Beta	T	Sig T
AGE	1.015396E-03	6.61324E-04	.07811	1.535	.1254
SEX	-.10358	.01032	-.33699	-10.038	.0000
MINORITY	-.05237	.01084	-.14157	-4.832	.0000
EDLEVEL	.03144	1.74805E-03	.59195	17.988	.0000
WORK	1.607508E-03	9.24066E-04	.09143	1.740	.0826
(Constant)	3.38530	.03323		101.866	.0000

17.37 Determining Important Variables

In multiple regression, one sometimes wants to assign relative importance to each independent variable. For example, you might want to know whether education is more important in predicting beginning salary than previous work experience. There are two possible answers, depending on which of the following questions is asked:

- How important are education and work experience when each one is used alone to predict beginning salary?
- How important are education and work experience when they are used to predict beginning salary along with other independent variables in the regression equation?

The first question is answered by looking at the correlation coefficients between salary and the independent variables. The larger the absolute value of the correlation coefficient, the stronger the linear association. Figure 17.34 shows that education correlates more highly with the log of salary than does previous work experience (0.686 and 0.040, respectively). Thus, you would assign more importance to education as a predictor of salary.

The answer to the second question is considerably more complicated. When the independent variables are correlated among themselves, the unique contribution of each is difficult to assess. Any statement about an independent variable is contingent upon the other variables in the equation. For example, the regression coefficient *(B)* for work experience is 0.0007 when it is the sole independent variable in the equation, compared to 0.00161 when the other four independent variables are also in the equation. The second coefficient is more than twice the size of the first.

17.38 Beta Coefficients

It is also inappropriate to interpret the *B*'s as indicators of the relative importance of variables. The actual magnitude of the coefficients depends on the units in which the variables are measured. Only if all independent variables are measured in the same units—years, for example—are their coefficients directly comparable. When variables differ substantially in units of measurement, the sheer magnitude of their coefficients does not reveal anything about relative importance.

One way to make regression coefficients somewhat more comparable is to calculate *Beta* weights, which are the coefficients of the independent variables when all variables are expressed in standardized (*Z*-score) form (see Figure 17.36b). The Beta coefficients can be calculated directly from the regression coefficients using

$$BETA_k = B_k\left(\frac{S_k}{S_Y}\right)$$

Equation 17.38

where S_k is the standard deviation of the *k*th independent variable.

However, the values of the Beta coefficients, like the *B*'s, are contingent on the other independent variables in the equation. They are also affected by the correlations of the independent variables and do not in any absolute sense reflect the importance of the various independent variables.

17.39 Part and Partial Coefficients

Another way of assessing the relative importance of independent variables is to consider the increase in R^2 when a variable is entered into an equation that already contains the other independent variables. This increase is

$$R^2_{change} = R^2 - R^2_{(i)}$$ **Equation 17.39a**

where $R^2_{(i)}$ is the square of the multiple correlation coefficent when all independent variables except the *i*th are in the equation. A large change in R^2 indicates that a variable provides unique information about the dependent variable that is not available from the other independent variables in the equation. The signed square root of the increase is called the *part correlation coefficient.* It is the correlation between Y and X_i when the linear effects of the other independent variables have been removed from X_i. If all independent variables are uncorrelated, the change in R^2 when a variable is entered into the equation is simply the square of the correlation coefficient between that variable and the dependent variable.

The output in Figure 17.39 shows that the addition of years of education to an equation that contains the other four independent variables results in a change in R^2 of 0.266 ($.51593^2$). The square of the part coefficient tells only how much R^2 increases when a variable is added to the regression equation. It does not indicate what proportion of the unexplained variation this increase constitutes. If most of the variation had been explained by the other variables, a small part correlation is all that is possible for the remaining variable. It may therefore be difficult to compare part coefficients.

Figure 17.39 Zero-order, part, and partial correlation coefficients

```
Variable(s) Entered on Step Number
   5..     EDLEVEL    EDUCATIONAL LEVEL

Multiple R              .78420
R Square                .61498              R Square Change      .26619
Adjusted R Square       .61086              F Change          323.55404
Standard Error          .09559              Signif F Change      .0000

F =       149.50125         Signif F = 0.0

------------- Variables in the Equation --------------

Variable        Correl  Part Cor   Partial          F   Sig F

AGE            -.04780    .04404    .07080      2.357   .1254
SEX            -.54802  -.28792   -.42090    100.761   .0000
MINORITY       -.17284  -.13860   -.21799     23.349   .0000
WORK            .03994    .04990    .08015      3.026   .0826
EDLEVEL         .68572    .51593    .63934    323.554   .0000
(Constant)                                  10376.613   .0000
```

A coefficient that measures the proportional reduction in variation is

$$Pr_i^2 = \frac{R^2 - R^2_{(i)}}{1 - R^2_{(i)}}$$ **Equation 17.39b**

The numerator is the square of the part coefficient; the denominator is the proportion of unexplained variation when all but the *i*th variable are in the equation. The signed square root of Pr_i^2 is the *partial correlation coefficient.* It can be interpreted as the correlation between the *i*th independent variable and the dependent variable when the linear effects of the other independent variables have been removed from both X_i and Y. Since the denominator of Pr_i^2 is always less than or equal to 1, the part correlation coefficient is never larger in absolute value than the partial correlation coefficient.

Plots of the residuals of Y and X_i, when the linear effects of the other independent variables have been removed, are a useful diagnostic aid. They are discussed in Section 17.49.

17.40 Variance of the Estimators

The variability of the estimated regression coefficients must also be considered in evaluating the relative importance of the independent variables. Coefficients with large standard errors are unreliable and may differ markedly from sample to sample. It is a dangerous practice to identify variables as important for prediction based only on their significant individual t values.

When the independent variables are correlated among themselves, the parameter estimates are correlated as well. High intercorrelations among the variables can affect the regression estimates in several ways. The estimated variance of the regression coefficient for the ith independent variable is

$$S^2_{B_i} = \frac{S^2}{(1 - R_i^2)(N - 1)S_i^2} \qquad \textbf{Equation 17.40}$$

Here, R_i^2 is the squared multiple correlation when the ith independent variable is considered the dependent variable and the regression equation between it and the other independent variables is calculated. A large value of R_i^2 indicates that the ith independent variable is almost a linear function or combination of the other independent variables. The proportion of variability not explained by the other variables is, as before, $1-R_i^2$. This quantity is usually called the *tolerance* of the variable. As can be seen from Equation 17.40, for a fixed sample size and standard error S, the smaller the tolerance the larger the standard error of the coefficient. Small tolerance values can also cause computational problems for regression solutions. SPSS/PC+ displays the tolerances as shown in Figure 17.40a.

Figure 17.40a Tolerances

Variable	Tolerance
AGE	.31792
SEX	.72998
MINORITY	.95839
WORK	.29784
EDLEVEL	.75966

The variance-covariance matrix and the correlation matrix of the parameter estimates are shown in Figure 17.40b. The entries on the diagonal are the variances of the coefficients. The correlations are displayed above the diagonal and the covariances are displayed below. Note the high correlations between the coefficients for work experience and age (−0.80753). Very small and very large values are displayed in scientific notation. The exponent follows the letter "E." For example, 4.373E−07 is 0.0000004373.

Figure 17.40b Variance-covariance output

Var-Covar Matrix of Regression Coefficients (B)
Below Diagonal: Covariance Above: Correlation

	AGE	SEX	MINORITY	WORK	EDLEVEL
AGE	4.3735E-07	-.28621	-.00722	-.80753	.00388
SEX	-1.953E-06	1.0647E-04	.10271	.38399	.40581
MINORITY	-5.175E-08	1.1485E-05	1.1744E-04	-.04290	.13519
WORK	-4.935E-07	3.6613E-06	-4.296E-07	8.5390E-07	.18683
EDLEVEL	4.4798E-09	7.3196E-06	2.5610E-06	3.0178E-07	3.0557E-06

17.41 Building a Model

Our selection of the five variables to predict beginning salary has been arbitrary to some extent. It is unlikely that all relevant variables have been identified and measured. Instead, some relevant variables have no doubt been excluded, while others that were included may not be very important determinants of salary level. This is not unusual; one must try to build a model from available data, as voluminous or scanty as the data may be. Before considering several formal procedures for model building, we will examine some of the consequences of adding and deleting variables from regression equations. The SPSS/PC+ statistics for variables not in the equation are also described.

17.42 Adding and Deleting Variables

The first step in Figure 17.42 shows the equation and summary statistics when years of education is the sole independent variable and log of beginning salary is the dependent variable. Consider the second step in the same figure, when another variable, sex, is added. The value displayed as R Square Change in the second step is the change in R^2 when sex is added. R^2 for education alone is 0.47021, so R^2_{change} is 0.57598−0.47021, or 0.10577.

Figure 17.42 Adding a variable to the equation

```
Variable(s) Entered on Step Number
   1..      EDLEVEL    EDUCATIONAL LEVEL


Multiple R              .68572
R Square                .47021             R Square Change      .47021
Adjusted R Square       .46909             F Change          418.92011
Standard Error          .11165             Signif F Change      .0000

------------------ Variables in the Equation ------------------

Variable                  B          SE B         Beta         T  Sig T

EDLEVEL             .03642 1.77959E-03       .68572    20.468  .0000
(Constant)         3.31001      .02455                134.821  .0000

Variable(s) Entered on Step Number
   2..      SEX         SEX OF EMPLOYEE


Multiple R              .75893
R Square                .57598             R Square Change      .10577
Adjusted R Square       .57418             F Change          117.48552
Standard Error          .09999             Signif F Change      .0000

------------------ Variables in the Equation ------------------

Variable                  B          SE B         Beta         T  Sig T

EDLEVEL             .02984 1.70549E-03       .56183    17.498  .0000
SEX                -.10697 9.86858E-03      -.34802   -10.839  .0000
(Constant)         3.44754      .02539                135.806  .0000
```

The null hypothesis that the true population value for the change in R^2 is 0 can be tested using

$$F_{change} = \frac{R^2_{change}(N - p - 1)}{q(1 - R^2)} = \frac{(0.1058)\,(474\text{-}2\text{-}1)}{1(1\text{-}0.5760)} = 117.48 \qquad \textbf{Equation 17.42}$$

where N is the number of cases in the equation, p is the total number of independent variables in the equation, and q is the number of variables entered at this step. Sometimes, this is referred to as a *partial F test.* Under the hypothesis that the true change is 0, the significance of the value labeled F Change can be obtained from the F distribution with q and $N-p-1$ degrees of freedom.

The hypothesis that the real change in R^2 is 0 can also be formulated in terms of the β parameters. When only the ith variable is added in a step, the hypothesis that the change in R^2 is 0 is equivalent to the hypothesis that β_i is 0. The F value printed for the change in R^2 is the square of the t value displayed for the test of the coefficient, as shown in Figure 17.42. For example, the t value for sex from Figure 17.42 is −10.839. This value squared is 117.48, the value displayed for F Change.

When q independent variables are entered in a single step, the test that R^2 is 0 is equivalent to the simultaneous test that the coefficients of all q variables are 0. For example, if sex and age were added in the same step to the regression equation that contains education, the F test for R^2 change would be the same as the F test which tests the hypothesis that $\beta_{sex}=\beta_{age}=0$.

Entering sex into the equation with education has effects in addition to changing R^2. For example, note the decrease in magnitude of the regression coefficient for education from Step 1 to Step 2 (from 0.03642 to 0.02984) in Figure 17.42. This is attributable to the correlation between sex and level of education.

When highly intercorrelated independent variables are included in a regression equation, results may appear anomalous. The overall regression may be significant while none of the individual coefficients are significant. The signs of the regression coefficients may be counterintuitive. High correlations between independent variables inflate the variances of the estimates, making individual coefficients quite unreliable without adding much to the overall fit of the model. The problem of linear relationships between independent variables is discussed further in Sections 17.52 through 17.54.

17.43 Statistics for Variables Not in the Equation

When you have independent variables that have not been entered into the equation, you can examine what would happen if they were entered at the next step. Statistics describing these variables are shown in Figure 17.43. The column labeled Beta In is the standardized regression coefficient that would result if the variable were entered into the equation at the next step. The F test and level of significance are for the hypothesis that the coefficient is 0. (Remember that the partial F test and the t test for the hypothesis that a coefficient is 0 are equivalent.) The partial correlation coefficient with the dependent variable adjusts for the variables already in the equation.

Figure 17.43 Coefficients for variables not in the equation

------------- Variables not in the Equation -------------

Variable	Beta In	Partial	Min Toler	F	Sig F
WORK	.14425	.20567	.77382	20.759	.0000
MINORITY	-.12902	-.19464	.84758	18.507	.0000
AGE	.13942	.20519	.80425	20.659	.0000

From statistics calculated for variables not in the equation, you can decide what variable should be entered next. This process is detailed in Section 17.45.

17.44 The "Optimal" Number of Independent Variables

Having seen what happens when sex is added to the equation containing education (Figure 17.42), consider now what happens when the remaining three independent variables are entered one at a time in no particular order. Summary output is shown in Figure 17.44. Step 5 shows the statistics for the equation with all independent variables entered. Step 3 describes the model with education, sex, and work experience as the independent variables.

Figure 17.44 All independent variables in the equation

Step	MultR	Rsq	AdjRsq	F(Eqn)	SigF	RsqCh	FCh	SigCh		Variable	BetaIn	Correl
1	.6857	.4702	.4691	418.920	.000	.4702	418.920	.000	In:	EDLEVEL	.6857	.6857
2	.7589	.5760	.5742	319.896	0.0	.1058	117.486	.000	In:	SEX	-.3480	-.5480
3	.7707	.5939	.5913	229.130	0.0	.0179	20.759	.000	In:	WORK	.1442	.0399
4	.7830	.6130	.6097	185.750	0.0	.0191	23.176	.000	In:	MINORITY	-.1412	-.1728
5	.7842	.6150	.6109	149.501	0.0	.0019	2.357	.125	In:	AGE	.0781	-.0478

Examination of Figure 17.44 shows that R^2 never decreases as independent variables are added. This is always true in regression analysis. However, this does not necessarily mean that the equation with more variables better fits the

B Statistics Guide

population. As the number of parameters estimated from the sample increases, so does the goodness of fit to the sample as measured by R^2. For example, if a sample contains six cases, a regression equation with six parameters fits the sample exactly, even though there may be no true statistical relationship at all between the dependent variable and the independent variables.

As indicated in Section 17.10, the sample R^2 in general tends to overestimate the population value of R^2. Adjusted R^2 attempts to correct the optimistic bias of the sample R^2. Adjusted R^2 does not necessarily increase as additional variables are added to an equation and is the preferred measure of goodness of fit because it is not subject to the inflationary bias of unadjusted R^2. This statistic is shown in the column labeled AdjRsq in the output.

Although adding independent variables increases R^2, it does not necessarily decrease the standard error of the estimate. Each time a variable is added to the equation, a degree of freedom is lost from the residual sum of squares and one is gained for the regression sum of squares. The standard error may increase when the decrease in the residual sum of squares is very slight and not sufficient to make up for the loss of a degree of freedom for the residual sum of squares. The F value for the test of the overall regression decreases when the regression sum of squares does not increase as fast as the degrees of freedom for the regression.

Including a large number of independent variables in a regression model is never a good strategy, unless there are strong, previous reasons to suggest that they all should be included. The observed increase in R^2 does not necessarily reflect a better fit of the model in the population. Including irrelevant variables increases the standard errors of all estimates without improving prediction. A model with many variables is often difficult to interpret.

On the other hand, it is important not to exclude potentially relevant independent variables. The following sections describe various procedures for selecting variables to be included in a regression model. The goal is to build a concise model that makes good prediction possible.

17.45 Procedures for Selecting Variables

You can construct a variety of regression models from the same set of variables. For instance, you can build seven different equations from three independent variables: three with only one independent variable, three with two independent variables, and one with all three. As the number of variables increases, so does the number of potential models (ten independent variables yield 1,023 models).

Although there are procedures for computing all possible regression equations, several other methods do not require as much computation and are more frequently used. Among these procedures are forward selection, backward elimination, and stepwise regression. None of these variable selection procedures is "best" in any absolute sense; they merely identify subsets of variables that, for the sample, are good predictors of the dependent variable.

17.46 Forward Selection

In *forward selection*, the first variable considered for entry into the equation is the one with the largest positive or negative correlation with the dependent variable. The F test for the hypothesis that the coefficient of the entered variable is 0 is then calculated. To determine whether this variable (and each succeeding variable) is entered, the F value is compared to an established criterion. You can specify one of two criteria in SPSS/PC+. One criterion is the minimum value of the F statistic that a variable must achieve in order to enter, called *F-to-enter* (keyword FIN), with a default value of 3.84. The other criterion you can specify is the probability associated with the F statistic, called *probability of F-to-enter* (keyword PIN), with a default of 0.05. In this case, a variable enters into the equation only if the probability associated with the F test is less than or equal to the default 0.05 or the value you specify. By default, the probability of F-to-enter is the criterion used.

These two criteria are not necessarily equivalent. As variables are added to the equation, the degrees of freedom associated with the residual sum of squares decrease while the regression degrees of freedom increase. Thus, a fixed *F* value has different significance levels depending on the number of variables currently in the equation. For large samples, the differences are negligible.

The actual significance level associated with the *F*-to-enter statistic is not the one usually obtained from the *F* distribution, since many variables are being examined and the largest *F* value is selected. Unfortunately, the true significance level is difficult to compute since it depends not only on the number of cases and variables but also on the correlations between independent variables.

If the first variable selected for entry meets the criterion for inclusion, forward selection continues. Otherwise, the procedure terminates with no variables in the equation. Once one variable is entered, the statistics for variables not in the equation are used to select the next one. The partial correlations between the dependent variable and each of the independent variables not in the equation, adjusted for the independent variables in the equation, are examined. The variable with the largest partial correlation is the next candidate. Choosing the variable with the largest partial correlation in absolute value is equivalent to selecting the variable with the largest *F* value.

If the criterion is met, the variable is entered into the equation and the procedure is repeated. The procedure stops when there are no other variables that meet the entry criterion.

To include a specific number of independent variables in the equation, you can specify the number of steps and SPSS/PC+ selects only the first *n* variables that meet entry requirements. Another criterion that is always checked before a variable is entered is the tolerance, which is discussed in Section 17.54.

Figure 17.46a shows output generated from a forward-selection procedure using the salary data. The default entry criterion is PIN=0.05. In the first step, education (variable EDLEVEL) is entered since it has the highest correlation with beginning salary. The significance level associated with education is less than 0.0005, so it certainly meets the criterion for entry.

Figure 17.46a Summary statistics for forward selection

Step	MultR	Rsq	AdjRsq	F(Eqn)	SigF	RsqCh	FCh	SigCh		Variable	BetaIn	Correl
1	.6857	.4702	.4691	418.920	.000	.4702	418.920	.000	In:	EDLEVEL	.6857	.6857
2	.7589	.5760	.5742	319.896	0.0	.1058	117.486	.000	In:	SEX	-.3480	-.5480
3	.7707	.5939	.5913	229.130	0.0	.0179	20.759	.000	In:	WORK	.1442	.0399
4	.7830	.6130	.6097	185.750	0.0	.0191	23.176	.000	In:	MINORITY	-.1412	-.1728

To see how the next variable, SEX, was selected, look at the statistics shown in Figure 17.46b for variables not in the equation when only EDLEVEL is in the equation. The variable with the largest partial correlation is SEX. If entered at the next step, it would have an *F* value of approximately 117 for the test that its coefficient is 0. Since the probability associated with the *F* is less than 0.05, variable sex is entered in the second step.

Figure 17.46b Status of the variables at the first step

------------------ Variables in the Equation ------------------

Variable	B	SE B	Beta	F	Sig F
EDLEVEL	.03642	1.77959E-03	.68572	418.920	.0000
(Constant)	3.31001	.02455		18176.773	.0000

------------- Variables not in the Equation -------------

Variable	Beta In	Partial	Min Toler	F	Sig F
SEX	-.34802	-.44681	.87327	117.486	.0000
WORK	.22747	.30241	.93632	47.408	.0000
MINORITY	-.08318	-.11327	.98234	6.121	.0137
AGE	.15718	.20726	.92113	21.140	.0000

Once variable SEX enters at Step 2, the statistics for variables not in the equation must be examined (see Figure 17.43). The variable with the largest absolute value for the partial correlation coefficient is now years of work experience. Its *F* value is 20.759 with a probability less than 0.05, so variable work is entered in the next step. The same process takes place with variable minority and it is entered, leaving age as the only variable out of the equation. However, as shown in Figure 17.46c, the significance level associated with the age coefficient *F* value is 0.1254, which is too large for entry. Thus, forward selection yields the summary table for the four steps shown in Figure 17.46a.

Figure 17.46c The last step

------------- Variables not in the Equation -------------					
Variable	Beta In	Partial	Min Toler	F	Sig F
AGE	.07811	.07080	.29784	2.357	.1254

17.47 Backward Elimination

While forward selection starts with no independent variables in the equation and sequentially enters them, *backward elimination* starts with all variables in the equation and sequentially removes them. Instead of entry criteria, removal criteria are specified.

Two removal criteria are available in SPSS/PC+. The first is the minimum *F* value (FOUT) that a variable must have in order to remain in the equation. Variables with *F* values less than this *F-to-remove* are eligible for removal. The second criterion available is the maximum probability of *F*-to-remove (keyword POUT) a variable can have. The default FOUT value is 2.71 and the default POUT value is 0.10. The default criterion is POUT.

Figure 17.47a Backward elimination at the first step

------------------ Variables in the Equation ------------------								
Variable	B	SE B	Beta	Correl	Part Cor	Partial	F	Sig F
AGE	1.015396E-03	6.61324E-04	.07811	-.04780	.04404	.07080	2.357	.1254
SEX	-.10358	.01032	-.33699	-.54802	-.28792	-.42090	100.761	.0000
MINORITY	-.05237	.01084	-.14157	-.17284	-.13860	-.21799	23.349	.0000
EDLEVEL	.03144	1.74805E-03	.59195	.68572	.51593	.63934	323.554	.0000
WORK	1.607508E-03	9.24066E-04	.09143	.03994	.04990	.08015	3.026	.0826
(Constant)	3.38530	.03323					10376.613	.0000

Look at the salary example again, this time constructing the model with backward elimination. The output in Figure 17.47a is from the first step, in which all variables are entered into the equation. The variable with the smallest partial correlation coefficient, age, is examined first. Since the probability of its *F* (0.1254) is greater than the default POUT criterion value of 0.10, variable AGE is removed.

Figure 17.47b Backward elimination at the last step

Variable	B	SE B	Beta	Correl	Part Cor	Partial	F	Sig F
SEX	-.09904	9.90104E-03	-.32223	-.54802	-.28733	-.41933	100.063	.0000
MINORITY	-.05225	.01085	-.14125	-.17284	-.13828	-.21700	23.176	.0000
EDLEVEL	.03143	1.75056E-03	.59176	.68572	.51577	.63827	322.412	.0000
WORK	2.753243E-03	5.45823E-04	.15659	.03994	.14489	.22685	25.444	.0000
(Constant)	3.41195	.02838					14454.046	.0000

The equation is then recalculated without AGE, producing the statistics shown in Figure 17.47b. The variable with the smallest partial correlation is MINORITY. However, its significance is less than the 0.10 criterion, so backward elimination stops. The equation resulting from backward elimination is the same as the one from forward selection. This is not always the case, however. Forward- and backward-selection procedures can give different results, even with comparable entry and removal criteria.

17.48 Stepwise Selection

Stepwise selection of independent variables is really a combination of backward and forward procedures and is probably the most commonly used method. The first variable is selected in the same manner as in forward selection. If the variable fails to meet entry requirements (either FIN or PIN), the procedure terminates with no independent variables in the equation. If it passes the criterion, the second variable is selected based on the highest partial correlation. If it passes entry criteria, it also enters the equation.

From this point, stepwise selection differs from forward selection: the first variable is examined to see whether it should be removed according to the removal criterion (FOUT or POUT) as in backward elimination. In the next step, variables not in the equation are examined for entry. After each step, variables already in the equation are examined for removal. Variables are removed until none remain that meet the removal criterion. To prevent the same variable from being repeatedly entered and removed, PIN must be less than POUT (or FIN greater than FOUT). Variable selection terminates when no more variables meet entry and removal criteria.

In the salary example, stepwise selection with the default criteria results in the same equation produced by both forward selection and backward elimination (see Figure 17.48).

Figure 17.48 Stepwise output

```
Listwise Deletion of Missing Data

Equation Number 1    Dependent Variable..   LOGBEG

Beginning Block Number  1.  Method:  Stepwise


Variable(s) Entered on Step Number
   1..     EDLEVEL   EDUCATIONAL LEVEL


Multiple R              .68572
R Square                .47021
Adjusted R Square       .46909
Standard Error          .11165

F =       418.92011         Signif F =  .0000

------------------ Variables in the Equation ------------------

Variable                B           SE B        Beta          F  Sig F

EDLEVEL            .03642 1.77959E-03       .68572    418.920  .0000
(Constant)        3.31001      .02455                18176.773  .0000


------------- Variables not in the Equation -------------

Variable      Beta In  Partial  Min Toler          F  Sig F

SEX          -.34802  -.44681     .87327    117.486  .0000
WORK          .22747   .30241     .93632     47.408  .0000
MINORITY     -.08318  -.11327     .98234      6.121  .0137
AGE           .15718   .20726     .92113     21.140  .0000

             * * * * * * * * * * * * * * * * * * * * * * *

Variable(s) Entered on Step Number
   2..     SEX        SEX OF EMPLOYEE


Multiple R              .75893
R Square                .57598
Adjusted R Square       .57418
Standard Error          .09999

F =       319.89574         Signif F = 0.0


------------------ Variables in the Equation ------------------

Variable                B           SE B        Beta          F  Sig F

EDLEVEL            .02984 1.70549E-03       .56183    306.191  .0000
SEX               -.10697 9.86858E-03      -.34802    117.486  .0000
(Constant)        3.44754      .02539                18443.284  .0000

                                                      CONTINUED...
```

Figure 17.48 *continued*

```
------------- Variables not in the Equation -------------

Variable      Beta In  Partial  Min Toler          F  Sig F

WORK           .14425   .20567     .77382     20.759  .0000
MINORITY      -.12902  -.19464     .84758     18.507  .0000
AGE            .13942   .20519     .80425     20.659  .0000

         * * * * * * * * * * * * * * * * * * * * * *

Variable(s) Entered on Step Number
   3..     WORK       WORK EXPERIENCE

Multiple R           .77066
R Square             .59391
Adjusted R Square    .59132
Standard Error       .09796

F =      229.13001        Signif F = 0.0

------------------ Variables in the Equation ------------------

Variable                 B        SE B       Beta          F  Sig F

EDLEVEL             .03257 1.77493E-03     .61321    336.771  .0000
SEX                -.09403      .01008   -.30594     87.099  .0000
WORK         2.536163E-03 5.56642E-04     .14425     20.759  .0000
(Constant)         3.38457      .02845             14150.645  .0000

------------- Variables not in the Equation -------------

Variable      Beta In  Partial  Min Toler          F  Sig F

MINORITY      -.14125  -.21700     .75967     23.176  .0000
AGE            .07633   .06754     .29839      2.149  .1433

         * * * * * * * * * * * * * * * * * * * * * *

Variable(s) Entered on Step Number
   4..     MINORITY  MINORITY CLASSIFICATION

Multiple R           .78297
R Square             .61304
Adjusted R Square    .60974
Standard Error       .09573

F =      185.74958        Signif F = 0.0

------------------ Variables in the Equation ------------------

Variable                 B        SE B       Beta          F  Sig F

EDLEVEL             .03143 1.75056E-03     .59176    322.412  .0000
SEX                -.09904 9.90104E-03   -.32223    100.063  .0000
WORK         2.753243E-03 5.45823E-04     .15659     25.444  .0000
MINORITY           -.05225      .01085   -.14125     23.176  .0000
(Constant)         3.41195      .02838             14454.046  .0000

------------- Variables not in the Equation -------------

Variable      Beta In  Partial  Min Toler          F  Sig F

AGE            .07811   .07080     .29784      2.357  .1254

End Block Number   1   PIN =     .050 Limits reached.
```

The three procedures do not always result in the same equation, though you should be encouraged when they do. The model selected by any method should be carefully studied for violations of the assumptions. It is often a good idea to develop several acceptable models and then choose among them based on interpretability, ease of variable acquisition, parsimony, and so forth.

17.49 Checking for Violation of Assumptions

The procedures discussed in Sections 17.17 through 17.22 for checking for violations of assumptions in bivariate regression apply in the multivariate case as well. Residuals should be plotted against predicted values as well as against each independent variable. The distribution of residuals should be examined for normality.

Several additional residual plots may be useful for multivariate models. One of these is the partial regression plot. For the *j*th independent variable, it is obtained by calculating the residuals for the dependent variable when it is predicted from all the independent variables excluding the *j*th and by calculating the residuals for the *j*th independent variable when it is predicted from all of the other independent variables. This removes the linear effect of the other independent variables from both variables. For each case, these two residuals are plotted against each other.

A partial regression plot for educational level for the regression equation that contains work experience, minority, sex, and educational level as the independent variables is shown in Figure 17.49a. (Summary statistics for the regression equation with all independent variables are displayed in the last step of Figure 17.73.)

Figure 17.49a Partial regression plot from PLOT

```
                  PLOT OF RES1 WITH RES2
         +---+----+----+----+----+----+----+---+
      .6+                              1       +
         |                                     |
         |                     1               |
R        |                      11             |
e        |                     1 11 1111       |
s     .3+               1            31    1   +
i        |                  2  32541           R
d        |          1     21 2 22B221          |
u        |      1       1111382516812  1       |
a        |              1364681 232121 1       |
l      0+   12142212213DB4CB 86221        1    +
         |   242512454SI9E7718212 1            |
         | 2 3  3312AA1298FM2 1121             |
         |          22    142                  |
         R  1 1            1                   |
    -.3+                                       +
         +---+----+----+----+----+----+----+---+
       -6.75     -2.25      2.25      6.75
             -4.5        0         4.5

                        Residual

474 cases plotted. Regression statistics of RES1 on RES2:
Correlation  .63827 R Squared  .40739  S.E. of Est     .09542 Sig.  .0000
Intercept(S.E.)      .00000(   .00438)  Slope(S.E.)     .03143(    .00174)
```

Several characteristics of the partial regression plot make it a particularly valuable diagnostic tool. The slope of the regression line for the two residual variables (0.03143) is equal to the coefficient for the EDLEVEL variable in the multiple regression equation after the last step (Step 4 in Figure 17.73). Thus, by examining the bivariate plot, you can conveniently identify points that are "influential" in the determination of the particular regression coefficient (see Sections 17.23 through 17.25). The correlation coefficient between the two residuals, 0.638, is the partial correlation coefficient coefficient discussed in Section 17.39. The residuals from the least-squares line in Figure 17.49a are equal to the residuals from the final multiple regression equation, which includes all the independent variables.

The partial regression plot also helps you assess inadequacies of the selected model and violations of the underlying assumptions. For example, the partial regression plot of educational level does not appear to be linear, suggesting that an additional term, such as years of education squared, might also be included in the model. This violation is much easier to spot using the partial regression plot than the plot of the independent variable against the residual from the equation with all independent variables. Figures 17.49b and 17.49c show the residual scatterplot and partial regression plot produced by the REGRESSION procedure. Note that the nonlinearity is much more apparent in the partial regression plot.

Figure 17.49b Residual scatterplot from REGRESSION

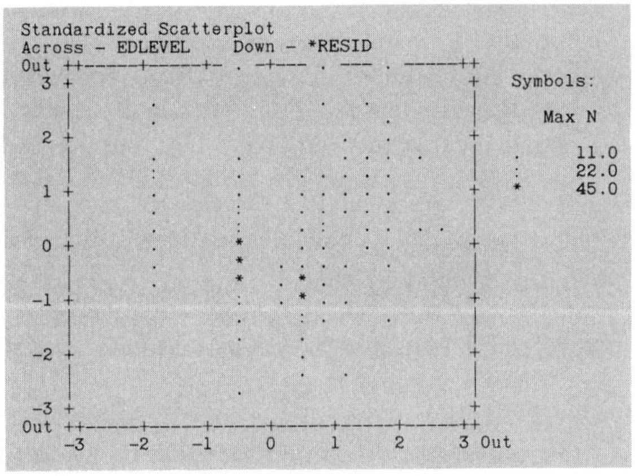

Figure 17.49c Partial regression plot from REGRESSION

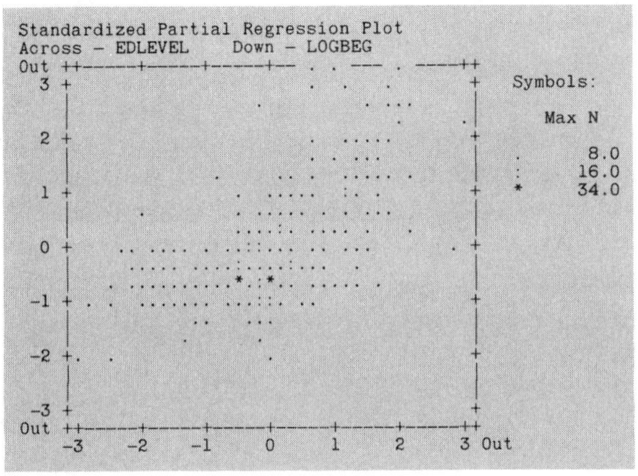

Figure 17.49d contains the summary statistics when the number of years of education squared is included in the multiple regression equation. The multiple R^2 increases from 0.61 (Step 4 in Figure 17.73) to 0.71, a significant improvement.

Figure 17.49d The regression equation with education squared

```
Variable(s) Entered on Step Number  1..    WORK       WORK EXPERIENCE
                                    2..    MINORITY   MINORITY CLASSIFICATION
                                    3..    SEX        SEX OF EMPLOYEE
                                    4..    EDLEVEL    EDUCATIONAL LEVEL
                                    5..    ED2

Multiple R            .84302        Analysis of Variance
R Square              .71068                        DF      Sum of Squares      Mean Square
Adjusted R Square     .70759        Regression       5             7.89331          1.57866
Standard Error        .08286        Residual       468             3.21345           .00687

                                    F =     229.91286         Signif F = 0.0
```

The casewise plot of cases with residuals greater than 3 in absolute value is shown in Figure 17.49e. An additional statistic shown with this plot is the centered leverage (*LEVER), which is a measure of the influence of a point. The centered leverage ranges from $-1/N$ to $(N-1)/N$, where N is the number of observations. The mean value for the centered leverage is N/p, where p is the number of independent variables in the equation. The Mahalanobis' distance previously described is obtained by multiplying the leverage value by $N-1$. A leverage of $-1/N$ identifies a point with no influence on the fit, while a point with a leverage of $(N-1)/N$ indicates that a degree of freedom has been devoted to fitting the data point. See Belsley, Kuh, and Welsch (1980) for further details.

Figure 17.49e Casewise plot of outliers

```
Casewise Plot of Standardized Residual

Outliers = 3.    *: Selected   M: Missing

       -6.           -3.  3.           6.
 Case #  0:.............:  :.............:0     *PRED    *RESID    *LEVER    *MAHAL    *COOK D
     2   .               ..           *  .    3.9303     .4499     .0042    2.0100      .0316
    56   .               ..      *       .    4.1478     .3572     .0210    9.9449      .0751
    67   .               ..      *       .    3.9303     .3484     .0042    2.0100      .0190
   122   .               ..       *      .    3.9465     .3758     .0103    4.8777      .0437
   283   .               ..*             .    3.8276     .2516     .0066    3.1029      .0136
   402   .               ..       *      .    3.7578     .3725     .0141    6.6804      .0565
   415   .               ..      *       .    3.8888     .3577     .0106    5.0249      .0406

    7 Outliers found.
```

17.50 Interpreting the Equation

The multiple regression equation estimated above suggests several findings. Education appears to be the best predictor of beginning salary, at least among the variables included in this study (Figure 17.47a). The sex of the employee also appears to be important. Women are paid less than men since the sign of the regression coefficient is negative (men are coded 0 and women are coded 1). Years of prior work experience and race are also related to salary, but when education and sex are included in the equation, the effect of experience and race is less striking.

Do these results indicate that there is sex discrimination at the bank? Not necessarily. It is well recognized that all education is not equally profitable. Master's degrees in business administration and political science are treated quite differently in the marketplace. Thus, a possible explanation of the observed results is that women enter areas that are just not very well paid. Although this may suggest inequities in societal evaluation of skills, it does not necessarily imply discrimination at the bank. Further, many other potential job-related skills or qualifications are not included in the model. Also, some of the existing variables, such as age, may make nonlinear as well as linear contributions to the fit. Such contributions can often be approximated by including new variables that are simple functions of the existing one. For example, the age values squared may improve the fit.

17.51 Statistics for Unselected Cases

As previously noted, a model usually fits the sample from which it is derived better than it fits the population. A sometimes useful strategy for obtaining an estimate of how well the model fits the population is to split the sample randomly into two parts. One part is then used to estimate the model, while the remaining cases are reserved for testing the goodness of fit.

It is also sometimes interesting to split the data on some characteristics of the sample. For example, you can develop the salary equation for males alone and then apply it to females to see how well it fits. For example, Figure 17.51 shows histograms of residuals for males (denoted as selected cases) and females (unselected cases). Note that the females' salaries are too large when predicted from the male equation since most of the residuals are negative. The multiple R

for the females is 0.45596, which is smaller than the 0.73882 for males (stepwise selection was used).

Figure 17.51 Histograms for males (selected) and females (unselected)

```
Histogram - Standardized Residual
- Selected Cases
 N Exp N       (* = 1 Cases,     . : = Normal Curve)
 3   .20   Out ***
 3   .40  3.00 ***
 0  1.01  2.67 .
 1  2.30  2.33 *.
 2  4.71  2.00 **   .
 5  8.63  1.67 *****    .
11 14.16  1.33 ***********   .
19 20.81  1.00 *******************  .
26 27.40   .67 **************************.
35 32.32   .33 *******************************:***
31 34.15  0.0  *******************************   .
30 32.32  -.33 ****************************** .
42 27.40  -.67 **************************:***************
31 20.81 -1.00 ********************:**********
10 14.16 -1.33 **********    .
 5  8.63 -1.67 *****    .
 3  4.71 -2.00 *** .
 0  2.30 -2.33   .
 1  1.01 -2.67 :
 0   .40 -3.00
 0   .20   Out

Histogram - Standardized Residual
- Unselected Cases
 N Exp N       (X = 1 Cases,     . : = Normal Curve)
 0   .17   Out
 0   .33  3.00
 0   .84  2.67 .
 0  1.93  2.33   .
 0  3.94  2.00     .
 1  7.22  1.67 X      .
 1 11.85  1.33 X          .
 3 17.42  1.00 XXX             .
 4 22.94   .67 XXXX                 .
10 27.06   .33 XXXXXXXXXX               .
14 28.59  0.0  XXXXXXXXXXXXXX               .
32 27.06  -.33 XXXXXXXXXXXXXXXXXXXXXXXXXX:XXXXX
32 22.94  -.67 XXXXXXXXXXXXXXXXXXXXXX:XXXXXXXXX
52 17.42 -1.00 XXXXXXXXXXXXXXXX:XXXXXXXXXXXXXXXXXXXXXXXXXXXXXXXXXXX
22 11.85 -1.33 XXXXXXXXXXX:XXXXXXXXXX
13  7.22 -1.67 XXXXXX:XXXXXX
12  3.94 -2.00 XXX:XXXXXXXX
 8  1.93 -2.33 X:XXXXXX
 7   .84 -2.67 :XXXXXX
 1   .33 -3.00 X
 4   .17   Out XXXX
```

17.52 Problems of Multicollinearity

Preceding sections deal with the consequences of correlated independent variables in regression analysis. The estimates of the β's and the sum of squares attributable to each variable are dependent on the other variables in the equation. Variances of the estimators also increase when independent variables are interrelated. This may result in a regression equation with a significant R^2, although virtually none of the coefficients is significantly different from 0. If any independent variable is a perfect linear combination of other independent variables, the correlation matrix is *singular* and a unique, unbiased least-squares solution does not exist.

Although situations involving singularities do occur, they are not as common as those involving near-singularities—variables that are almost linear combinations of other independent variables. These variables are often called *multicollinear*.

17.53 Methods of Detection

Multicollinearities can be detected in several ways. Large coefficients in the correlation matrix always signal the presence of multicollinearity. However, multicollinearity can exist even when none of the correlation coefficients is very large.

One of the most frequently used indicators of interdependency between variables is the tolerance (see Section 17.40). If the variable has a large R^2—or equivalently a small tolerance—when it is predicted from the other independent variables, a potentially troublesome situation exists. Not only are the variances of the estimators inflated, but computational problems can occur.

17.54 SPSS/PC+ and Multicollinearity

In the SPSS/PC+ REGRESSION procedure, various steps are taken to warn you of multicollinearity. Before an independent variable is entered into the equation, its tolerance with other independent variables already in the equation is calculated.

It is possible for a variable not in the equation to have an acceptable tolerance level but when entered to cause the tolerance of other variables already in the equation to become unacceptably small (Berk, 1977; Frane, 1977). Thus, the tolerances of all the variables in the equation are recomputed at each step. If either the tolerance of the variable or the tolerance of any variable already in the equation is less than 0.01, a warning is issued and the variable is not entered unless the default TOLERANCE criterion has been altered.

In SPSS/PC+, you can display both the tolerance of a variable and the minimum tolerance of all independent variables in the equation if the variable were entered.

17.55 RUNNING PROCEDURE REGRESSION

The REGRESSION procedure provides five equation-building methods: forward selection, backward elimination, stepwise selection, forced entry, and forced removal. The subcommands for residual analysis help detect influential data points, outliers, and violations of the regression model assumptions.

17.56 Building the Equation

To build a simple regression model, you must specify three required subcommands: a VARIABLES subcommand that names the variables to be analyzed, a DEPENDENT subcommand that indicates the dependent variable, and a METHOD subcommand that names the method to be used. For example, to build the simple bivariate model of beginning salary and current salary discussed earlier in the chapter, specify

```
REGRESSION VARIABLES=SALBEG SALNOW
  /DEPENDENT=SALNOW
  /METHOD=ENTER SALBEG.
```

The beginning (SALBEG) and current (SALNOW) salaries are named, with the latter specified as the dependent variable. The ENTER keyword enters beginning salary into the equation. The output produced from this command is shown in Figures 17.3b, 17.10, and 17.11a.

17.57 The VARIABLES and DEPENDENT Subcommands

The VARIABLES subcommand lists all variables needed for the regression analysis, including those that are named on the DEPENDENT and METHOD subcommands. There can be only one VARIABLES subcommand per REGRESSION command, and it must be specified before the DEPENDENT and METHOD subcommands.

For example, to run both the bivariate and multivariate examples that have been developed in this chapter, specify

```
REGRESSION VARIABLES=SALBEG SALNOW LOGBEG
          EDLEVEL SEX WORK MINORITY AGE
  /DEPENDENT=SALNOW
  /METHOD=ENTER SALBEG
  /DEPENDENT=LOGBEG
  /METHOD=ENTER EDLEVEL TO AGE.
```

The first DEPENDENT subcommand defines a single equation with SALNOW as the dependent variable, and the METHOD subcommand requests that SALBEG be entered into the equation. The second DEPENDENT subcommand defines another equation, with LOGBEG as the dependent variable. The associated METHOD subcommand requests that the variables EDLEVEL to AGE be entered into the equation. The TO convention for naming consecutive variables used in the second METHOD subcommand refers to the order in which the variables are named on the VARIABLES subcommand. See Figures 17.36a and 17.36b for the output from the second equation.

Usually only one variable is named as a dependent variable on a DEPENDENT subcommand. If you name more than one dependent variable, SPSS/PC+ develops an equation for the first variable, then for the second, and so on.

17.58 The METHOD Subcommand

At least one METHOD subcommand must immediately follow each DEPENDENT subcommand, specifying the method to be used in developing the regression equation. The available methods are

FORWARD *Forward variable selection.* Variables are entered one at a time based on entry criteria (Section 17.46).

BACKWARD *Backward variable elimination.* All variables are entered and then removed one at a time based on removal criteria (Section 17.47).

STEPWISE *Stepwise variable entry and removal.* Variables are examined at each step for entry or removal (Section 17.48).

ENTER *Forced entry.* The variables named are entered in a single step. The default variable list is all independent variables.

REMOVE (varlist) *Forced removal.* The variables named are removed in a single step. REMOVE must have an accompanying variable list.

TEST (varlist) *Test indicated subsets of independent variables.* TEST offers an easy way to test a variety of models using R^2 change and its test of significance as the criterion for the "best" model. TEST must have an accompanying variable list.

A variable list is required with the REMOVE and TEST keywords and is optional for the other METHOD keywords. The default variable list for methods FORWARD, BACKWARD, STEPWISE, and ENTER includes all variables named on the VARIABLES subcommand that are not named on the preceding DEPENDENT subcommand. For example, to request the backward-elimination method discussed in Section 17.47, specify

```
REGRESSION VARIABLES=LOGBEG EDLEVEL SEX WORK MINORITY AGE
  /DEPENDENT=LOGBEG
  /METHOD=BACKWARD.
```

You can specify multiple METHOD subcommands. For example, you might want to force one variable into the equation first and then enter the remaining variables in a forward-selection fashion, as in

```
REGRESSION VARIABLES=LOGBEG EDLEVEL SEX WORK MINORITY AGE
  /DEPENDENT=LOGBEG
  /METHOD=ENTER EDLEVEL
  /METHOD=FORWARD SEX TO AGE.
```

17.59 The CRITERIA Subcommand

You can control the statistical criteria by which REGRESSION chooses variables for entry into or removal from an equation with the CRITERIA subcommand. Place the CRITERIA subcommand after the VARIABLES subcommand and before the DEPENDENT subcommand. A CRITERIA subcommand affects any subsequent DEPENDENT and METHOD subcommands and remains in effect until overridden with another CRITERIA subcommand.

The CRITERIA keywords are

DEFAULTS *PIN(0.05), POUT(0.10), and TOLERANCE(0.01).* These are the defaults if no CRITERIA subcommand is specified. If criteria have been changed, DEFAULTS restores the default values.

PIN(value) *Probability of* F*-to-enter.* Use to override the default value of 0.05.

POUT(value) *Probability of* F*-to-remove.* Use to override the default value of 0.10.

FIN(value) F*-to-enter.* The default value is 3.84. FIN and PIN are mutually exclusive.

FOUT(value) F*-to-remove.* The default value is 2.71. FOUT and POUT are mutually exclusive.

TOLERANCE(value) *Tolerance.* The default value is 0.01. All variables must pass both tolerance and minimum tolerance tests before entering the equation. (See Sections 17.40, 17.53 and 17.54.)

MAXSTEPS(n) *Maximum number of steps.* For the STEPWISE method, the default is twice the number of independent variables. For the FORWARD and BACKWARD methods, the default maximum is the number of variables meeting the PIN and POUT or FIN and FOUT criteria. The MAXSTEPS value applies to the total model. The default value for the total model is the sum of the maximum number of steps over each method in the model.

For example, to change stepwise entry and removal criteria to FIN and FOUT and use their default values of 3.84 and 2.71, respectively, specify

```
REGRESSION VARIABLES=LOGBEG EDLEVEL SEX WORK MINORITY AGE
  /CRITERIA=FIN,FOUT
  /DEPENDENT=LOGBEG
  /METHOD=STEPWISE.
```

17.60 The STATISTICS Subcommand

By default, REGRESSION displays the four sets of statistics described for keywords R, ANOVA, COEFF, and OUTS below. These statistics are shown in Figures 17.3b, 17.10, and 17.11a for the bivariate equation, and in Figures 17.36a and 17.36b for the multivariate equation. You can specify exactly which statistics you want displayed by any of the following keywords on the STATISTICS subcommand.

DEFAULTS R, *ANOVA, COEFF, and OUTS.* These statistics are displayed when the STATISTICS subcommand is omitted or if no keywords are specified on the subcommand. If you specify statistics keywords on a STATISTICS subcommand, the default statistics will not appear unless you specify them explicitly, either individually, or with the DEFAULTS keyword.

ALL *All statistics except* F, *LINE, and END.*

R *Multiple* R. Displays multiple R, R^2, adjusted R^2, and the standard error. (See Figure 17.10.)

ANOVA *Analysis of variance table.* Displays degrees of freedom, sums of squares, mean squares, F value for multiple R, and the observed significance level of F. (See Figure 17.11a.)

CHA *Displays change in* R^2 between steps, F value for change in R^2, and significance of F. (See Figure 17.42.)

BCOV *Variance-covariance matrix.* Displays a matrix with covariances above the diagonal, correlations below the diagonal, and variances on the diagonal. (See Figure 17.40b.)

XTX *Sweep matrix.*

COND *Condition number bounds.* Prints the lower and upper bounds for the condition number of the submatrix of the sweep matrix that contains independent variables already entered. (Berk, 1977)

COEFF *Statistics for variables in the equation.* Displays regression coefficient B, standard error of B, standardized coefficient Beta, t value for B, and two-tailed significance level of t.

OUTS *Statistics for variables not in the equation* that have been named on the VARIABLES subcommand. Statistics are Beta if the variable were entered, t value for Beta, significance level of t, partial correlation with the dependent variable controlling for variables in the equation, and minimum tolerance.

ZPP *Zero-order, part, and partial correlation.* (See Figure 17.39.)

CI *Confidence intervals.* Displays the 95% confidence interval for the unstandardized regression coefficient. (See Figure 17.8.)

SES *Approximate standard error of the standardized regression coefficients.* (Meyer and Younger, 1976)

TOL *Tolerance.* Displays tolerance for variables in the equation and, for variables not in the equation, the tolerance a variable would have if it were the only variable entered next. (See Figure 17.40a.)

F F *value for* B *and significance of* F. Displayed instead of *t* for COEFF and OUTS. (See, for example, Figures 17.46b, 17.46c, 17.47a, and 17.47b.)

LINE *Summary line for each step in step methods.* Displays a single summary line for each step in BACKWARD, FORWARD, or STEPWISE methods and the default or requested statistics at the end of each method block (BACKWARD, FORWARD, STEPWISE, ENTER, REMOVE, or TEST).

HISTORY *Step history.* Displays a summary report with a summary line for each method (ENTER, REMOVE, or TEST, if the equation changes) or step if the method entails steps (FORWARD, BACKWARD, or STEPWISE). If history is the only statistic requested, COEFF is displayed for the final equation. (See Figures 17.44 and 17.46a.)

END *One summary line per step or method block.* Displays a summary line per step for BACKWARD, FORWARD, or STEPWISE, and one summary line per block for ENTER, REMOVE, or TEST, if the equation changes.

The STATISTICS subcommand must appear before the DEPENDENT subcommand that initiates the equation and remains in effect until overridden by another STATISTICS subcommand. For example, to produce the output in Figure 17.8, specify

```
REGRESSION VARIABLES=SALBEG SALNOW
  /STATISTICS=CI
  /DEPENDENT=SALNOW
  /METHOD=ENTER SALBEG.
```

To produce the output for the multivariate example shown in Figure 17.42, specify

```
REGRESSION VARIABLES=LOGBEG EDLEVEL SEX WORK MINORITY AGE
  /STATISTICS=R CHANGE COEFF
  /DEPENDENT=LOGBEG
  /METHOD=ENTER EDLEVEL
  /METHOD=ENTER SEX.
```

17.61 The ORIGIN Subcommand

The regression model contains a constant term. You can use the ORIGIN subcommand to suppress this term and obtain regression through the origin. The NOORIGIN subcommand, which is the default, requests that equations include a constant term.

Place the ORIGIN or NOORIGIN subcommand between the VARIABLES subcommand and the DEPENDENT subcommand for the equation. For example,

```
REGRESSION VARIABLES=SALBEG SALNOW,EDLEVEL
  /DEPENDENT=SALNOW
  /METHOD=ENTER SALBEG
    /ORIGIN
    /DEPENDENT=SALBEG
    /METHOD=ENTER EDLEVEL.
```

requests two equations, the first with a constant term (the default) and the second with regression through the origin.

There are no specifications for the ORIGIN and NOORIGIN subcommands. Once specified, the ORIGIN subcommand remains in effect until NOORIGIN is requested.

17.62 The SELECT Subcommand

Use the SELECT subcommand to select a subset of cases for computing the regression equation. Only selected cases contribute to the correlation coefficients and to the regression equation. Residuals and predicted values are calculated and reported separately for both selected and unselected cases. The SELECT subcommand can precede or immediately follow the VARIABLES subcommand and is in effect for the entire REGRESSION command. The form of the SELECT subcommand is

/SELECT= varname relation value

The *relation* can be EQ, NE, LT, LE, GT, or GE.

For example, to generate separate residuals histograms for males and females based on the equation developed for males alone (SEX=0), as shown in Figure 17.51, specify

```
REGRESSION SELECT SEX EQ 0
  /VARIABLES=LOGBEG EDLEVEL SEX WORK MINORITY AGE
  /DEPENDENT=LOGBEG
  /METHOD=STEPWISE
  /RESIDUALS=HISTOGRAM.
```

17.63 The MISSING Subcommand

Use the MISSING subcommand to specify the treatment of cases with missing values. If the MISSING subcommand is omitted, a case with user- or system-missing values for any variable named on the VARIABLES subcommand is excluded from the computation of the correlation matrix on which all analyses are based. The MISSING subcommand can precede or immediately follow the VARIABLES subcommand and is in effect for the entire REGRESSION command.

The available keywords are

LISTWISE *Delete cases with missing values listwise.* Only cases with valid values for all variables listed on the VARIABLES subcommand are included in analyses. If INCLUDE is also specified, only cases with system-missing values are deleted listwise. LISTWISE is the default.

PAIRWISE *Delete cases with missing values pairwise.* Cases with complete data on the pair of variables being correlated are used to compute the correlation coefficient. If INCLUDE is also specified, only cases with system-missing values are deleted pairwise.

MEANSUBSTITUTION *Replace missing values with the variable mean.* All cases are used for computations, with the mean of a variable substituted for missing observations. If INCLUDE is also specified, user-missing values are included in the computation of the means and only system-missing values are substituted.

INCLUDE *Include all cases with user-missing values.* Only cases with system-missing values are excluded.

17.64 The DESCRIPTIVES Subcommand

You can request a variety of descriptive statistics with the DESCRIPTIVES subcommand. These statistics are displayed for all variables specified on the VARIABLES subcommand, regardless of which variables you specify for computations. Descriptive statistics are based on all valid cases for each variable if you have specified PAIRWISE or MEANSUB on the MISSING subcommand. Otherwise, only cases that are included in the computation of the correlation matrix are used. If you specify the DESCRIPTIVES subcommand without any keywords, the statistics listed for keyword DEFAULTS are displayed. If you name any statistics on DESCRIPTIVES, only those explicity requested are displayed.

The following descriptive statistics are available:

DEFAULTS *MEAN, STDDEV, and CORR.* This is the default if DESCRIPTIVES is specified without any keywords.

MEAN *Variable means.*

STDDEV *Variable standard deviations.*

VARIANCE *Variable variances.*

CORR *Correlation matrix.*

SIG *One-tailed significance levels for the correlation coefficients.*

BADCORR *Correlation matrix only if some coefficients cannot be computed.*

COV *Covariance matrix.*

XPROD *Cross-product deviations from the mean.*

N *Number of cases used to compute the correlation coefficients.*

ALL *All descriptive statistics.*

For example, to produce the correlation matrix shown in Figure 17.34, specify

```
REGRESSION DESCRIPTIVES=CORR
  /VARIABLES=LOGBEG EDLEVEL SEX WORK MINORITY AGE
  /DEPENDENT=LOGBEG
  /METHOD=ENTER EDLEVEL TO AGE.
```

17.65 Analyzing Residuals

Once you have built an equation, REGRESSION can calculate twelve temporary variables containing several types of residuals, predicted values, and related measures. You can use these variables to detect outliers and influential data points and to examine the regression assumptions described in Sections 17.17 through 17.22.

The following temporary variables are available for the analysis of residuals.

PRED *Unstandardized predicted values.* (See Section 17.13.)

ZPRED *Standardized predicted values.* (See Section 17.13.)

SEPRED *Standard errors of the predicted values.* (See Section 17.14.)

RESID *Unstandardized residuals.* (See Section 17.18.)

ZRESID *Standardized residuals.* (See Section 17.18.)

SRESID *Studentized residuals.* (See Section 17.18.)

MAHAL *Mahalanobis' distance.* (See Section 17.24.)

ADJPRED *Adjusted predicted values.* (See Section 17.25.)

DRESID *Deleted residuals.* (See Section 17.25.)

SDRESID *Studentized deleted residuals.* (See Section 17.25.)

COOK *Cook's distances.* (See Section 17.25.)

LEVER *Leverage values.* (See Section 17.49.)

Residuals analysis is specified with four subcommands: RESIDUALS, CASEWISE, PARTIALPLOT, and SCATTERPLOT. You can specify these subcommands in any order, but you cannot specify more than one of each per equation, and they must immediately follow the last METHOD subcommand that completes an equation. The residuals subcommands affect only the equation they follow. Requesting any residuals analysis always produces descriptive statistics on at least four of the temporary variables (PRED, ZPRED, RESID, and ZRESID).

All variables are standardized before plotting. If an unstandardized version of a variable is requested, the standardized version is plotted.

17.66 The RESIDUALS Subcommand

Use the RESIDUALS subcommand to obtain the statistics and plots listed below. Specifying the RESIDUALS subcommand without any specifications produces the display described for keyword DEFAULTS. If any keywords are specified on RESIDUALS, *only* the displays for those keywords are produced.

DEFAULTS *HISTOGRAM(ZRESID), NORMPROB(ZRESID), OUTLIERS plots(ZRESID), SIZE(SMALL), and DURBIN.* These plots are produced if RESIDUALS is specified without any specifications.

HISTOGRAM(tempvars) *Histogram of standardized temporary variables named.* The default temporary variable is ZRESID. Other variables that can be plotted are PRED, RESID, ZPRED, DRESID, ADJPRED, SRESID, and SDRESID. (See Figure 17.22a.)

NORMPROB(tempvars) *Normal probability (P-P) plot of standardized values.* The default variable is ZRESID. Other variables that can be plotted are PRED, RESID, ZPRED, and DRESID. (See Figure 17.22b.)

SIZE(plotsize) *Plot sizes.* The plot size can be specified as SMALL or LARGE. The default is LARGE if the display width is at least 120 and the page length is at least 55.

OUTLIERS(tempvars) *The ten most extreme values for the temporary variables named.* The default temporary variable is ZRESID. Other variables can be RESID, DRESID, SRESID, SDRESID MAHAL, and COOK. (See Figure 17.24.)

DURBIN *Durbin-Watson test statistic.* (See Section 17.21.)

ID(varname) *Identification labels for casewise and outlier plots.* Cases are labeled with values of the variable named after the ID keyword. By default, the plots are labeled with the sequential case number. ID also labels the CASEWISE list of cases. (See Figures 17.23 and 17.24.)

POOLED *Pooled plots and statistics when the SELECT subcommand is in effect.* All cases in the active file are used. The default is separate reporting of residuals statistics and plots for selected and unselected cases.

For example, to produce the output shown in Figures 17.22a, 17.22b, and 17.24, specify

```
REGRESSION VARIABLES=SALBEG SALNOW
  /DEPENDENT=SALNOW
  /METHOD=ENTER SALBEG
  /RESIDUALS=HISTOGRAM(SRESID) NORMPROB
             OUTLIERS(MAHAL) ID(SEXRACE).
```

17.67 The CASEWISE Subcommand

You can display a casewise plot of one of the temporary variables accompanied by a listing of the values of the dependent and the temporary variables. The plot can be requested for all cases or limited to outliers. Specifying the CASEWISE subcommand without keywords produces the output listed for DEFAULTS.

The following may be specified on the CASEWISE subcommand.

DEFAULTS *OUTLIERS(3), PLOT(ZRESID), DEPENDENT, PRED, and RESID.* This is the default if CASEWISE is specified without any keywords.

OUTLIERS(value) *Limit plot to outliers greater than or equal to the standardized absolute value of the plotted variable.* The default value is 3. (See Figure 17.23.)

ALL *Include all cases in the casewise plot.* Produces a plot of all cases, including outliers. The keyword OUTLIERS is ignored when ALL is specified.

PLOT(tempvar) *Plot the standardized values of the temporary variable named.* The default variable is ZRESID. The other variables that can be plotted are RESID, DRESID, SRESID, and SDRESID. (See Figure 17.23.)

varlist *List values of the DEPENDENT and temporary variables named.* Any temporary variable, including LEVER, can be listed. The defaults are DEPENDENT (the dependent variable), PRED, and RESID. (See Figures 17.16 and 17.23.)

For example, to produce the casewise plot shown in Figure 17.16, specify

```
REGRESSION VARIABLES=SALBEG SALNOW
  /DEPENDENT=SALNOW
  /METHOD=ENTER SALBEG
  /RESIDUALS=ID(SEXRACE)
  /CASEWISE=ALL DEPENDENT PRED RESID SEPRED.
```

To plot outliers whose absolute values are equal to or greater than 3 based on ZRESID, you need only specify the CASEWISE subcommand. To base the plot on Studentized residuals and label it with an ID variable, as shown in Figure 17.23, specify

```
REGRESSION VARIABLES=SALBEG SALNOW
  /DEPENDENT=SALNOW
  /METHOD=ENTER SALBEG
  /RESIDUALS=ID(SEXRACE)
  /CASEWISE=PLOT(SRESID).
```

If you request more variables than will fit on the page width set either with the SET WIDTH command or the WIDTH subcommand in REGRESSION, your output will be truncated. (see Section 17.72).

17.68 The SCATTERPLOT Subcommand

Use the SCATTERPLOT subcommand to generate scatterplots for the variables in the equation. You must name at least one pair of variables on the SCATTERPLOT subcommand. Optionally, you can specify the SIZE keyword to control the size of the plots. All scatterplots are standardized.

The specifications for SCATTERPLOT are

(varname,varname) *The pair of variables to be plotted.* Available variables are PRED, RESID, ZPRED, ZRESID, DRESID, ADJPRED, SRESID, SDRESID, and any variable named on the VARIABLES subcommand. Temporary variables should be preceded by an asterisk on this subcommand.

SIZE(plotsize) *Plot sizes.* Plotsize can be SMALL or LARGE. The default is SMALL.

The first variable named inside the parentheses is plotted on the vertical (*Y*) axis, and the second is plotted on the horizontal (*X*) axis. For example, to generate the scatterplot shown in Figure 17.20, specify

```
REGRESSION VARIABLES=SALBEG SALNOW
  /DEPENDENT=SALNOW
  /METHOD=ENTER SALBEG
  /SCATTERPLOT=(*SRESID,*PRED).
```

To produce a scatterplot for SRESID and PRED based on the logarithmic transformation of both the dependent and independent variables, as shown in Figure 17.30a, use the SCATTERPLOT subcommand above along with the following transformation commands:

```
COMPUTE LOGBEG=LG10(SALBEG).
COMPUTE LOGNOW=LG10(SALNOW).
REGRESSION VARIABLES=LOGBEG,LOGNOW
  /DEPENDENT=LOGNOW
  /METHOD=ENTER LOGBEG
  /SCATTERPLOT=(*SRESID,*PRED).
```

To produce more than one scatterplot, simply add pairs of variable names in parentheses, as in

```
/SCATTERPLOT=(*SRESID,*PRED)(SALBEG,*PRED)
```

17.69 The PARTIALPLOT Subcommand

Use the PARTIALPLOT subcommand to generate partial residual plots. Partial residual plots are scatterplots of the residuals of the dependent variable and an independent variable when both variables are regressed on the rest of the independent variables.

If no variable list is given on the PARTIALPLOT subcommand, a partial residual plot is produced for every independent variable in the equation. Plots are displayed in descending order of the standard error of *B*. All plots are standardized.

The specifications on the PARTIALPLOT subcommand are

varlist *Independent variables to be used in partial residual plot.* At least two independent variables must be in the equation for a partial residual plot to be produced. You can specify the keyword ALL to obtain the default plots for every independent variable in the equation.

SIZE(plotsize) *Plot sizes.* The plot size can be specified as SMALL or LARGE. The default plotsize is SMALL.

17.70 The SAVE Subcommand

Use the SAVE subcommand to save any or all of the 12 temporary variables described in Section 17.65. The format is the name of the temporary variable followed by a valid variable name in parentheses, as in

```
GET FILE='BANK.SYS'.
REGRESSION VARIABLES=SALBEG, SALNOW
  /DEPENDENT=SALNOW
  /METHOD=ENTER SALBEG
  /SAVE=SEPRED(SE).
PLOT CUTPOINTS=EVERY(20) /SYMBOLS='.'
     /PLOT=SE WITH SALBEG.
```

This example saves the standard errors of the predicted values with variable name SE. Then the PLOT procedure is used to plot the standard errors against the values of the independent variable SALBEG. Figure 17.14a shows the plot.

17.71 The READ and WRITE Subcommands

Procedure REGRESSION can read and write matrix materials, which can be processed more quickly than cases. Use the WRITE subcommand to write matrix materials to a file. You can write default matrix materials or specify the materials you want to write, including variable means, standard deviations, variances, a correlation or covariance matrix, and the number of cases used to compute the correlations or covariances. You can then use the READ subcommand to read the matrix materials into REGRESSION for additional analysis.

The READ subcommand can also read matrix materials written by other procedures, such as CORRELATION, or entered as data in free or fixed format. See Command Reference: Regression—Matrix Materials for complete instructions on using matrix materials with REGRESSION.

17.72 The WIDTH Subcommand

You can use the WIDTH subcommand to control the width of the display produced by the REGRESSION procedure. The default is the width specified on the SET command. The WIDTH subcommand in REGRESSION overrides the width specified on SET.

Changing the width may affect the appearance of your output. A smaller page width limits the number of statistics that can be displayed in a summary line and may also cause casewise output to be truncated (see Section 17.67). Specifying a smaller page width may also reduce the size of scatter and normal-probability plots in the residuals output.

17.73 Annotated Example

To produce the stepwise variable selection example discussed in Section 17.48, specify

```
DATA LIST FILE='EMPLOYEE.DAT'
  /  ID 1-4 SALBEG 6-10 SEX 12 TIME 14-15
     AGE 17-20 (2) SALNOW 22-26 EDLEVEL 28-29
     WORK 31-34 (2) JOBCAT MINORITY 36-37.
VAR LABELS  ID 'EMPLOYEE CODE'/
            SALBEG 'BEGINNING SALARY'/
            SEX 'SEX OF EMPLOYEE'/
            TIME 'JOB SENIORITY'/
            AGE 'AGE OF EMPLOYEE'/
            SALNOW 'CURRENT SALARY'/
            EDLEVEL 'EDUCATIONAL LEVEL'/
            WORK 'WORK EXPERIENCE'/
            JOBCAT 'EMPLOYMENT CATEGORY'/
            MINORITY 'MINORITY CLASSIFICATION'.
VALUE LABELS   SEX  0'MALES' 1'FEMALES'/
               JOBCAT 1 'CLERICAL' 2 'OFFICE TRAINEE'
                      3 'SECURITY OFFICER'
                      4 'COLLEGE TRAINEE' 5 'EXEMPT EMPLOYEE'
                      6 'MBA TRAINEE' 7 'TECHNICAL'/
            MINORITY  0 'WHITE' 1 'NONWHITE'.
MISSING VALUES SALBEG,TIME TO EDLEVEL,JOBCAT (0)/SEX,MINORITY (9).
FORMATS  SALBEG SALNOW (COMMA6.0).
COMPUTE LOGBEG=LG10(SALBEG).
COMPUTE LOGNOW=LG10(SALNOW).
REGRESSION VARIABLES=LOGBEG,EDLEVEL,SEX,WORK,MINORITY,AGE/
           STATISTICS=R,COEFF,OUTS,F/
           DEPENDENT=LOGBEG/
           METHOD=STEPWISE.
FINISH.
```

- The DATA LIST command defines the data file and variable names and gives the column locations for each variable.
- The VARIABLE LABELS and VALUE LABELS commands supply descriptive labels for the variables and their values.

- The MISSING VALUES command assigns the value 0 as missing for SALBEG, TIME TO EDLEVEL, and JOBCAT, and the value 9 as missing for SEX and MINORITY.
- The FORMATS command assigns a comma display and write format to the variables SALBEG and SALNOW.
- The COMPUTE commands create the new variables LOGBEG and LOGNOW.
- The REGRESSION command asks for a stepwise regression of the named variables with LOGBEG as the dependent variable. It also asks for the statistics *R*, COEFF, OUTS, and *F*.

The output from this example is shown in Figure 17.48.

Contents

Chapter 18 Reporting Results

Statistics Guide

Case listings and descriptive statistics are basic tools for studying and presenting data. You can obtain case listings with LIST, frequency counts and descriptive statistics with FREQUENCIES and DESCRIPTIVES, and subpopulation statistics with MEANS. Each of these procedures uses a format designed to make the information clear; but if that format isn't what you need for presentation, there is little you can do to change it. REPORT gives you the control you need over data presentation.

REPORT is a formatting tool. It allows you to present case listings and summary statistics (including frequencies) in report format. Your report can be one page long, or it can be hundreds of pages long. With REPORT's subcommands and keywords you can specify the variables you want to summarize or list and organize them into subgroups. You can calculate summary statistics on the report variables and also calculate cross-variable statistics that are unavailable in other procedures, such as the ratio of two means.

18.1 BASIC REPORT CONCEPTS

This section introduces you to the two basic elements of a report: its contents, and the organization of those contents. To illustrate these concepts (and others in this chapter), we'll use information from a retail company's personnel file. The file contains data such as employees' names, salaries, employment grades, length of time in grade, overall length of employment, store branch and department, shift, and so forth.

18.2 Report Contents: Summaries and Listings

Reports contain summary statistics calculated for groups of cases, listings of individual cases, or a combination of both statistics and listings.

18.3 Summary Reports

Summary reports display summary statistics but do not display case listings. The summary information consists of the statistic or statistics you request for the report variables. Each statistic prints in a separate row on the report, referred to as a *summary line*. The cells within the summary line display the statistical values. Figure 18.3 shows a summary report presenting means as the summary statistics.

Figure 18.3 Summary personnel report

```
Personnel Data                                                                          PAGE    1

                    Tenure     Tenure
                      in         in
Division     Age    Company    Grade     Salary--Annual
----------   -----  -------    -------   --------------

Carpeting

Mean        30.75     4.04      3.31          $11,754

Appliances

Mean        31.11     3.81      3.54          $12,508

Furniture

Mean        36.87     4.79      4.08          $13,255

Hardware

Mean        36.20     4.60      4.57          $17,580
```

18.4 Listing Reports

Listing reports list individual cases. The case listings comprise the values and/or labels recorded for each of the report variables. In addition to listing cases, a listing report can display as many summary statistics as a summary report. Figure 18.4a presents a report that contains case listings, and 18.4b shows one that contains both listings and summaries.

Figure 18.4a Personnel report with case listings

Personnel Data PAGE 1

Division	Age	Tenure in Company	Tenure in Grade	Salary--Annual
Carpeting	22.00	3.92	3.08	$10,900
	27.00	3.67	2.17	$9,200
	23.00	3.92	3.08	$10,900
	35.00	6.00	5.33	$19,500
	36.00	3.83	3.25	$10,000
	24.00	4.00	3.25	$10,000
	44.00	4.83	4.33	$15,690
	33.00	3.75	3.25	$10,000
	27.00	4.33	3.17	$10,000
	33.00	2.67	2.67	$9,335
	35.00	3.50	3.00	$15,520
	30.00	4.08	3.08	$10,000
Appliances	42.00	6.50	6.50	$18,000
	26.00	2.92	2.08	$8,000
	24.00	3.17	3.17	$8,975
	21.00	2.67	2.67	$8,700
	38.00	5.00	4.42	$28,300
	33.00	3.42	2.92	$8,900
	30.00	2.67	2.67	$7,500
	32.00	2.92	2.92	$8,900
	34.00	5.08	4.50	$15,300

Figure 18.4b Personnel report with case listings and summaries

Personnel Data PAGE 1

Division	Age	Tenure in Company	Tenure in Grade	Salary--Annual
Carpeting	22.00	3.92	3.08	$10,900
	27.00	3.67	2.17	$9,200
	23.00	3.92	3.08	$10,900
	35.00	6.00	5.33	$19,500
	36.00	3.83	3.25	$10,000
	24.00	4.00	3.25	$10,000
	44.00	4.83	4.33	$15,690
	33.00	3.75	3.25	$10,000
	27.00	4.33	3.17	$10,000
	33.00	2.67	2.67	$9,335
	35.00	3.50	3.00	$15,520
	30.00	4.08	3.08	$10,000
Mean	30.75	4.04	3.31	$11,754
Appliances	42.00	6.50	6.50	$18,000
	26.00	2.92	2.08	$8,000
	24.00	3.17	3.17	$8,975
	21.00	2.67	2.67	$8,700
	38.00	5.00	4.42	$28,300
	33.00	3.42	2.92	$8,900
	30.00	2.67	2.67	$7,500
	32.00	2.92	2.92	$8,900
	34.00	5.08	4.50	$15,300
Mean	31.11	3.81	3.54	$12,508

18.5 Report Organization

A report is organized into columns and rows. Each column is defined by a variable. The columns on the left are the *break variables*—those whose values divide the report into subgroups. In Figures 18.3, 18.4a, and 18.4b, DIVISION is the break variable. The other columns, on the right, are the *report variables*—those whose values are listed or summarized. The order of the columns is the order in which you specify the variables.

The rows consist of case listings or summary statistics. The order of the rows is the order of the cases in the file. (See Section 18.14 for more information about the order of cases.)

18.6 BUILDING THE REPORT

Four basic subcommands determine the appearance of reports:

FORMAT Determines the report's general page layout and whether cases are listed. If it is specified, FORMAT precedes all other subcommands on the REPORT command.

VARIABLES Specifies the *report variables* that are listed and/or summarized in columns in the report. The order in which you list variables on the VARIABLES subcommand determines the order in which columns appear on the report. The VARIABLES subcommand is required.

BREAK Specifies the *break variable(s)* that break the report rows into subgroups. Break variables always print as the leftmost columns on the report.

SUMMARY Names a statistic or set of statistics to calculate.

The VARIABLES subcommand is required on every REPORT command. In addition, a listing report requires the FORMAT subcommand with the LIST keyword, while a summary report requires the BREAK and SUMMARY subcommands.

Each subcommand, when used, must follow the order above: FORMAT followed by VARIABLES, BREAK, and SUMMARY. You can specify multiple BREAK and SUMMARY subcommands.

REPORT has other subcommands, but the four above are the ones you'll use most often. They are described in more detail below.

18.7 Choosing Formats and Obtaining Listings

Use the FORMAT subcommand to specify whether AUTOMATIC or MANUAL default formats are implemented. The AUTOMATIC keyword, new in SPSS/PC+ V2.0, facilitates report design by automatically implementing the basic format features you are most likely to use. MANUAL implements the defaults that were used in earlier versions of SPSS/PC+. MANUAL is the default. See REPORT in Part C for a table that compares the default settings for AUTOMATIC and MANUAL.

The FORMAT subcommand also determines whether a report contains summaries or listings, or both. Summaries are the default. To produce a listing report, add the keyword LIST to the FORMAT subcommand. For example,

```
REPORT FORMAT=AUTOMATIC LIST /
```

The keyword LIST applies to all variables named on the VARIABLES subcommand.

18.8 Specifying Report Variables

The required VARIABLES subcommand determines the variables that appear in the report columns. The minimum VARIABLES specification is a list of variables, as in

```
VARIABLES=LNAME AGE TENURE JTENURE
```

which instructs REPORT to list or summarize data for the variables LNAME, AGE, TENURE, and JTENURE.

18.9 Defining Break Groups

The BREAK subcommand specifies the variable whose values break the report rows into subgroups. For example, the REPORT subcommand

```
BREAK=DIVISION /
```

specifies a subgroup for each value of DIVISION.

The BREAK subcommand is required on all summary reports but is optional on listing reports. If you do not want to break a listing report into subgroups, omit the BREAK subcommand.

The REPORT command does not organize data itself, so you must first use the SORT command to organize cases into the break groups you intend to specify on the BREAK subcommand (see Section 18.14).

18.10 Requesting Summary Statistics

To request statistics for subgroups of cases on either a summary or listing report, specify the statistic on a SUMMARY subcommand immediately after a BREAK subcommand. You cannot use a SUMMARY subcommand without a corresponding BREAK subcommand, and the BREAK subcommand must precede its associated SUMMARY subcommands.

For example, the commands

```
SORT CASES BY DIVISION.
REPORT FORMAT=AUTOMATIC /
       VARS=AGE TENURE JTENURE SALARY /
       BREAK=DIVISION /
       SUMMARY=MEAN.
```

request the mean values for AGE, TENURE, JTENURE, and SALARY at each break of the variable DIVISION, as shown in Figure 18.3.

For a list of the summaries available in REPORT, see the REPORT command in Part C.

18.11 Adding Statistics

To add more summaries, you simply add more SUMMARY subcommands. For example, the following commands

```
SORT CASES BY DIVISION.
REPORT FORMAT=AUTOMATIC /
       VARS=AGE TENURE JTENURE SALARY /
       BREAK=DIVISION /
       SUMMARY=MEAN /
       SUMMARY=MIN /
       SUMMARY=MAX.
```

request the minimum and maximum values in addition to the means for variables AGE, TENURE, JTENURE, and SALARY. Each summary appears on its own line on the report.

18.12 Adding Break Levels

Beyond breaking data into subgroups, you might want to further subdivide it into multiple break levels, adding break columns on the left of the report. To do this, specify multiple BREAK subcommands. Successive BREAK subcommands group the data within preceding BREAK subcommands.

Before specifying a report with multiple break levels, use SORT CASES to sort the break variables into the order you intend to use them (see Section 18.14). Then specify successive BREAK subcommands, as in:

```
SORT CASES BY DIVISION STORE.
REPORT FORMAT=AUTOMATIC /
       VARS=AGE TENURE SALARY /
       BREAK=DIVISION /
       SUMMARY=MEAN /
       SUMMARY=VALIDN /
       BREAK=STORE /
       SUMMARY=MEAN.
```

Figure 18.12 shows a portion of the report. The DIVISION and STORE columns are defined by the break variables. Only the means are shown for each level of store branch within division. Both means and the number of valid cases are shown for each division.

Figure 18.12 Personnel report with multiple breaks

```
Personnel Data                                                                    PAGE   1

                  Branch              Tenure
Division          Store       Age       in      Salary--Annual
                                     Company
----------        ---------   -----  --------   --------------

Carpeting         Suburban

                  Mean        26.75    4.37        $12,625

                  Downtown

                  Mean        32.75    3.87        $11,318

Mean                          30.75    4.04        $11,754
N                                12      12             12
```

When you specify a SUMMARY subcommand, REPORT calculates the statistic you request for the specified break level, only. If you want the same statistic for multiple break levels, you must request it at each desired level (see Section 18.32 for a shorthand way of doing this). If you don't want any summary statistics at a given break level, omit the SUMMARY subcommand at that level.

18.13 Breaks Defined by More Than One Variable

To combine break groups in a single column, list multiple variables on the same BREAK subcommand. For example, the commands

```
TITLE Personnel Data.
SORT CASES BY DIVISION SHIFT.
REPORT FORMAT=AUTOMATIC /
       VARS=AGE TENURE SALARY /
       BREAK=DIVISION SHIFT 'Division' 'and' 'Shift' /
       SUMMARY=MEAN.
```

combine break variables DIVISION and SHIFT in the same column. By default, REPORT uses the heading from the first variable on the BREAK subcommand. To include both variable names, another column heading is specified in apostrophes (see Section 18.20).

Figure 18.13 shows that REPORT displays the value labels of both break variables in the same column, with the specified column heading, and calculates the mean for AGE, TENURE, and SALARY any time the value of either DIVISION or SHIFT changes. (Only two divisions are shown in FIGURE 18.13.) Note that REPORT does not calculate overall means for each division. To obtain means by division, specify DIVISION and SHIFT on separate BREAK subcommands (see Section 18.12).

Figure 18.13 Report with stacked break variables

```
Personnel Data                                                      PAGE   1

Division              Tenure
and                     in
Shift         Age     Company    Salary--Annual
----------   -----    --------   --------------

Carpeting
First

Mean         30.57      3.80        $11,649

Carpeting
Second

Mean         24.00      4.00        $10,000

Carpeting
Weekend

Mean         32.75      4.48        $12,375

Appliances
First

Mean         31.17      3.75        $13,062

Appliances
Second

Mean         42.00      6.50        $18,000

Appliances
Weekend

Mean         25.50      2.67         $8,100
```

18.14 Preparing Data for REPORT

REPORT does not organize the data itself as it reads the cases and computes the summaries. Instead, it simply reads cases in the order they reside in the file, listing them (if you request a listing) while keeping track of information you requested for summaries.

For example, when REPORT calculates a statistic, it calculates it for the subgroup defined by the first value of the break variable. When the value of the break variable changes in the data, REPORT calculates and prints the statistics you requested. It then prints the next value of the break variable and resumes reading cases until the value of the break variable changes again. It prints the requested summary statistics every time it reads a change in the value for the break variable.

Therefore, before you run REPORT, the data must be organized in the file in such a way that all cases with the same value for the break variable reside together. It doesn't matter if the values are in ascending, descending, or some other order—only that all the cases in the same group are together.

To organize data in the file, use the SORT command immediately before the REPORT command. For example, the command

```
SORT BY DIVISION.
```

organizes the data in Figure 18.3 by the subgroups Carpeting, Appliances, Furniture, and Hardware. If you wanted to further subdivide these divisions according to the subgroupings of other variables, you would name additional variables on the SORT command. For example, the command

```
SORT BY DIVISION STORE.
```

groups the data into the divisions of Carpeting, Appliances, Furniture, and Hardware and then organizes the data within each of the divisions according to the subgroups of the variable STORE, which are Suburban and Downtown.

18.15 Trial Runs

Because REPORT is so flexible and its output can have so many components, you may want to experiment with the report layout before you run the final report. If you have a large data file, you can experiment with a subset of the data until you obtain your intended format.

To process a subset of the data, use one of the following techniques:

- Use the N command to limit the number of cases read. The system does not attempt to read more cases than the number specified on the N command. For example, if the file contains 10,000 cases and you specify 10 on the N command, REPORT reads only the first 10 cases.
- If the REPORT contains break variables, and the labeling and spacing of breaks are important considerations, you can obtain a subset of the entire file containing cases from several breaks by using the SAMPLE command.

18.16 REFINING THE REPORT

In addition to specifying the basic report structure as described above, REPORT gives you the ability to refine various report components. These include the following:

- Titles and footnotes
- Margins and alignment
- Column contents and labeling
- Horizontal spacing
- Summary statistics
- String variables
- Missing values

These topics are discussed in Sections 18.17 through 18.36.

18.17 Adding Titles and Footnotes

The optional TITLE and FOOTNOTE subcommands in REPORT enable you to place titles and footnotes on the left, in the center, and on the right of each page of a report. Enclose each line of the title or footnote in apostrophes, separating lines by a comma or a space. To include an apostrophe in a title, either enclose the string in quotation marks or use double apostrophes not separated by a space. For example,

```
SORT CASES BY DIVISION.
REPORT FORMAT=AUTOMATIC LIST /
       VARS=LNAME AGE TENURE SALARY /
       BREAK=DIVISION /
       SUMMARY=MEAN /
       TITLE= 'Personnel Report' "Employees' Profile".
```

specifies a two line title. *Personnel Report* prints on the first title line, and *Employee's Profile* prints beneath it on the second title line. The quotation marks enclosing the second title line assure the apostrophe will print in the title.

If you don't specify a title, SPSS/PC+ uses its system title as a default when the REPORT width is greater than or equal to the system title's width plus 12. To specify a blank title line, type a space between apostrophes. There is no default footnote.

To specify the left, right, or center positions for either titles or footnotes, use the following conventions:

```
TITLE=
   LEFT   '     '   Left-justified title
   RIGHT  '     '   Right-justified title
   CENTER '     '/  Centered title

FOOTNOTE=
   LEFT   '     '   Left-justified note
   RIGHT  '     '   Right-justified note
   CENTER '     '/  Centered note
```

Using TITLE or FOOTNOTE with no reference to a position is equivalent to specifying the center position. However, if you specify a left or right position on the same subcommand, you must use the keyword CENTER to specify the center position. To specify multiple title and footnote lines in any given position, specify each line in apostrophes. Do not repeat the positional keywords. For example,

```
TITLE=LEFT   'Personnel Report' 'As of January 1, 1987'
      RIGHT 'ACME Products' '2201 LaSalle Park'
            'Chicago, Illinois 60611'
```

produces a two-line left title and a three-line right title. Notice that LEFT, RIGHT, and CENTER are not separated by slashes.

Titles and footnotes print on each page of a multiple-page report. The LEFT, RIGHT, and CENTER alignments are relative to the report's margins. If you specify a title or footnote that is wider than the report width, REPORT generates an error message.

You can specify titles and footnotes anywhere after the FORMAT subcommand (except between BREAK and SUMMARY) as long as all title subcommands precede all footnote subcommands. There is no fixed limit to the number of title or footnote lines you can define on a report.

Three special arguments are available in titles and footnotes:

)PAGE *Print the page number right-justified in a five-character field.*

)DATE *Print the current date in the form* dd/mmm/yy *right-adjusted in a nine-character field.*

)variable *Print this variable's value label in this relative position.*

You can use all three arguments in as many titles and footnotes as you like on a single report. The following are typical examples:

```
TITLE=LEFT  'Personnel Report' 'Prepared on )DATE' /
      RIGHT 'Page )PAGE' /
FOOTNOTE=RIGHT 'Regional Manager: )MGRNM' /
```

The left title includes the current date, and the right title includes the current page numbers. The footnote includes the value label for MGRNM.

Each variable you specify with *)variable* must be one you've defined in the active file, though it does not need to be a variable you've included as a column on your report. If you specify a variable that has no value label, the value itself will print, formatted according to its print format. You cannot specify a system variable or a variable you create with the STRING subcommand (see Section 18.33).

One label or value from each variable specified in a *)variable* argument prints on every page of the report. The label REPORT prints for each varies from page to page and is chosen from cases determined as follows:

- If a new page starts with a case listing, REPORT takes the labeled value from the first case listed.
- If a new page starts with a BREAK line, REPORT takes the labeled value from the first case of the new break group.
- If a new page starts with a summary line, REPORT takes the labeled value from the last case of the break group being summarized.
- If you specify the same variable in both a title and a footnote, REPORT takes the title value from the top case on the page, and the footnote value from the last case on the page.

You cannot use variables named DATE or PAGE in the *)variable* argument because they will only print the current date or a page number. If you want to use a variable named DATE or PAGE, change the variable's name with the MODIFY VARIABLES command before you use it in the *)variable* argument.

18.18 Margins and Report Alignment

Report margins determine the maximum width allowed for a report. (The finished report may be narrower, however.) REPORT uses a default left margin of 1, and a default right margin equal to your system's defined width. To change the margins, use the MARGINS keyword on the FORMAT subcommand (see REPORT in Part C).

Report alignment refers to a report's position relative to its defined margins. For example, report margins of (1,80) allow for an 80-character report. If the finished report is only 50 characters wide, the report's alignment refers to whether the report is left-justified against column 1 of the margins, right-justified against column 80, or centered between columns 1 and 80. The default is left-justified. To change the default, use the ALIGNMENT keyword on the FORMAT subcommand (see REPORT in Part C).

18.19 Column Contents and Labeling

Sections 18.20 through 18.24 discuss options for controlling the contents, layout, and labeling of columns.

18.20 Column Headings

Each column in a report has a heading. By default, REPORT uses the variable label for the heading. If there is no label, REPORT uses the variable name.

REPORT wraps default column headings within their column widths, using as many lines as necessary and attempting to split lines meaningfully at spaces. With FORMAT=AUTOMATIC, REPORT centers each column heading within the width of its column. However, when value labels or string values exceed the width of the longest word in the heading, REPORT left-justifies the heading.

To specify a column heading, enclose the heading in apostrophes or quotation marks following the variable name on either the VARIABLES or

BREAK subcommands. With automatic format, REPORT does not wrap a heading you specify; it prints the heading exactly as you arrange it in your specification. To print the heading on multiple lines, enclose each line within its own set of apostrophes or quotation marks. To include an apostrophe in the heading, use quotation marks around the heading. For example,

```
SORT CASES BY DIVISION
REPORT FORMAT=AUTOMATIC LIST
      /VARIABLES=LNAME 'Employee'
                 AGE "Employee's Age" TENURE JTENURE
      /BREAK=DIVISION 'Company' 'Division'
      /SUMMARY=MEAN
```

produces the report in Figure 18.20 (only the first few lines of the report are shown). Each heading is centered, except for Company Division and Employee. Because DIVISION and LNAME are string variables with values wider than the longest word in the column heading, the column headings are left-justified.

Figure 18.20 Column headings

```
Personnel Data                                                                        PAGE   1

                                                 Tenure     Tenure
Company                                            in         in
Division     Employee        Employee's Age     Company     Grade
----------   -------------   --------------     -------     -------

Carpeting    Ford                 27.00           3.67       2.17
             Cochran              22.00           3.92       3.08
             Hoawinski            23.00           3.92       3.08
             Gates                24.00           4.00       3.25
             Mulvihill            30.00           4.08       3.08
             Lavelle              27.00           4.33       3.17
             Mahr                 33.00           2.67       2.67
```

18.21 Column Widths and Spacing

When you don't specify a column width for a variable, REPORT determines a default width, using the *larger* of the following for each variable:

- The widest print format in the column, whether it is a variable print format or a summary print format.
- The width of any temporary variable you define with REPORT's STRING subcommand.
- If you assign a column heading, the length of the longest title line in that heading. For example, VARS = TENURE 'Tenure in Company' 'Measured' 'in Months' specifies 'Tenure in Company' as the longest line in a three-line heading.
- When no column heading is specified, the length of the longest word in the variable label, or the length of the variable name. (If FORMAT=MANUAL, variable labels are not evaluated.)
- If you specify (LABEL), the length of the variable's longest value label. (With FORMAT=AUTOMATIC, (LABEL) is the default for break variables.) REPORT reads value labels from the dictionary, so it's possible for you to exclude a value from your report, yet still have that value's label determine the width of a column. (If FORMAT=MANUAL, REPORT uses the length of the variable's longest value label, up to 20 characters; 20 is the largest value it uses for this criterion with MANUAL format.)

To override the default, specify a column width in parentheses following the variable name on the VARIABLES or BREAK subcommand, as in:

```
REPORT FORMAT = AUTOMATIC /
       VARIABLES = LNAME (10) AGE /
       BREAK = DIVISION (5) /
       SUMMARY = MEAN.
```

Intercolumn Spacing. REPORT subtracts the combined column widths of the break and report variables from the REPORT margins. It then divides the result by the number of columns minus 1. It uses this value or 4, whichever is least, as the space between each column.

Automatic Fit. When the above criteria result in a report that is too wide for the report margins, AUTOMATIC format shrinks the report. It performs the following steps, stopping as soon as the report fits within the margins:

- It reduces intercolumn spacing incrementally until it reaches a minimum intercolumn space of 1. It will never reduce it to 0.
- It shortens strings, begining with the longest string from a string variable at least 15 characters wide. It shortens the string as much as needed—up to 40% of its length. If necessary it repeats the step, using different string variables. It will not shorten the same string twice.

REPORT does not implement the *automatic fit* criteria unless FORMAT= AUTOMATIC.

18.22 Labels versus Values

Frequently when you run a listing report you want to see value labels for the report variables rather than the values themselves. For example, if you want to run a personnel report that includes an employee's sex, you would prefer to see the labels Female and Male rather than the values 1 and 2 on the report. By specifying the keyword (LABEL) after the variable name on the VARIABLES subcommand, you instruct REPORT to print labels rather than values for that variable.

For example, to produce a listing report that shows the sex, age, job grade, and work shift for each employee, you would use the following commands:

```
SORT CASES BY LNAME
REPORT FORMAT=AUTOMATIC LIST
      /VARS=LNAME SEX(LABEL) AGE JOBGRADE(LABEL) SHIFT(LABEL)
```

The keyword (LABEL) generates a descriptive report, as you can see in Figure 18.22 (only part of the report is shown).

Figure 18.22 Personnel report with keyword (LABEL)

```
                                                                                   PAGE   1
Personnel Data

Last Name       Sex       Age    Job Grade            Shift
-------------   ------   -----   ------------------   ------

Baker           Male     42.00   Sales Staff          Weekend
Blount          Male     41.00   Sales Staff          Weekend
Carlyle         Female   40.00   Sales Staff          Weekend
Cochran         Female   22.00   Sales Staff          First
Cochran         Male     39.00   Sales Staff          Second
Dan             Male     36.00   Sales Staff          Weekend
Farkas          Male     37.00   Sales Staff          First
Ford            Female   27.00   Support Staff        First
Ford            Female   36.00   Sales Staff          Weekend
Gates           Female   24.00   Sales Staff          Second
Golden          Female   42.00   Sales Staff          First
Gonzales        Female   42.00   Support Staff        First
Hoawinski       Female   23.00   Sales Staff          First
Jacobesen       Male     44.00   Supervisory Staff    Second
Johnson         Female   42.00   Sales Staff          Second
```

If the variable list contains a set of inclusive variables implied by the keyword TO, (LABEL) applies to the entire set of variables in the list. However, (LABEL) cannot be implied for a set of variables named individually. For example, in

```
/VARIABLES=V1 TO V5(LABEL)
```

(LABEL) applies to all variables implied by V1 TO V5. But in

```
/VARIABLES=V1 V2 V3 V4 V5(LABEL)
```

(LABEL) applies only to V5.

18.23 Stacking Report Variables

When FORMAT=LIST, REPORT permits you to stack report variables together in a single column by linking them with + signs on the VARIABLES subcommand. For example,

```
TITLE Personnel Data.
SORT CASES BY DIVISION LNAME.
REPORT FORMAT=AUTOMATIC LIST/
       VARS=LNAME
       TENURE + STORE(LABEL) 'Tenure' 'and' 'Location'
       AGE SALARY/
       BREAK=DIVISION/
       SUMMARY=MEAN.
```

stacks the variables TENURE and STORE in a single column, as shown in Figure 18.23 (only one division is shown). LNAME, AGE, and SALARY are unaffected. The stacked variables each start a new line on the report and are listed in the order they are defined on the VARIABLES subcommand. REPORT will not split the values for a single case across page breaks in the report.

Figure 18.23 Report with stacked report variables

```
Personnel Data                                                                      PAGE   1

                                Tenure
                                 and
Division       Last Name       Location     Age     Salary--Annual
----------     -----------     --------     ----    --------------

Carpeting      Cochran             3.92    22.00           $10,900
                               Suburban
               Dan                 3.83    36.00           $10,000

               Ford                3.67    27.00            $9,200
                               Suburban
               Gates               4.00    24.00           $10,000
                               Downtown
               Hoawinski           3.92    23.00           $10,900
                               Suburban
               Jones               4.83    44.00           $15,690
                               Downtown
               Katz                3.75    33.00           $10,000
                               Downtown
               Lavelle             4.33    27.00           $10,000
                               Downtown
               Mahr                2.67    33.00            $9,335
                               Downtown
               McAndrews           3.50    35.00           $15,520
                               Downtown
               Mulvihill           4.08    30.00           $10,000
                               Downtown
               Tygielski           6.00    35.00           $19,500
                               Suburban

Mean                               4.04    30.75           $11,754
```

REPORT reads values from all the stacked variables when it determines column widths. However, REPORT uses the default heading from the first variable on the stacked list (if you don't specify a heading), and it uses only values from the first variable on the stacked list to calculate summaries.

18.24 Creating Empty Columns

You can add space between columns by creating a dummy variable on the VARIABLES subcommand. Since a dummy variable has no values, its column will be blank. To insert this blank column between other variables' columns, specify the variable name followed by the keyword (DUMMY) between the names of the variables whose columns are to be separated by the blank column. For example,

```
/VARIABLES = AGE TENURE XX(DUMMY) (7) ' ' SALARY
```

places a blank column 7 spaces wide between the columns for TENURE and SALARY. The blank title specified between apostrophes overrides the default column heading.

The space created by dummy variables is also useful for holding the results of composite functions (see Section 18.31).

18.25 Horizontal Spacing

The keyword LIST on FORMAT has an optional argument, (n), that controls the spacing of cases in listing reports. By default, there are no blank lines between cases. An integer value in parentheses specifies a blank line after every *n* cases. For example, LIST(1) produces a double-spaced listing and LIST(3) lists cases in sets of three. The spacing option makes it easier to read reports with many cases.

The following commands generate the report shown in Figure 18.25 (only part of the report is shown):

```
SORT CASES BY LNAME
REPORT FORMAT=AUTOMATIC LIST(3)
      /VARS=LNAME AGE TENURE JTENURE
```

To insert a blank line between several variables, that are stacked in one column (see Stacking Report Variables, above) either specify (DUMMY) after a non-existent variable name, or specify a blank space between two plus signs.

Figure 18.25 Personnel report with optional spacing

```
Personnel Data                                                    PAGE    1

                          Tenure     Tenure
                            in         in
Last Name          Age    Company    Grade
--------------   ------   -------    ------

Baker             42.00     5.25      3.75
Blount            41.00     5.25      3.75
Carlyle           40.00     6.00      6.00

Cochran           22.00     3.92      3.08
Cochran           39.00     5.50      5.50
Dan               36.00     3.83      3.25

Farkas            37.00     4.42      3.67
Ford              27.00     3.67      2.17
Ford              36.00     4.50      3.67
```

Other keywords in REPORT allow you to control spacing between titles and column headings, between headings and listings, between individual rows within the body of the report, and between the body of the report and footnotes. These keywords are listed in Table 18.37 and described under REPORT in Part C.

18.26 Summary Statistics

REPORT gives you precise control over the types of summaries to be printed and their appearance on the report. Sections 18.27 through 18.32 describe summary specifications.

18.27 Selecting Variables for Statistics

A statistic specified on the SUMMARY subcommand is calculated for all the variables named on the VARIABLES subcommand. If you want the statistic to be calculated only for selected report variables, name the desired variables in parentheses on the SUMMARY subcommand, as in

```
SORT CASES BY DIVISION.
REPORT FORMAT=AUTOMATIC LIST /
       VARIABLES=LNAME AGE TENURE JTENURE SALARY /
       BREAK=DIVISION /
       SUMMARY=MEAN(JTENURE, SALARY).
```

You cannot use the TO keyword on SUMMARY to imply a set of variables.

18.28 Summaries Across Break Groups (Report Totals)

The keyword (TOTAL) on the BREAK subcommand calculates summary statistics for all the cases on the report. These totals print at the end of the report. You can specify (TOTAL) for both listing and summary reports.

To generate a report that has break level summaries plus total summaries, specify (TOTAL) after the variable named on the BREAK subcommand that precedes the summary whose totals you want, as in:

```
SORT CASES BY STORE DIVISION.
REPORT FORMAT=AUTOMATIC /
       VARIABLES=AGE TENURE JTENURE SALARY /
       BREAK=STORE(TOTAL) /
       SUMMARY=VALIDN (AGE) /
       BREAK=DIVISION /
       SUMMARY=MEAN.
```

The first set of BREAK and SUMMARY subcommands calculates the number of employees who work at each branch store and the total number of employees who work for the company by counting those that have valid values for AGE. The second set calculates a mean for each division but does not calculate a mean for the entire company because (TOTAL) is not specified at this break level. Figure 18.28a shows the report.

Figure 18.28a Summary report with totals

Personnel Data PAGE 1

Branch Store	Division	Age	Tenure in Company	Tenure in Grade	Salary—Annual
Suburban	Carpeting				
	Mean	26.75	4.37	3.42	$12,625
	Appliances				
	Mean	30.20	4.05	3.77	$14,395
	Furniture				
	Mean	35.29	4.71	4.32	$12,975
	Hardware				
	Mean	32.00	4.33	4.33	$22,500
N		17			
Downtown	Carpeting				
	Mean	32.75	3.87	3.25	$11,318
	Appliances				
	Mean	32.25	3.52	3.25	$10,150
	Furniture				
	Mean	38.25	4.86	3.86	$13,500
	Hardware				
	Mean	37.25	4.67	4.62	$16,350
N		24			
TOTAL					
N		41			

When the statistics you want REPORT to calculate for the totals differ from those you want it to calculate at the various break levels, specify (TOTAL) on a BREAK subcommand and do not name a variable; then use as many SUMMARY subcommands as you need to specify the summary totals you want REPORT to calculate. Name your break variables on subsequent BREAK subcommands. For example, when you specify

```
SORT CASES BY DIVISION.
REPORT FORMAT=AUTOMATIC /
       VARIABLES=AGE TENURE JTENURE SALARY /
       BREAK=(TOTAL) /
       SUMMARY=VALIDN (AGE) /
       BREAK=DIVISION /
       SUMMARY=MEAN.
```

REPORT calculates the total number of employees in the company but does not calculate the number of employees within each division. It calculates a mean for each division but does not calculate a mean for the entire company (see Figure 18.28b).

Figure 18.28b Totals that differ from break level summaries

Personnel Data PAGE 1

Division	Age	Tenure in Company	Tenure in Grade	Salary—Annual
Carpeting				
Mean	30.75	4.04	3.31	$11,754
Appliances				
Mean	31.11	3.81	3.54	$12,508
Furniture				
Mean	36.87	4.79	4.08	$13,255
Hardware				
Mean	36.20	4.60	4.57	$17,580
N	41			

You can specify (TOTAL) on a BREAK subcommand that does not name a variable to obtain totals for a listing report that does not contain break levels. Use as many SUMMARY subcommands as you need to specify the totals you want REPORT to calculate, as in:

```
SORT CASES BY LNAME.
REPORT FORMAT=AUTOMATIC LIST/
       VARIABLES=LNAME AGE TENURE JTENURE SALARY /
       BREAK=(TOTAL) /
       SUMMARY=MEAN /
       SUMMARY=VALIDN (JTENURE).
```

Each statistic specified on a SUMMARY subcommand is calculated for all the cases in the report. Though the file is sorted by LNAME to alphabetize the listing, SORT CASES is optional since the report has no break variables.

Figure 18.28c Listing report with totals

Personnel Data PAGE 1

Last Name	Age	Tenure in Company	Tenure in Grade	Salary—Annual
.				
.				
.				
Sedowski	30.00	2.67	2.67	$7,500
Shavilje	32.00	2.92	2.92	$8,900
Snolik	34.00	5.08	4.50	$15,300
Syms	32.00	4.33	4.33	$22,500
Totman	41.00	4.50	3.50	$13,300
Tygielski	35.00	6.00	5.33	$19,500
Wajda	26.00	3.17	3.17	$8,975
Washington	32.00	4.42	3.67	$14,400
White	25.00	3.92	3.33	$11,000
Wilson	36.00	4.42	3.75	$14,000
Mean				
	33.73	4.34	3.79	$13,179
N				
			41	

Figure 18.28c shows the last lines of the report. The mean values are those for the entire 41 employees. The summary titles *Mean* and *N* begin in the left-most column and are left-justified. The values for each statistic print one row beneath their title, allowing you to specify a summary title that spans as far as the right margin of the report (see Section 18.29).

18.29 Summary Titles

By default, REPORT uses its own summary titles (listed in Part C) on a report to identify each statistic you request on the SUMMARY subcommand. It displays the summary title in the column corresponding to the break being summarized, aligning the titles with the labels or values in the column. The title prints in mixed case or upper case, depending upon your system's default specifications.

You can change the default summary title by enclosing a one-line title in apostrophes or quotation marks, as in

```
SUMMARY=STDDEV 'Standard Deviation'
```

Use leading blanks to indent summary titles within the break column.

If you specify a summary title wider than its break column, the title extends into the break column to its right. If the width of the available break columns is insufficient to display the full summary title, the title is truncated.

REPORT allows you to move summary titles between break columns. To do so, specify in parentheses the number of the break column in which you want the summary title to appear, counting the first specified BREAK subcommand as 1. For example, the commands

```
REPORT FORMAT=AUTOMATIC LIST /
       VARIABLES=LNAME AGE TENURE JTENURE SALARY /
       BREAK=STORE /
       SUMMARY=MEAN(JTENURE)
       BREAK=DIVISION /
       SUMMARY=MEAN(SALARY) 'Average Salary' (1)
```

display the title Average Salary in the column for the first break, STORE, rather than in the column for DIVISION.

Although summary titles can only be one line long, when you use multiple SUMMARY subcommands you can continue a summary title from one summary line to another. For example,

```
SUMMARY=SUM 'Sums and Averages' /
SUMMARY=MEAN 'Based on 1986 5% Sample'
```

produces a two-line summary with a title continuing from the first to the second line. You can also specify a blank title for any summary line.

18.30 Print Formats for Summary Statistics

Every summary function has a default display format. For example, the PERCENT function uses a width of 6 places, including one decimal place. For the default formats, see REPORT in Part C.

You can specify COMMA and DOLLAR formats when they are not the default, or PLAIN to override those formats when they are the default. You can also specify an alternative number of decimal digits by enclosing the desired number in a separate set of parentheses. For example, in the subcommand

```
SUMMARY = MEAN(SALARY(COMMA)(2))
```

SUMMARY displays the mean of SALARY without a dollar sign and with two decimal digits.

If the column is not wide enough to display the specified decimal digits for a given function, REPORT displays fewer decimals. REPORT uses scientific notation or displays asterisks if the column is not wide enough to display the integer portion of the number.

18.31 Composite Functions

A composite function operates on simple aggregate statistics and their arguments to produce a single result. For example, say you want to run a listing report that shows the average age, tenure, and salary of employees, and also reflects what the average salary will be when next year's 7% across the board salary increase takes effect. The following commands produce such a report:

```
SORT CASES BY DIVISION LNAME
REPORT FORMAT=AUTOMATIC LIST /
       VARS=LNAME AGE TENURE SALARY /
       BREAK=DIVISION /
       SUMMARY=MEAN /
       SUMMARY=MULTIPLY(MEAN(SALARY)1.07)  '7% Raise'
```

Note that the summary title is specified after the final parentheses enclosing the arguments, not between the function name (MULTIPLY) and its arguments.

Figure 18.31 shows one division from the report. Note that the composite function computes only one result.

Figure 18.31 Report with composite function

Personnel Data

Division	Last Name	Age	Tenure in Company	Salary—Annual
Carpeting	Cochran	22.00	3.92	$10,900
	Dan	36.00	3.83	$10,000
	Ford	27.00	3.67	$9,200
	Gates	24.00	4.00	$10,000
	Hoawinski	23.00	3.92	$10,900
	Jones	44.00	4.83	$15,690
	Katz	33.00	3.75	$10,000
	Lavelle	27.00	4.33	$10,000
	Mahr	33.00	2.67	$9,335
	McAndrews	35.00	3.50	$15,520
	Mulvihill	30.00	4.08	$10,000
	Tygielski	35.00	6.00	$19,500
Mean		30.75	4.04	$11,754
7% Raise				$12,577

By default, REPORT prints the result of the composite function calculation in the column defined by the first variable in the function that is also named on the VARIABLES subcommand. To move the result to another variable's column, specify that variable's name in parentheses after you completely define the composite function's argument. For example,

```
SUMMARY=MULTIPLY(MEAN(SALARY)1.07)  (LNAME)
```

places the result in the column defined by LNAME. Without (LNAME), this summary would appear in the column for SALARY. Unlike simple functions, the composite function result can be placed in any report column, including those defined by dummy variables or string variables.

It is often advisable to specify the format and number of decimal digits you want to print. Specify print formats for composites within parentheses following the name of the variable in whose column the result is to be printed. For example, in the subcommand

```
SUMMARY=MULTIPLY(MEAN(SALARY)1.07)  (SALARY(2))
```

(SALARY(2)) specifies that the increased average salary be displayed with 2 decimal places in the SALARY column.

You can use any numeric SPSS/PC+ variables, not just REPORT variables, as arguments to composite functions. You do not have to name a variable on the VARIABLES subcommand to use it in a composite function. You cannot use a composite function as an argument to a composite function. You can only use simple functions, variables, and constants.

There are three basic reasons for using composite functions:

1 To place a summary statistic in a column other than the one for which it is calculated. For example, the sum of VARA is normally printed in the column corresponding to VARA. It can, however, be placed in any other column by using a composite such as

```
SUMMARY=ADD(SUM(VARA))  (VARB)
```

which adds the sum of VARA to nothing and places it in VARB's column. The SUBTRACT, GREAT, LEAST, and AVERAGE composite functions can achieve the same result.

2 To manipulate statistics and change the unit of analysis. Certain arithmetic operations between variables produce the same result at either the case level or the subpopulation level. You can subtract the sum of variable A from the sum of variable B and obtain the same answer as subtracting A from B for each case and summing the result. However, dividing the sum of A by the sum of B is not the same as dividing A by B for each case and averaging the results.

3 To manipulate variables not named on the VARIABLES subcommand, so that these variables are not allocated a column in the report. For example, you might want to adjust dollar figures by some index when printing means but not to allocate a column for the index.

For a complete list of the composite functions available in REPORT, see Part C.

18.32 Repeating Summary Specifications

The keyword PREVIOUS on SUMMARY references a set of SUMMARY subcommands defined for a previous BREAK subcommand. For example,

```
REPORT  FORMAT=LIST/
        VARIABLES=NAME AGE TENURE /
        BREAK=DIVISION /
        SUMMARY=MEAN /
        SUMMARY=VALIDN(AGE) /
        BREAK=STORE /
        SUMMARY=PREVIOUS.
```

prints means and the valid number of cases for AGE for each STORE within the DIVISION.

PREVIOUS accepts an optional argument in parentheses to point to the particular set of summaries to be copied. For example, PREVIOUS(1) copies all the SUMMARY subcommands applying to the first BREAK subcommand, and PREVIOUS(2) copies summary specifications for the second BREAK subcommand. No other specification can be used on a SUMMARY subcommand using PREVIOUS.

18.33 Using Strings

The STRING subcommand enables you to link together SPSS/PC+ variables and constants to create new temporary variables you can use in REPORT. You can link together both alphanumeric and numeric variables and also intermix them.

For example, the subcommand

```
STRING= PHONE (AREA '/' EXCH '-' NUM)
```

creates a string variable named PHONE, which comprises three variables and two constants.

- The STRING subcommand must precede the VARIABLES subcommand.
- New STRING variables are temporary and available to the REPORT procedure only. The name you assign them must be unique and must follow SPSS/PC+ variable-naming conventions.
- You cannot use the keyword TO to imply a list of variables on the STRING subcommand.

You can use STRING variables on both the VARIABLES and BREAK subcommands, specifying column headings for them as you do for any other variables.

18.34 A STRING Application

One use of the STRING subcommand is to separate report columns with a column of special characters, such as asterisks or vertical bars. For example,

```
TITLE Personnel Data.
REPORT FORMAT=AUTOMATIC LIST /
       STRING=FILL ('*') /
       VARS=LNAME AGE TENURE FILL(1) ' ' SALARY.
```

defines a string variable FILL with the value "*" for each case. Naming FILL between TENURE and SALARY on the VARIABLES subcommand prints a vertical column of asterisks between those two columns on a listing report. The (1) after FILL on VARIABLES specifies a column width of 1; the space between apostrophes specifies a blank column heading. Figure 18.34 shows the first few lines of the report.

Figure 18.34 Column of special characters

Personnel Data PAGE 1

Last Name	Age	Tenure in Company		Salary—Annual
Ford	27.00	3.67	*	$9,200
Cochran	22.00	3.92	*	$10,900
Hoawinski	23.00	3.92	*	$10,900
Gates	24.00	4.00	*	$10,000
Mulvihill	30.00	4.08	*	$10,000
Lavelle	27.00	4.33	*	$10,000
Mahr	33.00	2.67	*	$9,335
Katz	33.00	3.75	*	$10,000
Jones	44.00	4.83	*	$15.690

18.35 Handling Missing Values

The MISSING subcommand within REPORT controls the treatment of missing values. You can specify one MISSING subcommand per REPORT command. It must follow the VARIABLE subcommand and precede BREAK.

The MISSING subcommand has three options:

VAR *Treat missing values separately for each variable named on the VARIABLES subcommand.* Missing values are indicated in case listings but are not included in calculating summary statistics. MISSING=VAR is the default.

LIST(varlist n) *Eliminate any case with missing values on n or more of the variables named on the variable list.*

NONE *Handle all user-defined missing values as though they were not missing.* The keyword NONE applies to the entire set of variables named on the VARIABLES subcommand. It cannot be used to ignore missing-data indicators for some variables selectively.

For example, the subcommand

```
MISSING=LIST (AGE TENURE SALARY 2)
```

deletes any case with missing values on two or more of the variables AGE, TENURE and SALARY from case listings and from summaries. If a case is missing for just one or for none of the variables listed, it is not deleted from case listings but is deleted from summaries for those variables for which it is missing. If no number is specified, the default is 1.

MISSING specifications apply to REPORT-generated strings, as well as to other variables. If one variable in a REPORT-generated string is missing, the string is missing.

Only variables named on the VARIABLES and SUMMARY subcommands are checked for missing values. You must use a SELECT IF command to eliminate cases missing on break variables.

18.36 Changing the Missing-Value Indicator

REPORT uses a period to indicate missing values in case listings, break values, and summary statistics.

The MISSING keyword on the FORMAT subcommand enables you to change the default missing-value indicator from a period (.) to any other character, including a blank. To do so, specify any one-character symbol, including a blank, in apostrophes, as in

```
FORMAT=AUTOMATIC LIST MISSING 'm' /
```

18.37 SUMMARY

REPORT has a full range of keywords that allow you to control the format and layout of a report. Table 18.37 will help you determine the subcommands and keywords to use to achieve the results you want. You can then consult the REPORT in Part C to quickly locate full descriptions of each.

Table 18.37 Summary of REPORT keywords

Function	Subcommand	Keyword
Adjusting margins	FORMAT	MARGINS(l,r)
Aligning columns	FORMAT	ALIGN(LEFT,CENTER,RIGHT)
Aligning report contents	VARIABLES, BREAK	(OFFSET(n|CENTER))
Determining column contents	FORMAT	LIST NOLIST MISSING ' '
	VARIABLES, BREAK	(VALUE LABEL)
	BREAK	NAME NONAME
Moving summary titles	SUMMARY	( ) col number
Page lengths	FORMAT	LENGTH(*|n)
	BREAK	(PAGE(RESET))
Page numbers	FORMAT	PAGE1(n)
Shifting column headings	FORMAT	CHALIGN(TOP,BOTTOM) TSPACE(n)
	VARIABLES, STRING, BREAK	' ' literal
	VARIABLES, BREAK	(CENTER, LEFT, RIGHT)
Spacing between columns	FORMAT	COLSPACE (4 n)
	VARIABLES	(DUMMY)
Spacing between rows	FORMAT	TSPACE(n) CHDSPACE(n) LIST(n) BRKSPACE(n) SUMSPACE(n) FTSPACE(n)
	VARIABLES	(DUMMY)
	BREAK, SUMMARY	(SKIP(n))
Specifying column widths	VARIABLES, STRING, BREAK	(n) following varname
Totals on listing reports	BREAK	(TOTAL)
Underscores between listings and summaries	BREAK	UNDERSCORE
Underscoring column headings	FORMAT	UNDERSCORE(ON OFF)

Bibliography

Anderson, R., and S. Nida. Effect of physical attractiveness on opposite and same-sex evaluations. *Journal of Personality,* 46:3 (1978), 401–413.

Beard, C. M., V. Fuster, and L. R. Elveback. Daily and seasonal variation in sudden cardiac death, Rochester, Minnesota, 1950–1975. *Mayo Clinic Proceedings,* 57 (1982), 704–706.

Berk, K. N. Comparing subset regression procedures. *Technometrics,* 20 (1978), 1–6.

———. Tolerance and condition in regression computation. *Journal of the American Statistical Association,* 72 (1977), 863–866.

Black and Sherba. Contracting to problem solve to lose weight. *Behavior Therapy*, 14 (1983), 105–109.

Blalock, H. M. *Social Statistics.* New York: McGraw-Hill, 1979.

Borgatta, E. F., and G. W. Bohrnstedt. Level of measurement once over again. *Sociological Methods and Research,* 9:2 (1980), 147–160.

Cedercreutz, C. Hypnotic treatment of 100 cases of migraine. In F. H. Frankel and H. S. Zamansky, eds. *Hypnosis at Its Bicentennial.* New York: Plenum, 1978.

Churchill, G. A., Jr. *Marketing Research: Methodological Foundations.* Hinsdale, Il.: Dryden Press, 1979.

Conover, W. J. Some reasons for not using the Yates continuity correction on 2 × 2 contingency tables. *Journal of the American Statistical Association,* 69 (1974), 374–376.

Cook, R. D. Detection of influential observations in linear regression. *Technometrics,* 19 (1977), 15–18.

Davis, J. A. *General Social Surveys, 1972–1982: Cumulative Codebook.* Chicago: National Opinion Research Center, 1982.

Davis, H., and E. Ragsdale. Unpublished working paper. Chicago: University of Chicago, Graduate School of Business, 1983.

Dineen, L. C., and B. C. Blakesley. Algorithm AS 62: A generator for the sampling distribution of the Mann-Whitney *U* statistic. *Applied Statistics,* 22 (1973), 269–273.

Everitt, B. S. *The Analysis of Contingency Tables.* London: Chapman and Hall, 1977.

Fienberg, S. E. *The Analysis of Cross-Classified Categorical Data.* Cambridge: MIT Press, 1977.

Frane, J. W. Some simple procedures for handling missing data in multivariate analysis. *Psychometrika,* 41 (1976), 409–415.

———. A note on checking tolerance in matrix inversion and regression. *Technometrics,* 19 (1977), 513–514.

Goodman, L. A. Interactions in multidimensional contingency tables. *Annals of Mathematical Statistics,* 35 (1964), 632–646.

Goodman, L. A., and W. H. Kruskal. Measures of association for cross-classification. *Journal of the American Statistical Association,* 49 (1954), 732–764.

Greeley, A. M., W. C. McCready and G. Theisen. *Ethnic Drinking Subcultures.* New York: Praeger Publishers, 1980.

Haberman, S. J. *Analysis of Qualitative Data,* Vol. 1. London: Academic Press, 1978.

Hansson, R. O., and K. M. Slade. Altruism toward a deviant in city and small town. *Journal of Applied Social Psychology,* 7:3 (1977), 272–279.

Hoaglin, D. C., and R. E. Welsch. The hat matrix in regression and ANOVA. *American Statistician,* 32 (1978), 17–22.

Hocking, R. R. The analysis and selection of variables in linear regression. *Biometrics,* 32 (1976), 1–49.

Information Please Almanac Atlas and Yearbook, 1983. New York: Information Please Publishers, 1983.

Jonassen, C. T. and S. H. Peres. *Interrelationships of Dimensions of Community Systems.* Columbus: Ohio State Univeristy Press, 1960.

King, M. M., et al. Incidence and growth of mammary tumors induced by 7,12-dimethylbenz(a) anthracene as related to the dietary content of fat and antioxident. *Journal of the National Cancer Institute,* 63:3 (1979), 657–663.

Kleinbaum, D. G., and L. L. Kupper. *Applied Regression Analysis and Other Multivariable Methods.* North Scituate, Massachusetts: Duxbury Press, 1978.

Lee, E. T. *Statistical Methods for Survival Data Analysis.* Belmont, California: Lifetime Learning Publications, 1980.

Loether, H. J., and D. G. McTavish. *Descriptive and Inferential Statistics: An Introduction.* Boston: Allyn and Bacon, 1976.

Mantel, N. Comment and a suggestion on the Yates continuity correction. *Journal of the American Statistical Association,* 69 (1974), 378–380.

Meyer, L. S., and M. S. Younger. Estimation of standardized coefficients. *Journal of the American Statistical Association,* 71 (1976), 154–157.

Neter, J., and W. Wasserman. *Applied Linear Statistical Models.* Homewood, Illinois: Richard D. Irwin Inc., 1974.

Overall, J. E., and C. Klett. *Applied Multivariate Analysis.* New York: McGraw-Hill, 1972.

Paul, O., et al. A longitudinal study of coronary heart disease. *Circulation,* 28 (1963), 20–31.

Rabkin, S. W., F. A. Mathewson and R. B. Tate. Chronobiology of cardiac sudden death in men. *Journal of the American Medical Association,* 244:12, (1980), 1357–1358.

Roberts, H. V. *An Analysis of Employee Compensation.* Rpt. 7946, Center for Mathematical Studies in Business and Economics, University of Chicago: October 1979.

———. Statistical bases in the measurement of employment discrimination. In E. Robert Livernash, ed., *Comparable Worth: Issues and Alternatives.* Washington, D.C.: Equal Employment Advisory Council, 1980, 173–195.

Siegel, S. *Nonparametric Statistics for the Behavioral Sciences.* New York: McGraw-Hill, 1956.

Sigall, H., and N. Ostrove. Beautiful but dangerous: effects of offender attractiveness and nature of the crime on juridic judgment. *Journal of Personality and Social Psychology,* 31 (1975), 410–414.

Smirnov, N. V. Table for estimating the goodness of fit of empirical distributions. *Annals of Mathematical Statistics,* 19 (1948), 279–281.

Sneath, P. H. A., and R. R. Sokal. *Principles of Numerical Taxonomy.* London: Freeman, 1973.

Snedecor, G. W., and W. G. Cochran. *Statistical Methods.* Ames, Iowa: Iowa State University Press, 1967.

Somers, R. H. A new symmetric measure of association for ordinal variables. *American Sociological Review,* 27 (1962), 799–811.

Speed, M. F. Response curves in the one way classification with unequal numbers of observations per cell. *Proceedings of the Statistical Computing Section,* American Statistical Association, 1976.

SPSS Inc. *SPSS^X User's Guide,* 2nd ed. New York: McGraw-Hill, 1986.

SPSS Inc. *SPSS Statistical Algorithms.* Chicago: SPSS Inc., 1985.

Stevens, S. S. On the theory of scales of measurement. *Science,* 103 (1946), 677–680.

Theil, H. *Economics and Information Theory.* Chicago: Rand McNally, 1967.

Velleman, P. F., and R. E. Welsch. Efficient computing of regression diagnostics. *American Statistician,* 35 (1981), 234–242.

Winer, B. J. *Statistical Principles in Experimental Design.* New York: McGraw-Hill, 1971.

Wynder, E. L. Nutrition and cancer. *Federal Proceedings,* 35 (1976), 1309–1315.

Wyner, G. A. Response errors in self-reported number of arrests. *Sociological Methods and Research,* 9:2 (1980), 161–177.

Command Reference

Universals

Commands

Contents

Universals

Syntax

Using SPSS/PC+ means becoming familiar with its language. Every effort has been made to keep the language natural, consistent, and straightforward.

Syntax Diagrams

Each SPSS/PC+ command described in this Command Reference includes a syntax diagram that shows all the subcommands, keywords, and specifications allowed for that command. The syntax diagram is used to show all the specifications for a command. By remembering the following rules, you can use the syntax diagram as a quick reference for each command.

- Elements shown in capital letters are keywords.
- Elements in lower case describe specifications supplied by the user.
- Elements in boldface type are defaults. Some defaults are indicated with **.
- Special delimiters, such as parentheses, apostrophes, or quotation marks, are required where indicated.
- Elements enclosed in square brackets ([ ]) are optional. When brackets would confuse the format, they are omitted. The command description explains which specifications are required or optional.
- Braces ({ }) indicate a choice between elements.
- The word "varlist" stands for a list of variable names.
- The command terminator is not shown in the syntax diagram.

Command Order

There are few formal rules regarding the order in which SPSS/PC+ commands should be specified. Understanding how SPSS/PC+ works will make it easy to specify commands in the right order.

- Variables must be defined before they can be used in procedures or assigned missing values or labels. You must use at least one DATA LIST, GET, JOIN, AGGREGATE, IMPORT, or TRANSLATE command to define variables for a session. (You can also define variables using the DE command available with SPSS Data Entry II.) IF, COUNT, COMPUTE, and CREATE commands also define variables.
- The logical outcome of command processing frequently determines the order: although data transformations are not carried out until the data are read, the result is as though the commands were executed when encountered.

Commands

Commands are the instructions that you give SPSS/PC+ to initiate an action. The following rules apply to all SPSS/PC+ commands:

- Commands begin with a keyword that is the name of the command and often have additional specifications, such as subcommands and user specifications. Refer to the discussion of each command to see which subcommands and additional specifications are required.
- Each command ends with a command terminator. The default command terminator is a period (.).
- The command terminator must be the last non-blank character in a line.
- Commands can begin in any column of a command line and continue for as many lines as needed. The exception is the END DATA command, which must begin in the first column of the first line after the end of data.
- The maximum length of any line in a command is 80 characters, including the prompt and the command terminator.

- Spaces can be added or lines broken at any point where a single blank is allowed. The only exceptions are the END DATA command, which can have only one space between words, and TITLE and SUBTITLE specifications, which cannot be broken across two lines.
- Commands and any command specifications can be entered in upper and lower case. Commands, subcommands, keywords, and variable names are translated to upper case before processing. All user specifications, including labels and data values, preserve upper and lower case.
- Most two-word commands—such as BEGIN DATA, SORT CASES, or VALUE LABELS—can be abbreviated to their first word. END DATA is an exception: you must enter both words. If the first specification on the command, such as a variable name, begins with the same three letters as the second word of a two-word command, you must enter the second word of the command explicitly before entering the specification. If two commands begin with the same three letters, such as DATE (in SPSS/PC+ Trends) and DATA LIST, you must specify the second word of the two-word command.

Subcommands

Many commands include additional specifications called *subcommands* for locating data, handling data, and formatting the output display.

- Subcommands begin with a keyword that is the name of the subcommand. Some subcommands include additional specifications.
- A subcommand keyword is separated from its specifications, if any, by an equals sign. The equals sign is optional unless the first three characters of the specification conflict with a variable name.
- Most subcommands can be named in any order. However, some commands require a specific subcommand order.
- Subcommands are separated from one another by a slash.

Keywords

Keywords are words specially defined by SPSS/PC+ to identify commands, subcommands, functions, operators, and other specifications.

- Keywords, including commands and subcommands, can be truncated to the first three characters of each word. The only exception is the keyword WITH, which must be spelled in full.
- Some keywords are reserved and cannot be used as variable names. Logical operators (AND, OR, and NOT), relational operators (EQ, GE, GT, LE, LT, and NE), and ALL, BY, TO, and WITH are the reserved keywords.
- Keyword ALL refers to all user-defined variables in the active file.
- Keyword THRU between two values specifies a range. The range includes the specified values.
- Keyword TO between two variable names specifies an inclusive list of variables.

Values

Values refer to specifications in commands or the data points processed by SPSS/PC+.

- A number specified as an argument to a subcommand can be entered with or without leading zeros.
- Data values of numeric variables can be specified as integers or real numbers, with or without leading zeros.
- Whenever values of string (alphanumeric) variables are used in commands, they must be specified within apostrophes or quotation marks, including all blanks.
- String values in data files or entered between BEGIN DATA and END DATA commands do not need to be enclosed in special delimiters (see DATA LIST: Freefield Format for exceptions).
- Blanks within apostrophes or quotation marks are significant.

Delimiters

Delimiters are used to separate data values, keywords, arguments, and specifications.

- The blank is usually used to delimit one specification from another, except when another delimiter serves the same purpose or where the comma is required.
- Commas are required to separate arguments to functions. Otherwise, commas are generally valid substitutes for blanks and vice versa.
- Arithmetic operators (+, −, *, and /) serve as delimiters in expressions. Blanks before and after arithmetic operators are optional.
- Special delimiters include parentheses, apostrophes, quotation marks, the slash, the equals sign, and so forth. Blanks before and after special delimiters are optional.
- The slash is used primarily to separate subcommands and lists of variables. Although slashes are sometimes optional, entering them as shown in the syntax diagrams is good practice.
- The equals sign is used between a subcommand and its specifications, as in FILE='filename', and to show equivalence, as in old variable list=new variable list. Equals signs following subcommands are frequently optional, but it is best to enter them.

Strings

The term "string" is used to refer to alphanumeric data or specifications such as titles and labels.

- The values of string variables can contain alphabetical characters as well as numbers.
- String variables whose values contain 8 or fewer characters are referred to as *short string variables.* String variables that contain more than 8 characters are referred to as *long string variables.* The maximum number of characters in a long string is 255. Strings longer than 130 are essentially useless, however.
- Only short string variables or values can be used in SPSS/PC+ transformation commands.
- Each string specified in a command should be enclosed in a set of either apostrophes or quotation marks.
- String specifications cannot be broken across command lines.
- Missing values cannot be generated for string variables read from a raw data file, since any character is a legal string value.
- When a transformation command on a string variable yields a missing or undefined result, a blank value is assigned.
- Blank is therefore defined as the user-missing value for string variables created by transformation commands. If you compute one string variable equal to another, user-missing values will be converted to blanks (but will remain user-missing).

Files

SPSS/PC+ uses a number of files in its operation. This section provides an overview to the types of files you can use, as well as a discussion of the active file.

Types of Files

The files that you can use in a session are

active file *A file specially formatted for use in SPSS/PC+, containing the data and a data dictionary.* The active file is initially defined by the DATA LIST, GET, IMPORT, JOIN, or TRANSLATE command, or the SPSS Data Entry II DE command. It is available until you replace it with a new active file or until you enter FINISH. You can modify the active file using transformations, analyze it using any of the procedures, and save it with SAVE or EXPORT. You can replace it with a new active file using a DATA LIST, GET, JOIN,

AGGREGATE, TRANSLATE, DE, or IMPORT command. The active file is stored in SPSS/PC+ workspace and in temporary files on disk, and it cannot be named.

command file *A file that contains SPSS/PC+ commands.* You can use the INCLUDE command to process the commands in a command file as an alternative to entering commands interactively, or you can submit a command file directly from REVIEW.

data file *A file that contains only raw data.* The data file is the file you specify on the DATA LIST command (unless the data are inline). Data files can be arranged in fixed or freefield format. In fixed-format files, the values of each variable for each case are recorded in the same location on each record. In freefield format, the values of each variable are recorded in the same order but not necessarily in the same location (see DATA LIST: Fixed Format and DATA LIST: Freefield Format).

listing file *A file containing output from SPSS/PC+ procedures.* Filenames for listing files can be specified on the SET command.

log file *A file created by SPSS/PC+ that contains a copy of all the commands that are executed during a session.* Filenames for log files can be specified on the SET command. You can use a log file as a command file in subsequent sessions.

results file *A file that contains procedure data output.* This file can contain a rectangular data set produced by the WRITE command or matrix materials produced by CLUSTER, CORRELATION, DSCRIMINANT, FACTOR, MANOVA, ONEWAY, QUICK CLUSTER, or REGRESSION. Filenames for results files can be specified on the SET command.

scratch pad *A file used by REVIEW as a work area for building commands.* The scratch pad is used as a command file.

system file *A binary file that is a copy of the active file and is saved on disk for later use.* In later runs, processing a system file is considerably more efficient than recreating the active file from raw data. You create system files with the SAVE command and read system files with the GET command.

portable file *A portable ASCII file containing data and a dictionary.* A portable file can be created with the EXPORT command of SPSS/PC+, SPSS^X, or SPSS/PC Release 1.0 or 1.1. The mainframe SAS procedure PROC TOSPSS also creates a portable file. Portable files are read using the IMPORT command. Portable files are much less efficient to process than system files but have the advantage that they can be read by SPSS^X or SAS.

- All filenames must be enclosed in apostrophes.
- Unless otherwise specified, file specifications default to the current drive and directory.
- Command files, data files, and system files can be read from any drive and directory; system files, log files, and listing files can be written to any drive and directory. When using a file on another drive or directory, you must explicitly specify the drive or path in the file specification.
- The results file cannot be written to a directory other than the current directory.
- A portable file can neither be written to nor read from a directory other than the current directory. It can be written to a floppy drive.

SPSS/PC+ Active File

When SPSS/PC+ processes data definition commands, it builds an internal file called the *active file*. The active file contains data and an associated data dictionary (stored in memory) of variable names, variable and value labels, missing-value flags, and format specifications. Data in the active file are stored in temporary files kept on the current default drive or on the drive specified with SET WORKDEV. The active file data can be altered by transformation commands and is used as input for SPSS/PC+ procedures.

- The active file is intially defined with a DATA LIST, GET, JOIN, IMPORT, TRANSLATE, or SPSS Data Entry II DE command.
- The active file can contain up to 200 variables.
- Each 8-character portion of a long string variable counts toward the system limit of 200 variables. For example, a 20-character string variable counts as 3 toward the system limit of 200 variables.
- The active file is available for use with SPSS/PC+ commands until the FINISH command is entered or until a new active file is built.
- The data contained in the active file are always read when a procedure is executed.
- Data transformation commands cause the active file data to be altered prior to processing data for a procedure. When these commands are entered prior to a procedure, the data are passed twice: once to incorporate the instructions from the transformation commands, and then to perform the SPSS/PC+ procedure.
- The N command limits the number of cases processed from the active file. The SELECT IF command limits the number of cases processed based upon some logical criteria. SELECT IF and N permanently affect the number of cases in the active file.
- The SAMPLE and PROCESS IF commands affect the number of cases only for the next procedure.
- The FORMATS, VALUE LABELS, VARIABLE LABELS, and MISSING VALUES commands affect the active file dictionary, not the data. No transformation pass occurs.
- The WEIGHT command affects the value of the existing system variable, $WEIGHT. No transformation pass occurs.
- The active file is replaced with a new active file when a new DATA LIST, GET, DE, JOIN, IMPORT, TRANSLATE (with FROM), or AGGREGATE (with OUTFILE=*) command is issued. The SPSS/PC+ Trends commands CREATE and RMV also replace the active file. MODIFY VARS may replace the active file.

Variables

This section describes rules for defining variables in SPSS/PC+.

Variable-Naming Conventions

Variable-naming conventions are the rules used to establish variable names in the active file dictionary and to refer to variables in commands.

- Variable names have a maximum of eight characters. The first character of user-defined variables must be an alphabetical letter or the @ character.
- The period, underscore, and the characters $, #, and @ can be used *within* variable names. For example, A._$@#1 is a valid variable name.
- Variable names ending with a period should be avoided, since the period will be interpreted as a command terminator when the variable name comes at the end of a command line.
- SPSS/PC+ creates special system variables that begin with a dollar sign ($). Such variables cannot be named on the DATA LIST, GET, or IMPORT commands.
- Variable names can be established on the DATA LIST, COMPUTE, COUNT, and IF commands. They can be changed with MODIFY VARS.
- You can establish the names of a set of variables on DATA LIST using the TO convention. Specify a character prefix with a numeric suffix before and after the keyword TO. The prefix can be any valid name and the number suffixes can be any integers, so long as the first number is smaller than the second. Each variable name, including the number, must not exceed eight characters. For example, ITEM1 TO ITEM5 establishes five variables named ITEM1, ITEM2, ITEM3, ITEM4, and ITEM5.
- With the TO convention, leading zeros used in suffixes are included in the variable name. For example, V001 TO V100 establishes 100 variables, V001, V002, V003, . . . , V100. V1 TO V100 establishes 100 variables, V1, V2, V3, . . . , V100.

- The TO keyword can also be used on commands other than DATA LIST to refer to a set of consecutive variables on the active file. AVAR TO VARB refers to the variables AVAR and all other variables up to and including VARB on the active file. Use the DISPLAY command or the Variables menu ((Alt) (V)) in REVIEW to see the order of variables on the active file.
- Reserved keywords that cannot be used as variable names are

ALL	AND	BY	EQ	GE	GT	LE
LT	NE	NOT	OR	TO	WITH	

System Variables

SPSS/PC+ provides three system variables that are included as part of each case in the active file. The three variables are $DATE, $CASENUM, and $WEIGHT.

- System variables begin with a dollar sign ($).
- System variables cannot be named on a DATA LIST, GET, or IMPORT command.
- System variables cannot be named as target variables on COUNT, COMPUTE, and IF commands.
- $DATE and $CASENUM are established when data are read with a DATA LIST or IMPORT command.
- The value of $DATE is the date a case was read in on the DATA LIST command or created in SPSS Data Entry II.
- The value of $CASENUM is the sequence number of each case as it is read by SPSS/PC+ during execution of a DATA LIST or IMPORT command, or as it is created by SPSS Data Entry II.
- The value of $CASENUM remains unchanged, even after SORT, SELECT IF, SAVE, and GET commands, so you can always identify the original case number.
- The initial value of $WEIGHT for each case is 1.00.
- The value of $WEIGHT is changed using the WEIGHT transformation.
- When a file is written using SAVE, the current values of each system variable are written to the file.
- When a file is written using EXPORT, $WEIGHT is the only system variable written to the file.
- When a file is written using WRITE, $CASENUM and $WEIGHT can be included. $DATE cannot be included.

Variable Format

Values are stored internally in double precision. They are displayed or printed according to a format specification. This format specification is used to print values in procedures and to write data to other files. The format specification *does not* affect the precision of data values stored in memory. You can use the DISPLAY command specifying VARIABLES=ALL to see the format of each of your variables.

Variable formats have two components, the variable type and the variable width. Variables can be one of two types, numeric or string. Numeric variables can contain numbers, decimal points, and optional leading plus or minus signs. String variable values can contain numbers, letters, and punctuation characters (see Strings).

The format of a variable is defined on the DATA LIST command or is assigned by SPSS/PC+ on an IF, COUNT, or COMPUTE command. You can change the format of numeric variables with the FORMATS command. You cannot change the format of string variables.

Variable Type

- By default, the DATA LIST command assumes variables are numeric. String variables are indicated by the A format specification.
- Variables created by COMPUTE or IF commands are assigned a format type based on specifications in the assignment expression (see COMPUTE and IF).

Variable Width

- The width determines how the values are printed.
- Variables defined with DATA LIST using keyword FIXED use the column specifications to calculate the maximum width of data values. If a decimal place is implied on DATA LIST, one column is added to the width required to print the value to allow for the decimal point.
- Numeric variables read with DATA LIST FREE are assigned a print width of eight characters.
- String variables read with DATA LIST FREE are assigned a width of eight characters, including a decimal point and two decimal digits. If the actual width is greater than eight columns, you must specify the maximum number of characters after the format specification, as in NAME (A20).
- Numeric variables created with COMPUTE or IF are assigned a width of eight characters, including a decimal point and two decimal digits.
- String variables created with COMPUTE or IF are assigned a width based on the assignment expression. When the assignment expression creates a variable by equating it to an existing variable, the width of the new variable is equal to the width of the existing variable. When the assignment expression creates a new string variable by equating it to a string constant, the width of the new string variable has a width equal to the initial specification of the string constant.

Format Specifications

The FORMATS command allows you to change the print and write width, print additional decimal values, and add a dollar sign and commas to values of numeric variables.

- You cannot change the variable type from string to numeric or vice versa.
- The FORMATS command changes only the print and write formats, not the internal representation, of a variable.
- If you do not allow enough room to print the values of a variable, SPSS/PC+ prints the value without decimal values, commas, or dollar signs. When the value cannot be reasonably represented in the width provided, SPSS/PC+ rounds it, uses scientific notation, or prints asterisks (**) in the available space.

Transformation Expressions

Transformation expressions are used in COMPUTE, IF, PROCESS IF, and SELECT IF commands. The following sections describe the different types of operators and functions that can be used in transformation expressions.

Arithmetic Operators

Arithmetic operators are used with numeric variables in expressions on COMPUTE, IF, and SELECT IF commands. String variables and string constants cannot be used with arithmetic operators. The arithmetic operators are

+	Addition	–	Subtraction
*	Multiplication	/	Division
**	Exponentiation		

Syntax

- No two operators can appear consecutively.
- Arithmetic operators cannot be entered before or after relational and logical operators (see Logical Expressions).
- Blanks (not commas) can be inserted before and after an operator to improve readability.

Operations

- Arithmetic operators are executed after functions (see Functions).
- The order of operations is exponentiation, then multiplication and division, and then addition and subtraction.
- Operators at the same level are executed from left to right.
- Use parentheses to override the order of operation. Execution begins with the innermost set of parentheses and progresses out.
- If any variables or values in an arithmetic or logical expression are missing, the result is always system-missing. The only exceptions are values used as arguments to the missing-value functions VALUE, SYSMIS, and MISSING.
- A negative number raised to a noninteger power is system-missing.

Functions

Functions are used on COMPUTE, IF, and SELECT IF command expressions. All functions are available for use with numeric variables. Only the LAG function can be used with short string variables. No functions can be used with long string variables. For examples using functions, see COMPUTE, IF, and SELECT IF commands.

The expression that is transformed by a function is called the *argument*. Most functions have a variable name or a list of variable names as arguments. You can substitute constants for variable names. Arguments are always enclosed in parentheses. For example, to generate the square root of variable X, specify X as the argument to the SQRT function, as in SQRT(X).

Arguments can include arithmetic and exponential operators and numeric constants to form a complex expression. The expression VARA+VARB, enclosed in parentheses, forms an argument that can be used by most functions. You can use sets of parentheses in complex expressions used as arguments. For functions that take multiple arguments, such as a list of of variables, each argument is separated by a comma and the entire list of arguments is enclosed in parentheses.

- By default, functions in expressions are evaluated before arithmetic, relational, and logical operators. The default order of evaluation is numeric functions, then exponentiation, then arithmetic operators, then relational operators, and then logical operators.
- Operations at the same level are executed from left to right.
- Use parentheses to override the default order of operation. Execution begins with the innermost set of parentheses and progresses out.

Numeric Functions

- Numeric functions always return numbers (or the system-missing value whenever the result is indeterminate).
- All numeric functions take one argument enclosed in parentheses. The argument can be either a variable name or an expression.

ABS(arg) *Absolute value.* ABS(−4.7) is 4.7; ABS(4.7) is 4.7. The argument can be an expression or a variable name. Returns system-missing if the argument is missing.

RND(arg) *Round the absolute value to an integer and reaffix the sign.* RND(−4.7) is −5. The argument can be an expression or a variable name. Returns system-missing if the argument is missing.

TRUNC(arg) *Truncate to an integer.* TRUNC(−4.7) is −4. The argument can be an expression or a variable name. Returns system-missing if the argument is missing.

MOD10(arg) *Remainder (modulus) when the argument is divided by 10.* MOD10(198) is 8; MOD10(−198) is −8. The argument can be an expression or a variable name. Returns 0 if the argument is 0 and system-missing if the argument is missing.

SQRT(arg) *Square root.* The argument can be an expression or a variable name. Returns system-missing if the argument is negative or missing.

EXP(arg) *Exponential. e* is raised to the power of the argument. The argument can be an expression or a variable name. Returns system-missing if the argument is missing or if the argument produces a result too large to be represented.

LG10(arg) *Base 10 logarithm.* The argument can be an expression or a variable name. Returns system-missing if the argument is negative, 0, or missing.

LN(arg) *Natural or Naperian logarithm (base* e*).* The argument can be an expression or a variable name and must be positive and greater than 0. Returns system-missing if the argument is missing.

ARTAN(arg) *Arctangent* (alias ATAN). The argument can be an expression or a variable name. The result is given in radians. Returns system-missing if the argument is missing.

SIN(arg) *Sine.* The argument can be an expression or a variable name but it must yield radians. Returns system-missing if the argument is missing.

COS(arg) *Cosine.* The argument can be an expression or a variable name but it must yield radians. Returns system-missing if the argument is missing.

The arc sine function is not directly available but can be computed easily as the following:

```
COMPUTE ARCSINX = ARTAN(X/SQRT(1-X*X)).
```

Note: this identity is valid only if X is greater than −1 and less than 1.

Missing-Value Functions

- Each missing-value function takes one variable name enclosed in parentheses as an argument.

VALUE(arg) *Returns the numeric value of valid or user-missing data in a form which is not considered missing in expressions.* The argument must be a single variable name. Use this function to test for user-missing values in IF and SELECT IF logical expressions. For example, if 999 was declared as a user-missing value for VAR1, the expression IF (VALUE(VAR1) EQ 999) is evaluated as *true* for cases where VAR1 has that value; the expression IF (VAR1 EQ 999) would be evaluated as *missing,* since this expression contains a variable with a missing value.

MISSING(arg) *Return 1 (true) if the value is either system-missing or user-missing and return 0 (false) otherwise.* The argument must be a single variable name.

SYSMIS(arg) *Return 1 (true) if the value is system-missing and 0 (false) otherwise.* The argument must be a single variable name.

Cross-Case Function

LAG(arg) *The value of the variable one case before.* The argument must be a numeric variable. LAG(GNP) returns the value of GNP for the case before the current one. If you are selecting cases from a file, LAG returns the value for the case previously selected. Returns system-missing for the first case.

Random-Number Functions

- The seed value used for random number functions can be changed using the SEED specification on the SET command.

UNIFORM(arg) *A uniform pseudo-random number.* The random number is uniformly distributed with values varying between 0 and the value of the argument. The argument must be a numeric constant or variable name.

NORMAL(arg) *A normal pseudo-random number.* The random number is randomly distributed with a mean of 0 and a standard deviation equal to the argument. The argument must be a numeric constant or variable name. Returns system-missing if the argument is zero or negative.

Date Function

YRMODA(arg list) *Convert year, month, and day to a day number.* The number returned is the number of days since October 15, 1582 (day 1 of the Gregorian calendar).

- Arguments for YRMODA can be variables, constants, or any other type of numeric expression but must yield integers.
- Year, month, and day must be specified in that order.
- The first argument can be any year between 0 and 99, or between 1582 to 47516.
- If the first argument yields a number between 00 and 99, 1900 through 1999 is assumed.
- The month can range from 1 through 13. Month 13 yields the last day of the year—as in YRMODA(YEAR,13,0)—or the first month of the next year—as in YRMODA(YEAR,13,DAY).
- The day can range from 0 through 31. Day 0 is the last day of the previous month. For example, YRMODA(YEAR,MONTH + 1,0) is the last day of MONTH.
- Returns missing if any of the three arguments is missing.

- Returns missing if the arguments do not form a valid date after October 15, 1582.

Logical Expressions

A logical expression is an expression that can be evaluated as true, false, or missing, based upon conditions found in the data. Logical expressions can be simple logical relations among variables, or they can be complex logical tests involving variables, constants, functions, relational operators, and logical operators. Logical expressions can be used on the IF, SELECT IF, and PROCESS IF commands. However, logical expressions used with PROCESS IF can contain only the relational operators, not the logical operators.

Relational operators used in expressions are

EQ or =	Equal to	NE or ~= or <>	Not equal to
LT or <	Less than	LE or <=	Less than or equal to
GT or >	Greater than	GE or >=	Greater than or equal to

Logical operators are

AND or & OR or | NOT or ~

Syntax

- Parentheses enclosing a logical expression are required.
- Blanks (not commas) must separate the relational and logical operators from the expressions.
- String values must be enclosed in apostrophes or quotation marks.
- The operators EQ and NE must compare one string variable to another or one numeric value to another.
- Only strings of the same length are compared using EQ or NE.
- Long string values cannot be used in logical expressions or transformations.

Operations

- In expressions, functions and arithmetic operators are executed before relational and logical operators.
- Relational operators are executed before logical operators.
- Logical NOT is executed first, then AND, and then OR.
- Operators at the same level are executed from left to right.
- Use parentheses to override the order of operation. Execution begins with the innermost set of parentheses and progresses out.

Limitations

- The complexity of a logical expression is limited by available memory.

AND and OR

- AND returns true if both expressions are true.
- OR returns true if either expression is true.

NOT

- NOT reverses the outcome of the expression.
- NOT affects only the expression that immediately follows (unless otherwise indicated by parentheses.)

Logical Outcomes

- Logic is indeterminate and the outcome is missing if the expression on either side of a relational operator is missing.

Commands

C

AGGREGATE

```
AGGREGATE OUTFILE={'filename'}
                  {*         }

 [/PRESORTED]

 [/MISSING=COLUMNWISE]

 /BREAK=varlist [{(A)}] [varlist [{(A)}]] ...
                 {(D)}            {(D)}

 /aggvar ['label'] aggvar ['label'] ...
      =function(varlist[,arguments])

 [/aggvar ...]
```

Example:

```
AGGREGATE OUTFILE='AGGEMP.SYS'
/BREAK=LOCATN82 DEPT82
/COUNT=N
/AVGSAL AVGRAISE = MEAN(SALARY82 RAISE82)
/SUMSAL SUMRAISE = SUM(SALARY82 RAISE82)
/BLACKPCT 'Percentage Black' = PIN(RACE,1,1)
/WHITEPCT 'Percentage White' = PIN(RACE,5,5).
```

Overview

Procedure AGGREGATE creates a new active or system file from the current active file by aggregating groups of cases into single cases. The values of one or more variables in the active file define the case groups. A series of aggregate functions creates new variables that have one value for each case group. Each function operates on a *source variable* in the active file to create an *aggregated variable* for the new active or system file.

Defaults

By default, AGGREGATE excludes missing values from all aggregate variable calculations except those involving functions N, NU, NMISS, and NUMISS. Unless otherwise specified, AGGREGATE sorts the aggregated cases in ascending order of the values of the case grouping variables.

Tailoring

Output File. You can produce either a new active file or a system file with AGGREGATE. You can specify any legal DOS filename as the name of the system file.

Case Grouping. You can group cases according to the values of any one or more of the variables in your active file. Variables that you use to group cases are called *break variables*. A set of cases in the file with identical values for each break variable is called a *break group*. AGGREGATE calculates a single value for each new variable for each break group.

Sorting. You can sort the aggregated cases into either ascending or descending order of the values of each break variable. If the file is already sorted by the break variables, you can instruct AGGREGATE to skip this final pass through the file.

Aggregated Variables. You can create aggregated variables using any of 19 aggregate functions. Functions SUM, MEAN, and SD can take only numeric variables as arguments, but all other aggregate functions will accept both numeric and string variables.

Labels and Formats. You can specify variable labels for the aggregated variables on the AGGREGATE command. Variables created with functions MAX, MIN, FIRST, and LAST will assume the formats and value labels of their respective source variables. All other variables will assume the default print formats described under "Aggregate Functions" below.

Missing Values. By default, AGGREGATE declares a value of a new variable missing only if all values of the source variable are missing in the relevant break group. Optionally, you can declare a new variable missing if there is a single missing value for the source variable among any cases in the break group. You can also include user-missing values in the calculation of an aggregate function.

Syntax

- The minimum specification is the OUTFILE subcommand with a file name or asterisk, the BREAK subcommand with a single variable name, and an aggregate function that creates a new aggregated variable.
- The OUTFILE subcommand must be specified first.
- The aggregate functions must be specified last.
- If specified, the PRESORTED subcommand must precede the BREAK subcommand.
- Subcommands and variable definition statements must be separated by slashes.

Operations

- AGGREGATE causes the data to be read.
- The system files that AGGREGATE produces are, like all system files, binary files designed to be read and written by SPSS/PC+ only. They cannot be edited.
- When AGGREGATE produces a system file, the active file remains unchanged and is still available for analysis.
- AGGREGATE places the new system file in the current DOS directory.
- When AGGREGATE creates a new active file, it erases the old active file. Only the new active file is available for analysis.
- AGGREGATE includes the break variables in the file it creates.
- If the active file is already sorted in the order you want your aggregated file, use the PRESORTED subcommand. If you specify PRESORTED, a new aggregate case is created each time a different value or combination of values is encountered on variables named on the BREAK subcommand.

OUTFILE Subcommand

The OUTFILE subcommand specifies whether AGGREGATE should create a new system file or replace the active file.

- An asterisk following the equals sign tells AGGREGATE to replace the active file.
- A name in single quotes following the equals sign tells AGGREGATE to create a new system file with the given name.
- Names for system files must adhere to the DOS restrictions on filenames.

Example

```
AGGREGATE OUTFILE='AGGEMP.SYS'
/BREAK=LOCATN82
/AVGSAL = MEAN(SALARY82).
```

- This example creates a system file called AGGEMP.SYS in the current DOS directory.
- The new system file will contain variables LOCATN82 and AVGSAL.

BREAK Subcommand

The BREAK subcommand lists the *break variables.* Each unique combination of values of the break variables defines one break group. A *break group* consists of cases with identical values for each break variable.

- The variables named on the BREAK subcommand can be any combination of variables from the active file.

- Unless it encounters the PRESORTED subcommand (see below), AGGREGATE sorts cases after aggregating. By default, cases are sorted in ascending order of the values of the break variables.
- You can control the sort order by specifying an A (for ascending) or D (for descending) in parentheses after any break variables.
- The designations (A) and (D) apply to all preceding undesignated variables.
- AGGREGATE sorts first on the first-named variable, then on the second-named variable within the groups created by the first, and so on.
- Subcommand PRESORTED overrides all sorting specifications.

Example

```
/BREAK=LOCATN82 DEPT82 (A) TENURE (D)
```

- This subcommand names variables LOCATN82, DEPT82 and TENURE as the break variables.
- Cases are sorted in ascending order of LOCATN82, in ascending order of DEPT82 within LOCATN82, and in descending order of TENURE within LOCATN82 and DEPT82.

PRESORTED Subcommand

The PRESORTED subcommand indicates that cases in the active file are sorted according to the values of the break variables. This prevents AGGREGATE from sorting cases that have already been sorted.

- If specified, the PRESORTED subcommand must precede the BREAK subcommand.
- When you specify PRESORTED, SPSS/PC+ forms an aggregate case out of each group of *adjacent* cases with the same values for the break variable(s).
- If the active file is not sorted by the break variables in ascending order and PRESORTED is specified, a warning message is generated but the procedure is executed. Each group of adjacent cases with the same values for break variables form a case in the output file, which may produce multiple cases with the same values for the break variables. In this case, the output file will not be sorted by the break variables.

Example

```
/PRESORTED
/BREAK=LOCATN82 DEPT82
```

- In this example, the PRESORTED subcommand tells AGGREGATE that the cases are already sorted by variables LOCATN82 and DEPT82.
- AGGREGATE does not make an extra data pass to sort the cases.

Aggregate Functions

An aggregated variable is created by applying an aggregate function to a variable in the active file. The variable in the active file is the *source* variable, and the new aggregated variable is called the *target* variable.

- The simplest specification is a target variable list, followed by an equals sign, the function keyword, and a list of source variables.
- The number of target variables named must match the number of source variables.
- When several aggregate variables are defined at once, the first-named target variable is a function of the first-named source variable, the second-named target is a function of the second-named source, and so on.
- You can optionally specify variable labels for the target variables. A label is specified in single quotation marks after the target variable to which it applies. Value labels cannot be assigned on the AGGREGATE command.
- Print formats are automatically assigned to target variables according to the function it is based on (see list of functions below).
- If you are creating a system file, it will not be possible to change print formats or add value labels immediately. However, you can do so later by calling the system file into the active file, specifying the new labels and formats, and resaving.

The following functions are available:

SUM(varlist) *Sum across cases.* Dictionary formats are F8.2.

MEAN(varlist) *Mean across cases.* Dictionary formats are F8.2.

SD(varlist) *Standard deviation across cases.* Dictionary formats are F8.2.

MAX(varlist) *Maximum value across cases.* Complete dictionary information is copied from the source variables to the target variables.

MIN(varlist) *Minimum value across cases.* Complete dictionary information is copied from the source variables to the target variables.

PGT(varlist,value) *Percentage of cases greater than value.* Dictionary formats are F5.1.

PLT(varlist,value) *Percentage of cases less than value.* Dictionary formats are F5.1.

PIN(varlist,value1,value2) *Percentage of cases between value1 and value2 inclusive.* Dictionary formats are F5.1.

POUT(varlist,value1,value2) *Percentage of cases not between value1 and value2.* Cases where the source variable equals value1 or value2 are not counted. Dictionary formats are F5.1.

FGT(varlist,value) *Fraction of cases greater than value.* Dictionary formats are F5.3.

FLT(varlist,value) *Fraction of cases less than value.* Dictionary formats are F5.3.

FIN(varlist,value1,value2) *Fraction of cases between value1 and value2 inclusive.* Dictionary formats are F5.3.

FOUT(varlist,value1,value2) *Fraction of cases not between value1 and value2.* Cases where the source variable equals value1 or value2 are not counted. Dictionary formats are F5.3.

N(varlist) *Weighted number of cases in break group.* Dictionary formats are F7.0 for unweighted files and F8.2 for weighted files.

NU(varlist) *Unweighted number of cases in break group.* Dictionary formats are F7.0.

NMISS(varlist) *Weighted number of missing cases.* Dictionary formats are F7.0 for unweighted files and F8.2 for weighted files.

NUMISS(varlist) *Unweighted number of missing cases.* Dictionary formats are F7.0.

FIRST(varlist) *First nonmissing observed value in break group.* Complete dictionary information is copied from the source variables to the target variables.

LAST(varlist) *Last nonmissing observed value in break group.* Complete dictionary information is copied from the source variables to the target variables.

- Functions SUM, MEAN, and SD operate only on numeric source variables. All other functions accept short and long string variables as well as numeric ones.
- The N and NU functions do not require arguments. Without arguments, they return the number of weighted and unweighted cases in a break group. If you supply a variable list, they return the weighted and unweighted number of valid cases for the variables specified.

- For several functions, the argument includes values as well as a source variable designation. PGT, PLT, FGT, and FLT take one value; PIN, POUT, FIN, and FOUT take two values.
- You can use either blanks or commas to separate the components of an argument list.
- String values specified in an argument should be enclosed in single quotes.
- The order used in aggregation functions with string values is alphabetical.
- For PIN, POUT, FIN, and FOUT, the first value should be less than or equal to the second. If the first is higher, AGGREGATE automatically reverses them and prints a warning message.
- If the two values are equal, PIN and FIN calculate the percentages and fractions, respectively, of values equal to the argument. POUT and FOUT calculate the percentages and fractions, respectively, of values not equal to the argument.

Example

```
AGGREGATE OUTFILE='AGGEMP.SYS'
/BREAK=LOCATN82
/AVGSAL 'Average Salary' AVGRAISE = MEAN(SALARY82 RAISE82).
```

- This example defines two aggregate variables, AVGSAL and AVGRAISE.
- AVGSAL is the mean of SALARY82 for each break group, and AVGRAISE is the mean of RAISE82.
- The label 'Average Salary' is assigned to AVGRAISE.

Example

```
AGGREGATE OUTFILE=*
 /BREAK=DEPT
 /LOWVAC,LOWSICK = PLT (VACDAY SICKDAY,10).
```

This example assigns the percentage of cases with values less than 10 for VACDAY to LOWVAC and for SICKDAY to LOWSICK.

Example

```
AGGREGATE OUTFILE='GROUPS.SYS'
 /BREAK=OCCGROUP
 /COLLEGE = FIN(EDUC,13,16).
```

This example assigns the fraction of cases having 13 to 16 years of education to COLLEGE.

Example

```
AGGREGATE OUTFILE=*
 /BREAK=CLASS
 /LOCAL = PIN(STATE,'IL','IO').
```

This example creates variable LOCAL, which is the percentage of cases in each break group whose two-letter state code represents Illinois, Indiana, or Iowa. (The abbreviation for Indiana, IN, is between IL and IO in an alphabetical sort sequence.)

Missing Values

By default, AGGREGATE uses all nonmissing values of the source variable to calculate aggregated variables. An aggregated variable will have a missing value only if the source variable is missing for every case in the break group. You can alter the default missing-value treatment by using the MISSING subcommand or by specifying the inclusion of missing values on any function.

- The COLUMNWISE keyword is the only specification available on MISSING.
- MISSING=COLUMNWISE declares the value of an aggregated variable as missing if the source variable is missing for any case in the break group.
- The MISSING subcommand must precede the BREAK subcommand.
- The MISSING subcommand does not affect the calculation of the N, NU, NMISS, or NUMISS functions.

You can force a function to include user-missing values in its calculations by specifying a period after the function name.

- AGGREGATE ignores periods used with functions N, NU, NMISS, and NUMISS if these functions have no argument.
- User-missing values are treated as valid when these four functions are followed by a period and have a variable as an argument. NMISS.(AGE) gives the number of cases for which AGE has the system-missing value only.

The effect of specifying a period with N, NU, NMISS, and NUMISS is illustrated by the following:

N = N. = N(AGE) + NMISS(AGE) = N.(AGE) + NMISS.(AGE)

NU = NU. = NU(AGE)+ NUMISS(AGE) = NU.(AGE) + NUMISS.(AGE)

- The function N (the same as N. with no argument) yields a value for each break group that equals the number of cases with valid values plus the number of cases with user- or system-missing values.
- This in turn equals the number of cases with either valid or user-missing values plus the number with system-missing values.
- The same identities hold for the NU, NMISS, and NUMISS functions.

Example

```
AGGREGATE OUTFILE='AGGEMP.SYS'
/MISSING=COLUMNWISE
/BREAK=LOCATN82
/AVGSAL = MEAN(SALARY82).
```

- This example specifies that AVGSAL will be missing for an aggregate case if SALARY82 is missing for any case in the break group.

Example

```
AGGREGATE OUTFILE=*
 /BREAK=DEPT
 /LOVAC = PLT.(VACDAY,10).
```

- This function sets variable LOVAC to the percentage of cases within each break group with values less than 10 for VACDAY, even if some of those values are defined as user-missing.

Example

```
AGGREGATE OUTFILE='CLASS.AVG'
 /BREAK=GRADE
 /FIRSTAGE = FIRST.(AGE).
```

- This example assigns the first value of AGE in each break group to variable FIRSTAGE.
- If the first value of AGE in a break group is user-missing, that value will be assigned to FIRSTAGE. However, the value will retain its missing-value status since variables created with FIRST take their dictionary information from their source variables.
- Function LAST with a period operates in parallel fashion.

ANOVA

```
ANOVA [VARIABLES=] varlist BY varlist(min,max) [WITH varlist]
      [/[VARIABLES=] varlist ...]

      [/OPTIONS=option numbers]

      [/STATISTICS={statistic numbers}]
                   {ALL              }
```

Options:

1	Include user-missing values	7	Covariates with main effects
2	Suppress labels	8	Covariates after main effects
3	Suppress all interaction terms	9	Regression approach
4	Suppress three-way terms	10	Hierarchical approach
5	Suppress four-way terms	11	Narrow format
6	Suppress five-way terms		

Statistics:

1 MCA table
2 Unstandardized regression coefficients for covariates
3 Display cell means and counts

Example:

```
ANOVA VARIABLES=YVAR1,YVAR2 BY XVAR(1,3) ZVAR1,ZVAR2(1,2)
/OPTIONS=4
/STATISTICS=3.
```

Overview

Procedure ANOVA performs analysis of variance for factorial designs. The default is the full factorial model if there are five or fewer factors. Analysis of variance tests the hypothesis that the group means of the dependent variable are equal. The dependent variable is interval level, and one or more categorical variables define the groups. These categorical variables are termed *factors.* ANOVA also allows you to include continuous explanatory variables, termed *covariates.* Other SPSS/PC+ procedures that perform analysis of variance are ONEWAY and MEANS.

Defaults

By default, the model includes all interaction terms up to five-way interactions. In the default model, the sums of squares are decomposed using the classical experimental approach, in which covariates, main effects, and ascending orders of interaction are assessed separately in that order. The default display includes an analysis of variance table with variable labels, sums of squares, degrees of freedom, mean square, *F*, probability of *F* for each effect, and a count of valid and missing cases. By default, a case that has a missing value for any variable in an analysis list is omitted from the analysis.

Tailoring

Statistical Display. You can choose among three methods for controlling the order in which covariates and main effects are assessed. You can also select among three methods for decomposing the sums of squares, and you can pool interaction effects in the error term. In addition, you can request a labeled multiple classification analysis (MCA) table, a table of means of the dependent variable within groups formed by the factors, and unstandardized regression coefficients for the covariates.

Display Format. You can suppress the display of variable labels on all tables and the display of value labels on the MCA table.

Missing Values. You can include cases with user-missing values in the analysis.

Syntax

- The minimum specification is a single VARIABLES subcommand with an analysis list. The actual keyword VARIABLES may be omitted.
- The minimum analysis list specifies a list of dependent variables, the keyword BY, a list of factor variables, and the minimum and maximum integer values of the factors in parentheses.
- Subcommands can be specified in any order and must be separated by slashes.

Operations

- ANOVA causes the data to be read.
- A separate analysis of variance is performed for each dependent variable in an analysis list, using the same factors and covariates.
- With the exception of cell means and counts (Statistic 3), the output is always displayed in narrow format, regardless of the width defined on SET.

Limitations

- Maximum 5 VARIABLES subcommands.
- Maximum 1 each STATISTICS and OPTIONS subcommands.
- Maximum 5 dependent variables per VARIABLES subcommand.
- Maximum 10 factors per VARIABLES subcommand.
- Maximum 10 covariates per VARIABLES subcommand.
- Maximum 5 interaction levels.
- Maximum 25 value labels per variable displayed in the MCA table.
- The combined number of categories for all factors in an analysis list plus the number of covariates must be less than the sample size.
- Memory requirements for ANOVA are roughly proportional to the square of the product of the number of values for each independent variable.
- Both the number of categories in each factor and the number of interaction terms included in the model will determine the amount of workspace required.

Example

```
ANOVA VARIABLES=YVAR1,YVAR2 BY XVAR(1,3) ZVAR1,ZVAR2(1,2)
/OPTIONS=4
/STATISTICS=3.
```

- The VARIABLES subcommand specifies two three-way analyses of variance: YVAR1 by XVAR, ZVAR1, and ZVAR2; and YVAR2 by XVAR, ZVAR1, and ZVAR2.
- Variables ZVAR1 and ZVAR2 both have the values 1 and 2 included in the analysis.
- The OPTIONS subcommand pools all three-way interaction terms into the error sum of squares.
- The STATISTICS subcommand requests the display of a table of cell means of YVAR1 and YVAR2 within the combined categories of XVAR, ZVAR1, and ZVAR2.

VARIABLES Subcommand

The VARIABLES subcommand specifies the analysis list. The actual keyword VARIABLES may be omitted.

- Variables named before the BY keyword are dependent variables.
- Variables named after the BY in the analysis list are factor (independent) variables. All factors are used simultaneously in the analysis of variance requested by that analysis list.
- Every factor variable must have a value range indicating its minimum and maximum values. The values must be separated by a space or comma and enclosed in parentheses.
- Variables named after the keyword WITH are covariates.
- Each analysis list can include only one BY and one WITH keyword.
- Factor variables must be integer. Noninteger values for factors are truncated.
- Cases with values outside the range specified for a factor are excluded from the analysis.
- If two or more factors have the same value range, you can specify the value range once following the last factor to which it applies.
- You can specify a single minimum and maximum value range that encompasses the ranges of all factors in the list. However, this may reduce performance and cause memory problems if the specified range is larger than the actual range.

- You can specify multiple VARIABLES subcommands on a single ANOVA command. The slash between subcommands is required; the VARIABLES keyword is not.

Suppressing Interaction Effects

By default, all interaction effects up to and including fifth-order interaction effects are tested. You can suppress any of these higher-order interactions and pool them into the error (residual) sums of squares by specifying Option 3, 4, 5, or 6. When you specify any of these options, cell means corresponding to suppressed interaction terms are not displayed.

Option 3 *Suppress all interaction terms.*

Option 4 *Suppress three-way and higher-order interaction terms.*

Option 5 *Suppress four-way and higher-order interaction terms.*

Option 6 *Suppress five-way and higher-order interaction terms.*

Specifying Order of Entry of Covariates

By default, covariates are assessed first and main effects are assessed after adjusting for the covariates. To change this order, use Option 7 or 8.

Option 7 *Process covariates concurrently with main effects for factors.*

Option 8 *Process covariates after main effects for factors.* Option 8 is ignored when Option 9 is in effect.

Decomposing Sums of Squares

By default, each type of effect is assessed separately in the following order (unless Option 7 or 8 has been specified):

- Effects of covariates.
- Main effects of factors.
- Two-way interaction effects.
- Three-way interaction effects.
- Four-way interaction effects.
- Five-way interaction effects.

To change this order, specify Option 9 or 10.

Option 9 *Regression approach.* All effects are assessed simultaneously, with each effect adjusted for all other effects in the model. Option 9 overrides Options 7 and 8. Statistics 1 and 3 are not available with Option 9.

Option 10 *Hierarchical approach.* Factor main effects and covariate effects are assessed hierarchically. Factor main effects are adjusted only for factor main effects already assessed, and covariate effects are adjusted only for covariates already assessed. Factors are assessed in the order they are listed in the analysis list on the ANOVA command.

Table A shows how effects would be assessed under Options 9, 10 and the default for the following command:

```
ANOVA VARIABLES=Y BY A B C (0 3).
```

A Terms adjusted for under each option

Effect	Default	Option 9	Option 10
A	B,C	ALL OTHERS	NONE
B	A,C	ALL OTHERS	A
C	A,B	ALL OTHERS	A,B
AB	A,B,C,AC,BC	ALL OTHERS	A,B,C,AC,BC
AC	A,B,C,AB,BC	ALL OTHERS	A,B,C,AB,BC
BC	A,B,C,AB,AC	ALL OTHERS	A,B,C,AB,AC
ABC	A,B,C,AB,AC,BC	ALL OTHERS	A,B,C,AB,AC,BC

Summary of Analysis Methods

Table B describes the results obtained with various combinations of methods for controlling entry of covariates and decomposing the sums of squares.

B Summary of analysis methods

	Assessments between types of effects	Assessments within the same type of effect
Default	**Covariates** *then* **Factors** *then* **Interactions**	**Covariates:** Adjust for all other covariates **Factors:** Adjust for covariates and all other factors **Interactions:** Adjust for covariates, factors, and all other interactions of the same and lower orders
Option 7	**Factors** **Covariates** concurrently *then* **Interactions**	**Covariates:** Adjust for factors and all other covariates **Factors:** Adjust for covariates and all other factors **Interactions:** Adjust for covariates, factors, and all other interactions of the same and lower orders
Option 8	**Factors** *then* **Covariates** *then* **Interactions**	**Factors:** Adjust for all other factors **Covariates:** Adjust for factors and all other covariates **Interactions:** Adjust for covariates, factors, and all other interactions of the same and lower orders
Option 9	**Covariates** **Factors** **Interactions** simultaneously	**Covariates:** Adjust for factors, interactions, and all other covariates **Factors:** Adjust for covariates, interactions, and all other factors **Interactions:** Adjust for covariates, factors, and all other interactions
Option 10	**Covariates** *then* **Factors** *then* **Interactions**	**Covariates:** Adjust for preceding covariates **Factors:** Adjust for covariates and preceding factors **Interactions:** Adjust for covariates, factors, and all other interactions of the same and lower orders
Options 7 and 10	**Factors** **Covariates** concurrently *then* **Interactions**	**Factors:** Adjust only for preceding factors **Covariates:** Adjust for factors and preceding covariates **Interactions:** Adjust for covariates, factors, and all other interactions of the same and lower orders
Options 8 and 10	**Factors** *then* **Covariates** *then* **Interactions**	**Factors:** Adjust only for preceding factors **Covariates:** Adjust for factors and preceding covariates **Interactions:** Adjust for covariates, factors, and all other interactions of the same and lower orders

Multiple Classification Analysis

Multiple classification analysis is useful for displaying the results of analysis of variance when there are no significant interaction effects (see Andrews et al., 1973).

- For each category of each factor, the MCA table presents the unadjusted mean of the dependent variable expressed as a deviation from the grand mean; the deviation from the grand mean of the category mean adjusted for other factors; and the deviation from the grand mean of the category mean adjusted for both factors and covariates.

- For each factor, the complete MCA display displays the correlation ratio (eta) with the unadjusted deviations (the square of eta indicates the proportion of variance explained by all categories of the factor); a partial beta equivalent to the standardized partial regression coefficient that would be obtained by assigning the unadjusted deviations to each factor category and regressing the dependent variable on the resulting variables; and the parallel partial betas from a regression that includes covariates in addition to the factors, and the multiple R and R^2 from this regression.

To obtain an MCA table, specify Statistic 1 on the STATISTICS subcommand.

Statistic 1 *Multiple classification analysis.* The MCA display is affected by the form of analysis specified by Options 7 through 10 or their defaults. When covariates are specified, a complete MCA table can be obtained only in conjunction with Option 8, Options 8 and 10, or Options 7 and 10. With a model in which factors are not processed first, effects adjusted only for factors do not appear. The MCA table cannot be produced when Option 9 is in effect.

Statistical Display

You can request the following optional statistics on the STATISTICS subcommand:

Statistic 2 *Display unstandardized regression coefficients for covariates.* The regression coefficients are computed when the covariates are entered into the equation. Thus, their values depend on the design you specify with Options 7 through 10 or the defaults.

Statistic 3 *Display cell means and counts for the dependent variable.* A table is created by crossing all factors. Marginal means and counts as well as cell means and counts are displayed. With Options 3 through 6, cells corresponding to the suppressed interaction terms are not displayed. Statistic 3 is not available when Option 9 is in effect.

ALL *Display all statistics.* Includes display produced by Statistics 2 and 3.

Display Format

In the default display, the ANOVA table is labeled with the variable labels and the MCA table includes value labels. In addition, the table of cell means and counts (Statistic 3) uses the available width. To change these defaults, specify the following on the OPTIONS subcommand:

Option 2 *Suppress variable and value labels.*

Option 11 *Narrow format.* Restricts the display produced by Statistic 3 to narrow width regardless of the width defined on SET.

Missing Values

By default, a case that has a missing value for any variable named on a VARIABLES subcommand is omitted from all analyses requested by that subcommand. You can change the treatment of missing values by specifying the following on the OPTIONS subcommand:

Option 1 *Include user-missing values.* User-missing value specifcations are ignored. A case with a system-missing value for any variable named on a VARIABLES subcommand is omitted from the analyses requested by that subcommand.

Reference

Andrews, F., J. Morgan, J. Sonquist, and L. Klein. *Multiple classification analysis,* 2nd ed. Ann Arbor: University of Michigan, 1973.

BEGIN DATA—END DATA

```
BEGIN DATA
lines of data
END DATA
```

Example:

```
BEGIN DATA.
1  3424   274 ABU DHABI 2
3 39932    86 AMSTERDAM 4
3  8889   232 ATHENS
2  3424   294 BOGOTA    3
END DATA.
```

Overview

The BEGIN DATA command signals the beginning of data lines in a command file, and the END DATA command signals the end of data lines. Both BEGIN DATA and END DATA must be used when data are part of an SPSS/PC+ command file (inline data). BEGIN DATA and END DATA are also used for inline matrix data or matrix materials.

Syntax

- BEGIN DATA, the data, and END DATA must appear before the first SPSS/PC+ procedure.
- Data lines must *not* have a command terminator.
- The BEGIN DATA command must be entered immediately before the first line of inline data.
- The END DATA command must be entered immediately after the last line of data, beginning in the column 1.
- Only a single space is allowed between the words END and DATA.
- END DATA cannot be abbreviated to END. BEGIN DATA can be abbreviated to BEGIN (or BEG).
- Procedures and additional transformations can follow the END DATA command.

Operations

- When SPSS/PC+ encounters BEGIN DATA, it begins to read and process data on the next input line. All preceding transformation commands are processed as a file is built for use in SPSS/PC+ procedures.
- SPSS/PC+ continues to evaluate input lines as data until it encounters END DATA, at which point it begins evaluating input lines as SPSS/PC+ commands.
- No other SPSS/PC+ commands are recognized between BEGIN DATA and END DATA.
- You can use a file that contains BEGIN DATA, data lines, and END DATA with the INCLUDE command, provided you omit the FILE specification on your DATA LIST command (see INCLUDE).

Example

```
DATA LIST /XVAR 1 YVAR ZVAR 3-12 CVAR 14-22(A) JVAR 24.
BEGIN DATA.
1  3424   274 ABU DHABI 2
3 39932    86 AMSTERDAM 4
3  8889   232 ATHENS
2  3424   294 BOGOTA    3
4 11323   332 HONG KONG 3
3   323   232 MANILA    1
4  3234   899 CHICAGO   4
1 78998  2344 VIENNA    3
2  8870   983 ZURICH    5
END DATA.
MEANS XVAR BY JVAR.
```

- The DATA LIST command defines the names and column locations of the variables. The FILE subcommand is omitted because the data are inline.
- There are nine cases in the inline data.
- Each line of data is completed by pressing the enter key.
- The END DATA command begins in column 1, has only a single space between END and DATA, and signals the end of lines of data.

* (Comment)

```
* text
```

Example:

```
* CREATE A NEW VARIABLE AS A COMBINATION OF TWO OLD VARS.
COMPUTE XYVAR=0.
IF (XVAR EQ 1 AND YVAR EQ 1)XYVAR=1.
```

Overview

The comment facility allows you to insert text within your SPSS/PC+ job for the purpose of documentation.

Syntax

- The first line of a comment must begin with an asterisk (*) in column 1.
- Comment text can extend for multiple lines. Continuation comment lines can start in any column and do not begin with an asterisk.
- You cannot specify a continuation line if the preceding lines ends with a period. Instead, you must start a new comment with an asterisk.
- The comment text can contain any characters.
- Comments cannot be imbedded within data lines or within lines of multiple-line commands.
- A command terminator must be placed at the end of a comment. If you omit the command terminator, the first command following the comment is treated as a continuation of the comment.

Operations

- Comments are included in the command printback on the log file.

Example

```
* CREATE A NEW VARIABLE AS A COMBINATION OF TWO OLD VARIABLES.
COMPUTE XYVAR=0.
IF (XVAR EQ 1 AND YVAR EQ 1) XYVAR=1.
```

- The one-line comment will be included in the log file.

Example

```
DATA LIST /XVAR 1 YVAR ZVAR 3-12 CVAR 14-22(A) JVAR 24.
* THIS IS AN EXAMPLE OF A THREE LINE
COMMENT.  THE DATA BELOW ARE ENTERED
INTERACTIVELY.
BEGIN DATA.
1  3424  274 ABU DHABI 2
3 39932   86 AMSTERDAM 4
3  8889  232 ATHENS
2  3424  294 BOGOTA    3
4 11323  332 HONG KONG 3
3   323  232 MANILA    1
4  3234  899 CHICAGO   4
1 78998 2344 VIENNA    3
2  8870  983 ZURICH    5
END DATA.
MEANS XVAR BY JVAR
/OPTIONS=1.
FINISH.
```

- The comment text begins just after the DATA LIST command and extends for two additional lines.

COMPUTE

```
COMPUTE target variable=expression
```

Arithmetic Operators:

+	Addition	−	Subtraction
*	Multiplication	/	Division
**	Exponentiation		

Numeric Functions:

ABS	Absolute value	RND	Round
TRUNC	Truncate	MOD10	Modulus
SQRT	Square root	EXP	Exponential
LG10	Base 10 logarithm	LN	Natural logarithm
SIN	Sine	COS	Cosine
ATAN	Arctangent		

Missing-Value Functions:

VALUE	Treat user-missing as valid
SYSMIS	Return 1 if system-missing
MISSING	Return 1 if missing

Cross-case Function:

LAG Lag

Random-Number Functions:

UNIFORM	Uniform pseudo-random number
NORMAL	Normal pseudo-random number

Date Function:

YRMODA Date function

Example:

```
COMPUTE YVAR1=RND((YVAR/ZVAR)*100).
COMPUTE NEWSTRNG='maxwidth'.
```

Overview

The COMPUTE transformation creates a new variable or modifies the values of an existing variable for each case in your active file. The variable name on the left of the equals sign is the *target variable.* The variables, values, and specifications on the right side of the equals sign form an *assignment expression.*

Numeric Transformations

You can use both arithmetic operations and functions in the transformation of numeric variables. The assignment expression can include combinations of arithmetic operations, constants, and functions. Parentheses are used to indicate the order of operations and to enclose the argument for a function.

String Transformations

You can create and modify short string variables using COMPUTE. A variable name can be set equal to a string constant or to an existing string variable. The LAG function is available for cross-case transformation of short string variables. All other functions are available for numeric transformations only.

Syntax

- The target variable is named first.
- The equals sign is required.
- You cannot mix numeric and string variables in an expression.
- Each function takes on at least one argument enclosed in parentheses.
- For a complete discussion of each function and its argument, see Universals: Functions.

Numeric Variables

- Parentheses are used to indicate the order of transformations and to set off the arguments for a function.
- To evaluate unary minus, the minus sign and the variable or constant must be enclosed in parentheses.
- Numeric functions can take on expressions, enclosed in parentheses, as arguments.

- The arc sine function is not available directly but can be computed as shown in the examples below.
- The VALUE, SYSMIS, MISSING, and LAG functions take only one variable name enclosed in parentheses as an argument.
- UNIFORM takes a single value or variable name, enclosed in parentheses, as an argument. The resulting random variable ranges between 0 and the value of the argument.
- NORMAL takes a single positive value or a variable name, enclosed in parentheses, as an argument. The resulting random variable has a mean of 0 and a standard deviation equal to the argument. The result is system-missing when the argument is negative.
- YRMODA requires three arguments that represent a year, month, and day, in that order. The arguments must yield integer values and can be in the form of variables, constants, or a combination of variables and constants.

String Variables

- Only short string variables, values, and constants can be used in an expression. Long string variables cannot be used.
- String values and constants must be enclosed in apostrophes or quotation marks.
- LAG is the only function available for strings.
- The LAG function takes only one variable enclosed in parentheses as an argument.

Operations

- COMPUTE is a transformation and is executed before the data are read for the next procedure.
- If the target variable already exists, its values are replaced.
- If the target variable does not already exist, it is created as a new variable.
- Invalid syntax stops all processing of the COMPUTE command. New variables are not created and existing target variables remain unchanged.

Numeric Variables

- New numeric variables created with COMPUTE are assigned a dictionary format of a width of eight characters with two decimal places.
- Existing numeric variables transformed with COMPUTE retain their original dictionary formats.
- You can change the format of a numeric variable using the FORMAT command.
- If a case is system-missing on any variable named in an assignment expression, SPSS/PC+ returns the system-missing value for that case since the operation is indeterminate.
- You can include user-missing values in computations by using the VALUE function.
- The MISSING function returns the value 1 if the variable named has either a user-missing or system-missing value.
- The SYSMIS function returns the value 1 if the variable named has a system-missing value.
- The YRMODA function returns the number of days since the beginning of the Gregorian calendar (October 15, 1582). The arguments for YRMODA must yield integers.
- The LAG function returns the value of the previous case for the named variable. The first case will have a system-missing value for the target variable.
- The UNIFORM and NORMAL functions return values with decimal places. Use the TRUNC or RND function to change these values to integers.
- You can change the seed value used by UNIFORM and NORMAL with the SEED specification on SET.
- All expressions are evaluated in the following order: first functions, then exponentiation, and then arithmetic operations. You can change the order of operations by using parentheses.

String Variables

- A new string variable created by setting a variable name equal to a string constant is assigned a dictionary format equal to the width of the string constant.
- A new string variable created by setting a variable name equal to an existing string variable has the same dictionary format as the existing variable.
- Existing string variables transformed with COMPUTE retain their original dictionary formats.
- All subsequent transformations on a string variable must use the format defined when the variable is first named. All leading or trailing blanks must be specified.
- You cannot change the format of string variables.
- The LAG function returns the value of the previous case for the named variable. The first case will have a system-missing value for the target variable.

Limitations

- Only 1 variable can be created or transformed per COMPUTE command.
- The number of variables created with COMPUTE, COUNT, and IF plus the number defined on DATA LIST, IMPORT, and GET cannot exceed the system maximum of 200 variables.

Numeric Examples

The following examples illustrate the use of the COMPUTE command with numeric variables.

Arithmetic Operations

```
COMPUTE XVAR=25.
COM YVAR1=(YVAR/XVAR)*100.
```

- XVAR is initialized to 25 for all cases.
- YVAR1 is computed as the percentage YVAR is of XVAR.

Numeric Functions

```
COMPUTE WTCHANGE=ABS(WEIGHT1-WEIGHT2).
COMPUTE YVAR1=RND((YVAR/ZVAR)*100).
COMPUTE ARCSINX=ARTAN(X/SQRT(1-X*X)).
```

- WTCHANGE is computed as the absolute value of WEIGHT1 minus WEIGHT2.
- YVAR1 is computed as a percentage of ZVAR and is rounded to an integer using the RND function.
- ARCSINX is computed as the arc sine of X, using the ARTAN and SQRT functions. This trigonometric identity is valid only if X is greater than −1 and less than +1.

Missing Values

```
MISSING VALUE XVAR1 XVAR2 XVAR3 (0).
COMPUTE FVAR1=XVAR1 + XVAR2 + XVAR3.
COMPUTE FVAR4=VALUE(XVAR1) + VALUE(XVAR2) + VALUE(XVAR3).
COMPUTE FVARSM=SYSMIS(XVAR1) + SYSMIS(XVAR2) + SYSMIS(XVAR3).
COMPUTE FVARM=MISSING(XVAR1) + MISSING(XVAR2) + MISSING(XVAR3).
```

- The MISSING VALUE command declares the value 0 as missing for XVAR1, XVAR2, and XVAR3.
- FVAR1 is computed as the sum of three variables only for cases with valid values for all three variables. FVAR1 is assigned the system-missing value for a case if any variable in the assignment expression has a system-missing or user-missing value.
- The VALUE function overrides user-missing value declarations. Thus, FVAR4 is the sum of XVAR1, XVAR2, and XVAR3 for each case, including cases with value 0 (the user-missing value) for any of the three variables. Cases with system-missing values are not included.
- The SYSMIS function on the third COMPUTE returns value 1 if the variable is system-missing. Thus, FVARSM ranges from 0 to 3 for each case, depending on whether the variables XVAR1, XVAR2, and XVAR3 are system-missing for the case.
- The MISSING function on the fourth COMPUTE returns the value 1 if the variable named is system-missing or user-missing. Thus, FVARM ranges from

0 to 3 for each case, depending on whether variables XVAR1, XVAR2, and XVAR3 are user- or system-missing for that case.

- Alternatively, you could use the COUNT command to create variables FVARSM and FVARM.

Across-Case Operations

```
COMPUTE LVAR1=LAG(LVAR).
COMPUTE LVAR2=LAG(LVAR1).
```

- The LAG function on the first COMPUTE sets LVAR1 equal to the value of LVAR for the previous case.
- The second COMPUTE sets LVAR2 equal to the value of LVAR1 for the previous case. In effect, LVAR2 is equal to the value of LVAR for two cases previous.
- This example demonstrates the use of successive LAG transformations to perform multiple-case LAG operations.

Random-Number Functions

```
COMPUTE QVAR=UNIFORM(10).
COMPUTE SVAR=NORMAL(1.5).
```

- The first COMPUTE sets QVAR equal to a pseudo-random number from a distribution with values ranging between 0 and the specified value of 10.
- The second COMPUTE sets SVAR equal to a pseudo-random number from a distribution with a mean of 0 and a standard deviation of the specified value of 1.5.
- You can change the seed value of the pseudo-random-number generator with the SEED specification on SET.

Date Function

```
COMPUTE AGER=(YRMODA(1980,08,23)-
YRMODA(YRBIRTH,MOBIRTH,DABIRTH))/365.25.
```

- The YRMODA function converts the current date (in this example, August 23, 1980) and birthdate to a number of days. Birthdate is subtracted from current date and the remainder is divided by the number of days in a year to yield age in years.

String Examples

The examples below illustrate the use of the COMPUTE command with string variables.

Equivalence

```
COMPUTE NEWSTR=YVAR.
COMPUTE DAYVAR='TODAY'.
```

- The first COMPUTE creates a new variable NEWSTR with the same values as YVAR. The format of NEWSTR is the same as YVAR.
- The second COMPUTE creates a new variable DAYVAR with the value "TODAY" for each case. The width of DAYVAR is five characters.

LAG Operations

```
COMPUTE NEIGHBOR=LAG(HSHOLDER).
```

- COMPUTE creates a new short string variable NEIGHBOR as the value for HSHOLDER for the previous case. The first case will have the system-missing value for NEIGHBOR.

CORRELATION

```
CORRELATION [VARIABLES=] {varlist} [WITH varlist] [/varlist ...]
                         {ALL    }

              [/OPTIONS=option numbers]

              [/STATISTICS={statistic numbers}]
                           {ALL              }
```

Options:

1	Include user-missing values	4	Write count and correlation matrix
2	Exclude missing values pairwise	5	Display count and probability
3	Two-tailed probability		

Statistics:

1 Univariate mean, standard deviation, and count
2 Cross-product deviations and covariance

Example:

```
CORRELATION VARIABLES=WVAR XVAR YVAR
/VARIABLES=ZVAR1 TO ZVAR5 WITH ZVAR6 TO XVAR8
/OPTIONS=2 3
/STATISTICS=1.
```

Overview

Procedure CORRELATION (alias PEARSON CORR) produces Pearson product-moment correlations with one-tailed probabilities and, optionally, univariate statistics, covariances, and cross-product deviations. Other procedures that read and write correlation matrices are FACTOR (see *SPSS/PC+ Advanced Statistics)* and REGRESSION.

Defaults

By default, CORRELATION displays a rectangular matrix of correlation coefficients for each analysis list. The default display includes the coefficient and the number of valid cases used in the computation of the matrix. An asterisk (*) indicates that a coefficient has a one-tailed probability of less than 0.01, and two asterisks (**) indicate a probability of less than 0.001. Cases that have missing values for any variable in the matrix are excluded.

Tailoring

Statistical Display. You can display counts and probabilities in the matrix of coefficients, two-tailed probabilities, univariate statistics for each variable, and cross-product deviations and covariances for each pair of variables.

Writing Matrices. You can write out a square matrix containing correlation coefficients and the number of cases for use in other SPSS/PC+ procedures (see FACTOR in *SPSS/PC+ Advanced Statistics* and REGRESSION: Matrix Materials).

Missing Values. Missing-value options allow you to include cases with user-missing values in the computation of each coefficient. Alternatively, you can exclude cases with missing values on a pairwise basis.

Syntax

- The minimum specification is the VARIABLES subcommand with a single analysis list. The actual keyword VARIABLES can be omitted.
- Subcommands are separated by slashes.
- You can invoke CORRELATION with its alias, PEARSON CORR.

Operations

- CORRELATION causes the data to be read.
- A correlation of a variable with itself is displayed as 1.0000.
- A correlation that cannot be computed is displayed as a period (.).
- Correlation coefficients are displayed with four decimal places. Optional statistics are displayed with four decimal places where possible.
- Long or short string variables on an analysis list prevents execution of CORRELATION.
- The display uses the width set on the SET command.

Limitations

- The maximum number of variables that can be named on a CORRELATION command is the same as the system limit.
- Maximum 40 VARIABLES subcommands per CORRELATION command.
- Maximum 1 each of the OPTIONS and STATISTICS subcommands.
- Maximum 250 individual elements per CORRELATION command. Variable names, keywords, and special delimiters count as 1. Variables implied by the TO keyword do not count.

Example

```
CORRELATION VARIABLES=WVAR XVAR YVAR
/VARIABLES=ZVAR1 TO ZVAR5 WITH ZVAR6 TO ZVAR8
/OPTIONS=2 3
/STATISTICS=1.
```

- The first VARIABLES subcommand requests a square matrix of correlation coefficients among variables WVAR, XVAR, and YVAR.
- The second VARIABLES subcommand requests a rectangular correlation matrix in which variables ZVAR1 through ZVAR5 are the rows and ZVAR6 through ZVAR8 are the columns.
- Option 2 specifies pairwise deletion. All cases with valid values for the pair of variables used to compute a coefficient are included in the computation of that coefficient.
- Option 3 requests two-tailed probabilities.
- Statistic 1 asks for univariate statistics for all variables named on the VARIABLES subcommands.

VARIABLES Subcommand

The VARIABLES subcommand specifies the analysis list. The actual keyword VARIABLES can be omitted.

- An analysis list composed of a simple variable list produces a square matrix of correlations of each variable with every other variable.
- An analysis list that consists of a list of variables and keyword WITH produces a rectangular correlation matrix. Variables listed before WITH define the rows of the matrix and those listed after WITH define the columns.
- The keyword ALL can be used in an analysis list to refer to all user-defined variables.
- You can specify multiple VARIABLES subcommands on a single CORRELATION command. The slash between the subcommands is required; the keyword VARIABLES is not.

Statistical Display

By default, the correlation matrix and number of valid cases on which the matrix is based are displayed. One-tailed probabilities of less than 0.01 are indicated by an asterisk (*) and less than 0.001 by two asterisks (**). In addition, you can request the following on the OPTIONS and STATISTICS subcommands:

Option 3 *Two-tailed probability.* Two-tailed probabilities less than 0.01 are indicated by an asterisk (*) and less than 0.001 by two asterisks (**). With Options 3 and 5, the exact two-tailed probabilties are displayed.

Option 5 *Display count and probability.* The number of cases used to compute each coefficient and exact probability are displayed.

Statistic 1 *Univariate mean, standard deviation, and count.* Displays the mean, standard deviation, and number of nonmissing cases for each variable. Cases with missing values are excluded on a variable-by-variable basis regardless of the missing-value option in effect.

Statistic 2 *Cross-product deviations and covariance.*

ALL *Display all statistics.* Includes statistics available with Statistics 1 and 2. Specify ALL on the STATISTICS subcommand.

Writing Matrices

By default, matrices are written only to the display file. Optionally, you can write matrices to the results file named on the SET command for use in other procedures by specifying Option 4 on the OPTIONS subcommand.

- Any VARIABLES subcommand that contains the keyword WITH is ignored when matrices are written.
- With listwise deletion of cases with missing values (the default), the correlation matrix precedes a record containing the *n* used to compute all coefficients in the matrix.
- With pairwise deletion of cases with missing values (Option 2), each correlation matrix precedes a matrix of the *n*'s used to compute the coefficients.
- Correlation matrices are written with F10.7 format.
- Matrices of *n*'s are written with an F10.0 format.
- Each row of a matrix begins on a new record.
- Each record has a maximum of eight values.
- A matrix of coefficients followed by the number of cases is written for each VARIABLES subcommand in the order in which the VARIABLES subcommands are specified.
- If the results file named on the SET command is not empty when CORRELATION is executed, the contents of the file will be overwritten. Use the SET command immediately before the CORRELATION command to control the destination of the matrix that is being written.

Option 4 *Write count and correlation matrix.* The correlation matrix and number of cases used to compute each coefficient are written for each analysis list to the results file named on the SET command. Ignored for VARIABLES subcommands with the keyword WITH.

Example

```
SET RESULTS='SAVINGS.MAT'.
DATA LIST FILE='SAVINGS.DAT'/ VAR1 TO VAR10 1-20.
CORRELATION VARIABLES=VAR1 TO VAR10
/OPTIONS=4.
```

- This example writes one matrix of correlations followed by a single *n* to file SAVINGS.MAT.

Missing Values

By default, a case with a user- or system-missing value on any variable in a matrix is excluded from the computation of that matrix. Alternatively, you can specify the following missing-value treatments on the OPTIONS subcommand:

Option 1 *Include user-missing values.* Cases with user-missing values are considered valid when listwise or pairwise deletion of cases with missing values is performed.

Option 2 *Exclude missing values pairwise.* Cases with valid values for the variables used to compute a coefficient are included in the computation of that coefficient, regardless of whether the cases have missing values for other variables named on the analysis list.

COUNT

```
COUNT varname=varlist (value list) varlist (value list)
      [/varname=...]
```

Keywords available for numeric value lists:

```
LO  LOWEST  HI  HIGHEST  THRU  SYSMIS  MISSING
```

Example:

```
COUNT RVAR=XVAR,YVAR,ZVAR (2).
```


Overview

The COUNT transformation creates a numeric variable that, for each case, counts the occurrences of the same value (or list of values) across a list of variables. The new variable is called the target variable. The variables and values that are counted are the criterion variables and values. You can create the target variable from numeric or string variables or both.

Syntax

- The minimum specification is the target variable, an equals sign, a criterion variable, and a criterion value enclosed in parentheses.
- Only one target variable is allowed per specification.
- A variable can be specified more than once in the criterion variable list.
- The criterion variable list can include both string and numeric variables, provided they have separate value specifications.
- The TO keyword can be used to name consecutive criterion variables that have the same criterion value or values.
- You can specify more than one criterion value, separating each by a comma or space.
- String values must be enclosed in apostrophes.
- Keywords THRU, LOWEST (LO), and HIGHEST (HI) can be used in a numeric value list.
- SYSMIS counts system-missing values for numeric variables.
- MISSING counts both user- and system-missing values for numeric variables.
- You cannot specify any keywords with string variables.
- You can create more than one target variable on a single COUNT command by separating the specifications with a slash.

Operations

- COUNT is a transformation and is executed when the data are read for the next procedure.
- The target variable is numeric.
- The target variable is initialized to 0 for each case.
- If the target variable already exists, its previous values are replaced.
- Variables created with COUNT are assigned a dictionary format of eight columns with two decimal places.
- COUNT does not propagate missing values automatically. The target variable will never be system-missing. To declare missing-value flags, use the RECODE or MISSING VALUE command.

Limitations

- The number of variables created with COUNT combined with the number created with COMPUTE and IF and defined on DATA LIST, IMPORT, or GET cannot exceed the system maximum of 200 variables.

Example

```
COUNT RVAR=XVAR,YVAR,ZVAR (2).
```

- The value of RVAR for each case will be either 0, 1, 2, or 3, depending on the number of times the value 2 occurs across the three variables.
- RVAR is a numeric variable with a format of eight columns with two decimal places.

Example

```
COUNT QLOW=QVAR1 TO QVAR10 (LO THRU 0)
/QSYSMIS=QVAR1 TO QVAR10 (SYSMIS).
```

- Assuming there are 10 variables between and including QVAR1 and QVAR10 on the active file, QLOW ranges from 0 to 10, depending on the number of times a case has a negative or 0 value across variables QVAR1 to QVAR10.
- QSYSMIS ranges from 0 to 10, depending on how many system-missing values are encountered for QVAR1 to QVAR10 for each case.
- Both QLOW and QSYSMIS are numeric variables and have a format of eight columns with two decimal places.

Example

```
COUNT SVARC=AVAR,BVAR ('male  ') CVAR, DVAR, EVAR ('female').
```

- SVARC ranges from 0 to 5, depending on the number of times a case has a value of "male" for AVAR and BVAR and value "female" for CVAR, DVAR, and EVAR.
- SVARC is a numeric variable with a format of eight columns with two decimal places.

Example

```
COUNT MIXVAR=SEX (2) EVAL ('good').
```

- MIXVAR ranges from 0 to 2, depending on the number of times a case has value 2 for variable SEX and value "good" for variable EVAL.
- MIXVAR is a numeric variable with a format of eight columns with two decimal places.

CROSSTABS

```
CROSSTABS [TABLES=] {varlist} BY varlist [BY...]
                    {ALL    }
          [/[TABLES=] varlist...]

          [/OPTIONS=option numbers]

          [/STATISTICS={statistic numbers}]
                       {ALL              }
```

Options:

1	Include user-missing values	13	Suppress cell counts
2	Suppress all labels	14	Display expected frequencies
3	Display row percentages	15	Display chi-square residuals
4	Display column percentages	16	Display standardized chi-square residuals
5	Display two-way table total percentages	17	Display adjusted standardized chi-square residuals
6	Suppress value labels	18	Display all cell information
8	Order rows by descending value	19	Suppress values
12	Suppress tables		

Statistics:

1	Chi-square	6	Kendall's tau-b
2	Phi for 2 × 2 tables, Cramér's V for larger tables	7	Kendall's tau-c
		8	Gamma
3	Contingency coefficient	9	Somers' d
4	Lambda	10	Eta
5	Uncertainty coefficient	11	Pearson's r

Example:

```
CROSSTABS TABLES=WVAR XVAR BY YVAR BY ZVAR/TABLES=AVAR BY BVAR
/OPTIONS=3 4 14 15
/STATISTICS=1.
```

Overview

Procedure CROSSTABS (alias XTABS) produces tables showing the joint distribution of two or more variables that have a limited number of distinct values. The frequency distribution of one variable is subdivided according to the values of one or more variables. The unique combination of values for two or more variables defines a cell, the basic element of all tables. To analyze contingency tables using hierarchical log-linear models, use HILOGLINEAR in SPSS/PC+ Advanced Statistics.

Defaults

By default, CROSSTABS produces two-way to *n*-way crosstabulations for variables that have a limited number of numeric or string values. The default table consists of cell counts and is labeled with the names of the variables, the values of the variables, and their variable and value labels if these have been defined. Values of the row variable are displayed in ascending order. Cases that have user-missing or system-missing values for the variables that define a table are omitted from that table and the number of missing cases is reported.

Tailoring

Display Format. Display format options include ordering the table rows in descending value order, suppressing display of variable and value labels, suppressing display of values, and suppressing display of the table itself.

Cell Contents. You can include row, column, and total percentages, expected chi-square frequencies, and chi-square residuals in the cells of the table.

Statistical Display. Measures of association and significance tests are available for each subtable.

Missing Values. You can include user-missing values in the tables and in the calculation of statistics.

Syntax

- The minimum specification is a single TABLES subcommand with a tables list. The actual keyword TABLES may be omitted.
- The minimum tables list specifies a list of row variables, the keyword BY, and a list of column variables.
- Subcommands are separated by slashes.
- Subcommands can be specified in any order.
- You can invoke procedure CROSSTABS with the XTABS command.

Operations

- CROSSTABS causes the data to be read.
- If a long string variable is used, only the short-string portion is tabulated.
- Statistics are calculated separately for each two-way table or two-way subtable. Missing values are reported for the table as a whole.
- If you request only percentages and cell counts, percent signs are not displayed for percentages, and zero values for counts and percentages are represented by blanks.
- If you request percentages and any expected values or residuals, percent signs and zeros are displayed.
- Value labels for column variables are displayed in two lines with eight characters per line. Row value labels are displayed with 16 characters on one line.
- Percentages, expected values, and residuals are displayed with one decimal place.
- The display uses the width defined on the SET command.
- The BOXSTRING subcommand on SET controls the characters used in the table display.

Limitations

- The number of variables allowed per CROSSTABS command is the same as the system limit.
- Maximum 250 nonempty rows or columns are displayed for each variable.
- Maximum 20 TABLES subcommands lists per CROSSTABS command.
- Maximum 1 each OPTIONS and STATISTICS subcommands.
- Maximum 10 dimensions (9 BY keywords) per tables list.
- Maximum 250 value labels are displayed on any single table.

Example

```
CROSSTABS TABLES=WVAR XVAR BY YVAR BY ZVAR/TABLES=AVAR BY BVAR
/OPTIONS=3 4 14 15
/STATISTICS=1.
```

- The first tables list generates bivariate subtables of WVAR by YVAR for each value of ZVAR, followed by subtables of XVAR by YVAR for each value of ZVAR.
- The second tables list produces a single table in which AVAR is the row variable and BVAR is the column variable.
- The OPTIONS subcommand requests row and column percentages, expected cell frequencies, and residuals.
- The STATISTICS subcommand requests the chi-square statistic.

Example

```
XTABS TABLES=JOBCAT BY EDCAT BY SEX BY INCOME3.
```

- Assuming that SEX has values 1 and 2 and INCOME3 has values 1, 2, and 3, this tables list will produce a subtable of JOBCAT by EDCAT first for value 1 of SEX and value 1 of INCOME3, then for value 2 of SEX and value 1 of INCOME3, then for value 1 of SEX and value 2 of INCOME3, and so forth.
- This example uses the CROSSTABS alias, XTABS.

TABLES Subcommand

The TABLES subcommand specifies the tables lists. The actual keyword TABLES may be omitted.

- Variables named before the first BY in a tables list are row variables, and variables named after the first BY in a tables list are column variables.
- Variables named after the second (or subsequent) BY are control variables.
- Each subsequent use of the keyword BY in a tables list adds a new dimension to the tables requested and introduces a new order of control among the independent variables.
- You can name more than one variable in each dimension.
- You can use keyword ALL to include all user-defined variables in a dimension.
- When the tables list specifies two dimensions, tables are produced that crosstabulate the first variable before BY with each variable after BY, then the second variable before BY with each variable after BY, and so forth.
- When the tables list specifies more than two dimensions, a two-way subtable is produced for each combination of values of control variables.
- When the tables list specifies more than three dimensions, the value of the last variable mentioned changes the most slowly in determining the order in which the tables are displayed.
- You can specify multiple TABLES subcommands on a single CROSSTABS command. The slash between the subcommands is required; the keyword TABLES is not.

Display Format

By default, CROSSTABS displays values, variable labels, and value labels and orders rows from the lowest to the highest value. You can change these defaults by specifying the following on the OPTIONS subcommand:

Option 2 *Suppress all labels.* Neither variable nor value labels are displayed.

Option 6 *Suppress value labels.* Only variable labels are displayed on each table.

Option 8 *Order rows by descending value.*

Option 19 *Suppress values.* Values are displayed only if no value labels have been defined.

Cell Contents

By default, CROSSTABS table cells contain only counts. You can display additional information in the cells of each bivariate table or subtable. Optionally, you can control cell contents by specifying the following on the OPTIONS subcommand:

Option 3 *Display row percentages.* The cell count as a percentage of cases in the row is included in the cell.

Option 4 *Display column percentages.* The cell count as a percentage of cases in the column is included in the cell.

Option 5 *Display two-way table total percentages.* The cell count as a percentage of all cases in the subtable is included in the cell.

Option 13 *Suppress cell counts.* Unless you specify Option 3, 4, 5, 14, 15, 16, or 17, nothing is displayed. If you request both Options 13 and 18, Option 13 is ignored.

Option 14 *Display expected frequencies.* The expected cell count if the two variables in the subtable were statistically independent is included in the cell.

Option 15 *Display chi-square residuals.* The observed cell count minus the expected value is included in the cell.

Option 16 *Display standardized chi-square residuals.* The standardized residual is included in the cell. (See Haberman, 1978.)

Option 17 *Display adjusted standardized chi-square residuals.* The adjusted standardized residual is included in the cell. (See Haberman, 1978.)

Option 18 *Display all cell information.* Includes cell count; row, column, and total percentages; expected value; residual; standardized residual; and adjusted standardized residual. If you request both Options 13 and 18, Option 13 is ignored.

Statistical Display

By default, only the cell counts, marginal percentages, and number of missing cases are displayed for each table. You can request additional statistics by specifying the following on the STATISTICS subcommand:

Statistic 1 *Chi-square.* Fisher's exact test is computed using the rounded values of the cell entries when there are fewer than 20 cases in a 2 × 2 table that does not result from missing rows or columns in a larger table; Yates' corrected chi-square is computed for all other 2 × 2 tables.

Statistic 2 *Phi for 2 × 2 tables, Cramér's* V *for larger tables.*

Statistic 3 *Contingency coefficient.*

Statistic 4 *Lambda.*

Statistic 5 *Uncertainty coefficient.*

Statistic 6 *Kendall's tau*-b. CROSSTABS does not calculate this statistic for tables containing string variables.

Statistic 7 *Kendall's tau*-c. CROSSTABS does not calculate this statistic for tables containing string variables.

Statistic 8 *Gamma.* Zero-order gammas are displayed for 2-way tables and conditional gammas are displayed for the 2-way subtables of 3-way to 10-way tables. Statistic 8 is ignored with string variables.

Statistic 9 *Somers'* d, *symmetric and asymmetric.* Ignored for string variables.

Statistic 10 *Eta.* Ignored for string variables.

Statistic 11 *Pearson's* r. Ignored for string variables.

ALL *Display all statistics.*

Missing Values

By default, a case missing on any of the variables that define a table is not used either in the table display or in the calculation of the statistics. You can change the handling of cases with missing values by specifying the following on the OPTIONS subcommand:

Option 1 *Include cases with user-missing values.* Cases with user-missing values are displayed in tables and are included in the calculation of statistics.

Reference

Haberman, S. J. 1978. *Analysis of Qualitative Data,* Vol. 1. London: Academic Press.

DATA LIST: Fixed Format

```
DATA LIST [FILE='filename'] [FIXED] [TABLE]

 /varlist columns [{(0)}] [varlist columns ...]
                   {(n)}
                   {(A)}

  [/ ...] [/ ...]
```

Format	**Meaning**
(n)	Implied decimal places for numeric variables
(A)	String variable

Example:

```
DATA LIST /ID 1-3 SEX 5 (A) AGE 7-8 OPINION1 TO OPINION5 10-14.
```


Overview

The DATA LIST command assigns names to variables and provides information about the column location and format. Data can be inline (entered with SPSS/PC+ commands) or stored in an external file. The DATA LIST command with keyword FIXED defines data arranged in fixed format. In fixed format, the values for each variable are found in the same location on the same record for each case. Fixed format is the default for DATA LIST.

See DATA LIST: Freefield Format for defining data organized in freefield format, and DATA LIST: Matrix Materials for defining matrix materials.

- Use DATA LIST to read a data file containing ASCII data (numbers and other alphanumeric characters).
- Use the GET command, not DATA LIST, to read a *system file* created with the SAVE command (see GET and SAVE).
- Use the IMPORT command, not DATA LIST, to read a *portable file* created with the EXPORT command in SPSS/PC+ or SPSS^X (see IMPORT and EXPORT).

Defaults

By default, SPSS/PC+ assumes that data are inline, entered interactively or contained in a file named on the INCLUDE command. (The file named on INCLUDE must also contain the BEGIN DATA command; see BEGIN DATA and INCLUDE.) By default, all variables are assumed to be numeric without implied decimal places.

Tailoring

Data Source and Formats. You can use data from an external file. You can also define string variables and specify implied decimal places for numeric variables.

Summary Table. You can ask SPSS/PC+ to display a table that summarizes your variable definitions.

Syntax

- The minimum DATA LIST specification for fixed format is a slash followed by at least one variable name and its location.
- The keyword FIXED is optional.
- The keyword TABLE is optional.

Operations

- Variable names are stored in the active file dictionary.
- The order of the variables in the active file dictionary is the order in which they are defined on the DATA LIST command, not their sequence on the input data file. This order is important if you later use the TO convention.
- By default, variables are assumed to be numeric. Alphabetical and special characters, except the decimal point and leading plus and minus signs, are not valid numeric values and are set to system-missing if encountered in the data.
- Blanks to the left or right of a number in the default format are ignored; embedded blanks are invalid.

- The system-missing value is assigned to a completely blank field for numeric variables. The value assigned to blanks can be changed using the BLANKS specification on the SET command.
- Formats are stored in the active file dictionary and are used to print and write out the values. Use the FORMATS command to change formats of numeric variables defined on DATA LIST.

Limitations

- You cannot define more than 200 variables on a DATA LIST command.
- Each 8-character portion of a long string variable counts toward the 200-variable limit. For example, an 18-character long string counts as three short string variables (see Universals: Strings).
- The maximum length of an input record is 1024 characters.
- The maximum number of format "tokens," or syntactic elements, on the DATA LIST command is 600.

Example

```
DATA LIST /ID 1-3 SEX 5 (A) AGE 7-8 OPINION1 TO OPINION5 10-14.
BEGIN DATA.
001 m 28 12212
002 f 29 21212
003 f 45 32145
lines of data
128 m 17 11194
END DATA.
```

- The data are assumed to be inline because no data file is specified.
- The data are in fixed format (the default).
- Variable definitions start with ID in columns 1 through 3.
- Variable SEX is a short string variable in column 5.
- AGE is a two-column variable in columns 7 and 8.
- Variables OPINION1, OPINION2, OPINION3, OPINION4, and OPINION5 are named using the TO convention (see Universals: Variable-Naming Conventions). Each is a one-column variable, with OPINION1 located in column 10 and OPINION5 located in column 14.
- The BEGIN DATA and END DATA commands enclose the inline data. Note that the values of SEX are in lower-case characters and must be specified as such on subsequent commands.

FILE Subcommand

- The FILE subcommand is required when data are contained in an external data file. It must not be used when the data are contained in a file included with an INCLUDE command or when the data are inline (see INCLUDE and BEGIN DATA).
- The file specification must be enclosed in apostrophes.
- The file specification can be fully qualified, including drive, directory, filename, and extension.
- The FILE subcommand can be specified before or after the optional keyword FIXED.
- The FILE subcommand and keyword FIXED must be separated by at least one blank or comma.

Data on Floppy Diskettes

You can read an input data file from a floppy diskette.

- If you have two floppy diskette drives, insert the data diskette in B: and include the drive specification on the FILE subcommand, as in DATA LIST FILE= 'B:FLOPPY.DAT'.
- If you have one floppy diskette drive, you must start the system with the key diskette in that drive. When the security check is complete and the indicator light on the floppy drive has gone out, remove the key diskette and insert the data diskette. Enter the DATA LIST and other commands. SPSS/PC+ will

read the data from the floppy diskette when it needs them for a procedure. Later in the session, SPSS/PC+ may request that you reinsert the key diskette into drive A:.

Example

```
DATA LIST FIXED FILE='\INVENTORY\MARCH.DAT'
 /NUTS3 1-2 BOLTS35 3-5 NAILS3P 6-10.
```

- Keyword FIXED indicates that the variables to be defined are on a fixed-format data file.
- The FILE specification directs SPSS/PC+ to read data from the file MARCH.DAT in directory \INVENTORY.
- Three numeric variables, NUTS3, BOLTS35, and NAILS3P, are defined for use in subsequent SPSS/PC+ procedures.

Keyword TABLE

- Keyword TABLE displays a table summarizing the variable definitions supplied on DATA LIST FIXED. For each variable, the table displays the variable name, record number, starting column, ending column, format, width, and number of decimal places. The table also includes the number of records per case.
- TABLE must be specified before the variable definitions (before the first slash).
- The TABLE keyword is only valid on DATA LIST FIXED.

Record Specification

- Records are indicated on the DATA LIST command by a slash, followed by the variables to be defined from that record.
- The first slash indicates the first (or only) record.
- The second and any subsequent slashes tell SPSS/PC+ to skip to a new record.
- You must specify a slash for every record, even if no variables are being defined from that record.
- Variables from each record can be named in any order, regardless of their sequence on the data file.
- All variables to be read from one record must be defined before proceeding to the next record.

Example

```
DATA LIST FILE='SOCSUR82.DAT'
  / ID 1-7 SEX 15 AGE 16-18
  / ANOMIA 15 LIKEPOL 17
  // OPIN1 76 OPIN2 77
  /.
```

- The DATA LIST command defines data in fixed format from file SOCSUR82.DAT in the current directory.
- Three variables, ID, SEX, and AGE, are defined from the first record.
- ANOMIA and LIKEPOL are defined from the second record.
- The third data record for each case is skipped; no variables are defined.
- The fourth record contains two variables, OPIN1 and OPIN2.
- The fifth record is skipped.
- The data file contains a total of five records per case. The DATA LIST command defines seven variables from three of these records.

Variable Names

- Variable names can contain up to eight characters.
- All variable names must begin with a letter or the @ character. System variables (beginning with a $) cannot be defined on DATA LIST.
- You can name a list of variables using the TO convention. For more information on the TO convention and other variable-naming rules, see Universals: Variable-Naming Conventions.

Variable Locations

- Each variable name is followed by its column location.
- If the variable is one column wide, specify the number of the column. If the variable is two or more columns wide, specify the number of the first column followed by a dash (–) and the number of the last column.
- The same column locations can be used to define different variables.
- If several variables are recorded in adjacent columns on the same record and have the same width and format type, you can use an abbreviated format for specifying column location. First list all variable names, then the beginning column location of the first variable in the list, a dash, and the ending column location of the last variable in the list. SPSS/PC+ divides the total number of columns specified equally among the variables. If the number of columns do not divide equally, an error message is issued.

Example

```
DATA LIST FILE='AGES.DAT' TABLE / BIRTHDA 1-2 BIRTHMO 3-4
    BIRTHYR 5-8 BIRTHDAY 1-8 PRSNT1 TO PRSNT5 11-15
    CELEBRAT 65-68 CAKES 50-52 / WISHES 10-11.
```

- The DATA LIST command defines variables from the fixed-format file AGES.DAT in the current directory.
- Keyword TABLE generates a summary table of the names, formats, and locations of variables specified on the DATA LIST command.
- Three variables, BIRTHDA, BIRTHMO, and BIRTHYR, are read from the first eight columns on the first record. Variable BIRTHDAY is also read from the first eight columns as one variable.
- Variables PRSNT1, PRSNT2, PRSNT3, PRSNT4, and PRSNT5 are defined using the TO convention. Each of these variables is one column wide.
- Variable CELEBRAT is read from columns 65 through 68, and next CAKES is read from columns 50 through 52. The SPSS/PC+ active file dictionary will contain these variable names in the order they are defined on DATA LIST, even though this order differs from their order in the data file.
- Variable WISHES is read from columns 10 and 11 on the second data record.

Example

```
DATA LIST / LINENUM 1 ID 2-6 V1 TO V7 7-13
OPINREL OPINSEX OPINDRUG OPINRAR 15-18
LOCATN76 TO LOCATN83 20-35.
```

- The DATA LIST command defines inline data in fixed format.
- Variables V1, V2, V3, V4, V5, V6, and V7 are named using the TO convention. Each variable is one column wide.
- Four opinion variables, OPINREL, OPINSEX, OPINDRUG, and OPINRAR, are defined separately in columns 15 through 18. Each of these variables is one column wide.
- Eight location variables are defined by the LOCATN76 TO LOCATN83 specification. Each of these variables is two columns wide.

Variable Formats

- In the default format, variables are assumed to be numeric, either signed or unsigned integer or real numbers.
- String (alphanumeric) variables are indicated with an A in parentheses following the column specification.
- If a value is encountered which cannot be read according to the format type specified, it is assigned the system-missing value and a warning message is issued.

Numeric Formats

- When a decimal point is not actually coded in real data, the number of implied decimal places can be indicated in parentheses following the column specification.
- A coded decimal point in the data overrides the number of implied decimal places indicated on the DATA LIST command.

- The table below compares how values are interpreted for a four-column numeric variable when no decimal places are defined on DATA LIST and when two decimal places are defined.

Values in the data file	Default	Two defined decimal places
2001	2001	20.01
201	201	2.01
−201	−201	−2.01
2	2	.02
20	20	.20
2.2	2.2	2.2
.201	.201	.201
2 01	Undefined	Undefined

String Formats

- The values of string (alphanumeric) variables can contain any number, letter, or character, including special characters and embedded blanks. For further discussion of string variables, see Universals: Strings.
- String variables whose values contain eight characters or less are called *short string variables.*
- String variables with values longer than eight characters and up to 255 characters are called *long string variables.*

Example

```
DATA LIST FILE='\SPSSDAT\FILEX.DAT' TABLE
/XVAR 1 YVAR 10-15 ZVAR 3-9(2) // AVAR 25-30(A) BVAR 31-45(A).
```

- The data are defined from file FILEX.DAT in directory \SPSSDAT and are arranged in fixed format.
- Keyword TABLE generates a summary table of the variable names, formats, and locations specified on the command.
- Numeric variable XVAR is found in column 1 of the first record.
- The next variable defined is YVAR, found in columns 10 through 15.
- Variable ZVAR, found in columns 3 through 9, contains two implied decimal places, indicated by (2).
- No variables are defined on the second record for each case.
- Two variables are defined from the third record for each case.
- AVAR is a six-column short string variable.
- BVAR is a long string variable read from columns 31 through 45. BVAR counts as two variables toward the 200-variable system limit.

DATA LIST: Freefield Format

```
DATA LIST [FILE='filename'] FREE

      /variable [({A })] varlist
                 {Aw}
```

Format	Meaning
(Aw)	String of width w

Example:

```
DATA LIST FILE='MYFILE.DAT' FREE/XVAR YVAR.
```

Overview

The DATA LIST command assigns names to variables and provides information about their formats. The data can be inline or read from an external file. The DATA LIST command with keyword FREE identifies data arranged in freefield format. In freefield format, all variables are recorded in the same order for each case but not necessarily in the same column locations. Each value in the data file is separated by one or more blanks or by one comma.

See DATA LIST: Fixed Format for information on defining data in fixed format, and DATA LIST: Matrix Materials to define matrix materials.

Defaults

By default, SPSS/PC+ assumes that data are inline, entered interactively or contained in a file named on the INCLUDE command. (The file named on INCLUDE must also contain the BEGIN DATA command; see BEGIN DATA and INCLUDE.) All data values are assumed to be numeric.

Tailoring

You can use data stored on an external file. You can also define string variables.

Syntax

- The minimum specification for freefield data on DATA LIST is the keyword FREE, a slash, and at least one variable name.
- Variables must be named in the order they are entered on the data file.
- There is no record or column specification for freefield format.

Operations

- FREE can read freefield-format data with multiple cases recorded on one record or with one case recorded on more than one record.
- Variable names are stored in the active file dictionary.
- In the default format, variables are assumed to be numeric. Alphabetical and special characters, except the decimal point and leading plus and minus signs, are not valid numeric values and are set to the system-missing value.
- You can use BASIC conventions of delimiting data values. Two commas together or two commas separated by a blank indicate either numeric system-missing or a string blank.

Limitations

- You cannot define more than 200 variables on a DATA LIST command.
- The maximum length of an input record is 1024 characters.
- Each eight-character portion of a long string variable counts toward the 200-variable system limit. For example, an eleven-character string variable counts as two short string variables (see Universals: Strings).
- The maximum number of format tokens, or syntactic elements, on the DATA LIST command is 600.

Example

```
DATA LIST FREE/XVAR YVAR.
BEGIN DATA.
1 3 2 15 3 16 4
156
5 22 6 -3
END DATA.
```

- The DATA LIST command indicates inline data in freefield format.
- Two variables, XVAR and YVAR, are named.

- The values for the first case are 1 for variable XVAR and 3 for variable YVAR. The second case has values of 2 and 15 for XVAR and YVAR. The third case has values 3 and 16, the fourth case has values 4 and 156, and so on. Note that in freefield format, a single line of data can include values for more than one case. Also, the values for one case can be split across lines, as for the fourth case. Individual values cannot be split across lines.

FILE Subcommand

- The FILE subcommand is required when data are contained in an external file. It is not required when the data are included using an INCLUDE command or when the data are inline (see INCLUDE and BEGIN DATA).
- The file specification must be enclosed in apostrophes.
- The file specification can be fully qualified, including directory, filename, and extension.
- The FILE subcommand can be specified before or after keyword FREE.
- The FILE subcommand and keyword FREE must be separated by at least one blank or comma.

Example

```
DATA LIST FILE='MYFILE.DAT' FREE/XVAR YVAR.
```

- The freefield-format data file is read from MYFILE.DAT in the current directory.
- Two numeric variables are defined.

Variable Names

- Variable names can contain up to eight characters.
- All variable names must begin with a letter. System variables (beginning with $) cannot be defined on DATA LIST.
- Each variable name corresponds to one value per case.
- You can name a list of variables using the TO convention. For more information on the TO convention and other variable-naming rules, see Universals: Variable-Naming Conventions.

Example

```
DATA LIST FREE / ID VAR1 TO VAR7.
```

- The DATA LIST command indicates inline data in freefield format.
- Eight variables are defined: ID, VAR1, VAR2, VAR3, VAR4, VAR5, VAR6, and VAR7.

Variable Formats

- In DATA LIST with keyword FREE, formats can be specified for string variables only.
- All numeric variables are automatically assigned print and write formats of F8.2. Use the FORMATS command to specify any other format for numeric variables.
- All numeric variable digits are read and stored by SPSS/PC+.
- Numeric values with decimal points in the data preserve the decimal point and decimal digits.
- String variables are indicated by an A in parentheses after the variable name.
- By default, all string variables are assigned formats of A8 (width of eight characters). You can change the format of a string variable by specifying A and a width enclosed in parentheses, such as (A20).
- A format specification applies only to the variable immediately preceding it.
- If the string value in the data is longer than the specified length, the string is truncated and a warning message is printed. Thus, you must specify formats for long string variables.
- If the string in the data is shorter than the specified format, it is right-padded with blanks and no warning message is printed. All subsequent transformations require value specifications with the declared format width, including all padded values.

Example

```
DATA LIST FREE FILE='\MASTER\APRIL.DAT'
  /ID SEX (A1) NAME (A15) AGE TENURE ETHNIC (A).
FORMATS AGE (F2.0) TENURE (F3.1).
```

- The DATA LIST command defines data in freefield format from file APRIL.DAT in directory \MASTER.
- Six variables, ID, SEX, NAME, AGE, TENURE, and ETHNIC, are defined.
- ID, AGE, and TENURE are numeric variables.
- SEX is a defined as a one-column short string variable. NAME is defined as a long string variable up to 15 columns wide. ETHNIC is a short string variable with an assumed width of eight columns.
- The FORMATS command changes the print and write formats of numeric variables AGE and TENURE.

Entering Freefield Data

- Values are read sequentially in the order variables are named.
- A value cannot be split across records.
- One data value is separated from another by any number of blanks or by a single comma.
- You can use both commas and blanks to distinguish data values in a single data file.
- Any number of consecutive blanks (except blanks specified within a string value) are interpreted as one delimiter.
- When commas are used as delimiters, two consecutive commas or commas separated by a blank indicate a system-missing numeric value or a blank string value.
- A blank field for a variable that is not delimited by commas causes values from that point on to be assigned to the wrong variable.
- String values which contain embedded blanks or commas must be delimited by apostrophes or quotation marks. The delimiters are not read as part of the string value (see Universals: Strings).
- You can include an apostrophe in string values by delimiting the value with quotation marks. You can include quotation marks by delimiting the value with apostrophes.
- You cannot use commas or blanks within numeric values.
- If there are not enough values to complete the last case, a warning is issued and the incomplete case is dropped.

Example

```
DATA LIST FREE/AVAR BVAR STATE (A) MAYOR(A10).
BEGIN DATA.
7500000 20000000 'S DAKOTA' "O'LEARY" 22222000 55000000
'INDIANA' 'JONES' 120000000 56000000 'NEW YORK' 'ALDRIDGE'
-1.2 2222.223 'MAINE' 'BURNS'
END DATA.
FORMATS AVAR (COMMA15.2) BVAR(DOLLAR14.2).
DISPLAY VAR=ALL.
FREQ VAR=ALL.
LIST VAR=ALL.
```

- The DATA LIST command defines inline data in freefield format.
- AVAR and BVAR are defined as numeric variables. STATE is defined as a string variable with the default width of eight columns. MAYOR is declared as a long string variable with a width of 10 columns (A10).
- The BEGIN DATA command indicates the beginning of data lines.
- The first case has a value of 7500000 for AVAR. BVAR has the value 20000000. STATE has the value "S DAKOTA" enclosed in apostrophes to preserve the embedded blank. The value "O'LEARY" is enclosed in quotation marks to preserve the embedded apostrophe. The format for MAYOR indicates a width of 10 columns, so O'LEARY is right-padded with three blanks.
- The second case is split across two records. With freefield format, you can split cases but not individual values across records.
- Decimal values are included for AVAR and BVAR for the fourth case.

- The END DATA command indicates the end of inline data.
- The FORMATS command changes the print and write formats of numeric variables AVAR and BVAR from the default format (eight characters with two decimal places) to the specified formats (see FORMATS).
- The DISPLAY command shows the current print and write formats of the variables.
- The FREQUENCIES procedure produces tables for each of the variables. Because FREQUENCIES uses the internal representation of values in tables, the DOLLAR and COMMA formats do not appear on these tables.
- The LIST procedure produces a listing of the values of each variable. The DOLLAR and COMMA formats are preserved in the listing.

Example

```
DATA LIST FREE/AVAR BVAR STATE (A) MAYOR(A10).
BEGIN DATA.
7500000,20000000,'S DAKOTA',"O'LEARY",22222000,,
'INDIANA','JONES',120000000,,,'ALDRIDGE'
-1.2 2222.223 'MAINE' 'BURNS'
END DATA.
```

- This example shows the use of both commas and blanks to separate values.
- The two commas at the end of the first record indicate missing information for BVAR for the second case.
- In the second line of data, the three commas after value 120000000 indicate missing values for BVAR and STATE for the third case. A blank between the commas would also indicate missing information.
- The blanks after ALDRIDGE indicate the end of the value. The last line of data uses blanks as delimiters.

DATA LIST: Matrix Materials

```
DATA LIST [FILE='filename'] MATRIX [{FIXED}]
                                    {FREE }

          /varlist
```

Example:

```
DATA LIST MATRIX FILE='REG.MAT'
/SUICIDE ANOMIE AGE.
N 488.
REG VAR=AGE SUICIDE ANOMIE
/READ CORR
/DEP=SUICIDE
/METHOD=ENTER.
```

Overview

The DATA LIST command with keyword MATRIX provides variable names and a dictionary for matrix materials used as input in CLUSTER, FACTOR, ONEWAY, REGRESSION, and MANOVA. (For information on CLUSTER, FACTOR, and MANOVA, see *SPSS/PC+ Advanced Statistics.)* The matrix materials can include correlation coefficients, covariance coefficients, a matrix of *n*'s, or group distance measures. Matrix materials can be read in fixed or freefield format but must conform to the requirements of the individual procedures (see each procedure for details). The matrix input can be inline or read from an external file.

For information on reading individual casewise data, see DATA LIST: Fixed Format or DATA LIST: Freefield Format.

Syntax

- The minimum specification is DATA LIST with keyword MATRIX, followed by a slash and a list of variable names.
- The slash (/) between keyword MATRIX and the variable names is required.
- Variable names must be eight characters or less and must begin with a letter or the @ character.
- You must supply names for all variables that will be used by the next procedure. If you specify more names on DATA LIST MATRIX than are specified for the procedure, the extra names are ignored.
- The order in which variables are named determines their order in the new active file but has no relation to the contents of the matrix.
- Format types are meaningless for matrix materials and cannot be specified.
- The FILE subcommand is required when the matrix materials are contained in an external file. The file specification must be enclosed in apostrophes.
- The FILE subcommand can be specified before or after keyword MATRIX.
- The FILE subcommand and keyword MATRIX must be separated by at least one blank or comma.
- You can specify files in directories other than the current directory by using fully qualified file specifications.

Operations

- DATA LIST with keyword MATRIX defines variable names to be used in processing a variety of matrix materials arranged for specific procedures. Reading and interpreting the matrix are actually performed by the next procedure.
- Each procedure that can process matrix materials interprets them according to its own specifications. Thus the format of a matrix to be read by SPSS/PC+ depends on the procedure which will use it.
- Each procedure that reads matrix materials can accept its own matrix output in either FIXED or FREE format. Matrices entered directly and matrices that will be processed by a procedure other than the one that created them should normally be read in FREE format.
- DATA LIST with keyword MATRIX cannot read individual casewise data.
- You cannot use DATA LIST MATRIX with procedures that expect casewise data.

- You cannot perform any SPSS/PC+ transformations on matrix materials.
- The BASIC convention of using two commas in a row to indicate missing data is *not* supported with DATA LIST MATRIX.

Limitations

- You cannot define more than 200 variables on the DATA LIST command.
- The maximum number of format "tokens," or syntactic elements, on the DATA LIST command is 600.

Example

```
DATA LIST MATRIX FILE='REG.MAT'
 / SUICIDE ANOMIE AGE.
N 488.
REG VAR=AGE SUICIDE ANOMIE
 /READ CORR
 /DEP=SUICIDE
 /METHOD=ENTER.
```

- DATA LIST reads matrix materials from file REG.MAT in the current directory. The matrix was written to this file by procedure CORRELATION in a previous SPSS/PC+ session.
- The matrix materials are read in fixed format (the default). CORRELATION writes matrices in the same format as does REGRESSION. This matrix could also be read in freefield format.
- The variable names SUICIDE, ANOMIE, and AGE are defined for use in the REGRESSION command.
- The N command indicates that the matrix input is based on 488 cases. REGRESSION uses this information in computing significance tests.
- The VARIABLES subcommand on REGRESSION identifies the variables in the correlation matrix and their order.
- The READ subcommand on REGRESSION indicates that a correlation matrix will be read by the REGRESSION procedure.
- This example takes advantage of spelling permitted by three-character truncation of keywords.

Example

```
DATA LIST FREE MATRIX/AGE SUICIDE ANOMIE.
BEGIN DATA.
1.0 .5555555 .3333333
.5555555 1.0 .4555555
.3333333 .4555555 1.0
488
END DATA.
REGRESSION VARIABLES=AGE SUICIDE ANOMIE
 /READ CORR N
 /DEPENDENT=SUICIDE
 /METHOD=ENTER.
```

- The DATA LIST command specifies matrix materials in freefield format.
- Because no file is specified on DATA LIST, the matrix data are assumed to be inline.
- The variable names AGE, SUICIDE, and ANOMIE are defined for use in the REGRESSION command.
- The matrix materials are entered between the BEGIN DATA and END DATA commands. Each row vector begins on a new line. A final line contains a single number indicating the number of cases.
- The READ subcommand on REGRESSION indicates that a correlation matrix followed by the *n* (number of cases) will be read.

Matrix Data

The SPSS/PC+ procedures CORRELATION, CLUSTER, ONEWAY, FACTOR, and REGRESSION write matrix materials in a fixed format that automatically conforms to the requirements of the various procedures that read matrix materials. Some materials, such as factor matrices written by FACTOR and matrices written by ONEWAY and MANOVA, are specially formatted for a specific procedure. If you enter your own matrix materials, they must conform to these formats as well as to the requirements below.

- Matrix materials can be arranged in fixed or freefield format.
- A matrix written by an SPSS/PC+ procedure can always be read in fixed format by the same procedure.
- A matrix written by an SPSS/PC+ procedure can always be read in freefield format by any procedure which accepts that type of matrix. Factor matrices from FACTOR, and matrix materials from ONEWAY and MANOVA, are not accepted by other procedures.
- Each cell of the matrix must contain a value.
- Each element in a row in freefield format matrix materials is separated by at least one space or a comma.
- Each row of a matrix begins on a new line.
- Each type of matrix material begins on a new line.
- In fixed format, there is a maximum number of elements that can be entered in a row (see discussion of individual procedures). The format must conform to the requirements of the procedure that reads the matrix.
- If the elements for a vector do not fit in one row, the elements can be continued on the next row. Each row must be filled before continuing to the next in fixed format.
- Individual matrix elements cannot be split across input lines.
- Decimal points in the data must be entered explicitly. You cannot specify implied decimal places.
- For additional information on CLUSTER, FACTOR, and MANOVA, see *SPSS/PC+ Advanced Statistics.*

Matrix Input for Procedure ONEWAY

- Procedure ONEWAY reads matrix materials in either fixed or freefield format.
- If you use matrix materials in fixed format, you must specify Option 7 on the ONEWAY command.
- Each matrix cell entry has a width of 10 columns with up to 4 decimal places. You can enter up to 8 cells in each row.

Example

```
DATA LIST MATRIX  / SCORE METHOD.
BEGIN DATA.
7         7         7
    4.4286    7.5714    6.7143
    1.2724    1.3973     .9512
END DATA.
ONEWAY SCORE BY METHOD(1,3)
/OPTION 7.
```

- The DATA LIST command specifies inline matrix materials in the default fixed format. The active file dictionary contains two variable names, SCORE and METHOD.
- The data are arranged with each vector element occupying 10 columns. The vector of counts does not require decimal places.
- Option 7 on ONEWAY indicates that a matrix with a vector of counts, a vector of means, and a vector of standard deviations will be read (see ONEWAY).

Example

```
DATA LIST FREE MATRIX/YVAR XVAR.
BEGIN DATA.
65 95 181 82 40 37
2.6462 2.7737 4.1796 4.5610 4.6625 5.2297
6.2699
494
END DATA.
ONEWAY VARIABLES=YVAR BY XVAR(1,6)
/OPTIONS=8.
```

- The DATA LIST command specifies inline matrix materials in freefield format. Two variable names, YVAR and XVAR, are in the active file dictionary.
- The data are arranged to conform to the requirements of Option 8 in ONEWAY, with one row of counts, a row of means, an entry for the pooled variance estimate, and an entry for the degrees of freedom.
- Option 8 on the ONEWAY command indicates that matrix materials arranged as a vector of counts, a vector of means, the pooled variance estimate (a single

entry), and the degrees of freedom (a single entry) will be read (see ONEWAY).

Matrix Input for Procedure REGRESSION

- Procedure REGRESSION reads matrix materials in either fixed or freefield format.
- Materials arranged in fixed format must have 10 columns for each vector entry with up to 7 decimal places. You can enter up to 8 entries per line for each vector.
- Materials arranged in freefield format must conform to the requirements described above under "Matrix Data."

Example

```
DATA LIST MATRIX/AGE SUICIDE ANOMIE.
BEGIN DATA.
1.0       0.555555  0.333333
0.555555  1.0       0.455555
0.333333  0.455555  1.0
488
END DATA.
REGRESSION VARIABLES=AGE SUICIDE ANOMIE
 /READ CORR N
 /DEPENDENT=SUICIDE
 /METHOD=ENTER.
```

- The DATA LIST command defines inline matrix materials in fixed format (the default). Three variable names are defined for the active file dictionary.
- The data are arranged to conform to the requirements of the REGRESSION procedure. Each row vector has 3 entries, each with a width of 10 columns. The last entry is the number of cases.
- The READ subcommand on REGRESSION reads the matrix of correlation coefficients and a value for the number of cases (see REGRESSION: Matrix Materials).

Example

```
DATA LIST MATRIX FREE / X1 X2 X3 X4 X5 Y.
BEGIN DATA.
35.0825 2.7315 40.9060 3.1405 25.0690 6.2550
5.8171 .4541 25.8985 9.6254 1.3138 .6543
33.8381250 .5079382 113.502300 51.9130250 2.5499079 2.7898921
.5079382 .2062029 2.1302800 1.0036676 .2998858 .0584658
113.502300 2.1302800 670.734846 206.202997 1.7372853 15.7105368
51.9130250 1.0036676 206.202997 92.6479839 2.3183584 5.1585079
2.5499079 .2998858 1.7372853 2.3183584 1.7260832 .1064316
2.7898921 .0584658 15.7105368 5.1585079 .1064316 .4281316
20
END DATA.
VAR LABELS X1 'STAFF SALARIES PER PUPIL'
 X2 '6TH GRADE PER CENT WHITE-COLLAR FATHERS'
 X3 'SES COMPOSITE'
 X4 'MEAN TEACHER VERBAL TEST SCORE'
 X5 '6TH GRADE MEAN MOTHER EDUCATION'
 Y 'VERBAL MEAN TEST SCORE, ALL 6TH GRADERS'.
REGRESSION READ=COV MEAN STDDEV N
 /DES DEF
 /VAR=Y,X1 TO X5
 /CRI TOL(.0001)
 /STATS ALL
 /DEP Y
 /ENT.
```

- The DATA LIST command specifies inline matrix materials in freefield format. Six variable names are specified for the active file dictionary.
- The data are arranged to conform to the requirements of the REGRESSION procedure. The first six rows form a covariance matrix, with each row starting on a new line. The seventh row forms a vector of means for each variable named. The eighth row is a vector of standard deviations. The last entry is the number of cases.
- The READ subcommand on REGRESSION indicates that matrix materials with a covariance matrix, a vector of means, a vector of standard deviations, and an entry for the number of cases will be read (see REGRESSION: Matrix Materials).

DESCRIPTIVES

```
DESCRIPTIVES [VARIABLES=] {varlist}
                          {ALL    }

             [/OPTIONS=option numbers]

             [/STATISTICS={statistic numbers}]
                          {ALL              }
```

Options:

1	Include user-missing values	6	Serial format
2	Suppress variable labels	7	Narrow format
3	Save Z scores on active file	8	Suppress variable names
5	Exclude missing values listwise		

Statistics:

1	Mean	9	Range
2	Standard error of mean	10	Minimum
5	Standard deviation	11	Maximum
6	Variance	12	Sum
7	Kurtosis	13	Mean, standard deviation, minimum, and maximum
8	Skewness		

Example:

```
DESCRIPTIVES VARIABLES=YVAR ZVAR, AVAR1 TO AVAR5,
           BETATEST, IOTATEST
/STATISTICS=6 13
/OPTIONS=5.
```

Overview

Procedure DESCRIPTIVES computes univariate statistics, including the mean, standard deviation, minimum, and maximum, for numeric variables. Because it does not sort values into a frequency table, DESCRIPTIVES is an efficient means of computing descriptive statistics for continuous variables. Other procedures that display descriptive statistics include FREQUENCIES and MEANS.

Defaults

The default table displays the variable name, variable label, mean, standard deviation, minimum, maximum, and number of cases with valid values on a single line for each variable. All cases with valid values for a variable are included in the calculation of statistics for that variable. The display uses the width set on the SET command.

Tailoring

Display Format. You can display statistics in serial format and restrict the width to narrow format regardless of the width defined on SET. DESCRIPTIVES also offers control over the display of variable labels and variable names.

Statistical Display. Optional statistics include the standard error of the mean, variance, kurtosis, skewness, range, and sum. DESCRIPTIVES does not compute the median or mode (see FREQUENCIES).

Z Scores. You can compute *Z* scores (standardized deviation scores from the mean) and add these to the active file as new variables.

Missing Values. You can include cases with user-missing values in the calculation of statistics. Optionally, you can exclude cases with missing values listwise.

Syntax

- The minimum specification is the VARIABLES subcommand with a list of variables. The actual keyword VARIABLES may be omitted.
- Subcommands are separated by slashes.

Operations

- DESCRIPTIVES causes the data to be read.
- If the STATISTICS subcommand is used, only those statistics explicitly requested are displayed.
- If a string variable is specified on the variable list, a warning is issued and no statistics are displayed for that variable.

- The available width and the statistics and options requested determine whether the statistics are displayed in tabular or serial form.
- If there is insufficient width to display the statistics requested, DESCRIPTIVES first truncates the variable label and then adopts serial format.
- If there is insufficient memory available to calculate statistics for all variables requested, DESCRIPTIVES truncates the variable list.
- Statistics that will fit within the allotted columns are displayed with two decimal places.

Limitations

- There is no fixed limit on the number of variables named or implied on DESCRIPTIVES.
- Maximum 1 each of the VARIABLES, OPTIONS, and STATISTICS subcommands.

Example

```
DESCRIPTIVES VARIABLES=YVAR ZVAR, AVAR1 TO AVAR5,
             BETATEST, IOTATEST
/STATISTICS=6 13
/OPTIONS=5.
```

- This example requests statistics for all the variables named or implied by the TO keyword.
- The STATISTICS subcommand requests the variance (Statistic 6) and the defaults: mean, standard deviation, minimum, and maximum (Statistic 13).
- Option 5 specifies that cases with missing values for any variable on the variable list will be omitted from the calculation of statistics for all variables.

Example

```
DESCRIBE VAR=RAGE RINC81.
```

- Because no STATISTICS subcommand is included, only the mean, standard deviation, minimum, and maximum for RAGE and RINC81 will be displayed.
- This example takes advantage of spelling permitted by three-character truncation of keywords.

VARIABLES Subcommand

The VARIABLES subcommand names the variables to be included in the table. The actual keyword VARIABLES may be omitted.

- You can use keyword ALL to refer to all user-defined variables on the active file.
- Variables named more than once appear in the display more than once.

Z Scores

The *Z* score transformation standardizes variables to the same scale, producing new variables with a mean of 0 and a standard deviation 1. These variables are added to the active file. There are two methods for requesting *Z* scores. One is to use Option 3:

Option 3 *Add Z scores to the active file for all variables on the DESCRIPTIVES command.* SPSS/PC+ forms variable names for the new variables, using wherever possible the letter Z and the first seven characters of the old variable name.

Alternatively, you can obtain *Z* scores for any variable by specifying a new variable name in parentheses after the variable on the variables list.

- You must specify new names individually; a *list* in parentheses is not recognized.
- SPSS/PC+ creates variable labels for the new *Z* score variables created with either method.
- Whenever *Z* scores are added to the file, a table is displayed showing the names of the new variables and of the original variables from which they were created.

- If you specify Option 3 and also enter variable names in parentheses for some variables, *Z* scores are calculated for all variables, using your names where you have supplied them and forming new names where you have not.
- If the new variables cause you to exceed the 200-variable limit, SPSS/PC+ prints an error message and does not process the DESCRIPTIVES command.

Example

```
DESCRIPTIVES VAR1 VAR2 SCORE (STDSCORE) INCOME
 / OPTIONS 3.
```

- *Z* scores are produced for four variables: VAR1, VAR2, SCORE, and INCOME.
- Since a name is specified only for one of the four variables, the other three are assigned names by the program.
- The variables ZVAR1, ZVAR2, STDSCORE, and ZINCOME are added to the end of the active file.

Display Format

By default, DESCRIPTIVES displays the statistics and a 40-character variable label for each variable on one line. If the statistics requested do not fit within the available width, DESCRIPTIVES first truncates the variable label and then uses serial format. Serial format provides larger field widths and permits more decimal places for very large or very small numbers than does the default format.

Optionally, you can request the following on the OPTIONS subcommand:

Option 2 *Suppress variable labels.*

Option 6 *Serial format.* The requested statistics are displayed below each variable name. This option is forced if the number of statistics requested does not fit within the available width.

Option 7 *Narrow format.* The display width is restricted to 79 columns regardless of the width defined on SET.

Option 8 *Suppress variable names.* The variable name will be displayed only if there is no variable label.

Statistical Display

DESCRIPTIVES automatically calculates the mean, standard deviation, minimum, and maximum for all variables in the variable list. The valid count on which statistics are based is always displayed. You can obtain additional statistics by specifying the following on the STATISTICS subcommand. If the STATISTICS subcommand is included, only statistics specifically requested are displayed.

Statistic 1 *Mean.*

Statistic 2 *Standard error of mean.*

Statistic 5 *Standard deviation.*

Statistic 6 *Variance.*

Statistic 7 *Kurtosis.* The standard error of the kurtosis is also displayed.

Statistic 8 *Skewness.* The standard error of the skewness is also displayed.

Statistic 9 *Range.*

Statistic 10 *Minimum.*

Statistic 11 *Maximum.*

Statistic 12 *Sum.*

Statistic 13 *Mean, standard deviation, minimum, and maximum.* This is the same as the default.

ALL *Display all statistics.*

Missing Values

By default, all cases with valid values for a variable are included in the calculation of statistics for that variable. You can alter the handling of cases with missing values by specifying the following on the OPTIONS subcommand:

Option 1 *Include cases with user-missing values.* Cases that have user-missing values will be included in the calculation of statistics for all variables named on the command.

Option 5 *Exclude cases with missing values listwise.* Cases missing on any variable named on the DESCRIPTIVES command are excluded from the calculation of statistics for all variables. The space reserved to display the valid counts for each variable is suppressed and the valid count is reported for the table as a whole.

DISPLAY

```
DISPLAY [{varlist}]
         {ALL    }
```

Example:

```
DISPLAY AVAR TO FVAR.
```

Overview

The DISPLAY command exhibits information about variables in the active file.

Syntax

- The minimum specification is simply the command keyword.
- When you specify a variable list, variable names can be separated by a comma or a space.
- You can specify keyword ALL to obtain detailed information on all variables in the active file.

Operations

- DISPLAY information is directed to output destinations. This includes, by default, the screen and the listing file SPSS.LIS. You can also have it sent to the printer or to an alternate listing file (see SET).
- DISPLAY with no specifications provides a list of all variables in the active file and their variable labels.
- When a list of variables or keyword ALL is specified, additional information is provided, including the variable name and label, value labels, missing-value flags, and variable type and width.

Example

```
GET 'WEATHER.SYS'.
DISPLAY.
```

- DISPLAY displays a list of variables and variable labels defined in the system file WEATHER.SYS.

Example

```
DISPLAY AVAR TO FVAR.
```

- DISPLAY exhibits detailed information on the variables between and including AVAR and FVAR in the active file.

EXECUTE

```
EXECUTE {path\filename{.ext}} ['parameters']
        {             {.EXE}}
        {DOS               }
```

Example:

```
EXECUTE '\SPSS\KERMIT.EXE'.
```

Overview

The EXECUTE command allows you to run other programs or execute DOS commands from within SPSS/PC+ and then return to your session at the point you left it. This is not intended to make SPSS/PC+ a regular operating environment: part of the system remains in memory, which limits the memory available for other programs. Rather, EXECUTE lets you run other programs briefly or execute DOS commands without having to reload SPSS/PC+ and recreate your active file and environment when you return.

SPSS/PC+ cannot control actions taken by other software that you invoke through EXECUTE and therefore cannot guarantee that any particular program will run safely. Read the cautions below before trying to use EXECUTE creatively.

Defaults

By default, the other program has approximately 128K of RAM in which to run. No parameters are passed to the other program.

Tailoring

RAM. You can allocate more RAM for use by other programs within your SPSS/PC+ session when you invoke SPSS/PC+ with the *SPSSPC* command from DOS.

Parameters. You can pass command-line parameters to the other program as if you were running it from DOS.

Syntax

- Specifications for EXECUTE consist of a file specification for an executable file with extension .EXE or .COM. The file specification should be in apostrophes.
- If you do not specify a file extension, the extension .EXE is assumed.
- Parameters in apostrophes following the file specification are passed to the program you invoke. For example, you could pass a filename to an editor.
- The keyword DOS is accepted as a synonym for COMMAND.COM, the DOS command processor. This provides access to any DOS command.
- The command *DOS* is accepted as a synonym for EXECUTE DOS. In this instance, the EXECUTE command itself is optional.
- To return from DOS to your SPSS/PC+ session, issue the DOS command *EXIT*.
- To return from any other program, issue the normal command used to leave that program.

Operations

- EXECUTE is an operation command and is performed immediately.
- SPSS/PC+ passes control to the program named on the EXECUTE command. By default, that program has approximately 128K of RAM available (see "Memory Considerations" below).
- All files used by SPSS/PC+ are closed before control is passed to the other program.
- When the program terminates, you return to the SPSS/PC+ session and may continue, provided that you have observed the limitations below.

Limitations

- The program invoked must not leave anything resident in memory when it exits.
- The program must not attempt to redefine any of the interrupt vectors in low memory on the PC.

- The number of files that the program can open will be smaller than the FILES specification in CONFIG.SYS because of files that remain allocated for use by SPSS/PC+.
- Temporary files used by SPSS/PC+ should not be deleted in the middle of a session. These include files named SPSS.SY1 and SPSS.SY2, which hold the active file at various times.
- System files that have been read by the GET command should not be deleted unless data transformation or selection has created a new active file on disk.
- Entering some DOS commands will make it impossible to resume your session (such as erasing SPSS/PC+ modules or a system file that will be needed by a future procedure).

See "Problems" below for further discussion of these limitations.

Example

```
DOS.
```

- This command is a synonym for EXECUTE DOS and executes the DOS command processor, COMMAND.COM.
- DOS will issue its command prompt, for example, **C:\MYFILES>**. You can then issue any DOS command but must remember that you are still inside an SPSS/PC+ session.
- To return to SPSS/PC+, issue the DOS *EXIT* command. You will again see the **SPSSPC:** prompt, and any active file you created earlier in your session will be available.

File Specification

To invoke a program from SPSS/PC+, specify an executable file with extension .COM or .EXE.

- You can omit the extension if it is .EXE.
- If the file is in a directory other than your current directory, you can include a path specification.
- Paths defined with the DOS *PATH* command are searched just as if you were naming the file from DOS.
- To execute a batch (.BAT) file, specify EXECUTE DOS and then invoke the batch file by name.
- Most programs can be invoked indirectly by executing DOS and running the program from DOS. This method slightly increases memory requirements and substantially increases the possibilities for confusion.

Example

```
EXECUTE '\SPSS\KERMIT.EXE'.
```

- This command runs KERMIT from within an SPSS/PC+ session. You can use KERMIT to log on to a mainframe and download a portable file created by the EXPORT procedure in SPSS^X.
- After logging off the mainframe and leaving KERMIT with the KERMIT *EXIT* or *QUIT* command, you can continue the SPSS/PC+ session and IMPORT the file you have downloaded.

Parameters

Any text following the file specification is passed to the program you invoke.

- Parameters specified after the file specification should be enclosed in apostrophes.

Example

This example shows how to use the DOS *PRINT* command to begin printing a lengthy listing file and then continue with your SPSS/PC+ session while the file prints. Prompts from DOS and SPSS/PC+ are shown in upper case and commands given by the user in lower case.

```
C:\MYFILES>print
NAME OF LIST DEVICE [PRN]:
RESIDENT PART OF PRINT INSTALLED
PRINT QUEUE IS EMPTY.

C:\MYFILES>spsspc
```

...beginning of SPSS/PC + session...

```
SPSS/PC:get file='bigfile.sys'.
SPSS/PC:set listing 'freq.lis'.
SPSS/PC:frequencies all.
```

...output from FREQUENCIES procedure...

```
SPSS/PC:dos.

C:\MYFILES>print freq.lis

       C:FREQ    :LIS IS CURRENTLY BEING PRINTED

C:\MYFILES>exit

SPSS/PC:crosstabs ....
```

- The DOS *PRINT* command is issued before entering SPSS/PC+ to install the print driver (see "Programs Remaining in Memory" below). Since no file is specified, nothing is printed and DOS issues the message **PRINT QUEUE IS EMPTY.**
- In the SPSS/PC+ session, the GET command specifies a system file, the SET command specifies a file (FREQ.LIS) for the output listing, and the FREQUENCIES command runs frequency distributions for all variables in the system file. If the frequency distributions were printed at this time (with SET PRINTER ON), you would have to wait for the printing to finish before continuing your analysis.
- The SPSS/PC+ command DOS (an abbreviation for EXECUTE DOS) invokes the DOS command processor, which responds with its prompt, in this case **C:\MYFILES>.**
- The DOS *PRINT* command prints the listing file containing the output from FREQUENCIES. The printing will take place as a "background" task while the session continues.
- The DOS *EXIT* command returns to SPSS/PC+.
- The SPSS/PC+ session can continue while DOS prints the frequency tables.

Problems

The following types of programs may make it impossible to return to your SPSS/PC+ session.

Programs Remaining in Memory

A program that leaves anything in RAM after exiting will cause SPSS/PC+ to abort if that program is *first* invoked from an SPSS/PC+ session with the EXECUTE command. Among the programs that can cause this problem are

- *The DOS PRINT command.* The first time you print a file, DOS loads a print driver into memory, where it remains. To use DOS PRINT during an SPSS/PC+ session, you must first print something prior to entering SPSS/PC+. This installs the print driver before SPSS/PC+ claims memory for its own use. As shown in the example above, the *PRINT* command without any filename will install the print driver.
- *The DOS MODE command.* Certain uses of the *MODE* command, including those needed to configure a printer, cause additional code to be loaded into memory (see your DOS manual). Do not attempt to do this when you are executing DOS from within an SPSS/PC+ session.
- *Desk-accessory programs.* Such programs, after being started, remain in memory and can be called up with a few keystrokes to perform notepad,

calendar, communications, and other functions. You can use these programs within SPSS/PC+ but cannot use the EXECUTE command to start them or you will abort the SPSS/PC+ session.

- *Keyboard utilities.* You must run a keyboard configuration utility before entering SPSS/PC+, not from within a session.

If you are uncertain whether any particular program remains in memory, run it before entering SPSS/PC+. You should then be able to use EXECUTE safely for the remainder of the session.

Programs Altering the Interrupt Vectors

Any program that alters the interrupt vectors stored in low memory of the PC and does not restore them before exiting will make it impossible for SPSS/PC+ to resume. Commercially available software is unlikely to do this.

Memory Considerations

Normally, SPSS/PC+ uses all available memory, beyond that occupied by SPSS/PC+ itself, as workspace. After obtaining this workspace, SPSS/PC+ will not give it up. Thus, only the 128K given up by SPSS/PC+ itself when you specify EXECUTE is available for use by another program.

If you intend to run a program that requires more than 128K, you must limit the amount of workspace SPSS/PC+ takes by means of a command-line *switch* specifying the size of the workspace SPSS/PC+ should request from DOS.

- The switch is specified when you invoke SPSS/PC+ from DOS with the *SPSSPC* command.
- The switch is entered as */S=nnnK,* where *nnnK* is the desired workspace size.
- A minimum of 20K is suggested for any productive work.
- To free up as much memory as possible at the expense of not being able to do large tasks in SPSS/PC+, enter the system with the DOS command *SPSSPC /S=20K.*
- If you specify a filename on the *SPSSPC* command, put the size switch after the filename, as in *SPSSPC MYDATA.DEF /S=20K.*

If you know that a program will require a certain amount of memory (more than 128K), and you want SPSS/PC+ to use as much as possible of the remaining memory for workspace, you can

1. Run SPSS/PC+ without the /S switch, so that it will request the maximum possible workspace.
2. Use the SHOW command to find out how large this maximum workspace is.
3. Run SPSS/PC+ again, using the /S switch to reduce the workspace sufficiently to leave the desired amount of memory free. Remember that your program will get about 128K of the SPSS/PC+ memory.

Example

To run a program requiring 192K with the EXECUTE command, you must reserve an additional 64K in addition to the 128K provided by SPSS/PC+. Suppose that the SHOW command reports a maximum workspace of 183K (this amount will vary from one machine to another). To reduce this by 64K, you would specify a workspace of 119K (or a bit less to be safe):

```
SPSSPC /S=115K
```

This will allow you to run SPSS/PC+ while leaving 192K for another program.

EXPORT

```
EXPORT OUTFILE='filename' [/KEEP={ALL    }] [/DROP=varlist]
                                 {varlist}

     [/RENAME=(old varlist=new varlist)...] [/MAP]

     [/DIGITS=number]
```

Example:

```
EXPORT OUTFILE='NEWDATA.POR'/RENAME=(V1 TO V3=ID, SEX,AGE)
/MAP.
```


Overview

EXPORT produces a portable ASCII data file and dictionary that can be read with the IMPORT command in SPSS/PC+ or SPSS^X. You can upload EXPORT files to a mainframe using KERMIT, provided that KERMIT is installed on both the IBM PC and the receiving mainframe computer (see Part F: Communications).

Defaults

EXPORT writes your active file, including all data and the data dictionary, to an external file. The dictionary contains variable and value labels, missing-value flags, and print formats for each variable. The portable file also contains the originating computer (IBM PC), the name and release number of SPSS/PC+, and the date and time the portable file was created.

Tailoring

You can save a subset of variables from your active file on the portable file and rename variables. You can also produce a record of all variables and their names on the exported file and specify the number of decimal digits of precision for the values of all numeric variables.

Syntax

- The minimum specification is the OUTFILE subcommand with a file specification enclosed in apostrophes.
- Subcommands can be named in any order and must be separated by a slash.

Operations

- EXPORT is a transformation and causes the data to be read.
- Portable files are written with 80-character record lengths.
- Portable files may contain some unprintable characters.
- The active file is still available for SPSS/PC+ transformations and procedures after the portable file is created.
- The system variables $CASENUM and $DATE are assigned when the file is read by IMPORT. EXPORT specifies the weighting variable on the portable file.

Limitations

- Maximum 200 variables on a portable file. Each 8-character portion of a long string variable counts as 1 toward this limit.
- You may not have enough available memory on your PC to write a large portable file. Use the DROP or KEEP subcommand to exclude extraneous variables from the portable file.

Example

```
EXPORT OUTFILE='NEWDATA.POR'/RENAME=(V1 TO V3=ID,SEX,AGE)/MAP.
```

- The portable file is written to NEWDATA.POR in the current directory.
- Variables V1, V2, and V3 are renamed ID, SEX, and AGE for the portable file. Their names remain V1, V2, and V3 in the SPSS/PC+ active file. None of the other variables written to the portable file are renamed.
- The MAP subcommand requests a listing of the variables in the portable file.

OUTFILE Subcommand

The OUTFILE subcommand specifies the filename of the portable file.

- The filename must be enclosed in apostrophes.
- The file must be in the current directory. You cannot direct portable files to other directories.
- You can direct the portable file to another drive, including the A: drive. Replace the key diskette in the A: drive after the EXPORT command is finished.

Example

```
EXP OUT='SALDATA.POR'.
```

- The complete active file is written to file SALDATA.POR.
- This example takes advantage of spelling permitted by three-character truncation of keywords.

DROP and KEEP Subcommands

Use the DROP and KEEP subcommands to save a subset of variables on the portable file.

- DROP excludes a variable or list of variables from the portable file. All variables not named are included in the portable file.
- KEEP includes a variable or list of variables on the portable file. All variables not named are excluded.
- Variables can be specified on DROP and KEEP in any order.
- With the DROP subcommand, the order of variables in the portable file is the same as their order on the active file.
- With the KEEP subcommand, the order of variables in the portable file is the order they are named on KEEP. Thus, you can also use KEEP to reorder variables in the portable file.
- You can use both DROP and KEEP on the same EXPORT command, provided they do not name any of the same variables.
- You can use the TO keyword to specify a group of consecutive variables on the active file.
- The active file is not affected by DROP or KEEP.

Example

```
EXPORT OUTFILE='NEWSUM.POR'/DROP=DEPT79 TO DEPT81.
```

- The portable file is written to NEWSUM.POR in the current directory.
- Variables between and including DEPT79 and DEPT81 on the active file are excluded from the portable file.
- All other variables are saved on the portable file.

RENAME Subcommand

Use the RENAME subcommand to rename variables being written to the portable file. The renamed variables retain their variable and value labels, missing-value flags, and print formats assigned in the SPSS/PC+ job.

- To rename a variable, specify the name of the variable in the active file, an equals sign, and the new name.
- The equals sign is required.
- You can specify lists of variables on both sides of the equals sign. The number of variables on both sides must be the same, and the entire specification must be enclosed in parentheses.
- You can use the TO convention for both variable lists (see Universals: Variable-Naming Conventions).

Example

```
EXPORT OUTFILE='NEWSUM.POR'/DROP=DEPT79 TO DEPT81
    /RENAME=(DEPT82,SALARY82=DEPT,SALARY).
```

- The RENAME subcommand renames DEPT82 and SALARY82 to DEPT and SALARY.
- DEPT and SALARY retain the variable and value labels, missing-value flags, and print formats assigned to DEPT82 and SALARY82.

MAP Subcommand

If you use the RENAME, DROP, or KEEP subcommands to tailor your file, you may find it helpful to produce a listing of your changes with the MAP subcommand.

- The MAP subcommand can be specified as often as you wish.
- The MAP subcommand produces a listing of all actions taken up to that point.
- When the MAP subcommand is specified last, it produces a listing of the contents of the portable file.

Example

```
EXPORT OUTFILE='NEWSUM.POR'/DROP=DEPT79 TO DEPT81/MAP
    /RENAME DEPT82=DEPT SALARY82=SALARY/MAP.
```

- The first MAP subcommand produces a listing of the variables in the file after the DROP subcommand has dropped the specified variables.
- The RENAME subcommand renames DEPT82 and SALARY82.
- The second MAP subcommand shows the variables on the file after renaming. Since this is the last subcommand, the listing will show the variables as they are written on the portable file.

DIGITS Subcommand

The DIGITS subcommand lets you specify the degree of precision for all values of noninteger numeric variables written to the portable file.

- The DIGITS subcommand has the general form DIGITS=*n*, where *n* is the number of digits of precision you want. The default is 10 digits.
- The DIGITS subcommand applies to all numbers for which rounding is required.
- You cannot specify different degrees of precision for different variables. Thus, DIGITS should be set according to the requirements of the variable that needs the most precision.

Example

```
EXPORT OUTFILE='NEWSUM.POR'/DROP=DEPT79 TO DEPT81
    /RENAME=(DEPT82,SALARY82=DEPT,SALARY)/MAP/DIGITS=4.
```

- The DIGITS subcommand guarantees the accuracy of values to four significant digits.
- For example, 12.34567890876 will be rounded to 12.35.

FINISH

```
FINISH
```

Overview

The FINISH command terminates an SPSS/PC+ session and returns control to DOS.

Syntax

- The minimum specification is simply the command keyword. FINISH has no additional specifications.
- The commands BYE, EXIT, and STOP are accepted as aliases for FINISH.

Operations

- FINISH causes SPSS/PC+ to stop reading commands.
- Any commands following FINISH in an INCLUDE file are ignored.

Example

```
DATA LIST FILE='NEW.DAT'/NAME 1-15(A) V1 TO V15 16-30.
LIST.
FINISH.
```

- The DATA LIST and LIST commands are executed.
- The SPSS/PC+ session is ended with the FINISH command.

FORMATS

```
FORMATS varlist (format) [varlist ...]
```

Format	Meaning
Fw.d	Numeric of width w and d decimal places
COMMAw.d	Numeric with commas and decimal places
DOLLARw.d	Numeric with dollar sign, commas, and decimal places

Examples:

```
FORMATS VARA (F4.2) VARB (DOLLAR9) VARC (COMMA7).
FORMATS VARD VARE VARF (F3.2).
```

Overview

The FORMATS command allows you to change the print and write formats of numeric variables. You can change the print width, specify additional decimal digits, and add commas and dollar signs. The values used in computations by SPSS/PC+ are not affected by the print and write FORMATS specification.

Syntax

- The syntax for specifying formats is Fw.d, COMMAw.d, or DOLLARw.d, where *w* specifies the total number of columns, including decimal point, commas, and dollar sign, and *d* specifies the number of decimal places. For example, to display the number 6543210 as $6,543,210, you must specify a dollar format of at least (DOLLAR10). The *d* specification is optional.
- Format specifications are enclosed in parentheses.
- You can specify format types for more than one variable on a single FORMATS command.
- You cannot use FORMATS with string variables.

Operations

- FORMATS is a transformation and is executed when the data are read for the next procedure.
- The FORMATS specification is saved on the active file dictionary.
- The FORMATS command specifies how values are printed in SPSS/PC+ procedures and how data values are written using the WRITE command.
- The formats specified on FORMATS are in effect for the duration of the session or until the variable is given a new FORMATS specification.
- Print and write formats are retained in system files (see SAVE) and portable files (see EXPORT).
- When a COMMA or DOLLAR format is incorrectly assigned, SPSS/PC+ attempts to display the value without commas or a dollar sign. If you have not allowed enough columns for printing a numeric value, SPSS/PC+ prints asterisks. The values in the active file are unchanged.

Limitations

- Some procedures are unable to print wide format values.

Example

```
DATA LIST / VARA 1-4 (3) VARB 6-10 VARC 12-17.
BEGIN DATA.
155  10500 429813
4309 25000 389213
6256 18750 35946
END DATA.
LIST.
FORMATS VARA (F4.2) VARB (DOLLAR9) VARC (COMMA7).
LIST VAR=ALL.
```

- The DATA LIST command indicates inline fixed-format data and defines three variables. VARA is four columns wide with three implied decimal places. VARB is five columns wide, and VARC is six columns wide.

- The first LIST command prints out all variables using the dictionary formats defined by the DATA LIST command. The results of the first LIST command are shown below:

```
VARA  VARB   VARC

 .155 10500 429813
4.309 25000 389213
6.256 18750  35946
```

- The FORMATS command defines new print formats for each variable. VARA is printed in four columns (including decimal point) with two decimal digits. VARB is printed in nine columns including dollar signs and commas. VARC is printed in seven columns including commas.
- The second LIST command prints out each of the variables using the new formats. The results of this LIST procedure are shown below:

```
VARA       VARB     VARC

 .15    $10,500 429,813
4.31    $25,000 389,213
6.26    $18,750  35,946
```

Example

```
DATA LIST FREE/ VARA VARB VARC.
BEGIN DATA.
155 10500 429813 4309 25000 389213 6256 18750 35946
END DATA.
LIST.
FORMATS VARA (F4.0) VARB (DOLLAR11.2) VARC (COMMA9.1).
LIST VAR=ALL.
```

- The DATA LIST command identifies inline data in freefield format. By default, variables VARA, VARB, and VARC have print and write formats of eight columns with two decimal places.
- The first LIST command shows the default formats of the three variables:

```
    VARA      VARB      VARC

  155.00  10500.00  429813.0
 4309.00  25000.00  389213.0
 6256.00  18750.00  35946.00
```

- The FORMATS command specifies a print format of four columns with no decimal digits for VARA; eleven columns, including a dollar sign, commas, and two decimal digits for VARB; and nine columns, including commas and one decimal place for VARC.
- The LIST command uses the new print formats and is shown below:

```
VARA           VARB      VARC

 155   $10,500.00 429,813.0
4309   $25,000.00 389,213.0
6256   $18,750.00  35,946.0
```

FREQUENCIES

```
FREQUENCIES [VARIABLES=]{varlist}
                        {ALL    }

[/FORMAT=[{CONDENSE}] [{NOTABLE }] [NOLABELS]
          {ONEPAGE }   {LIMIT(n)}
         [{DVALUE}] [DOUBLE] [NEWPAGE]]
          {AFREQ }
          {DFREQ }

[/MISSING=INCLUDE]

[/BARCHART=[MINIMUM(n)] [MAXIMUM(n)] [{FREQ(n)   }]]
                                      {PERCENT(n)}

[/HISTOGRAM=[MINIMUM(n)] [MAXIMUM(n)] [{FREQ(n)   }]]
                                       {PERCENT(n)}

           [{NONORMAL}] [INCREMENT(n)]]
            {NORMAL  }

[/HBAR=same keywords as HISTOGRAM]

[/NTILES=n]  [/PERCENTILES=value list]

[/STATISTICS=[DEFAULT] [MEAN] [STDDEV] [MINIUM] [MAXIMUM]
             [SEMEAN] [VARIANCE] [SKEWNESS] [SESKEW] [RANGE] [MODE]
             [KURTOSIS] [SEKURT] [MEDIAN] [SUM] [ALL] [NONE]]
```

Example:

```
FREQUENCIES VARIABLES=XVAR YVAR ZVAR1 TO ZVAR5
/FORMAT=NOTABLE
/STATISTICS=ALL
/HISTOGRAM.
```

Overview

FREQUENCIES produces tables of frequency counts and percentages for the values of individual variables.

Defaults

The default output is a table that displays counts for each value of a variable, the counts percentaged over all cases and over all cases with nonmissing values, and the cumulative percentage over all cases with nonmissing values. Values of numeric variables are ordered from lowest to highest; values of string variables are ordered alphabetically. The number of cases with user-missing and system-missing values are displayed. The table is labeled with variable and value labels if these have been defined. Each page displays as many entire single-spaced tables as will fit.

Tailoring

Display Format. You can suppress all tables or only tables for variables with more than a specified number of categories. You can specify one table per page, a condensed format that fits more values in each table, or double spacing, and you can suppress label printing. Table contents can be ordered by values in descending order, by frequencies in ascending order, or by frequencies in descending order.

Statistical Display. You can display percentiles or ntiles and other optional statistics for each numeric variable. Available statistics include mean, median, mode, standard deviation, variance, skewness, kurtosis, and sum.

Plots. Histograms are available for numeric variables, and bar charts are available for numeric or string variables.

Missing Values. You can include user-missing values in statistical calculations and plots.

Syntax

- The minimum specification is a list of variables.
- Subcommands can be named in any order.
- Subcommands must be separated by slashes.

- Keyword order on subcommands is unimportant.
- VARIABLES names the variables to be tabulated.
- All optional subcommands apply to all variables named on the VARIABLES subcommand.
- Specify subcommands VARIABLES, FORMAT, MISSING, and STATISTICS only once.
- Specify only one of the following subcommands at a time: BARCHART, HISTOGRAM, and HBAR. If you specify any two of these together, HBAR is assumed.
- You can use subcommands PERCENTILES and NTILES more than once. Multiple requests for the same percentiles are consolidated.

Operations

- FREQUENCIES causes the data to be read.
- FREQUENCIES tabulates numeric variables with or without decimal values, short string variables, and the short string portion of long string variables.
- Variables are tabulated in the order that they are mentioned on the VARIABLES subcommand. If a variable is mentioned more than once, it will be tabulated more than once.
- Percentages in tables are displayed with one decimal place. Statistics are displayed with three decimal places.
- If a requested ntile or percentile cannot be calculated, a period (.) is displayed.
- The display always uses narrow format regardless of the width defined on SET.
- The HISTOGRAM subcommand on the SET command controls the character used to draw histograms.

Limitations

- The maximum number of variables on a FREQUENCIES command is the same as the system limit.
- The maximum unique observed values over all variables depends on available workspace and on available labels space (up to 5000 characters for all labels combined).

Example

```
FREQUENCIES VARIABLES=XVAR YVAR ZVAR1 TO ZVAR5
/FORMAT=NOTABLE
/STATISTICS=ALL
/HISTOGRAM.
```

- This example requests FREQUENCIES for XVAR, YVAR, and all variables between and including ZVAR1 and ZVAR5.
- The FORMAT subcommand suppresses the display of frequency tables for all variables named on the VARIABLES subcommand.
- The STATISTICS subcommand requests that all statistics available be displayed for each variable named on the VARIABLES subcommand.
- The HISTOGRAM subcommand requests a histogram for each variable named on the VARIABLES subcommand.

VARIABLES Subcommand

VARIABLES names the variables to be tabulated and is the only required subcommand. The actual keyword VARIABLES can be omitted.

- You can use keyword ALL to refer to all user-defined variables in the active file.

FORMAT Subcommand

By default, FREQUENCIES displays as many single-spaced tables with complete labeling information as fit within the page length. The default table is ordered by ascending value (numeric variables) or in alphabetical order (string variables). Use FORMAT with the keywords listed below to change the defaults.

- If you specify FORMAT, only defaults that you explicitly alter are changed.

Table Formats

CONDENSE *Condensed format.* Displays counts in three columns without value labels and with valid and cumulative percentages rounded to integers. Overrides ONEPAGE.

ONEPAGE *Conditional condensed format.* Uses condensed format for tables that would otherwise require more than one page.

NEWPAGE *Each table starts on a new page.*

NOLABELS *No value labels.*

DOUBLE *Double-space frequency tables.*

Table Order

AFREQ *Sort categories in ascending order of frequency.* Ignored when HISTOGRAM, HBAR, NTILES, or PERCENTILES are requested.

DFREQ *Sort categories in descending order of frequency.* Ignored when HISTOGRAM, HBAR, NTILES, or PERCENTILES are requested.

DVALUE *Sort categories in descending order of values (numeric variables)* or in reverse alphabetical order (string variables). Ignored when HISTOGRAM, HBAR, NTILES, or PERCENTILES are requested.

Table Suppression

LIMIT(n) *No frequency tables with more than* n *categories.* The number of missing and valid cases and requested statistics are displayed for suppressed tables.

NOTABLE *No frequency tables.* The number of missing and valid cases are displayed for suppressed tables. Overrides LIMIT.

BARCHART Subcommand

BARCHART produces a bar chart for each variable named on the VARIABLES subcommand. By default, the horizontal axis for each bar chart is scaled in frequencies and the interval width is determined by the largest frequency count for the variable being plotted. Bar charts are labeled with value labels or with the value if no value label is defined.

- If you omit the BARCHART subcommand, no bar charts are displayed.
- If no keywords are specified on the BARCHART subcommand, the default bar charts are displayed.
- Only defaults you explicitly alter with keywords are changed.

MIN(n) *Lower bound* below which values are not plotted.

MAX(n) *Upper bound* above which values are not plotted.

FREQ(n) *Horizontal axis scaled in frequencies,* where optional *n* is the maximum. With no *n* or a too-small *n*, FREQUENCIES chooses 10, 20, 50, 100, 200, 500, 1000, 2000, and so forth, depending on the largest category. This is the default.

Example

```
FREQ VAR=JVAR/ BAR=MAX(10).
```

- This command requests a bar chart with values through 10.
- This example takes advantage of spelling permitted by three-character truncation of keywords.

HISTOGRAM Subcommand

HISTOGRAM displays a plot for each numeric variable named on the VARIABLES subcommand. If there are no specifications on HISTOGRAM, the horizontal axis of each histogram is scaled in frequencies and the interval width is determined by the largest frequency count of the variable being plotted.

- If you omit the HISTOGRAM subcommand, no histograms are displayed.
- If no keywords are specified, the default histograms are displayed.
- Only defaults you explicitly alter with keywords are changed.

MIN(n) *Lower bound* below which values are not plotted.

MAX(n) *Upper bound* above which values are not plotted.

PERCENT(n) *Horizontal axis scaled in percentages,* where optional *n* is the preferred maximum. With no *n* or a too-small *n*, FREQUENCIES chooses 5, 10, 25, 50, or 100, depending on the largest category.

FREQ(n) *Horizontal axis scaled in frequencies,* where optional *n* is the scale. With no *n* or a too-small *n,* FREQUENCIES chooses 10, 20, 50, 100, 200, 500, 1000, 2000, and so forth, depending on the largest category. This is the default.

INCREMENT(n) *Interval width,* where *n* is the size of the interval. Overrides the default number of intervals on the vertical axis, which depends on the system page length. For a variable that ranges from 1 to 100, INCREMENT(2) produces 50 intervals with 2 values each.

NORMAL *Superimpose a normal curve.* Based on all valid values for the variable, including values excluded by MIN and MAX.

Example

```
FREQS VAR=VARZ/HIST=NORMAL INCREMENT(4).
```

- This example requests a histogram with a superimposed normal curve and an interval width of 4.
- This example takes advantage of spelling permitted by three-character truncation of keywords.

HBAR Subcommand

HBAR produces a plot for each numeric and string variable named on the VARIABLES subcommand. For numeric variables, HBAR produces a bar chart if the number of categories fits within the page length (see SET). Otherwise, HBAR produces a histogram. HBAR produces bar charts for short string variables and for the short-string portion of long string variables, regardless of the number of values.

By default, the horizontal axis of each plot is scaled in frequencies and the interval is determined by the largest frequency count. All keyword specifications for HISTOGRAM and BARCHART work with HBAR.

PERCENTILES Subcommand

PERCENTILES displays the value below which the specified percentage of cases falls.

- There are no default percentiles.
- If you omit the PERCENTILES subcommand, no percentiles are displayed.

Example

```
FREQUENCIES VARIABLES=VARZ/ PERCENTILES=10 25 33.3 66.7 75.
```

- This example requests the values for percentiles 10, 25, 33.3, 66.7, and 75 for VARZ.

NTILES Subcommand

NTILES calculates the percentages that divide the distribution into the specified number of categories and displays the values below which the requested percentages of cases fall.

- There are no default ntiles for the NTILES subcommand.
- If you omit the NTILES subcommand, ntiles are not displayed.

Example

```
FREQUENCIES VARIABLES=VARZ/ NTILES=4.
```

- This example requests quartiles (percentiles 25, 50, and 75) for VARZ.

STATISTICS Subcommand

The STATISTICS subcommand controls the display of statistics. By default, cases with missing values are excluded from the calculation of statistics.

- If you use the STATISTICS subcommand without any specifications, it produces the mean, standard deviation, minimum, and maximum (the same as produced by the keyword DEFAULT).
- If you use the STATISTICS subcommand with any specifications, only statistics requested are displayed.

- If you omit the STATISTICS subcommand, no statistics are displayed.

The following can be specified on the STATISTICS subcommand:

MEAN *Mean.*

SEMEAN *Standard error of the mean.*

MEDIAN *Median.* Ignored with AFREQ or DFREQ on the FORMAT subcommand.

MODE *Mode.*

STDDEV *Standard deviation.*

VARIANCE *Variance.*

SKEWNESS *Skewness.*

SESKEW *Standard error of the skewness statistic.*

KURTOSIS *Kurtosis.*

SEKURT *Standard error of the kurtosis statistic.*

RANGE *Range.*

MINIMUM *Minimum.*

MAXIMUM *Maximum.*

SUM *Sum.*

DEFAULT *Mean, standard deviation, minimum, and maximum.*

ALL *Display all available statistics.*

NONE *No statistics.*

Example

```
FREQS VAR=AGE/STATS=DEF MODE.
```

- The keyword DEFAULT on the STATISTICS subcommand produces the mean, standard deviation, minimum, and maximum of AGE.
- The keyword MODE requests the mode of AGE.
- This example takes advantage of spelling permitted by three-character truncation of keywords.

MISSING Subcommand

By default, both user- and system-missing values are labeled as missing in the table but are not included in the valid and cumulative percentages, in the calculation of descriptive statistics, or in bar charts and histograms. Use the MISSING subcommand and keyword to alter the handling of missing values.

INCLUDE *Include cases with user-missing values.* Cases with user-missing values will be included in the statistics and plots.

GET

```
GET [FILE={'SPSS.SYS'**}] [/DROP=varlist]
          {'filename'  }
```

**Default if subcommand is omitted.

Example:

```
GET FILE='NEWDATA.SYS'.
```

Overview

GET reads an SPSS/PC+ system file produced by the SAVE command. A system file contains data and a dictionary with variable and value labels, missing-value flags, and print formats for each variable on the system file. GET also reads older system files produced by either Release 1.0 or 1.1 of SPSS/PC.

Defaults

By default, GET retrieves data and dictionary information for all variables on the system file SPSS.SYS in the current directory.

Tailoring

You can retrieve system files other than SPSS.SYS. You can also retrieve only a subset of variables from a system file.

Syntax

- The minimum specification is simply the command keyword.
- Subcommands are separated by an optional slash.

Operations

- GET causes the dictionary of the system file to be read.
- Data from the system file are copied into a separate active file on disk only if you use the DROP subcommand, or if you enter data transformation or selection commands. If neither of these circumstances applies, SPSS/PC+ will read data from the system file repeatedly when executing procedures.
- No declaration that a system file is either compressed or uncompressed is needed.
- System files are designed to be read by SPSS/PC+ only and should not be edited.
- The dictionary and data from the file named on the GET command are used in subsequent procedures.
- A file saved with weighting in effect maintains the values of variable $WEIGHT. For a discussion of turning off weights, see WEIGHT.
- The order of cases in a system file depends on their order at the time the file was saved. The values of $CASENUM are those from the raw data file, prior to any selection (see SELECT IF) or sorting (see SORT).
- The variable counts displayed by GET treat each 8-character portion of a long string variable as one variable. For example, a 17-character long string counts as three variables.

Example

```
GET.
```

- The GET command reads system file SPSS.SYS from the current directory.

FILE Subcommand

Use the FILE subcommand to specify a system file other than the default SPSS.SYS.

- The only specification on FILE is the name of the file.
- The file specification must be enclosed in apostrophes.
- You can specify files residing in other directories by supplying a fully qualified filename (see Universals: Files).

Example

```
GET FIL='\KLDIR\SALDATA.NOV'.
```

- The system file SALDATA.NOV is retrieved from directory KLDIR.
- This example takes advantage of spelling permitted by three-character truncation of keywords.

DROP Subcommand

Use the DROP subcommand to retrieve a subset of variables from the system file. DROP specifies a variable or list of variables to be dropped. All variables not named on the DROP subcommand will be included in the active file. The SPSS/PC+ system file on disk is not affected by specifying DROP on the GET command.

- You can specify variables on the DROP subcommand in any order.
- You can use the TO keyword to specify a group of consecutive variables on the SPSS/PC+ system file.

Example

```
GET FILE='NEWSUM.SYS'/DROP=DEPT79 TO DEPT81, SALARY81, HIRE.
```

- The system file retrieved is NEWSUM.SYS from the current directory.
- All variables between and including DEPT79 and DEPT81, as well as SALARY81 and HIRE, are excluded from the active file.
- All other variables are retained on the active file.

IF

```
IF (logical expression) target variable=assignment expression
```

Relational Operators:

EQ or =	Equal to	NE or ~= or <>	Not equal to
LT or <	Less than	LE or <=	Less than or equal to
GT or >	Greater than	GE or >=	Greater than or equal to

Logical Operators:

AND or & OR or | NOT or ~

Missing-Value Functions:

SYSMIS Returns 1 (true) if value is system-missing
MISSING Returns 1 (true) if value is system- or user-missing
VALUE Returns the value itself, ignoring user-missing flags

Example:

```
IF (SEX EQ 'F') EEOVAR=QUOTA+SXVAR.
IF (SYSMIS(QVAR)) RVAR=0.
IF (ABS(A-C) LT 100) INT=100.
IF (VALUE(AGE) = 99 OR VALUE(SEX) = 9) GROUP=9.
```

Overview

The IF command conditionally executes a single COMPUTE-like transformation based upon logical conditions found in the data. The transformation can create a new variable or modify the values of an existing variable for each case in your active file. You can create or modify the values of numeric and short string variables.

The IF command has three components. An expression enclosed in parentheses sets up the logical criteria and is called the *logical expression* (see Universals: Logical Expressions). The *target variable* (the one to be modified or created) is named next. Following the target variable is an equals sign and the *assignment expression.* The target variable's values are modified according to the assignment expression.

A series of IF commands testing the same variable can often be replaced by a single RECODE command, which is more efficient.

Syntax

- The minimum specification is a logical expression, followed by a target variable, an equals sign, and an assignment expression.
- Parentheses around the logical expression are required.
- The equals sign is required.
- Parentheses can also be used within the logical expression to specify the order of operations.
- The logical expression can contain string variables, numeric variables, or both.
- At least one relation, SYSMIS function, or MISSING function must be included in the logical expression.
- A relation includes a variable name, the relational operator, and a value or variable.
- Relations cannot be abbreviated: (A EQ 2 OR 5) is invalid.
- String values used in relations must be specified in quotes and must include any leading or trailing blanks. Lowercase letters are considered distinct from uppercase letters.
- A relation cannot compare a string variable to a numeric value or variable, or vice versa.
- Both the logical expression and the assignment expression can use arithmetic operations and functions allowed in COMPUTE transformations (see COMPUTE and Universals: Functions).

Operations

- IF is a transformation and is executed when the data are read for the next procedure.
- Each IF command is evaluated independently.

- The logical expression is evaluated as true or false. The assignment is executed only if the logical expression is true.
- If the logical expression is false or if one of the variables used in the logical expression is system- or user-missing, the assignment is not made. Existing target variables remain unchanged; new numeric variables are assigned the system-missing value and new string variables are set to blanks.
- Logical expressions are evaluated in the following order: first numeric functions, then exponentiation, then arithmetic operations, then relations, and finally logical operators. You can change the order of operations using parentheses.
- For assignment expressions, the order of evaluation is numeric functions, then exponentiation, and then arithmetic operators.
- Relational and logical operators cannot be used in assignment expressions.
- Numeric variables created with IF are initially set to the system-missing value. Short string variables created with IF are initially set to a blank value of the specified width.
- Numeric variables created with IF are assigned a print format of eight characters with two decimal places.
- String variables created with IF are assigned a width equal to the number of characters used in the initial assignment.

Limitations

- The number of variables created with IF, COMPUTE, and COUNT plus the number defined on DATA LIST, IMPORT, or GET cannot exceed the system maximum of 200 variables.

Example

```
IF (XVAR EQ 5) YVAR=3.
```

- Numeric variable YVAR is set to 3 for cases where XVAR equals 5 (the expression is true).
- When the expression is false or missing, the value of YVAR remains unchanged. If YVAR has not been previously defined, it contains the system-missing value.

Example

```
IF (SEX EQ 'F') EEOVAR=QUOTA+SVAR.
```

- The logical expression tests string variable SEX for the value "F."
- When the expression is true (when SEX equals F), the value of numeric variable EEOVAR is assigned the value of QUOTA plus SVAR. Both QUOTA and SVAR must be previously defined numeric variables.
- When the expression is false or missing (for example, if SEX equals f), the value of EEOVAR remains unchanged. If EEOVAR has not been previously defined, it contains the system-missing value.

Example

```
COMPUTE SVAR=0.
IF ((QVAR-RVAR) LE 7))SVAR=QVAR**2.
IF (ABS(A-C) LT 100)INT=100.
```

- COMPUTE assigns SVAR the value 0.
- The logical expression tests whether QVAR minus RVAR is less than or equal to 7. If it is, the value of SVAR is assigned the value of QVAR squared. Otherwise, the value of SVAR remains at 0.
- The second IF command tests whether the absolute value of variable A minus variable C is less than 100. If it is, INT is assigned the value 100. Otherwise, the value is unchanged or, if INT has not been previously defined, system-missing.

Example

```
IF (SYSMIS(QVAR)) RVAR=0.
COM VALID=0.
IF (NOT(SYSMIS(VARA))) VALID=1.
```

- The first IF command tests whether QVAR is system-missing. If it is, RVAR is assigned the value 0. Otherwise RVAR is unchanged, or system-missing if RVAR has not been previously defined.
- COMPUTE assigns variable VALID a value of 0.
- The next IF command tests whether VARA is not system-missing. For each case where VARA contains a valid value (is not system-missing), the value of VALID is set to 1. For each case that contains a system-missing value for VARA, the value of VALID equals 0.
- The example takes advantage of spelling permitted by three-character truncation of keywords.

Example

```
IF (STATE EQ 'IL' AND CITY EQ 13) COST=COST + .07 * COST.
```

- The logical expression tests whether STATE equals IL and CITY equals 13.
- If the logical expression is true, numeric variable COST is assigned the original value of COST plus 7% of the original value of COST (1.07 × COST).
- For any other value of STATE or CITY, the value of COST remains unchanged.

Example

```
IF (VALUE(VARA) GT 0) QVAR=AVAR*BVAR.
```

- The logical expression tests whether the value of VARA is greater than 0. If it is, QVAR is assigned the value of AVAR times BVAR. This test will be true even if VARA is user-missing, provided that it is greater than 0.
- For values less than or equal to 0, QVAR remains unchanged.
- For system-missing values, QVAR remains unchanged.

Example

```
IF (QVAR EQ 'ok') AVAR='fine'.
```

- The new string variable AVAR is set to "fine" when the value of QVAR is "ok."
- AVAR has format width of four characters.
- When QVAR is not equal to "ok," AVAR is defined as a four-column blank field.

Example

```
IF (RECV GT DUE OR (REVNUES GE EXPNS AND BALNCE GT
0))STATUS='SOLVENT'.
```

- The IF command specifies a complex logical expression.
- First, SPSS/PC+ tests whether REVNUES is greater than or equal to EXPNS and whether BALNCE is greater than 0.
- Second, SPSS/PC+ evaluates if RECV is greater than DUE.
- If either of these expressions is true, STATUS is assigned the value "SOLVENT."
- If both expressions are false, STATUS remains unchanged. If STATUS has not been previously defined, it is defined as a seven-column blank field.

IMPORT

```
IMPORT FILE='filename' [/KEEP={ALL    }] [/DROP=varlist]
                              {varlist}

     [/RENAME=(old varlist=new varlist)...] [/MAP]
```

Example:

```
IMPORT FILE='NEWDATA.POR'/RENAME=(ID,SEX,AGE=V1 TO V3)/MAP.
```

Overview

IMPORT reads a portable ASCII data file and dictionary produced by the EXPORT command in SPSS/PC, SPSS/PC+, SPSS^X, or by PROC TOSPSS in SAS. You can download files from SPSS^X on a mainframe to your IBM PC using KERMIT, provided KERMIT is installed on both the mainframe and the PC (see Part F: Communications).

Defaults

IMPORT reads all data and dictionary information for all cases on the portable file. The data dictionary contains the variable and value labels, missing-value flags, and print formats for each variable on the portable file. The file also contains a message with the name, release, and version of the originating software, and the date and time the portable file was created. When the file originates from SPSS^X, it also includes the file label and the name of the originating installation. SPSS/PC+ assumes that the file is in the current directory.

Tailoring

You can import a subset of variables from the portable file, and you can rename the imported variables. You can also obtain a listing of the imported variables. You can read a portable file from a floppy disk.

Syntax

- The minimum specification on IMPORT is a file specification.
- Subcommands can be named in any order and must be separated by a slash.

Operations

- IMPORT causes SPSS/PC+ to read the data from the portable file.
- Most IBM/PC editors are unable to read a portable file.
- The portable data file and dictionary become the SPSS/PC+ active file.
- You cannot import a portable file directly from a directory other than your current directory on the hard disk. You can import a portable file from a floppy diskette.
- The SPSS/PC+ active file has a more restrictive dictionary than mainframe versions of SPSS^X. SPSS/PC+ IMPORT changes the dictionary to conform to internal conventions. In particular, if the SPSS^X file defined a range of missing values for a variable, only the bottom of the range is defined in SPSS/PC+.
- A file saved with weighting in effect automatically uses the case weights when the file is read.

Limitations

- You may not have enough available memory on your PC to read a portable file produced by SPSS^X on a mainframe. When you produce portable files on a mainframe for use on a PC, you should include only the variables you need for particular SPSS/PC+ jobs.
- You can import a file with up to 200 variables. Each 8-character portion of a long string variable counts as one variable.

Example

```
IMPORT FILE='NEWDATA.POR'/RENAME=(ID,SEX,AGE=V1 TO V3)/MAP.
```

- The portable file is read from NEWDATA.POR in the current directory.
- Variables ID, SEX, and AGE are renamed V1, V2, and V3 in the active file. Their names are unchanged on the portable file on NEWDATA.POR. None of the other variables read into the SPSS/PC+ active file are renamed.
- The MAP subcommand requests a listing of variables on the active file.

FILE Subcommand

The FILE subcommand specifies the name of the file that contains the portable file.

- The filename must be enclosed in apostrophes.
- You cannot specify a path on the IMPORT command. The portable file must be in your current directory, if any, for the drive on which it resides.
- You can use a portable file stored on a floppy disk.

Example

```
IMP FIL='SALDATA.NOV'.
```

- The portable file is read from SALDATA.NOV in the current directory.
- This example takes advantage of spelling permitted by three-character truncation of keywords.

DROP and KEEP Subcommands

Use the DROP and KEEP subcommands to import a subset of variables from the portable file.

- DROP excludes a variable or list of variables from the active file. All variables not named on DROP are included.
- KEEP names a variable or list of variables to be included in the active file. All variables not named are excluded.
- Variables can be specified on DROP or KEEP in any order.
- With the DROP subcommand, the order of variables remaining on the active file is the same as their order on the portable file.
- With KEEP, the order of variables on the active file is the order in which they are named on KEEP. Thus, you can use KEEP to reorder variables.
- You can use both DROP and KEEP on the same IMPORT command, provided they do not name any of the same variables.
- You can use the TO keyword to specify consecutive variables on the portable file.
- The portable file is not affected by specifying DROP or KEEP on the IMPORT command.

Example

```
IMPORT FILE='NEWSUM.EXP'/DROP=DEPT79 TO DEPT81.
```

- The portable file is read from NEWSUM.EXP in the current directory.
- Variables between and including DEPT79 and DEPT81 are excluded from the active file. All other variables are on the active file.

RENAME Subcommand

Use the RENAME subcommand to rename variables being read from the portable file. The renamed variables retain the variable and value labels, missing-value flags, and print formats assigned in the job where the portable file was created.

- To rename a variable, specify the name of the variable in the active file, an equals sign, and the new name.
- The equals sign is required.
- You can specify lists of variables on both sides of the equals sign. The number of variables on both sides must be the same, and the entire specification must be enclosed in parentheses.
- You can use the TO convention for both variable lists (see Universals: Variable-Naming Conventions).

Example

```
IMPORT FILE='NEWSUM.POR'/DROP=DEPT79 TO DEPT81
    /RENAME=(DEPT82,SALARY82=DEPT,SALARY).
```

- The RENAME subcommand renames DEPT82 and SALARY82 to DEPT and SALARY.
- DEPT and SALARY retain the variable and value labels, missing-value flags, and print formats assigned to DEPT82 and SALARY82.

MAP Subcommand

If you use the RENAME, DROP, or KEEP subcommands to tailor your file, you may find it helpful to produce a listing of your changes with the MAP subcommand.

- The MAP subcommand can be specified as often as you wish.
- The MAP subcommand produces a listing of all actions taken *up to that point.*
- When the MAP subcommand is last, it produces a listing of the contents of the new active file.

Example

```
IMPORT FILE='NEWSUM.POR'/DROP=DEPT79 TO DEPT81/MAP
    /RENAME(DEPT82=DEPT SALARY82=SALARY)/MAP.
```

- The first MAP subcommand produces a listing of the variables after DROP has excluded the specified variables.
- The RENAME subcommand renames DEPT82 and SALARY82.
- The second MAP subcommand shows the variables after renaming. Since this is the last subcommand specified, it displays the variables that are contained in the active file.

SPSS^X Portable Files

SPSS^X EXPORT command writes a portable file that is read by a number of different mainframe computers as well SPSS/PC+. SPSS/PC+ uses a more restrictive dictionary than SPSS^X. If you are creating a portable file in SPSS^X that will be read by SPSS/PC+, keep the following in mind:

- After reading 200 variables, SPSS/PC+ stops processing variables. You can avoid this problem by dropping variables when you create the portable file, so that the portable file contains 200 or fewer variables.
- SPSS/PC+ only allows one missing-value flag per variable and accepts only the first of multiple missing-value declarations. You can override missing-value declarations prior to creating the portable file with SPSS^X.
- SPSS/PC+ prints values in one of the following formats: DOLLARw.d, COMMAw.d, Aw, or Fw.d. SPSS^X offers additional formats, which SPSS/PC+ attempts to translate. If the translation doesn't meet your purposes, you can change the print format of variables with the FORMAT command.
- SPSS^X portable system files can also be produced from SAS (Statistical Analysis System) data sets with the SAS procedure TOSPSS. SPSS Inc. distributes TOSPSS to SPSS^X sites and will, on request, provide TOSPSS along with KERMIT to SAS sites where SPSS^X is not installed.

INCLUDE

```
{INCLUDE 'filename'}
{@filename         }
```

Example:

```
INCLUDE '\MASTER\SET.CMD'.
INCLUDE 'DEFINE.INC'.
@CODEBOOK.RUN.
```

Overview

The INCLUDE command allows you to execute SPSS/PC+ commands from a file. With INCLUDE, you can prepare an entire session with your editor and leave SPSS/PC+ to execute unattended, as in batch-type processing. You can also use INCLUDE to execute all the file definition commands (such as DATA LIST and labeling commands) and then execute analysis commands interactively.

Another use for INCLUDE is to execute a "profile" for your machine configuration. For example, you can create a file containing a SET command that specifies printer characters, a prompt, and page size, and then include that file in any SPSS/PC+ session. A "profile" like this will be executed *automatically* when you enter SPSS/PC+ if you have named it SPSSPROF.INI and saved it in either your current directory (the one from which you work) or the directory in which the SPSS/PC+ system is saved.

You can nest INCLUDE commands so that one set of included commands includes another set of commands. This "nesting" can go five levels deep.

Syntax

- The only specification for INCLUDE is a filename enclosed in single quotes.
- You can include a file from another directory by using a fully qualified filename.
- The character @ is accepted as an alias for the INCLUDE command. With this form, the filename does not need to be in single quotes, although it can be. The space between @ and the filename is optional.
- You can use more than one INCLUDE command in a session, either in a series or nested.
- **If you include a file of inline data, the first line of the data file must contain the BEGIN DATA command. The END DATA command can be specified as the last line of the included file or with your SPSS/PC+ commands.**

Operations

- INCLUDE is an operation command and is executed immediately.
- INCLUDE identifies a file containing SPSS/PC+ commands which may include inline data.
- By default, each command from the INCLUDE file is displayed on your screen as it is processed. You can suppress this display by using the INCLUDE OFF specification on SET.
- Both the INCLUDE command and commands from the included file are copied to the LOG file. The INCLUDE command is executable from the LOG file. The commands from the included file are prefaced with an open bracket ([). These commands are treated as comments and are not executable. Thus, if you use the LOG file in a subsequent session, the INCLUDE command is read from the LOG file and the included commands are read from the original file. The included commands are executed only once.
- If an INCLUDE file contains a FINISH command, the SPSS/PC+ session ends and you are returned to DOS. No subsequent SPSS/PC+ commands are processed.

Limitations

- SPSS/PC+ will process only up to five levels of nested included files at a time.
- Only the first 80 characters of an included file are read. For data files with a record length exceeding 80 characters, you must use the FILE subcommand on DATA LIST. Command files can never have a record length longer than 80 characters.

Example

```
INCLUDE '\MASTER\SET.CMD'.
INCLUDE 'DEFINE.INC'.
@CODEBOOK.RUN.
```

- The first INCLUDE command processes the commands in file SET.CMD in directory \MASTER. SET.CMD contains the following:

```
SET DISK=ON/LENGTH=59/BOXST='-|+'/BEEP OFF.
```

- The second INCLUDE command processes commands in file DEFINE.INC in the current directory. DEFINE.INC contains the following:

```
DATA LIST FILE='CURRENT.DAT' / MONTH 1-2 (A) DAY 3-4 TEMP 6-7
     PRESSURE 8-12 (2) WINSPED 13-14.
MISSING VALUE DAY (99)/WINSPED (-1).
INCLUDE 'TRANSFOR.INC'.
```

- DEFINE.INC includes another INCLUDE command, which processes commands in file TRANSFOR.INC from the current directory. TRANSFOR.INC contains some data transformation commands. This is an example of a nested INCLUDE.
- The next command (@) is the abbreviated form of INCLUDE. Since the quotation marks and the space separating the command from the filename are optional with this form, they are omitted here. Commands are read from file CODEBOOK.RUN in the current directory. CODEBOOK.RUN contains the following commands:

```
FREQ VAR=ALL/HBAR.
DESC VAR=ALL.
```

JOIN

```
[JOIN]  {MATCH}
        {ADD  }

 /{FILE }={file specification}
  {TABLE} {*                 }

 [/KEEP=varlist ]

 [/DROP=varlist ]

 [/RENAME (old varlist=new varlist) [(old varlist=...)]]

  /FILE=...
    ...

 [/FILE=...  ]
    ...

 [/MAP]

 [/BY=varlist]
```

Example:

```
JOIN MATCH FILE='PART1.SYS'
/ RENAME (CASEID=ID)
/FILE=*
/BY ID.
```

Overview

JOIN permits you to combine two or more system files. With keyword MATCH you can make parallel or nonparallel matches between different files or perform table lookups. With keyword ADD you can add more cases to an existing file. JOIN works with SPSS/PC+ system files created with the SAVE command or optionally with the current active file.

JOIN combines files to produce a new *active file.* Statistical procedures following JOIN use this combined file unless you replace it by building another active file. You must use the SAVE command if you want to write the combined file to disk as a system file.

Defaults

By default, all variables from all input files are included in the new active file.

Tailoring

Variable Selection. You can choose which variables from each input file will be retained on the new active file.

Variable Names. You can rename variables on each input file before joining the files. This permits you to join variables that are the same but whose names differ on different input files or to distinguish different variables whose names are the same on different input files.

Variable Map. You can request a map showing all the variables on the new active file, their order, and the input file(s) from which they came.

Syntax

- The minimum specification is keyword MATCH or ADD and two or more FILE subcommands specifying the names of files to be joined.
- With keyword MATCH, a TABLE subcommand can be used instead of one of the FILE subcommands.
- With MATCH or ADD, an asterisk can be substituted for one filename to indicate the active file.
- The first specification must be either MATCH or ADD.
- The command name JOIN is optional. MATCH is accepted as an alias for JOIN MATCH, and ADD is accepted as an alias for JOIN ADD.
- Subcommands are separated by slashes.

- The RENAME, KEEP, and DROP subcommands apply only to variables in the file named on the immediately preceding FILE or TABLE subcommand. They can be repeated after each FILE or TABLE subcommand.
- The BY subcommand can only be specified once. It must follow all other subcommands except MAP.
- The MAP subcommand can be placed anywhere and repeated as desired.

Operations

- JOIN causes all input files named on the FILE or TABLE subcommands to be read and builds a new active file that replaces any active file created earlier in the session.
- Keywords MATCH and ADD, which are mutually exclusive, determine the type of processing.
- If the active file is named as an input file, any N and SAMPLE commands you have specified are applied to the active file before files are joined.
- The PROCESS IF command has no effect on JOIN.

Limitations

- Maximum 5 files total can be specified on the FILE and TABLE subcommands.
- Maximum 1 TABLE subcommand specifying 1 table file.
- Maximum 10 variables on the BY subcommand. Each 8-character portion of a long string variable counts as 1 toward this limit.
- JOIN creates only uncompressed active files.

Keyword MATCH

In general, use MATCH to combine two or more files containing the same cases but different variables.

- Unless you specify otherwise, the new active file will contain all variables from all input files.
- The new active file will contain all cases that are in any of the input files named on FILE subcommands.
- When two or more files have a variable with the same name, values in the resulting file are taken from the file named first in the JOIN command, even if they are missing in that file. Dictionary information for that variable is taken from the first file for which the variable has either value labels, a variable label, or a declared missing value.
- Cases that are absent from one of the input files will be assigned missing values for variables that exist only in that file.
- If you do not enter a BY subcommand, SPSS/PC+ performs a "parallel" or sequential match, combining the first case from each file, then the second case from each file, and so on, without regard to any identifying values that may be present.
- The BY subcommand specifies that cases should be joined according to a common value on one or more "key" variables present in all input files. All input files must be sorted in ascending order of the key variables.
- When you use the BY subcommand with MATCH, one of the input files can be specified on a TABLE subcommand to indicate that it is a *table lookup* file.

Example

```
MATCH FILE='PART1.SYS'
/FILE='PART2.SYS'
/FILE=*.
```

- MATCH is used here as an alias for JOIN MATCH.
- This example combines three files (the active file and two system files) in a parallel match. Cases are combined according to their order in each file.
- The new active file will contain as many cases as are in the largest of the three input files.
- If the same variable name is used in more than one input file, data are taken from the file listed first: PART1.SYS, then PART2.SYS, and then the active file.

BY Subcommand with MATCH

When used with MATCH, the BY subcommand specifies one or more identification or "key" variables that determine which cases are to be combined. BY is required unless all input files are to be matched simply according to the order of cases.

- BY must be entered after all other file specifications. Only the MAP subcommand, which requests optional output, can follow BY.
- Specifications for the BY subcommand consist of the names of one or more *key variables*. The maximum is 10 keys.
- Key variables must be present in all input files.
- All input files must be sorted by the key variable(s), in ascending order. If necessary, use SORT CASES before JOIN.
- Missing values on key variables are handled like any other values.
- String variables are permitted on the BY subcommand. Each 8-character portion of a long string variable counts as 1 toward the limit of 10 key variables. A 16-character string counts as 2 toward the limit, while a 17-character string counts as 3.
- Unmatched cases are assigned system-missing values (in numeric variables) or blanks (in string variables) for variables from the files that do not contain a match.
- A message informs you if two or more cases from any input file have identical values for the key variable(s). Only one such message is generated, regardless of the number of duplicates encountered, since duplicate keys are expected in many applications.

Table Lookup

A *table file* for a MATCH operation is an input file whose cases are not to be added to the output file. Instead, variables from a table file are added to whatever cases are taken from the other input file(s).

- A table file is identified on the TABLE subcommand.
- The BY subcommand naming the key variable or variables according to which data can be "looked up" is required to process a table file.
- Only one table file can be specified on any JOIN command.
- A table file cannot contain cases with duplicate values for the key variable(s).

Duplicate Cases

Duplicate cases have the same values for the key variable(s) named on the BY subcommand. If no BY subcommand is specified, the question of duplicate cases does not arise.

- Duplicate cases are permitted in any input files except table files.
- When there is no table file, the first case in each file from a group of duplicates is joined with the first matching case (if any) from the other files; the second duplicate case is matched with a second matching duplicate, if any, and so on. In effect, a parallel (sequential) match is performed within groups of duplicate cases. Unmatched cases are assigned system-missing values (in numeric variables) or blanks (in string variables) for the variables from the absent files.
- When a table file is specified, data from the table file are added to *all* cases in the other files with matching values for the BY variable(s).

Keyword ADD

In general, use keyword ADD to combine two or more files containing different cases but the same variables.

- Unless you specify the BY subcommand (see below), the resulting file contains all cases from the first-named input file first, followed by all cases from the second input file, and so on.
- The number of cases in the new active file is the sum of the number of cases in all the input files. No cases are combined.
- The cases in the new active file contain all variables that appear in *any* input files. Cases from files without all the variables will have system-missing values (numeric variables) or blanks (string variables) for the extra variables.
- You cannot use the TABLE subcommand to specify an input file for ADD.

Example

```
JOIN ADD FILE='JAN.SYS'
/FILE='FEB.SYS'
/FILE='MAR.SYS'
/FILE=*.
```

- This example combines cases from three system files saved previously on disk and from the current active file.
- Cases from JAN.SYS will be first in the new active file, followed by cases from FEB.SYS, then MAR.SYS, and then cases from the file that was active prior to the JOIN command.

BY Subcommand with ADD

If all input files are sorted by some variable, you can use the BY subcommand to interleave cases from the different files according to the values of that variable.

- All input files must be sorted in ascending order of the BY variable.
- You can specify up to 10 variables on the BY subcommand. Each 8-character portion of a long string variable counts as 1 toward this limit.
- A message informs you if two or more cases from any input file have identical values for the key variable(s). Only one such message is generated, regardless of the number of duplicates encountered, since duplicate keys are expected in many applications.
- Cases with identical values on the BY variable(s) are arranged in the order you name their input files.

Example

```
ADD FILE='SAMPLE1.SYS'
/FILE=*
/BY DATE.
```

- This example uses ADD as an alias for JOIN ADD.
- Cases from the system file SAMPLE1.SYS are interleaved with cases from the current active file in ascending order of their values for DATE.
- If cases have the same value for DATE, those from SAMPLE1.SYS will precede those from the current active file.

FILE Subcommand

The FILE subcommand identifies each input file (except a table file) for either a MATCH or an ADD operation.

- A maximum of five files total can be specified on the FILE and TABLE subcommands.
- Specifications on the FILE subcommand consist of either a filename in apostrophes or an asterisk to refer to the current active file.
- With MATCH, the order in which files are named on FILE subcommands determines the order of variables on the new active file. In addition, the order in which files are named determines which input file is used as the source for variables that can be taken from more than one input file (they are taken from the file named first).
- With ADD, the order in which files are named determines the order of cases in the new file (unless the BY subcommand is used).

TABLE Subcommand

One of the input files for a MATCH can be specified as a table file by entering its name on the TABLE subcommand instead of the FILE subcommand.

- A table file contributes variables, not cases, to the new active file.
- The BY subcommand is required when you specify a table file.
- A table file cannot contain duplicate cases (cases for which the key variable(s) have identical values).
- Variables from a table file are added to *all* of the cases from other file(s) which match on the key variable(s).

Example

```
JOIN MATCH FILE=* / TABLE='MASTER.SYS' / BY EMP_ID.
```

- This command adds variables from the system file MASTER.SYS to the current file, matching cases by the variable EMP_ID.

- No new cases are added to the current file as a result of the table lookup.
- Cases whose value for EMP_ID is not included in the table MASTER.SYS are assigned system-missing values for variables taken from the table.

RENAME Subcommand

Use the RENAME command to rename variables on the input files *before* they are processed by JOIN.

- Input system files are not changed on disk; only the copy of the file being joined is affected.
- RENAME applies only to the immediately preceding FILE or TABLE subcommand. To rename variables from more than one input file, enter a RENAME subcommand after each FILE or TABLE subcommand.
- Specifications for RENAME consist of a left parenthesis, a list of old variable names, an equals sign, a list of new variable names, and a right parenthesis. The two variable lists must have the same number of variables.
- You can enter more than one such specification, each enclosed in its parentheses, on a single RENAME subcommand.
- You can enter more than one RENAME subcommand for a single input file. Renaming variables is *not* cumulative: "old' variable names are always those on the input files, regardless of any previous RENAME subcommands.
- RENAME takes effect immediately. Any KEEP and DROP subcommands entered prior to a RENAME must use the old names, while KEEP and DROP subcommands entered after a RENAME must use the new names.
- The new name must be unique within that input file.
- All specifications within a single set of parentheses take effect simultaneously: the specification RENAME (A,B = B,A) is legal and swaps the names of the two variables.
- You can use RENAME to correct a situation where a key variable has different names on different input files. Since BY must be entered last, it always uses the new name of a key variable.
- The TO convention for variable lists is *not* valid on the RENAME subcommand.

Example

```
JOIN MATCH FILE='UPDATE.SYS'
/ RENAME=(NEWPHONE, NEWID = PHONE, ID)
/FILE='MASTER.SYS'
/BY ID.
```

- This example matches a master system file (MASTER.SYS) with an update system file (UPDATE.SYS).
- Two variables on UPDATE.SYS are renamed prior to the match. NEWPHONE is renamed PHONE to combine it with variable PHONE on the master file. NEWID is renamed ID so that it will have the same name as the identification variable in the master file and can be used on the BY subcommand.
- The BY subcommand ensures that only cases with the same ID will be joined.

DROP and KEEP Subcommands

Use DROP and KEEP to include only a subset of variables on the new active file. DROP specifies a set of variables to exclude, and KEEP specifies a set of variables to retain. One or the other will probably be easier to use, depending on how many variables you are excluding or keeping.

- Specifications consist of a list of variables on the input file separated by spaces or commas.
- DROP and KEEP apply only to the immediately preceding FILE or TABLE subcommand.
- The TO convention for variable lists is *not* valid on DROP or KEEP.
- KEEP does not affect the order of variables on the new active file. Variables that are kept retain their original order.

Example

```
MATCH FILE='THIS.SYS'
/RENAME (ONE_VAR,TWO_VAR = VAR1,VAR2)
/KEEP=ID,VAR1,VAR2
/FILE='THAT.SYS'
/KEEP=ID,ONE,TWO
/RENAME (ONE,TWO = VAR1,VAR2)
/BY ID.
```

- This example uses MATCH as an alias for JOIN MATCH.
- For the first file (THIS.SYS), the KEEP subcommand follows RENAME and therefore must use the new names.
- For the second file (THAT.SYS), the KEEP subcommand precedes the RENAME and therefore must use the old names.
- After the RENAME subcommands, each input file has variables named VAR1 and VAR2. These variables will therefore be combined, taking data from THIS.SYS for all cases present in that file and from THAT.SYS for cases absent in the THIS.SYS.

Example

```
JOIN ADD FILE='SAMPLE1.SYS'
/RENAME (SCALE=SCALE1)
/FILE='SAMPLE2.SYS'
/RENAME (SCALE=SCALE2)
/FILE=*
/RENAME (SCALE=SCALE3)
/DROP=TEMP1 TEMP2 TEMP3
/BY DATE.
```

- In this example RENAME is used to give *different* names to variables that would otherwise be combined.
- The DROP subcommand lists each variable individually since the TO convention is not valid.
- The new active file will be sorted by DATE, assuming that each of the three input files is already sorted by DATE.

MAP Subcommand

The MAP subcommand produces a map showing which variables are on the new active file and from what file or files they may be taken.

- More than one MAP subcommand can be entered as part of a single JOIN command.
- MAP can be specified anywhere on the JOIN command. However, the map will reflect only the subcommands that precede the MAP subcommand.
- To obtain a map of the new active file in its final state, specify MAP last.

LIST

```
LIST [[VARIABLES=]{ALL**   }]
                  {varlist}

  [/CASES=[FROM {1**}] [TO {eof**}] [BY {1**}] ]
                {n  }      {n    }      {n  }

  [/FORMAT=[{UNNUMBERED**}] [{WRAP**}] [WEIGHT] ]
            {NUMBERED    }    {SINGLE}
```

**Default if subcommand is omitted.

Example:

```
LIST VARIABLES=XVAR AVAR /CASES=FROM 10 TO 100 BY 2.
```

Overview

The LIST procedure displays the values of variables for cases in the active file.

Defaults

By default, all user-defined variables are listed for all cases in the active file. The listing for each case uses as many lines as needed.

Tailoring

You can specify a list of variables and request that the listing be numbered. You can also limit the listing to a particular sequence of cases. In addition, you can list values of system variables for each case in the active file and limit each case listing to a single line.

Syntax

- The minimum specification is simply the command keyword.
- All subcommands are optional.
- A subcommand or its abbreviation may not be recognized as such if a variable name is identical to it, unless the subcommand is followed by an equals sign.
- Subcommands can be specified in any order and are separated by slashes.

Operations

- LIST causes the data to be read.
- LIST uses the dictionary print formats (see Universals: Variable Format).
- If a value is longer than the format, the decimal portion will be rounded. If that fails, asterisks (*) are printed.
- The LIST display uses the width specified on SET.
- If a long string variable cannot be listed within the page width, it is truncated.
- Values of the variables listed for a case are always separated by at least one blank.
- System-missing values are listed as a period (.).
- For case listings that fit on one line, the column width for each variable is determined by the length of the variable name or the format, whichever is greater. If the variable names do not fit on one line, they are printed vertically.
- If case listings require more than one line, they are wrapped. LIST displays a table illustrating the location of the variables in the listing and prints the name of the first variable in each line at the beginning of the line.
- Each execution of LIST begins at the top of a new page.
- System variables will not be printed unless explicitly requested.

Limitations

- Maximum 1 each of the VARIABLES, CASES, and FORMAT subcommands.

Example

```
LIST VARIABLES=XVAR AVAR /CASES=FROM 10 TO 100 BY 2.
```

- This example produces a list of every second case starting with Case 10 and stopping at 100 for variables XVAR and AVAR.

VARIABLES Subcommand

By default, all user-defined variables in the active file are listed. The optional VARIABLES subcommand allows you to limit the listing to variables you specify or to specify the default explicitly with keyword ALL.

- Variables named on VARIABLES must already exist.
- Variables are listed in the order they are named.
- If a variable is named more than once it is listed more than once.
- The actual keyword VARIABLES is optional.
- You cannot enter the VARIABLES subcommand without specifications.

ALL *List all user-defined variables.* Variables are listed in the order in which they appear in the active file. This is the default if the VARIABLES subcommand is omitted.

CASES Subcommand

By default, all cases in the active file are listed. Use the CASES subcommand to limit the number and pattern of cases listed.

- If you omit CASES or include it with no specifications, all cases in the file are listed.
- Defaults that you do not change remain in effect.

The following may be specified on CASES:

FROM n *The case number of the first case to be listed.* The default is 1.

TO n *Upper limit on the cases to be listed.* The default is the end of the active file. CASES 100 is interpreted as CASES TO 100.

BY n *Increment used to choose cases for listing.* The default is 1.

FORMAT Subcommand

The default display does not number cases and uses more than one line per case if necessary. Use the optional FORMAT subcommand to change the defaults.

- If you omit the FORMAT subcommand or use it without any specifications, the default display is produced.
- Defaults that you do not change remain in effect.

The following specifications are available for FORMAT:

NUMBERED *Include the sequence number of each case in the listing.* The sequence number is displayed to the left of the values listed. The default is UNNUMBERED.

SINGLE *Limit each case listing to one line.* If the variables requested do not fit on a single line, LIST is not executed. The default is WRAP.

WEIGHT *List the value of the case's weight in the active file.*

MEANS

```
MEANS [TABLES=] {varlist} BY varlist [BY varlist ...] [/varlist...]
                {ALL    }

      [/OPTIONS=option numbers]

      [/STATISTICS={statistic numbers}]
                   {ALL               }
```

Options:

1 Include user-missing values
2 Exclude cases with user-missing dependent values
3 Suppress all labels
5 Suppress group counts
6 Display group sums
7 Suppress group standard deviations
8 Suppress value labels
9 Suppress independent variable names
10 Suppress independent variable values
11 Suppress group means
12 Display group variances

Statistics:

1 One-way analysis of variance
2 Test of linearity

Example:

```
MEANS TABLES=VAR1 TO VAR5 BY GROUP
/OPTIONS=9
/STATISTICS=1.
```

Overview

MEANS displays means, standard deviations, and group counts for a dependent variable within groups defined by one or more independent variables. Other SPSS/PC+ procedures that display univariate statistics are FREQUENCIES and DESCRIPTIVES.

Defaults

By default, MEANS displays means, standard deviations, and number of cases. The default table is labeled with the variable name and label of the dependent and independent variables. Groups are labeled with the variable name, variable label, values, and value labels of the independent variables. Cases that have missing values on any variables that define a table are excluded from the statistics calculated for that table.

Tailoring

Display Format. You can suppress the printing of all variable and value labels, value labels only, names of independent variables, and values of independent variables.

Statistical Display. Statistical display options allow you to display group sums and variances or to suppress group counts, group standard deviations, and group means. A one-way analysis of variance and a test of linearity are also available.

Missing Values. You can include cases with user-missing values on the dependent or independent variables. Alternatively, you can include cases with user-missing values on the independent variable only.

Syntax

- The minimum specification is a TABLES subcommand with a tables list. The actual keyword TABLES may be omitted.
- The minimum tables list specifies a dependent variable, the keyword BY, and an independent variable.
- Subcommands are separated by slashes.

Operations

- MEANS causes the data to be read.
- MEANS displays requested univariate statistics for the population as a whole and for each value of the first independent variable defined for the table in addition to statistics for groups.
- If an independent variable is a long string, only the short-string portion is used to identify groups in the analysis.

- Specifying a string variable as a dependent variable on any tables list stops execution of the MEANS procedure.
- Statistics are displayed with four decimal places where possible.
- The display uses the width defined on the SET command. If the statistics requested cannot fit within the available width, the command is not executed. You can use the OPTIONS subcommand to tailor the display to fit within the defined width.

Limitations

- The number of variables allowed per MEANS command is the same as the system limit.
- Maximum 250 tables can be produced.
- Maximum 30 TABLES subcommands.
- Maximum 1 each of the OPTIONS and STATISTICS subcommands.
- Maximum 6 dimensions (5 BY keywords) per TABLES subcommand.
- Maximum 200 value labels are displayed on any single table.

Example

```
MEANS TABLES=VAR1 TO VAR5 BY GROUP
/OPTIONS=9
/STATISTICS=1.
```

- In this example, the TABLES subcommand specifies that VAR1 through VAR5 are the dependent variables. GROUP is the independent variable.
- Assuming variables VAR2, VAR3, and VAR4 lie between between VAR1 and VAR5 on the active file, five tables are produced: VAR1 by GROUP; VAR2 by GROUP; VAR3 by GROUP, and so on.
- Option 9 suppresses the printing of variable name GROUP.
- Statistic 1 requests one-way analysis of variance tables of VAR1 through VAR5 by GROUP.

Example

```
MEA VARA BY VARB BY VARC/VAR1 VAR2 BY VAR3 VAR4 BY VAR5.
```

- This command contains two TABLES subcommands that omit the optional TABLES keyword.
- The first tables list requests one table. Statistics are produced for VARA within groups defined by each combination of values of VARB and VARC.
- The second tables list requests four tables: VAR1 by VAR3 by VAR5; VAR1 by VAR4 by VAR5; VAR2 by VAR3 by VAR5; and VAR2 by VAR4 by VAR5.
- This example takes advantage of spelling permitted by three-character truncation of keywords.

TABLES Subcommand

The TABLES subcommand specifies a tables list. The actual keyword TABLES may be omitted.

- The dependent variable is named first and must be numeric. The independent variables follow the BY keyword and can be numeric or string.
- You can specify more than one dependent variable in a tables list.
- Each use of the keyword BY in a tables list adds a dimension to the tables requested and introduces a new order of control among the independent variables.
- You can specify more than one independent variable in each dimension of a tables list.
- A table is built for each dependent variable by each combination of independent variables across dimensions.
- Each combination of values of the independent variables defined for a table defines a group.
- The order in which independent variables are displayed is the same as the order in which they are named. The values of the first independent variable defined for the table appear in the left-most column of the table and change most slowly in the definition of groups.

- You can use keyword ALL in each dimension to refer to all user-defined variables.
- You can specify multiple TABLES subcommands on a single MEANS command. The slash between the subcommands is required; the keyword is not.

Display Format

By default, MEANS displays the variable name and variable labels of both independent and dependent variables at the beginning of each table. Within the table, groups defined by the independent variables are identified by variable name, values, and value labels. Specify the following on the OPTIONS subcommand to change these defaults:

Option 3 *Suppress all labels.* No variable or value labels are displayed for either the independent or dependent variables.

Option 8 *Suppress value labels.* No value labels are printed for the independent variables.

Option 9 *Suppress independent variable names.*

Option 10 *Suppress independent variable values.*

Statistical Display

By default, MEANS displays means, standard deviations, and number of cases. Statistics are displayed for groups and for the entire population. Specify the following on the OPTIONS and STATISTICS subcommands to change these defaults:

Option 5 *Suppress group counts.* The number of cases in each group is not displayed.

Option 6 *Display group sums.*

Option 7 *Suppress group standard deviations.*

Option 11 *Suppress group means.*

Option 12 *Display group variances.*

Statistic 1 *One-way analysis of variance including eta and eta*^{2}. The analysis of variance is performed for the first independent variable defined for the table only.

Statistic 2 *Test of linearity.* Produces a one-way analysis of variance in which the between-groups sum of squares is subdivided into linear and nonlinear components. Pearson's r and r^2 are displayed as part of the test of linearity. The analysis of variance is performed for the first independent variable defined for the table only, and the test of linearity is ignored if the independent variable is a string.

ALL *Display all statistics.* Produces the same display as Statistic 2. Specify ALL on the STATISTICS subcommand.

Example

```
MEANS TABLES=INCOME81 BY AGECAT BY SEX
/STATISTICS=1.
```

- This example requests statistics for INCOME81 for groups defined by values of SEX within AGECAT.
- Statistic 1 requests an analysis of variance of INCOME81 by AGECAT.

Missing Values

By default, MEANS excludes cases that have missing values for any variables that define a table. You change the handling of missing values by specifying the following on the OPTIONS subcommand:

Option 1 *Include cases with user-missing values.* Cases with user-missing values on the independent or dependent variable are included.

Option 2 *Exclude cases with user-missing dependent values.* Cases with user-missing values for the dependent variable are excluded. Cases with user-missing values for independent variables are included.

MISSING VALUE

```
MISSING VALUE  {varlist}([value]) [[/]varlist ...]
               {ALL    }
```

Example:

```
MISSING VALUE XVAR (8)/ YVAR ZVAR (0)/ AVAR ('     ').
```

Overview

The MISSING VALUE command declares values for numeric and short string variables as user-missing. These values can then be treated specially in data transformations, statistical calculations, and case selection. By default, user-missing values are treated the same as the system-missing values. System-missing values are assigned automatically when no legal value can be assigned, as when input data for a numeric field are blank or when an illegal calculation is requested.

Syntax

- The minimum specification is a single variable followed by a value in parentheses.
- Each variable can have only one user-missing value.
- You can declare the same value as missing for more than one variable by specifying a variable list followed by the value in parentheses.
- You can declare different values as missing for different variables by specifying separate variable lists for each value. The slash between different specifications is optional.
- The missing-value specification must correspond to the variable type (numeric or string).
- You cannot assign missing values to long strings or system variables.
- Variable lists must have either all numeric or all string variables.
- Missing values for short string variables must be enclosed in apostrophes or quotation marks (see Universals: Strings).
- Keyword ALL can be used to refer to all user-defined variables in the active file provided the variables are either all numeric or all string.
- The TO keyword can be used to refer to consecutive variables on the active file.
- The missing-value specification for short string variables must include any leading or trailing blanks.
- More than one MISSING VALUE command can be specified per session.

Operations

- MISSING VALUE is a transformation command and is executed when the data are read for the next procedure.
- If a variable is mentioned more than once on one or more MISSING VALUE commands before a procedure, only the last specification is used.
- Missing-value specifications can be changed between procedures. New declarations replace previous ones.
- A variable list followed by an empty value specification (an empty set of parentheses) deletes any missing-value declarations for those variables.
- Missing-value declarations are saved in system files (see SAVE) and portable files (see EXPORT).

Example

```
MISSING VALUE XVAR (8)/ YVAR ZVAR (0)/ AVAR ().
```

- Value 8 is declared missing for numeric variable XVAR.
- Value 0 is missing for numeric variables YVAR and ZVAR.
- Any previously declared missing values for AVAR are deleted by the empty value specification.

Example

```
MIS VAL NAME1 TO NAME7 ('    ')/LIKE1 TO DLIKE7 (0).
```

- Blanks are declared missing for the variables between and including NAME1 and NAME7. All these variables must be string variables four columns wide.
- The value 0 is declared missing for the variables between and including LIKE1 and DLIKE7. All variables in this list must be numeric.
- This example takes advantage of spelling permitted by three-character truncation of keywords.

MODIFY VARS

This command available as of SPSS/PC+ V2.0.

```
MODIFY VARS [/REORDER = [{FORWARD**}] [{POSITIONAL**}] [(varlist)]]
                         {BACKWARD }   {ALPHA       }   (ALL**   )

            [/DROP = varlist]      [/KEEP = {varlist}]
                                   [        {ALL**  }]

            [/RENAME = (old varlist = new varlist) [(...)] ]

            [/MAP]
```

**Default if subcommand is omitted.

Example:

```
MODIFY VARS /RENAME (V1 TO V3 = ID GROUP AGE) /MAP.
```

Overview

The MODIFY VARS command lets you change the names or the order of the variables in your active file or drop variables from your active file.

Syntax

- The minimum specification is any one subcommand.
- Subcommands can be entered in any order and must be separated by slashes.
- MAP is the only subcommand that can be entered more than once.
- Variable lists on all subcommands refer to the original variable names (before renaming) and the original variable order (before reordering).
- You cannot reorder, rename, or drop a system variable ($CASENUM, $WEIGHT, or $DATE).
- You cannot specify both DROP and KEEP.

Operations

- MODIFY VARS causes the data to be read if (and only if) you specify DROP, KEEP, or REORDER.
- Specifications on all subcommands use the original variable names and reference the original variable order. Thus MODIFY VARS subcommands are *not* cumulative.
- Reordered or renamed variables retain variable labels, value labels, and missing-value and format specifications.
- MODIFY VARS affects your current active file. It does not affect any permanent disk file unless you use SAVE or some other command to write the file to disk.

Example

```
MODIFY VARS /RENAME (V1 TO V3 = ID GROUP AGE) /MAP.
```

- This example renames V1 to ID, V2 to GROUP, and V3 to AGE.
- SPSS/PC+ displays a map of the variable dictionary after renaming the three variables.

REORDER Subcommand

The REORDER subcommand allows you to change the order of the variables in your active file. This is useful in

- Grouping related variables so you can refer to all of them together on other commands with the TO keyword.
- Simplifying the tabulation of variables in alphabetic or some other order.

The REORDER subcommand allows you to specify the *direction* of ordering, the *type* of ordering, and the *variables* to be ordered. You can enter one, two, or all three kinds of specification. If you enter more than one, you must enter the direction first, then the type, and then the variables.

Specify the direction of ordering with:

FORWARD *For type POSITIONAL, the current order of variables in the active file. For type ALPHA, alphabetic order.* This is the default.

BACKWARD *For type POSITIONAL, the opposite of the current order of variables in the active file. For type ALPHA, reverse alphabetic order.*

Specify the type of ordering with:

POSITIONAL *The existing order of variables in the active file.* This is the default.

ALPHA *Alphabetic order, according to the variable names in the input file.* (The RENAME subcommand has no effect on this order, even if it precedes the REORDER subcommand.) For variable names with numeric suffixes, such as V1 TO V100, ALPHA refers to the numeric order of the suffixes rather than to strict lexicographic order; thus V9 precedes V10.

Specify the variables to be reordered in parentheses or specify ALL.

- The default is ALL.
- If you name one or more variables on REORDER, those variables are placed at the beginning of the new active file in the specified order. Variables not named are placed at the end of the new active file and retain their previous order among themselves.

Example

```
MODIFY VARS REORDER ALPHA (MAINE TO HAWAII) / MAP.
```

- This example arranges all of the variables from MAINE to HAWAII into alphabetic order.

After a variable list in parentheses, you can specify another direction and/or type of ordering and another variable list in parentheses, provided that the variable lists do not overlap. Variables named in the first set of parentheses precede those named in the second set of parentheses. Variables not named at all come at the end of the new active file.

Example

```
MODIFY VARS REORDER (ID SSNUMBER) ALPHA (GROUP TO SALARY).
```

- This example places ID and SSNUMBER at the beginning of the new active file. Since neither direction nor type of ordering is specified, the defaults FORWARD and POSITIONAL are used. Whichever of the two comes first in the old file will come first in the new file.
- After ID and SSNUMBER, all variables from GROUP to SALARY are arranged in alphabetic order by variable name.
- Any other variables in the file will come at the end of the new file, retaining their current order.

RENAME Subcommand

The RENAME subcommand allows you to assign new names to one or more variables in the active file.

- Specifications for RENAME consist of a list of existing variable names, an equals sign, and a list of new variable names (which must have as many names as the first list). All of these specifications must appear in parentheses.
- You can enter more than one set of parentheses containing old names, an equals sign, and new names on a RENAME subcommand.
- You cannot enter more than one RENAME subcommand on a MODIFY VARS command.
- All of the new names on a RENAME subcommand are assigned simultaneously. You can interchange variable names, as shown below.
- You can create a series of new names with numerical suffixes using the keyword TO, as shown below.
- SPSS/PC+ does *not* have to read the data to assign new names to variables.

Example

```
MODIFY VARS RENAME (PLAN, ALTERNAT = ALTERNAT, PLAN)
 (TASKA TO TASKJ = OLDJOB1 TO OLDJOB10)
 (ALT1 TO ALT10 = TASKA TASKB TASKC TASKD TASKE TASKF TASKG
                  TASKH TASKI TASKJ)
 /MAP.
```

- This command switches the names of the two variables PLAN and ALTERNAT.
- Existing variables from TASKA to TASKJ are renamed as OLDJOB1, OLDJOB2, ..., OLDJOB10.
- Existing variables from ALT1 to ALT10 are renamed as TASKA, TASKB, ..., TASKJ. Note that TO cannot be used on this list of new variable names, since the new names do not have numeric suffixes.

KEEP Subcommand

The KEEP subcommand lets you specify which variables to retain in your active file. Reducing the number of variables in the active file speeds processing, saves the disk space used (temporarily) to hold the active file, and enables you to create additional variables without exceeding the system limit.

- After KEEP, enter the names of the variables you want to retain in the active file.
- KEEP is an alternative to DROP. Use whichever is easier, but not both.

DROP Subcommand

The DROP subcommand lets you specify which variables to drop from your active file. Reducing the number of variables in the active file speeds processing, saves the disk space used (temporarily) to hold the active file, and enables you to create additional variables without exceeding the system limit.

- After DROP, enter the names of the variables you want to drop from the active file.
- DROP is an alternative to KEEP. Use whichever is easier, but not both.

MAP Subcommand

The MAP subcommand displays a table showing the names and order of the variables in your file at the current point in the MODIFY VARS specifications and before you issued the MODIFY VARS command.

Use MAP to verify that MODIFY VARS has produced the results you expect.

- MODIFY VARS subcommands that you enter *after* MAP are not reflected in the table.
- You can use MAP more than once to see the state of the active file at different points in the MODIFY VARS specifications.
- You should normally place MAP at the end of the command, to see the "final" state of the new active file.

Example

```
MODIFY VARS DROP ERR_1 TO ERR_40
 /RENAME (VSCORE MSCORE = SCOREV SCOREM)
 /MAP.
MODIFY VARS /REORDER (ID DEPT) ALPHA (AREA TO EVAL87)
 /MAP.
```

- Variables between ERR_1 and ERR_40 are dropped from the active file.
- The two variables VSCORE and MSCORE are renamed SCOREV and SCOREM. At this point, a table shows the results of the deletion and renaming of variables.
- The second MODIFY VARS command places ID and DEPT as the first two variables in the new file, followed by the variables between AREA and EVAL87 in alphabetic order, followed by any other variables in the file in their current order.
- A final table showing the names and order of variables in the new active file is displayed.

Two MODIFY VARS commands were required to alphabetize the variables using their new names, since the REORDER subcommand uses the names as they existed prior to the command.

N

```
N n [ESTIMATED]
```

Example:

```
N 100.
```

Overview

The N command limits the number of cases in the active file to the first *n* cases. You can also use N with keyword ESTIMATED to provide SPSS/PC+ with information about the estimated number of cases in your data file. SPSS/PC+ uses the estimate to optimize use of memory. In this case, the number of cases processed is *not* limited to the estimated number.

Syntax

- The number of cases must be a positive integer.
- The N command keyword and the integer must be separated by at least one space.
- N can be entered at any point in an SPSS/PC+ session and can be used more than once.
- The keyword ESTIMATED is specified after the integer estimate of the number of cases.

Operations

- N is a transformation command and is executed when the data are read for the next procedure.
- N limits the number of cases analyzed by all subsequent procedures.
- Without the N command, SPSS/PC+ processes all cases.
- N controls the building of cases, not the reading of individual data records.
- SPSS/PC+ stops processing input data when N is reached. If keyword ESTIMATED is used, all input data are processed.
- Any SAMPLE, PROCESS IF, or SELECT IF commands are executed before cases are counted toward the limit specified on N.
- If N is specified more than once in an SPSS/PC+ job, each N command places a limit on the number of cases available for later procedures only if a new active file is created by data transformations. Otherwise, the eliminated cases are still on the active file and can be restored for subsequent procedures using the N command with a larger number.
- You cannot increase the size of the active file by specifying a value for N greater than the number of cases written to the active file. An active file is written when a raw data file is processed by a DATA LIST command and again when data transformations are processed.
- The keyword ESTIMATED allows SPSS/PC+ to optimize the allocation of memory for some procedures.

Example

```
N 100.
```

- This example limits the number of cases in the active file to the first 100 cases.

Example

```
DATA LIST FILE='INVENT.DAT'/ITEM1 TO ITEM30 1-60.
N 400 ESTIMATED.
FREQ VAR=ITEM24.
N 23.
SELECT IF (ITEM11 EQ 8).
LIST VAR=ITEM1, ITEM12 TO ITEM18, ITEM24.
```

- The DATA LIST command defines 30 variables in the file INVENT.DAT in the current directory.
- The N command with keyword ESTIMATED tells SPSS/PC+ to allocate memory for processing approximately 400 cases.
- The FREQUENCIES procedure produces a frequency table for ITEM24 that includes all cases.

- The second N command limits the number of cases in the active file to 23 after selecting cases that have a value of 8 for variable ITEM11, as specified on the subsequent SELECT IF command.
- The SELECT IF command causes a new active file to be created. Therefore, cases for which ITEM11 does not equal 8 and all cases beyond the 23rd case are lost to the session (unless the original active file is recreated with a new DATA LIST).
- LIST produces a listing of the values of ITEM1, ITEM24, and all variables between and including ITEM12 and ITEM18 for the remaining cases in the active file.

NPAR TESTS

```
NPAR TESTS  [CHISQUARE=varlist [(lo,hi)]]

  [/EXPECTED={EQUAL**       }]
             {f1,f2,  ...fn}

  [/K-S ({UNIFORM [,lo,hi]})=varlist]
         {NORMAL [,m,sd]   }
         {POISSON [,m]     }

  [/RUNS ({MEAN  })=varlist]
          {MEDIAN}
          {MODE  }
          {value }

  [/BINOMIAL [({.5})]=varlist ({value1,value2})]
               {p }            {value         }

  [/MCNEMAR=varlist [WITH varlist]]

  [/SIGN=varlist [WITH varlist]]

  [/WILCOXON=varlist [WITH varlist]]

  [/COCHRAN=varlist]

  [/FRIEDMAN=varlist]

  [/KENDALL=varlist]

  [/MEDIAN [(value)]=varlist BY var (value1,value2)]

  [/M-W=varlist BY var (value1,value2)]

  [/K-S=varlist BY var (value1,value2)]

  [/W-W=varlist BY var (value1,value2)]

  [/MOSES[(n)]=varlist BY var (value1,value2)]

  [/K-W=varlist BY var (value1,value2)]

  [/OPTIONS=option numbers]

  [/STATISTICS=statistic numbers]
```

**Default if subcommand is omitted.

Options:

1 Include user-missing values
2 Exclude missing values listwise
3 Sequential pairing of variables for two related samples
4 Random sampling

Statistics:

1 Mean, maximum, minimum, standard deviation, and count
2 Quartiles and count

Example:

```
NPAR TESTS K-S(UNIFORM)=AVAR/ K-S(NORMAL,0,1)=BVAR.
```

Overview

Procedure NPAR TESTS is a collection of nonparametric tests that make minimal assumptions about the underlying distribution of data. All of these tests are described in Siegel (1956).

The tests available in NPAR TESTS can be grouped into three broad categories, based on how the data are organized: one-sample tests, related-samples tests, and independent-samples tests. A one-sample test analyzes one variable. A test for related samples compares two or more variables for the same set of cases. An independent-samples test analyzes one variable grouped by categories of another variable.

The one-sample tests available in procedure NPAR TESTS are

- BINOMIAL.
- CHISQUARE.
- K-S (Kolmogorov-Smirnov).
- RUNS.

Tests for two related samples are

- MCNEMAR.
- SIGN.
- WILCOXON.

Tests for *k* related samples are

- COCHRAN.
- FRIEDMAN.
- KENDALL.

Tests for two independent samples are

- M-W (Mann-Whitney).
- K-S (Kolmogorov-Smirnov).
- W-W (Wald-Wolfowitz).
- MOSES.

And the tests for *k* independent samples are

- K-W (Kruskal-Wallis).
- MEDIAN.

Tests are described below in alphabetical order.

Defaults

There are no default tests; each test must be requested by its subcommand keyword. By default, cases with missing values are deleted on a test-by-test basis within subcommands.

Tailoring

Statistical Display. In addition to the tests, you can request univariate statistics, quartiles, and counts for all variables named on the command. You can also control the pairing of variables in two-related-samples tests.

Random Sampling. NPAR TESTS must store cases in memory for tests that use ranks. You can use random sampling when there is not enough space to store all cases.

Missing Values. You can include cases with user-missing values in all tests. Optionally, you can exclude cases with missing values for any variable named on any subcommand from all tests.

Syntax

- The minimum specification is a single test subcommand and its arguments.
- Each test subcommand specifies a test and a list of variables to be tested. Some tests require additional specifications. CHISQUARE has an optional subcommand.

- The OPTIONS and STATISTICS subcommands are optional. Each can be specified only once per NPAR TESTS command.
- You can request any or all tests, and you can specify a test subcommand more than once on a single NPAR TESTS command.
- Subcommands must be separated by slashes.
- You can use keyword ALL in any variable list to refer to all user-defined variables in the active file.
- Keyword WITH controls pairing of variables in two-related-samples tests.
- Keyword BY introduces the grouping variable in two- and *k*-independent-samples tests.

Operations

- NPAR TESTS causes the data to be read.
- The display always uses narrow format.
- Specifying a string variable on any subcommand will stop execution of NPAR TESTS.
- When ALL is used, requests for tests of variables with themselves are ignored and a warning is printed.

Limitations

- The amount of memory required is directly proportional to the number of cases being analyzed.
- Maximum 1 each OPTIONS and STATISTICS subcommands.
- Maximum 100 subcommands per NPAR TESTS command.
- The maximum number of variables is the same as the system limit.
- The maximum range of values on the CHISQUARE subcommand is 200.

BINOMIAL Subcommand

```
NPAR TESTS BINOMIAL [({.5})]=varlist[({value,value})]
                      {p }              {value      }
```

BINOMIAL tests whether the observed distribution of a dichotomous variable is the same as that expected from a specified binomial distribution. By default, each variable named is assumed to have only two values, and the distribution of each variable named is compared to a binomial distribution with *p* (the proportion of cases expected in the first category) equal to 0.5. The default display includes the number of valid cases in each group, the test proportion, and the two-tailed probability of the observed proportion.

Syntax

- The minimum specification is a list of variables to be tested.
- To change the default 0.5 test proportion, specify a value in parentheses immediately after the BINOMIAL subcommand keyword.
- A single value in parentheses following the variable list is used as a cutting point. Cases with values equal to or less than the cutting point form the first category; the remaining cases form the second.
- If two values appear in parentheses after the variable list, cases with values equal to the first value form the first category and cases with values equal to the second value form the second category.
- If no values are specified, the variables must be dichotomous.

Operations

- The proportion observed in the first category is compared to the test proportion. Then, the probability of the observed proportion occurring given the test proportion and a binomial distribution is computed.
- If the test proportion is the default (0.5), a two-tailed probability is displayed. For any other test proportion, a one-tailed probability is displayed. The output label will always say two-tailed, however.
- A test statistic is calculated for each variable.

Example

```
NPAR TESTS BINOMIAL(.667)=YVAR(0,1).
```

- This example requests the one-tailed probability that, for cases having value 0 or 1 for YVAR, the proportion with value 0 is greater or less than 0.667.

CHISQUARE Subcommand

```
NPAR TESTS CHISQUARE=varlist [(lo,hi)] [/EXPECTED={EQUAL**     }]
                                                 {f1,f2,... fn}
```

The CHISQUARE (alias CHI-SQUARE) one-sample test tabulates a variable into categories and computes a chi-square statistic based on the differences between observed and expected frequencies. By default, equal frequencies are expected in each category. The display includes the frequency distribution, expected frequencies, residuals, chi-square, degrees of freedom, and probability.

Syntax

- The minimum specification is a list of variables to be tested.
- Optionally, you can specify a value range in parentheses following the variable list.
- You can also specify expected proportions with the EXPECTED subcommand.
- If you use the EXPECTED subcommand to specify unequal expected frequencies, you must specify a value greater than 0 for each observed category of the variable or the keyword EQUAL.
- The expected frequencies are specified in ascending order of category value.
- You can use the notation *n*f* to indicate that frequency *f* is expected for *n* consecutive categories.
- Specifying keyword EQUAL on the EXPECTED subcommand has the same effect as omitting the EXPECTED subcommand.
- EXPECTED applies to all variables named on the CHISQUARE subcommand.
- Use multiple CHISQUARE and EXPECTED subcommands to specify different expected proportions for variables.
- You can request CHISQUARE with its alias CHI-SQUARE.

Operations

- If no range is specified for the variables to be tested, each distinct value encountered defines a category.
- If a range is specified, integer-valued categories are established for each value within the range. Noninteger values are truncated before classification. Cases with values outside the specified range are excluded.
- EXPECTED values are interpreted as proportions, not absolute values. Values are summed, and then each value is divided by the total to calculate the proportion of cases expected in the corresponding category.
- A test statistic is calculated for each variable named.

Example

```
NPAR TESTS CHISQUARE=AVAR (1,5)/ EXPECTED= 12, 3*16, 18.
```

- This example requests the chi-square test for values 1 through 5 of variable AVAR.
- The observed frequencies for variable AVAR are compared with the hypothetical distribution of 12/78 occurrences of value 1; 16/78 occurrences each of values 2, 3, and 4; and 18/78 occurrences of value 5.

COCHRAN Subcommand

```
NPAR TESTS COCHRAN=varlist
```

COCHRAN calculates Cochran's *Q*, which tests whether the distribution of values of *k* dichotomous variables is the same for all the variables. The display shows the frequency distribution for each variable, degrees of freedom, and probability.

Syntax

- The minimum specification is a list of two variables.
- Variables must be dichotomous and must be coded with the same two values.

Operations

- A $2 \times k$ contingency table (category vs. variable) is constructed for dichotomous variables and the proportions for each variable are computed.
- Cochran's *Q* statistic has approximately a chi-square distribution.
- A single test comparing all variables is performed.

Example

```
NPAR TESTS COCHRAN=RVAR1 TO RVAR3.
```

- This example tests whether the distribution of values 0 and 1 for RVAR1, RVAR2, and RVAR3 is the same.

FRIEDMAN Subcommand

```
NPAR TESTS FRIEDMAN=varlist
```

FRIEDMAN tests whether *k* related samples have been drawn from the same population. The display shows the mean rank for each variable, number of valid cases, chi-square, degrees of freedom, and probability.

Syntax

- The minimum specification is a list of two variables.
- Variables should be at least at the ordinal level of measurement.

Operations

- The values of *k* variables are ranked from 1 to *k* for each case and the mean rank is calculated for each variable over all cases.
- The test statistic has approximately a chi-square distribution.
- A single test statistic comparing all variables is calculated.

Example

```
NPAR TESTS FRIEDMAN=SVAR TVAR UVAR
/STATISTICS = 1.
```

- This example tests variables SVAR, TVAR, and UVAR, and requests univariate statistics for all three.

K-S Subcommand (One Sample)

```
NPAR TESTS K-S({NORMAL  [mean,stddev]})=varlist
               {POISSON [mean]       }
               {UNIFORM [min,max]    }
```

The K-S (alias KOLMOGOROV-SMIRNOV) one-sample test compares the cumulative distribution function for a variable with a uniform, normal, or Poisson distribution and tests whether the distributions are homogeneous. The parameters of the test distribution can be specified; the defaults are the observed parameters. The display shows the number of valid cases, parameters of the test distribution, most-extreme absolute, positive, and negative differences, K-S *Z*, and two-tailed probability for each variable.

Syntax

- The minimum specification is a distribution keyword and a list of variables. The distribution keywords are

NORMAL *Normal distribution.* Default parameters are the observed mean and standard deviation.

POISSON *Poisson distribution.* The default parameter is the observed mean.

UNIFORM *Uniform distribution.* Default parameters are the observed minimum and maximum values.

- The distribution keyword and its optional parameters must be enclosed within parentheses.
- The distribution keyword must be separated from its parameters by blanks or commas.
- You can request K-S with its alias KOLMOGOROV-SMIRNOV.

Operations

- The Kolmogorov-Smirnov *Z* is computed from the largest difference in absolute value between the observed and test distribution functions.
- The K-S probability levels assume that the test distribution is specified entirely in advance. The distribution of the test statistic and resulting probabilities change when the parameters of the test distribution are estimated from the sample. No correction is made.
- For a mean of 100,000 or larger, a normal approximation to the Poisson distribution is used.
- A test statistic is calculated for each variable.

Example

```
NPAR TESTS K-S(UNIFORM)=AVAR/ K-S(NORMAL,0,1)=BVAR.
```

- The first K-S subcommand compares the distribution of AVAR with a uniform distribution that has the same range as AVAR.

- The second K-S subcommand compares the distribution of BVAR with a normal distribution with a mean of 0 and a standard deviation of 1.

K-S Subcommand (Two Sample)

```
NPAR TESTS K-S=varlist BY variable(value1,value2)
```

K-S (alias KOLMOGOROV-SMIRNOV) tests whether the distribution of a variable is the same in two independent samples defined by a grouping variable. The test is sensitive to any difference in median, dispersion, skewness, and so forth, between the two distributions. The display shows the count in each group, the largest absolute, positive, and negative differences between the two groups, K-S *Z*, and the two-tailed probability for each variable.

Syntax

- The minimum specification is a test variable, the keyword BY, a grouping variable, and a pair of values in parentheses.
- The test variable should be at least at the ordinal level of measurement.
- Cases with the first value form one group and cases with the second value form the other. The order in which values are specified determines which difference is the largest positive and which is the largest negative.
- You can request K-S with its alias KOLMOGOROV-SMIRNOV.

Operations

- The observed cumulative distributions for both groups and the maximum positive, negative, and absolute differences are computed.
- Cases with values other than those specified for the grouping variable are excluded.
- A test statistic is calculated for each variable named before BY.

Example

```
NPAR TESTS K-S=YVAR1 YVAR2 BY NVAR(0,1).
```

- This example specifies two tests. The first compares the distribution of YVAR1 for cases with value 0 for NVAR with the distribution of YVAR1 for cases with value 1 for NVAR.
- A parallel test is calculated for YVAR2.

K-W Subcommand

```
NPAR TESTS K-W=varlist BY variable(value1,value2)
```

K-W (alias KRUSKAL-WALLIS) tests whether *k* independent samples defined by a grouping variable are from the same population. The display shows the number of valid cases, mean rank of the variable in each group, chi-square, probability, and chi-square and probability after correcting for ties.

Syntax

- The minimum specification is a test variable, the keyword BY, a grouping variable, and a pair of values in parentheses.
- Every value in the range defined by the pair of values forms a group.
- You can request K-W with its alias KRUSKAL-WALLIS.

Operations

- Cases from the *k* groups are ranked in a single series and the rank sum for each group is computed.
- Kruskal-Wallis *H* has approximately a chi-square distribution.
- Cases with values other than those specified for the grouping variable are excluded.
- A test statistic is calculated for each variable named before BY.

Example

```
NPAR TESTS K-W=YVAR BY IVAR(0,4).
```

- This example tests YVAR for groups defined by values 0 through 4 of IVAR.

KENDALL Subcommand

```
NPAR TESTS KENDALL=varlist
```

KENDALL tests whether *k* related samples are from the same population. *W* is a measure of agreement among judges where each case is one judge's rating of several entities (variables). The display includes the mean rank for each variable, valid count, *W*, chi-square, degrees of freedom, and probability.

Syntax

- The minimum specification is a list of two variables.

Operations

- The values of the k variables are ranked from 1 to k for each case and the mean rank is calculated for each variable over all cases.
- Kendall's W and a corresponding chi-square statistic are calculated, correcting for ties.
- W ranges between 0 (no agreement) and 1 (complete agreement).
- A single test statistic is calculated for all variables.

Example

```
DATA LIST /XVAR1 TO XVAR5 1-10.
BEGIN DATA.
2 5 4 5 1
3 3 4 5 3
3 4 4 6 2
2 4 3 6 2
END DATA.
NPAR TESTS KENDALL=ALL.
```

- This example tests four judges (cases) on five entities (variables XVAR1 through XVAR5).

M-W Subcommand

```
NPAR TESTS M-W=varlist BY variable(value1,value2)
```

M-W (alias MANN-WHITNEY) compares two independent samples defined by a grouping variable on a single test variable. The test statistic uses the rank of each case to test whether the groups are drawn from the same population. The display shows the mean rank of the variable within each group, the valid count for each group, the Mann-Whitney U, Wilcoxon W (the rank sum of the smaller group), the two-tailed probability of U (or W), the Z statistic, and the two-tailed probability of Z corrected for ties.

Syntax

- The minimum specification is a test variable, the keyword BY, a grouping variable, and a pair of values in parentheses.
- Cases with the first value form one group and cases with the second value form the other. Order is unimportant.
- You can request M-W with its alias MANN-WHITNEY.

Operations

- Cases are ranked in order of increasing size, and test statistic U—the number of times a score from Group 1 precedes a score from Group 2—is computed.
- For fewer than than 30 cases, an exact significance level is computed.
- For more than 30 cases, U is transformed into a normally distributed Z statistic.
- Cases with values other than those specified for the grouping variable are excluded.
- A test statistic is calculated for each variable named before BY.

Example

```
NPAR TESTS M-W=YVAR BY XVAR(1,2).
```

- This example tests YVAR based on the two groups defined by values 1 and 2 of XVAR.

MCNEMAR Subcommand

```
NPAR TESTS MCNEMAR=varlist [WITH varlist]
[/OPTIONS=3]
```

MCNEMAR tests whether the changes in proportions are the same for pairs of dichotomous variables. The display shows the 2 × 2 contingency table, number of valid cases, and two-tailed probability for each pair of variables.

Syntax

- The minimum specification is a list of two variables.
- Variables must be dichotomous and must have the same two values.
- Without keyword WITH, each variable pair in the list is tested.
- With keyword WITH, each variable before WITH is tested with each variable following WITH.
- With Option 3 and no WITH, the first variable is paired with the second, the second with the third, the third with the fourth, and so on.
- With Option 3 and WITH, the first variable before WITH is paired with the first variable after WITH, the second variable before WITH with the second variable after WITH, and so on.

- With Option 3 and WITH, the number of variables specified before and after WITH must be the same.

Operations

- A 2 × 2 table is constructed for each pair of dichotomous variables and a chi-square statistic is computed for cases having different values for the two variables.
- If fewer than 10 cases change values from the first variable to the second variable, the binomial distribution is used to compute the probability.

Example

```
NPAR TESTS MCNEMAR=YVAR1 YVAR2 YVAR3.
```

- This example performs the MCNEMAR test on variable pairs YVAR1 and YVAR2, YVAR1 and YVAR3, and YVAR2 and YVAR3.

MEDIAN Subcommand

```
NPAR TESTS MEDIAN [(value)]=varlist BY variable(value1,value2)
```

MEDIAN tests whether the median of a variable is the same in *k* independent samples defined by a grouping variable. For each variable, the display shows a table of the number of cases greater than and less than or equal to the median in each category of the grouping variable, the median, chi-square, degrees of freedom, and probability. By default, the median tested is calculated from all cases included in the test.

Syntax

- The minimum specification is a single test variable, the keyword BY, a grouping variable, and two values in parentheses.
- If the first grouping value is less than the second, every value in the range defined by the pair of values forms a group and a *k*-sample test is performed.
- If the first value is greater than the second, two groups are formed using the two values and a two-sample test is performed.
- To override the default median, specify a median value in parentheses following the MEDIAN subcommand keyword.

Operations

- A 2 × *k* contingency table is constructed with counts of the number of cases greater than the median and less than or equal to the median for the *k* groups.
- For more than 30 cases, a chi-square statistic is computed.
- For 30 or fewer cases, Fisher's exact procedure (one-tailed) is used instead of chi-square.
- For a two-sample test, cases with values other than the two specified are excluded.
- A test statistic is calculated for each variable named before BY.

Example

```
NPAR TESTS MEDIAN(8.4)=YVAR BY XVAR(1,2)/ MEDIAN=YVAR BY XVAR(1,2)
    /MEDIAN=YVAR BY ZVAR(1,4)/ MEDIAN=YVAR BY ZVAR(4,1).
```

- The first two MEDIAN subcommands test variable YVAR grouped by values 1 and 2 of variable XVAR. The first test specifies a median of 8.4 and the second uses the observed median.
- The third MEDIAN subcommand requests a four-samples test, dividing the sample into four groups based on values 1, 2, 3, and 4 of variable ZVAR.
- The last MEDIAN subcommand requests a two-samples test, grouping cases based on values 1 and 4 of ZVAR and ignoring all other cases.

MOSES Subcommand

```
NPAR TESTS MOSES[(n)]=varlist BY variable(value1,value2)
```

The MOSES test of extreme reactions tests whether the range of an ordinal variable is the same in a control group and a comparison group defined by a grouping variable. For each variable tested, the display includes the count in the two groups, the number of outliers removed, the span of the control group before and after outliers are removed, and the one-tailed probability of the span with and without outliers. By default, 5% of the cases are trimmed from each end of the range of the control group to remove outliers.

Syntax

- The minimum specification is a test variable, the keyword BY, a grouping variable, and two values in parentheses.
- The test variable must be at least at the ordinal level of measurement.

- The first value of the grouping variable defines the control group, and the second value defines the comparison group.
- You can override the default 5% of cases to be trimmed from each end of the control group by specifying a value in parentheses following the subcommand keyword MOSES. This value represents an actual number of cases, not a percentage.

Operations

- Scores from the groups are arranged in a single ascending sequence.
- The span of the control group is computed as the number of cases in the sequence containing the lowest and highest control score.
- No adjustments are made for tied cases.
- Cases with values other than those specified for the grouping variable are excluded.
- A test statistic is calculated for each variable named before BY.

Example

```
NPAR TESTS MOSES=YVAR BY NVAR(0,1)/ MOSES=YVAR BY NVAR(1,0).
```

- The first MOSES subcommand tests YVAR using value 0 of NVAR to define the control group and value 1 for the experimental group. The second MOSES subcommand reverses the experimental and control groups.

RUNS Subcommand

```
NPAR TESTS RUNS({MEAN  })=varlist
                {MEDIAN}
                {MODE  }
                {value }
```

RUNS tests whether the sequence of values of a dichotomized variable is random. The display includes the test value (cutting point used to dichotomize the variable tested), number of runs, number of cases below the cutting point, number of cases equal to or greater than the cutting point, and test statistic *Z* with its one-tailed probability.

Syntax

- The minimum specification is a cutting point in parentheses followed by a test variable.
- The cutting point can be specified by an exact value or one of the keywords MEAN, MEDIAN, or MODE.

Operations

- All variables tested are treated as dichotomous: values less than the cutting point form one category, and values equal to or greater than the cutting point form the other category.
- A test statistic is calculated for each variable named.

Example

```
NPAR TESTS RUNS(MEDIAN)=XVAR/ RUNS(24.5)=XVAR/ RUNS(1)=ZVAR.
```

- This example performs three runs tests. The first two test variable XVAR, first using the median and then using 24.5 as the cutting point.
- The third test is for variable ZVAR, with value 1 specified as the cutting point.

SIGN Subcommand

```
NPAR TESTS SIGN=varlist [WITH varlist]
[/OPTION=3]
```

SIGN tests whether the number of positive and negative differences between two paired ordinal variables in a two-related-samples test is equal. The display includes the number of positive differences, number of negative differences, number of ties, and the two-tailed binomial probability.

Syntax

- The minimum specification is a list of two variables.
- Variables should be at least at the ordinal level of measurement.
- Without keyword WITH, each variable in the list is paired with every other variable in the list.
- With keyword WITH, each variable before WITH is paired with each variable after WITH.
- With Option 3 and no WITH, the first variable is paired with the second, the second with the third, the third with the fourth, and so on.

- With Option 3 and WITH, the first variable before WITH is paired with the first variable after WITH, the second variable before WITH with the second variable after WITH, and so on.
- With Option 3 and WITH, the number of variables specified before and after WITH must be the same.

Operations

- The positive and negative differences between the pair of variables are counted. Ties are ignored.
- The probability is taken from the binomial distribution if 25 or fewer differences are observed. Otherwise, the probability comes from the *Z* distribution.
- Under the null hypothesis for large sample sizes, *Z* is approximately normally distributed with a mean of 0 and a variance of 1.

Example

```
NPAR TESTS SIGN=NVAR1,MVAR1 WITH NVAR2,MVAR2
/OPTION=3.
```

- In this example, NVAR1 is tested with NVAR2, and MVAR1 is tested with MVAR2.

W-W Subcommand

```
NPAR TESTS W-W=varlist BY variable(value1,value2)
```

W-W (alias WALD-WOLFOWITZ) tests whether the distribution of a variable is the same in two independent samples. A runs test is performed with group membership as the criterion. The display includes the number of valid cases in each group, the number of runs, *Z*, and the one-tailed probability of *Z*. If ties are present, the minimum and maximum number of ties possible, their *Z* statistics, and one-tailed probabilities are displayed.

Syntax

- The minimum specification is a single test variable, the keyword BY, a grouping variable, and two values in parentheses.
- Cases with the first value form one group and cases with the second value form the other. Order is unimportant.
- You can request W-W with its alias WALD-WOLFOWITZ.

Operations

- Cases are combined from both groups and ranked from lowest to highest.
- A runs test is performed using group membership as the criterion.
- For ties involving cases from both groups, both the minimum and maximum number of runs possible are calculated.
- For a sample size of 30 or less, the exact one-tailed probability is calculated.
- For a sample size greater than 30, the normal approximation is used.
- Cases with values other than those specified for the grouping variable are excluded.
- Test statistics are calculated for each variable named before BY.

Example

```
NPAR TESTS W-W=YVAR BY NVAR(0,1).
```

- This example ranks cases from lowest to highest based on their values for YVAR. A runs test on the group variable is done.

WILCOXON Subcommand

```
NPAR TESTS WILCOXON=varlist [WITH varlist]
[/OPTION=3]
```

WILCOXON tests the hypothesis that there are no differences between two paired populations of ordered-metric scores. The test takes into account the magnitude of the differences between two paired variables. The display includes the number of positve and negative differences and their respective means, the number of ties, the valid count, *Z*, and the probability of *Z*.

Syntax

- The minimum specification is a list of two variables.
- Without keyword WITH, each variable is paired with every other variable in the list.
- With keyword WITH, each variable before WITH is paired with each variable after WITH.

- With Option 3 and no WITH, the first variable is paired with the second, the second with the third, the third with the fourth, and so on.
- With Option 3 and WITH, the first variable before WITH is paired with the first variable after WITH, the second variable before WITH with the second variable after WITH, and so on.
- With Option 3 and WITH, the number of variables specified before and after WITH must be the same.

Operations

- The differences between the pair of variables are counted, the absolute differences ranked, the positive and negative ranks summed, and the test statistic Z computed from the positive and negative rank sums.
- Under the null hypothesis for large sample sizes, Z is approximately normally distributed with a mean of 0 and a variance of 1.

Example

```
NPAR TESTS WILCOXON=VARA VARB VARC
/OPTIONS=3.
```

- This example pairs VARA and VARB and then VARB and VARC.

Statistical Display

The following options and statistics are available in NPAR TESTS. Option 3 can be specified only for two-related-samples tests (MCNEMAR, SIGN, and WILCOXON).

Option 3 *Sequential pairing of variables for two related samples.* With a simple variable list, sequential pairs are tested. With keyword WITH, the first variable before WITH is tested with the first variable after WITH, and so on. The variable lists on both sides of WITH must have the same number of variables.

Statistic 1 *Mean, maximum, minimum, standard deviation, and count.*

Statistic 2 *Quartiles and count.* Values corresponding to the 25th, 50th, and 75th percentiles and the number of valid cases are displayed.

Example

```
NPAR TESTS MCNEMAR=XVAR YVAR ZVAR
/OPTION=3.
```

- This example tests XVAR with YVAR and YVAR with ZVAR, but not XVAR with ZVAR, as is the case without Option 3.

Random Sampling

Option 4 *Random sampling.* Use if there is insufficient memory. Ignored for the RUNS subcommand.

Missing Values

By default, cases with missing values are deleted on a test-by-test basis. For subcommands specifying several tests, each test is evaluated separately.

Option 1 *Include user-missing values.* Cases with user-defined missing values are included in all tests requested on the NPAR TESTS command.

Option 2 *Exclude missing values listwise.* Cases missing on any variable named on any subcommand are excluded from all tests.

Reference

Siegel, S. *Nonparametric Statistics for the Behavioral Sciences.* New York: McGraw-Hill, 1956.

ONEWAY

```
ONEWAY [VARIABLES=]varlist BY varname(min,max)

   [/POLYNOMIAL=n]

   [/CONTRAST=coefficient list] [/CONTRAST=...]

   [/RANGES={SNK           }] [RANGES=...]
            {BTUKEY        }
            {TUKEY         }
            {ranges values }
            {LSD           } [({.05  })]
            {DUNCAN        }   {alpha}
            {MODLSD        }
            {SCHEFFE       }

   [/OPTIONS=option numbers]

   [/STATISTICS={statistic numbers}]
                {ALL               }
```

Options:

1 Include user-missing values
2 Exclude missing values listwise
3 Suppress variable labels
4 Write matrix of counts, means, and standard deviations
6 Use value labels as group labels
7 Read matrix of counts, means, and standard deviations
8 Read matrix of counts, means, pooled variance, and degrees of freedom
10 Harmonic mean of all group sizes as sample sizes in range tests

Statistics:

1 Group descriptive statistics
2 Fixed- and random-effects statistics
3 Homogeneity-of-variance tests

Example:

```
ONEWAY VARIABLES=YVAR BY XVAR(1,4).
```

Overview

Procedure ONEWAY produces a one-way analysis of variance for an interval-level dependent variable by one numeric independent variable that defines the groups for the analysis. Other SPSS/PC+ procedures that perform analysis of variance are MEANS, ANOVA, and MANOVA (in Advanced Statistics). Some tests not included in the other procedures are available as options in ONEWAY.

Defaults

By default, ONEWAY produces a labeled table for each dependent variable by the independent variable. The table contains the between and within groups sums of squares, mean squares, and degrees of freedom. The *F* ratio and the probability of *F* for the test are displayed. Cases that are missing on both variables used in each test are excluded from the calculation of the test statistics.

Tailoring

Trends, Contrasts, and Ranges. You can partition the between-groups sums of squares into linear, quadratic, cubic, and higher-order trend components. You can specify up to 10 contrasts to be tested with the *t* statistic. You can also specify seven different range tests for comparisons of all possible pairs of group means, or "multiple comparisons."

Display Format. You can suppress the display of variable labels. You can also label groups with the value labels of the independent variable.

Statistical Display. In addition to the default display, you can obtain means, standard deviations, and other descriptive statistics for each group. Fixed- and random-effects statistics as well as several tests for homogeneity of variance are also available. The harmonic mean of all group sizes can be used as the sample size for each group in range tests.

Writing and Reading Matrices. ONEWAY can write out or read a matrix of group frequencies, means, and standard deviations for use in subsequent analyses. ONEWAY also reads matrix materials consisting of group frequencies, means, pooled variance, and degrees of freedom for the pooled variance.

Missing Values. You can include cases with user-missing values in the analysis or omit cases with missing values for any variable in the analysis list from all calculations.

Syntax

- The minimum specification is the VARIABLES subcommand with a single analysis list. The actual keyword VARIABLES may be omitted.
- The minimum analysis list specifies a dependent variable, the keyword BY, an independent variable, and the minimum and maximum values of the independent variable in parentheses.
- The VARIABLES subcommand must be specified first. Other subcommands can be specified in any order.
- Subcommands must be separated by a slash.

Operations

- ONEWAY causes the data to be read.
- Noninteger values for the independent variable are truncated.
- Cases with values outside the range specified for the independent variable are omitted from the analysis.
- Specifying a string variable as an independent or dependent variable stops execution of ONEWAY.
- The display uses the width defined on SET.
- If SPSS/PC+ encounters more than one each of the POLYNOMIAL, OPTIONS, or STATISTICS subcommands, it uses the last one of each type.

Limitations

- Maximum 100 dependent variables.
- Maximum 1 independent variable.
- The number of categories for the independent variable is limited only by available memory.
- Maximum 1 VARIABLES subcommand.
- Maximum 1 POLYNOMIAL subcommand.
- Maximum 10 CONTRAST subcommands.
- Maximum 10 RANGES subcommands.
- Maximum 1 OPTIONS subcommand.
- Maximum 1 STATISTICS subcommand.
- Contrasts tests are not performed if the range of values for the independent variable exceeds 50.
- Range tests are not performed if there are more than 50 nonempty categories.
- Range tests are not performed on less than 3 groups.

Example

```
ONEWAY VARIABLES=YVAR BY XVAR(1,4).
```

- This example names YVAR as the dependent variable and XVAR as the independent variable with a minimum value of 1 and a maximum value of 4.

VARIABLES Subcommand

The VARIABLES subcommand specifies the analysis list. The actual keyword VARIABLES may be omitted.

- An analysis list specifies a dependent variable list, the keyword BY, an independent variable, and the minimum and maximum values of the independent variable in parentheses.
- There can be only one VARIABLES subcommand and it must be specified before any of the optional subcommands.
- All variables named must be numeric.

- The minimum and maximum values of the independent variable must be separated by a comma or a space and enclosed in parentheses. These values must be integers.

POLYNOMIAL Subcommand

The POLYNOMIAL subcommand partitions the between-groups sums of squares into linear, cubic, quadratic, or higher-order trend components. The display is an expanded analysis of variance table that provides the degrees of freedom, sums of squares, mean square, *F*, and probability of *F* for each partition.

- The value specified on POLYNOMIAL denotes the highest-degree polynomial to be used.
- The polynomial value must be a positive integer less than or equal to 5. If the polynomial specified is greater than the number of groups, the highest-degree polynomial possible is assumed.
- With balanced designs, ONEWAY computes the sums of squares for each order polynomial from weighted polynomial contrasts, using the group code as the metric. These contrasts are orthogonal.
- With unbalanced designs and equal spacing between groups, ONEWAY also computes sums of squares using the unweighted polynomial contrasts. These contrasts are not orthogonal.
- The deviation sums of squares are always calculated from the weighted sums of squares (Speed, 1976).
- Only one POLYNOMIAL subcommand can be specified per ONEWAY command.

Example

```
ONEWAY VARIABLES=WELL BY EDUC6 (1,6)
/POLYNOMIAL=2.
```

- This example requests an analysis of variance of WELL by EDUC6 with second-order (quadratic) polynomial contrasts.

CONTRAST Subcommand

Use the CONTRAST subcommand to specify a priori contrasts to be tested by the *t* statistic. Contrasts are specified as a vector of coefficients, where each coefficient corresponds to a category of the independent variable. The display for each contrast list is the value of the contrast, the standard error of the contrast, the *t* statistic, the degrees of freedom for *t*, and the two-tailed probability of *t*. Both pooled- and separate-variance estimates are displayed.

- A contrast must be specified or implied for every group in the range specified for the independent variable, even if the group is empty. If the number of contrast values is less than the number of groups, contrast values of 0 are assumed for the remaining groups.
- Only one set of contrast coefficients can be specified per CONTRAST subcommand. Additional contrasts on a single CONTRAST subcommand are ignored.
- You can use the notation $n*c$ to indicate that coefficient c is repeated n times.
- Coefficients are assigned to empty and nonempty groups defined by ascending integer values of the independent variable.
- Trailing coefficients of 0 do not need to be expressed.
- A warning is issued when sets of contrasts do not sum to 0.

Example

```
ONEWAY VARIABLES=YVAR BY XVAR(1,4)
/CONTRAST = -1 -1 1 1
/CONTRAST = -1 0 0 1
/CONTRAST = -1 0 .5 .5.
```

- The first CONTRAST subcommand contrasts the combination of the first two groups with the combination of the last two groups.

- The second CONTRAST subcommand contrasts the first group with the last group.
- The third CONTRAST subcommand contrasts the first group with the combination of the third and fourth groups.

Example

```
ONEWAY VARIABLES=YVAR BY XVAR(1,4)
/CONTRAST = -1 1 2*0
/CONTRAST = -1 1 0 0
/CONTRAST = -1 1.
```

- All three CONTRAST subcommands specify the the same contrast coefficients for a four-group analysis. The first group is contrasted with the second group in all three cases.
- The first CONTRAST uses the n^*c notation and the last CONTRAST omits the trailing zero coefficients.

RANGES Subcommand

Each RANGES subcommand specifies one of seven different tests for multiple comparisons between means. The RANGES display always includes multiple comparisons between all groups. Nonempty group means are sorted in ascending order, with asterisks indicating significantly different groups. In addition, homogeneous subsets are calculated for balanced designs. The means of the groups included within a subset are not significantly different.

- By default, the range tests use sample sizes of the two groups being compared. This is equivalent to using the harmonic mean of the sample size of the two groups begin compared. You can use Option 10 to change this default.
- The default alpha for all tests is 0.05. For some tests, you can specify a different alpha.

The tests available on the RANGES subcommand are

LSD(p) *Least-significant difference.* Alpha can be specified between 0 and 1. The default is 0.05.

DUNCAN(p) *Multiple range test.* Alpha can be specified as 0.01, 0.05, and 0.10 only. The default is 0.05. DUNCAN uses 0.01 if the alpha specified is less than 0.05; 0.05 if the alpha specified is greater than or equal to 0.05 but less than 0.10; 0.10 if the alpha specified is greater than or equal to 0.10; and 0.05 if no alpha is specified.

SNK *Student-Newman-Keuls.* Alpha is 0.05.

BTUKEY *Tukey's alternate procedure.* Alpha is 0.05.

TUKEY *Honestly significant difference.* Alpha is 0.05.

MODLSD(p) *Modified LSD.* Alpha can be specified between 0 and 1. The default is 0.05.

SCHEFFE(p) *Scheffé's test.* Alpha can be specified between 0 and 1. The default is 0.05.

Alternatively, you can use any other type of range by specifying range values:

- Specify the range values separated by commas or blanks.
- Up to $k-1$ range values can be specified in ascending order, where k is the number of groups and where the range value times the standard error of the combined subset is the critical value.
- If less than $k-1$ values are specified, the last value specified is used for the remaining range values.
- You can use the notation n^*r to indicate that the range r is repeated n times.
- To use a single critical value for all subsets, specify one range value.

Example

```
ONEWAY VARIABLES=WELL BY EDUC6 (1,6)
/RANGES=SNK
/RANGES=SCHEFFE (.01).
```

- This example requests two different range tests. The first uses the Student-Newman-Keuls test and the second uses Scheffé's test with an alpha of 0.01.

Example

```
ONEWAY VARIABLES=WELL BY EDUC (1,6)
/RANGES=2.81, 3.34, 3.65, 3.88, 4.05.
```

- The RANGES subcommand specifies five range values.

Harmonic Means

By default, range tests use the harmonic mean of the sizes of the two groups being compared. Use Option 10 on the OPTIONS subcommand to change this default.

Option 10 *Use the harmonic mean of* all *group sizes as the sample size for each group in range tests.* If Option 10 is used for unbalanced designs, ONEWAY determines homogeneous subsets for all range tests.

Display Format

The default display labels groups as GRP1, GRP2, and so forth, and includes variable labels. You can change these defaults by specifying the following on the OPTIONS subcommand:

Option 3 *Suppress variable labels.*

Option 6 *Use value labels for group labels.* Use the first eight characters from the value labels of the independent variable as group labels.

Statistical Display

By default, ONEWAY displays the between- and within-groups sums of squares, mean squares, degrees of freedom, the F ratio, and the probability of F for the test. You can obtain additional statistics by specifying the following on the STATISTICS subcommand:

Statistic 1 *Group descriptive statistics.* Displays the count, mean, standard deviation, standard error, minimum, maximum, and 95% confidence interval for each group for each dependent variable.

Statistic 2 *Fixed- and random-effects statistics.* Displays the standard deviation, standard error, and 95% confidence interval for the fixed-effects model and the standard error, 95% confidence interval, and estimate of between-component variance for the random-effects model.

Statistic 3 *Homogeneity-of-variance tests.* Displays Cochran's C, Bartlett-Box F, and Hartley's F max.

ALL *All statistics.*

Writing Matrices

ONEWAY writes matrix materials that it can read in subsequent analyses.

- To write matrix materials, specify Option 4 on the OPTIONS subcommand.
- Matrix materials are written to the results file named on the SET command (the default is SPSS.PRC).
- If the results file is not empty when ONEWAY is executed with Option 4, the contents of the file are overwritten. Use the SET command to specify a different results file.

Option 4 *Write a matrix containing vectors of counts, means, and standard deviations.* For each dependent variable, Option 4 writes a vector of group frequencies, followed by a vector of group means and a vector of group standard deviations. Vectors are written 80 characters per line with each vector beginning on a new line. The format for the frequencies vector is F10.2. The format for the means and standard deviation vectors is F10.4. There is thus a maximum of eight values per line.

Example

```
SET RESULTS='WELL.MAT'.
DATA LIST FILE='GSS80.DAT'
 / WELL 2-3 EDUC 4-5.
RECODE EDUC (0 THRU 8=1) (9 10 11=2) (12=3) (13 14 15=4)
   (16=5) (17 THRU 20=6).
ONEWAY VARIABLES=WELL BY EDUC (1,6)
 / OPTIONS=4.
```

- The SET command defines file WELL.MAT in the current directory as the results file.
- Option 4 writes group counts, means, and standard deviations for WELL by EDUC to the results file using the format supplied by ONEWAY.

Reading Matrices

You can read matrix materials in fixed or free format by specifying Option 7 or 8 on the OPTIONS subcommand. The general conventions for matrix materials are described in DATA LIST: Matrix Materials.

- If you specify Option 7 or 8 in ONEWAY, you must first use a DATA LIST command specifying matrix materials.
- All variables named on the ONEWAY analysis list must be named on the DATA LIST command.
- The analysis list on the ONEWAY command must be the same as the analysis list that was used when the matrix was written.
- There must be an entry for each group in the vectors of counts, means, and (with Option 7) standard deviations. Entries should be in ascending order of the values of the independent variable.
- Each user-generated vector must begin on a new line or record and can be entered in either fixed or freefield format.

Option 7 *Read matrix materials containing vectors of counts, means, and standard deviations.* ONEWAY expects a vector of counts for each group, followed by a vector of group means and a vector of group standard deviations like those written by Option 4.

Option 8 *Read matrix materials containing vectors of counts and means, followed by the pooled variance and the degrees of freedom.* ONEWAY expects a vector of counts for each group, followed by a vector of means, followed by the a record containing the pooled variance (within-groups mean square) and another record containing the (within-groups) degrees of freedom for the pooled variance. If the degrees of freedom are omitted, then $n-k$ degrees of freedom are assumed, where n is the number of cases and k is the number of groups. Statistics 1, 2, and 3, and the separate variance estimate for contrasts named on the CONTRAST command are not available.

Example

```
DATA LIST FREE MATRIX
 /WELL EDUC.
BEGIN DATA.
65 95 181 82 40 37
2.6462 2.7737 4.1796 4.5610 4.6625 5.2297
6.2699
494
END DATA.
ONEWAY VARIABLES=WELL BY EDUC(1,6)
/OPTIONS=8.
```

- The DATA LIST command specifies matrix input in freefield format.
- The counts for the six analysis groups (the six categories of EDUC, in this example) are on the first line of matrix input, the means for the six groups are on the second line, the within-groups mean square is on the third line, and the within-groups degrees of freedom are on the fourth line.
- Each vector to be read begins on a new line.
- Option 8 requests that matrix materials be read consisting of vectors of counts and means, plus the pooled variance and degress of freedom.

Missing Values

By default, cases with missing values on either the independent or dependent variable are excluded from the test. You can change the handling of cases with missing values by specifying one of the following on the OPTIONS subcommand:

Option 1 *Include cases with user-missing values.* Cases with user-missing values are included in the analyses.

Option 2 *Exclude cases with missing values listwise.* Cases that have missing values for any of the variables named in the analysis list are excluded from all analyses.

Reference

Speed, M. F. Response Curves in the One Way Classification with Unequal Numbers of Observations per Cell. *Proceedings of the Statistical Computing Section, American Statistical Association,* 1976.

PLOT

```
PLOT [MISSING=[{PLOTWISE**}] [INCLUDE]]
               {LISTWISE  }

  [/HSIZE={38**}]  [/VSIZE={16**}]
          {n   }           {n   }

  [/CUTPOINT={EVERY({1**})}]
             {      {n  } }
             {value list  }

  [/SYMBOLS={ALPHANUMERIC**                          }]
            {NUMERIC                                 }
            {'symbols'[,'overplot symbols']          }
            {X'hexsymbs'[,'overplot hexsymbs']}

  [/HORIZONTAL=['label'] [STANDARDIZE] [REFERENCE(vector)]]
                         [MIN(min)] [MAX(max)]
                         [UNIFORM]

  [/VERTICAL=['label'] [STANDARDIZE] [REFERENCE(vector)]]
                       [MIN(min)] [MAX(max)]
                       [UNIFORM]

  [/FORMAT={DEFAULT**        }]
           {CONTOUR[({10})]}
           {        {n }    }
           {OVERLAY          }
           {REGRESSION       }

  [/TITLE='title']

  /PLOT={varlist} WITH varlist [(PAIR)] [BY varname] [;varlist...]
        {ALL    }

  [/PLOT=...]
```

**Default if subcommand is omitted.

Example:

```
PLOT FORMAT=OVERLAY/ SYMBOLS='MD'/ VSIZE=12/ HSIZE=60
/TITLE='Marriage and Divorce Rates'
/VERTICAL='Rates per 1000 population'
/HORIZONTAL='Year' REFERENCE (1918, 1945) MIN (1900) MAX (1983)
/PLOT=MARRATE DIVRATE WITH YEAR.
```

Overview

Procedure PLOT produces two-dimensional line-printer plots, including simple bivariate scatterplots, scatterplots with a control variable, contour plots, and overlay plots. You can also request bivariate regression statistics. You can choose up to 36 ASCII or graphics symbols for the plot.

Defaults

By default, PLOT produces bivariate scatterplots within the page size specified on the SET command. Plots include all cases with valid values for both variables in the plot, with each symbol representing the count at that display position. The vertical and horizontal axes are labeled with variable labels. Default plot titles use either the names of the variables or type of plot requested.

Tailoring

Types of Plots. You can introduce a control variable for bivariate scatterplots or request regression plots with or without a control variable, contour plots, or overlay plots.

Plot Format. You can specify a title for the plot, and you can scale and label the horizontal and vertical axes. You can request reference lines and plot standardized variables. You can also control the plot size and specify plotting symbols and the frequency they represent.

Missing Values. You can include cases with user-missing values and request that cases with missing values be deleted listwise from all plots named on the PLOT subcommand.

Syntax

- The minimum specification is the PLOT subcommand with a variable or variable list for the vertical (Y) axis, the keyword WITH, and a variable or variable list for the horizontal (X) axis.
- The PLOT subcommand can be specified more than once on a PLOT command.
- PLOT must be the last subcommand specified.
- Subcommands MISSING, VSIZE, HSIZE, CUTPOINT, and SYMBOLS apply to all plots requested and can be specified only once within any PLOT command. They must be entered *before* the final PLOT subcommand.
- Subcommands HORIZONTAL, VERTICAL, FORMAT, and TITLE can be specified more than once and apply only to the following PLOT subcommand.
- Subcommands are separated by slashes.

Operations

- PLOT causes the data to be read.
- The default plot frame size depends on the page size specified on SET. The HSIZE and VSIZE subcommands override the default plot size.
- A longer page length can produce longer default plots within the same width. A wider page does not produce a wider default plot unless the page length is changed accordingly.

Limitations

- No limitation on the number of plots requested.
- No limitation on the number of variables on a PLOT command.
- Maximum 20 overlay plots per FORMAT subcommand.
- Maximum 1 control variable per PLOT subcommand.
- Maximum 60 characters per TITLE subcommand.
- Maximum 36 symbols per SYMBOLS subcommand.
- Maximum 35 cutpoints per CUTPOINT subcommand.
- Maximum 10 reference points on each HORIZONTAL and VERTICAL subcommand.
- Maximum 40 characters per label on each HORIZONTAL and VERTICAL subcommand.
- Maximum 35 contour levels for each CONTOUR plot.

Example

```
PLOT FORMAT=OVERLAY/ SYMBOLS='MD'/ VSIZE=12/ HSIZE=60
/TITLE='Marriage and Divorce Rates'
/VERTICAL='Rates per 1000 population'
/HORIZONTAL='Year' REFERENCE (1918, 1945) MIN (1900) MAX (1983)
/PLOT=MARRATE DIVRATE WITH YEAR.
```

- This example produces an overlay plot of marriage and divorce rates by year.
- SYMBOLS selects the symbols M and D, respectively, for the two plots.
- VSIZE and HSIZE limit the vertical and horizontal axes to 12 lines and 60 columns, respectively.
- TITLE specifies a plot title, and VERTICAL provides a title for the vertical axis.
- HORIZONTAL provides a title for the horizontal axis. REFERENCE, MIN, and MAX provide reference lines at values 1918 and 1945 and minumum and maximum scale values on the horizontal axis.

PLOT Subcommand

The PLOT subcommand names the variables to be plotted on each axis. The PLOT subcommand can also name a control or contour variable.

- PLOT is the only required subcommand.
- You can enter multiple PLOT subcommands.
- No other subcommands can follow the last PLOT subcommand.
- Each plot list first specifies a list of variables to be plotted on the vertical axis, then the keyword WITH, and then a list of variables to be plotted on the horizontal axis.

- By default, PLOT creates separate plots for each combination of variables listed on the left side of WITH with variables on the right.
- Use semicolons to separate multiple plot lists on a single PLOT subcommand.
- Keyword ALL can be used to refer to all user-defined variables.
- Use keyword BY followed by a variable name to specify a control or contour variable.
- Only one control variable can be specified on any plot list.
- If a control variable is specified, PLOT uses the first character of the control variable's value label as the plot symbol. If value labels have not been specified, the first character of the value is used. The symbol $ indicates that more than one control value occurs in that display position.

You can also request special pairing of variables with the following keyword:

(PAIR) *Plot corresponding pairs of variables.* The first variable before WITH is plotted against the first variable after WITH, and so on.

Example

```
PLOT PLOT=MARRATE WITH YEAR AGE;
        BIRTHS DEATHS WITH INCOME1 INCOME2 (PAIR);
        DIVRATE WITH AGE BY YEAR.
```

- This PLOT subcommand contains three plot lists. The first requests a plot of MARRATE with YEAR and of MARRATE with AGE.
- The second uses the keyword (PAIR) to request two plots: BIRTHS with INCOME1 and DEATHS with INCOME2.
- The third requests a plot of DIVRATE with AGE using YEAR as a control variable. The value labels for YEAR will be used to obtain plotting characters in this control plot.

FORMAT Subcommand

The FORMAT subcommand controls the type of plot produced.

- FORMAT can be specified only once before each PLOT subcommand and applies only to plots requested on that PLOT subcommand.
- If the subcommand is omitted or DEFAULT is specified, bivariate scatterplots are displayed.
- Only one keyword can be specified on each FORMAT subcommand.

The available keywords are

DEFAULT *Bivariate scatterplot.* When there is no control variable on the plot list, each symbol represents the case count at that plot position. When a control variable is specified, each symbol represents the first character of the value label of the control variable.

OVERLAY *Overlay plots.* All bivariate plots on the next PLOT subcommand appear in one plot frame. PLOT selects a unique symbol for each plot to be overlaid, plus a symbol to represent multiple plot points in one display position.

CONTOUR(n) *Contour plot with* n *levels.* Contour plots use a continuous variable as the control variable and *n* successive symbols to represent lowest to highest levels of the variable. Specify the control variable after BY on the PLOT subcommand. This variable is recoded into *n* equal-width intervals. If the levels specification is omitted, the default of 10 is used; the maximum is 35. When more than one level of the contour variable occurs at the same plot position, PLOT displays the value of the highest level.

REGRESSION *Regression of vertical-axis variable on horizontal-axis variable.* The regression-line intercepts on each axis are marked with the letter R. When there is no control variable, each symbol represents the frequency of cases at that plot position. If a control variable is specified, regression statistics are pooled over all categories and each symbol represents the first character of the value labels of the control variable.

Plot Symbols

A wide range of alphabetical, numeric, and special ASCII graphic characters are available for use as PLOT symbols. Two subcommands control the display of symbols: the SYMBOLS subcommand controls the choice of plot symbols, and the CUTPOINT subcommand controls the frequencies represented by a symbol. SYMBOLS and CUTPOINT can each be specified only once and apply to all plots requested in a PLOT command. If you have more than one FORMAT subcommand within a PLOT command, the meaning of the plotting symbols can vary. The operation of SYMBOLS and CUTPOINT for each FORMAT specification is summarized below.

- *DEFAULT or REGRESSION plot, no control.* Each symbol represents the frequency of cases. Controlled by SYMBOLS and CUTPOINT.
- *DEFAULT or REGRESSION plot, control.* Each symbol represents one value of the control variable. SYMBOLS and CUTPOINT do not apply. The plot symbol is the first character of the control variable's value label or the first character of the actual value if no VALUE LABELS have been declared; the uniqueness of these symbols is not checked.
- *OVERLAY.* Each symbol represents one of the overlaid plots. SYMBOLS is applicable; CUTPOINT is not.
- *CONTOUR.* Each symbol represents one level of the contour variable. SYMBOLS is applicable; CUTPOINT is not.

CUTPOINT Subcommand

By default, each frequency in a frequency plot is represented by a different plot symbol, and successive plotting symbols represent an interval width of 1. Use the CUTPOINT subcommand to alter the categories represented by plot symbols for bivariate and regression plots.

- CUTPOINT can be specified only once and applies to all frequency plots on the PLOT command.
- If the subcommand is omitted, the default interval of width 1 is in effect.
- Only one specification can be given on CUTPOINT.

The following specifications are available:

EVERY(n) *Frequency intervals of width* n. Each plot symbol represents the specified frequency interval. The default is an interval width of 1.

(value list) *Each value defines a cutpoint.* Successive plot symbols are assigned to each cutpoint. Up to 35 cutpoints can be specified. Specify values separated by blanks or commas. The number of cutpoints is one less than the number of intervals.

Example

```
PLOT CUTPOINT=EVERY(2)/ PLOT=YVAR WITH XVAR.
PLOT CUTPOINT=(5,10,20)/ PLOT=YVAR WITH XVAR.
```

- In the first PLOT command, 1 or 2 cases on a display position are represented by a 1; 3 or 4 cases by a 2, and so forth.
- In the second PLOT command, 1 to 5 cases on a display position are represented by a 1; 6 to 10 cases by a 2; 11 to 20 cases by a 3; and 21 or more cases by a 4.

SYMBOLS Subcommand

The SYMBOLS subcommand defines the plotting symbols for bivariate scatterplots and bivariate regression, overlay, and contour plots. Successive symbols represent increasing frequencies in scatterplots or regression plots, successive subplots in overlay plots, and successive intervals in contour plots.

- SYMBOLS can be specified only once and applies to all plots requested except control plots.
- If the subcommand is omitted, the default alphanumeric symbol set is used.
- If SYMBOLS is specified, a table defining the plotting symbols is displayed.

The following specifications are available for SYMBOLS:

ALPHANUMERIC *Alphanumeric plotting symbols.* Includes the characters 1 through 9, A through Z, and *. Thirty-six or more cases at a position are represented by a *. This is the default.

NUMERIC *Numeric plotting symbols.* Includes the characters 1 through 9 and *. Ten or more cases at a plot position are represented by a *.

'symbols'[,'ovprnt'] *List of plot symbols.* Up to 36 symbols can be specified. Symbols are specified without any intervening blanks or commas. If specified, the list of overprint symbols is separated from the first symbol list by a comma or space. Indicate hexadecimal symbols on either list by specifying X before the list and enclosing the list in apostrophes. You can also specify any special ASCII graphic characters available on your terminal. See the SET command and your DOS manual for further reference.

- When overprint symbols are displayed (on the screen), only the symbols from the first list remain visible. Other characters in the plot may be erased. Overprinting will occur when the plot is printed.

Example

```
PLOT CUTPOINTS=EVERY(5)/SYMBOLS='.+O','  X'
/PLOT=YVAR BY XVAR.
```

- This example uses a period (.) to represent 5 or fewer cases at one point, a plus sign (+) to represent 6 to 10 cases at the same position, and a symbol overprinting O and X to represent 11 or more cases at one position. Note the leading blanks in the list of overprint symbols.
- On the screen, the symbols .+O + X will remain visible to represent 1–5, 5–10, and 11 or more cases, respectively. The overprint of O and X will appear on printed output.

VSIZE and HSIZE Subcommands

Use the VSIZE and HSIZE subcommands to specify the vertical and horizontal frame size for the plot, respectively.

- VSIZE and HSIZE can each be used only once per PLOT command and apply to all plots requested.
- The default size of a plot depends on the current page size. If the SET command defines a width of 79 and a length of 24, the default plot width is 38 columns and the default plot length is 16 rows.
- VSIZE and HSIZE each use a single integer as their only specification.
- VSIZE and HSIZE values do not include display lines for the plot frame itself or for auxiliary information such as titles, axis scale numbers, regression statistics, or symbol table.
- If VSIZE is greater than the length specified on SET, the symbol table and other information are displayed on another page.
- If HSIZE is greater than the width specified on SET, the plot wraps on the screen if there is insufficient width for the plot.

VERTICAL and HORIZONTAL Subcommands

The VERTICAL and HORIZONTAL subcommands control labeling and scaling for the vertical and horizontal axes, respectively.

- VERTICAL and HORIZONTAL can each be specified once before each PLOT subcommand.
- VERTICAL and HORIZONTAL apply only to plots requested by the next PLOT subcommand.
- If VERTICAL and HORIZONTAL are omitted, all defaults are in effect. If VERTICAL and HORIZONTAL are included, only those defaults explictly altered are changed.

The following keywords are available for both VERTICAL and HORIZONTAL:

'label' *Label for axis.* The label can contain up to 40 characters. A label that will not fit in the plot frame is truncated. The default is the variable label or the variable name if no variable label has been declared.

MIN (n) *Minimum axis value.* If you specify a minimum value greater than the observed minimum value, some points will not be included in the plot. The default is the minimum observed value.

MAX (n) *Maximum axis value.* PLOT may extend the scale value slightly in order to display integer scale values of equal width. The default is the maximum observed value.

UNIFORM *Uniform values on axis.* All plots on the PLOT subcommand will have the same value scale on the axis. You imply a uniform scale when you specify both MIN and MAX. If you specify UNIFORM, PLOT determines the minimum and maximum observed values across all variables on the PLOT subcommand.

REFERENCE(values) *Values at which reference lines will be drawn.* Specify the values separated by blanks or commas. The default is no reference lines.

STANDARDIZE *Standardize variables on the axes.* Standardized variables are useful for overlay plots of variables with different scales. The default is to plot observed values.

TITLE Subcommand

Use the TITLE subcommand to label plots.

- TITLE can be specified once before each PLOT subcommand.
- TITLE applies to all plots named on the next PLOT subcommand.
- The default title is either the names of the variables in a bivariate plot or the type of plot requested on FORMAT.
- The title can be up to 60 characters long.
- The rules for specifying titles follow the usual conventions for strings (see Universals: Strings).
- The title is truncated if it exceeds the width specified on the HSIZE subcommand.

MISSING Subcommand

Use the MISSING subcommand to control the handling of cases with missing values.

- The MISSING subcommand can be specified only once on each PLOT command and applies to all plots.
- By default, cases with system-missing or user-missing values for any variables in a plot are omitted from that plot.
- The keywords LISTWISE and PLOTWISE (the default) are alternatives. Either one may also be specified with INCLUDE.

The following keywords are available for the MISSING subcommand:

PLOTWISE *Delete cases with missing values plotwise.* The default. Cases with missing values for any variable in a plot are not included in that plot. In overlay plots, PLOTWISE applies separately to each overlaid plot in the frame, not to the full list specified on the PLOT subcommand.

LISTWISE *Delete cases with missing values listwise.* Cases with missing values for any variable named on the PLOT subcommand are deleted from all plots specified on that PLOT subcommand.

INCLUDE *Include cases with user-missing values.* Only cases with system-missing values are excluded according to the missing-value treatment specified (PLOTWISE, LISTWISE, or the default).

PROCESS IF

```
PROCESS IF [(]variable relational operator value[)]
```

Relational Operators:

EQ or = NE or ~ = or <>
LT or < LE or <=
GT or > GE or >=

Example:

```
PROCESS IF (TYPE EQ 1).
```

Overview

The PROCESS IF transformation temporarily designates cases for inclusion in the next procedure.

Syntax

- The specification is a simple logical expression that can be evaluated as true or false (see Universals: Logical Expressions).
- Parentheses enclosing the logical expression are optional.
- All six relational operators are permitted: EQ, NE, LT, LE, GT, GE. Symbolic forms (=, ~ = or <>, <, <=, >, >=) are also accepted.
- Only numeric or short string variables can be specified. Long string variables may not be used.
- String values must be enclosed in apostrophes or quotation marks.
- String values must match the length of the short string being tested.
- PROCESS IF can be entered anywhere in an SPSS/PC+ session, except between BEGIN DATA and END DATA.

Operations

- PROCESS IF is a temporary transformation and is executed when the data are read for the next procedure.
- PROCESS IF temporarily designates cases for inclusion in the next procedure.
- If the logical expression is true, the case is processed. If it is false or missing, the case is not processed in the next procedure.
- If multiple PROCESS IF commands are entered before a procedure, only the last one is in effect.
- PROCESS IF has no effect on SORT CASES, and is ignored if SORT CASES is the next procedure.

Limitations

- PROCESS IF can take only one simple expression; logical operators are not accepted.

Example

```
PROCESS IF (AGE GT 50).
```

- In the next procedure, only cases for which variable AGE is greater than 50 are included.

Example

```
PROCESS IF (SEX EQ 'MALE  ').
```

- SEX is a six-column left-justified variable. The trailing blanks must be included in the value specification.
- The next procedure uses cases in which the value of SEX is MALE with two trailing blanks.

RECODE

Numeric Recodes:

```
RECODE varlist (value list=value)...(value list=value)

       [/varlist...]
```

Input keywords available for numeric recodes are:

```
LO  LOWEST  HI  HIGHEST  THRU  MISSING  SYSMIS  ELSE
```

Output keywords available for numeric recodes are:

```
SYSMIS
```

String Recodes:

```
RECODE varlist ('string',['string'...]='string')

      [varlist...]
```

Input keywords available for string recodes are:

```
ELSE
```

Examples:

```
RECODE IVAR1 TO IVAR3 (0=1) (1=0) (2,3=-1) (9=9) (ELSE=SYSMIS).

RECODE STRINGVAR ('A','B','C'='A')('D','E','F'='B')(ELSE=' ').
```


Overview

The RECODE command changes the coding scheme of an existing numeric or short string variable on a value-by-value basis or for a range of values. Where it can be used, RECODE is much more efficient than the series of IF commands that produce the same transformation.

Syntax

- The variable(s) to be recoded must already exist.
- Each set of value specifications is enclosed in parentheses.
- Input values are specified first (to the left of the equals sign), followed by a single output value (to the right of the equals sign).
- More than one input value can be recoded to a single output value.
- Values in the value lists must be separated by blanks or commas.
- Only one output value per set of input values is allowed.
- Input values that are not mentioned remain unchanged, unless the ELSE keyword is used.
- ELSE refers to all input values not previously mentioned.
- The equals sign preceding the output value is required.
- You can recode more than one variable using the same value specifications by specifying a list of variables before the value specifications.
- More than one variable can be recoded differently in one command by specifying the variable and the value specifications for each transformation, separated by a slash.

Numeric Variables

- THRU specifies a value range, inclusive of specified end values.
- LOWEST and HIGHEST (LO and HI) specify the lowest and highest values encountered in the data.
- LOWEST and HIGHEST include user-missing values.
- LOWEST and HIGHEST do not include the system-missing value.
- ELSE includes all input values not already specified, including the system-missing value.
- MISSING specifies user- and system-missing values for recoding. MISSING is an input specification only.
- SYSMIS specifies the system-missing value and can be used as both an input and output value.

String Variables

- Only short string variables can be recoded. Long strings cannot be recoded.
- Values must be enclosed in apostrophes or quotation marks.
- Blanks are significant characters.
- Input and output values must be specified according to the format width of the variable (see Universals: Strings).

Operations

- RECODE is a transformation and is executed when the data are read for the next procedure.
- Recode value specifications are scanned left-to-right.
- A value is recoded only once per RECODE command.
- Invalid syntax stops processing of the RECODE command and cancels any recoding of variables named on the command.

Numeric Variables

- Blank fields for numeric variables are handled according to the SET BLANKS specification prior to recoding.
- When you recode a value that was previously defined as user-missing on the MISSING VALUE command, the output value is not missing but is the new, recoded value.

String Variables

- If the input or output value is shorter or longer than the format width defined for that variable, it is an error.

Limitations

- You can recode (and count using the COUNT command) approximately 400 values prior to a data pass.
- Invalid specifications on a RECODE command that result in errors stop all processing of that RECODE command.

Example

```
RECODE IVAR1 TO IVAR3 (0=1) (1=0) (2,3=-1) (9=9) (ELSE=SYSMIS)/
   QVAR(1 THRU 5=1)(6 THRU 10=2)(11 THRU HI=3)(ELSE=0).
```

- Values for the list of numeric variables between and including IVAR1 and IVAR3 are changed: input values 0 and 1 are switched to 1 and 0, respectively; 2 and 3 become −1; 9 remains 9; and any other value is changed to the system-missing value.
- Values for variable QVAR are also changed: input values 1 through 5 become 1; 6 through 10 become 2; 11 through the highest value become 3; and any other value, including system-missing, becomes 0.

Example

```
RECODE STRINGVAR ('A','B','C'='A')('D','E','F'='B')(ELSE=' ').
RECODE PET ('IGUANA', 'SNAKE ' = 'WILD  ').
```

- Values A, B, and C are changed to value A. Values D, E, and F are changed to value B. All other values are changed to a blank.
- Values IGUANA and SNAKE are changed to value WILD. The variable PET has a format width of six characters. Thus, values SNAKE and WILD include trailing blanks to total six characters.
- Note that each string value is enclosed within apostrophes.

REGRESSION

```
REGRESSION VARIABLES={varlist   }
                     {ALL       }
                     {(COLLECT) }

  [/STATISTICS=[DEFAULTS**] [R] [COEFF] [ANOVA] [OUTS]
               [ZPP] [CHA] [CI] [F] [BCOV] [SES] [TOL]
               [COND] [XTX] [HISTORY] [END] [LINE] [ALL]]

  [/CRITERIA=[DEFAULTS**] [TOLERANCE({.0001 })] [MAXSTEPS({2v})]
                                    {value}               {n }

           [{PIN({0.05 })}]     [{POUT({0.1  })}]
            {    {value} }       {     {value} }
            {FIN({3.84 })}       {FOUT({2.71 })}
            {    {value} }       {     {value} }

  [/{NOORIGIN**}]
    {ORIGIN    }

  /DEPENDENT=varlist

  /[METHOD=]{STEPWISE [= varlist]      } [/METHOD=...]
            {FORWARD [=varlist]        }
            {BACKWARD [=varlist]       }
            {ENTER [=varlist]          }
            {REMOVE=varlist            }
            {TEST=(varlist) (varlist)  }

  [/DESCRIPTIVES=[DEFAULTS] [MEAN] [STDDEV] [CORR]
                 [VARIANCE] [XPROD] [SIG] [N] [BADCORR]
                 [COV]      [ALL]   [NONE**]]

  [/SELECT={varname relation value}]

  [/MISSING={LISTWISE**        } [INCLUDE]]
            {PAIRWISE          }
            {MEANSUBSTITUTION  }

  [/WIDTH={value on SET**}]
          {n             }

  [/REGWGT=variable]
```

**Default if subcommand is omitted.

Example:

```
REGRESSION VARIABLES=X1 TO X7,Y
/DEPENDENT=Y
/METHOD=ENTER X1 X2 X3
/METHOD=ENTER X4 TO X7
/METHOD=STEPWISE.
```

Overview

Procedure REGRESSION calculates multiple regression equations and associated statistics and plots. Several methods for selecting variables into the equation are available. Statistics for analyzing residuals and influential observations are also available. Several types of plots, including partial residual plots, can be displayed.

Defaults

The minimum specification for REGRESSION is a list of variables from which a correlation matrix is computed (VARIABLES), a dependent variable which begins the building of an equation (DEPENDENT) and a method for selecting blocks of independent variables for the equation (METHOD). For each block of variables selected, the default display includes statistics on the equation (including R^2 and analysis of variance), on the variables in the equation (including regression coefficients), and on variables being considered that are not in the equation. The default display uses the width specified on SET. By default, all cases in the active file with valid values for all the variables named on the VARIABLES subcommand are used to compute the correlation matrix on which the regression equations are based. The default equations include a constant (intercept).

Tailoring

The options for procedure REGRESSION can be grouped into logically related operations or specifications.

Global-Control Subcommands. These optional subcommands should be named only once and apply to the entire REGRESSION command. DESCRIPTIVES requests descriptive statistics on the variables in the analysis. SELECT estimates the model based on a subset of cases. MISSING specifies the treatment of cases with missing values.

Equation-Control Subcommands. These optional subcommands control the calculation and display of each equation. STATISTICS controls the statistics displayed as the equations are built, CRITERIA specifies the criteria used by the variable selection method, and ORIGIN specifies whether regression is through the origin.

Display Format. The WIDTH subcommand controls the width of the display for REGRESSION only. It applies to all output from the REGRESSION command.

Analysis of Residuals. The optional subcommands that analyze and plot residuals, and optionally add new variables to the active file containing predicted values, residuals, or related information, are described in REGRESSION: Residuals. These subcommands apply to a specific final equation, as defined by a DEPENDENT subcommand after processing all METHOD subcommands.

Writing and Reading Matrices. The optional subcommands that read and write matrix materials are described in REGRESSION: Matrix Materials. These subcommands are global in scope.

Syntax

- The minimum specification is three subcommands and their specifications: VARIABLES, which lists the variables that will be used to compute the correlation matrix; DEPENDENT, which initiates equation(s) and specifies at least one dependent variable; and METHOD, which specifies a method to be used in selecting independent variables.
- The VARIABLES subcommand can be used only once and must be specified before the DEPENDENT and METHOD subcommands.
- The MISSING, DESCRIPTIVES, and SELECT subcommands can be entered anywhere except between a DEPENDENT subcommand and the immediately following METHOD subcommands. If any of these subcommands appears more than once on the REGRESSION command, the last one of each type is in effect for the entire REGRESSION command.
- CRITERIA, STATISTICS, and ORIGIN subcommands remain in effect for all subsequent equations, until replaced. These subcommands must not be placed between a DEPENDENT subcommand and its METHOD subcommand(s).
- More than one DEPENDENT subcommand can be used. An equation is estimated for each variable listed on a DEPENDENT subcommand.
- A DEPENDENT subcommand must be followed immediately by one or more METHOD subcommands.
- If no variable list is specified on METHOD, all variables named on VARIABLES but not on DEPENDENT will be considered for selection.
- The WIDTH subcommand can appear anywhere. The last WIDTH subcommand encountered is in effect for all REGRESSION displays.
- All subcommands must be separated by slashes.

Operations

- REGRESSION causes the data to be read.
- REGRESSION calculates a correlation matrix that includes all variables named on the VARIABLES subcommand. All equations requested on the REGRESSION command are calculated from the same correlation matrix.

- The MISSING, DESCRIPTIVES, and SELECT subcommands control the calculation of the correlation matrix and associated displays.
- The DEPENDENT subcommand and the METHOD subcommands that follow it control the building of an equation for each variable named on the DEPENDENT subcommand.
- The equation for each variable named on a DEPENDENT subcommand uses the same independent variables and methods.
- Multiple METHOD subcommands operate, in sequence, on the equations defined by the preceding DEPENDENT subcommand.
- All independent variables that pass the tolerance criterion are candidates for entry (see CRITERIA Subcommand).
- If the width set is less than 132, some statistics requested may not be displayed. The WIDTH subcommand within REGRESSION allows you to increase the display width and obtain all the statistics available.

Limitations

- The number of variables that can be named on the VARIABLES subcommand depends on the memory available.

Example

```
REGRESSION VARIABLES=X1 TO X7,Y
/DEPENDENT=Y
/METHOD=ENTER X1 X2 X3
/METHOD=ENTER X4 TO X7
/METHOD=STEPWISE.
```

- VARIABLES requests that a correlation matrix of variables X1 to X7 and Y be calculated for use by REGRESSION.
- DEPENDENT defines a single equation, with Y as the dependent variable.
- The first METHOD subcommand requests that X1 to X3 be entered into the equation.
- The second METHOD subcommand requests that X4 to X7 be added to the equation containing X1 to X3.
- The last METHOD subcommand requests that the entire equation be evaluated using the stepwise method.

Subcommand Order

The standard subcommand order for REGRESSION is shown in Figure A.

A Subcommand order

```
REGRESSION VARIABLES=...
/DESCRIPTIVES=...
/SELECT=...
/MISSING=...
/WIDTH=...
/READ=...
/WRITE=...
```

Equation Blocks

```
/STATISTICS=...
/CRITERIA=...
/ORIGIN
/NOORIGIN
/DEPENDENT=...
```

Method Blocks

```
/METHOD=...
```

```
/RESIDUALS=...
/CASEWISE=...
/SCATTERPLOT=...
/PARTIALPLOT=...
```

- Subcommands listed outside the boxes apply to all analyses performed by the REGRESSION command. Subcommands listed in the equation block apply to all methods used in estimating that equation.
- A REGRESSION command can include multiple equation blocks, and each equation block can contain multiple METHOD subcommands. These methods are applied, one after the other, to the estimation of the equation.
- The STATISTICS, CRITERIA, and ORIGIN/NOORIGIN subcommands must precede the DEPENDENT subcommand for the equation to which they apply.
- The RESIDUALS, CASEWISE, SCATTERPLOT, and PARTIALPLOT subcommands must follow the last METHOD subcommand in an equation block and apply only to the final equation after all METHOD subcommands have been processed.
- This order of subcommands can be different from the order shown in Figure A. However, any analysis that can be performed with procedure REGRESSION can be performed using this order, repeating equation blocks and METHOD subcommands as needed.

VARIABLES Subcommand

The required VARIABLES subcommand names all the variables in the analysis with either a variable list or a keyword.

- The minimum specification is a list of two variables or the keyword ALL or (COLLECT). There is no default variable list.
- There must be one and only one VARIABLES subcommand, and it must precede any DEPENDENT or METHOD subcommands.
- All variables named on the DEPENDENT and METHOD subcommands must be named or implied on the VARIABLES subcommand.
- You can name any user-defined variable in the active file.
- You can use the TO keyword in the variable list to refer to consecutive variables on the active file.
- The order of variables in the correlation matrix constructed by REGRESSION is the same as their order on the VARIABLES subcommand.
- If you use the keyword (COLLECT), the order of variables in the correlation matrix is the order in which they are first referred to on the DEPENDENT and METHOD subcommands.

You can specify either of the following keywords instead of a variable list:

ALL *Include all user-defined variables in the active file.*

(COLLECT) *Include all variables named on the DEPENDENT and METHOD subcommands.* If (COLLECT) is used, the METHOD subcommands must have variable lists.

Example

```
REGRESSION VARIABLES=(COLLECT)
/DEPENDENT=Y
/METHOD=STEP B C D E F G
/METHOD=ENTER H
/DEPENDENT=H
/METHOD=ENTER G.
```

- (COLLECT) requests that the correlation matrix include Y, B, C, D, E, F, G, and H.
- The first DEPENDENT subcommand defines a single equation in which Y is the dependent variable.
- The first METHOD subcommand requests that the block of variables B to G be considered for inclusion using a stepwise procedure.
- The second METHOD subcommand specifies that variable H be added to the equation, provided that it satisfies the default tolerance criterion.
- A second DEPENDENT subcommand requests an equation in which H is the dependent variable.
- Variable G will be entered into this equation, provided that it satisfies the default tolerance criterion.

DEPENDENT Subcommand

The required DEPENDENT subcommand specifies a list of variables and requests that an equation be built for each.

- The minimum specification is a single variable. There is no default variable list.
- You can specify more than one DEPENDENT subcommand. Each must be followed by at least one METHOD subcommand.
- All variables named on a DEPENDENT subcommand must have been previously named or implied on the VARIABLES subcommand.
- The keyword TO on a DEPENDENT subcommand refers to the order variables are named on the VARIABLES subcommand. If VARIABLES=(COLLECT) was specified, TO refers to the order of variables on the active file.
- If the DEPENDENT subcommand names more than one variable, an equation is built for each using the same independent variable(s) and methods.
- No variable named on the DEPENDENT subcommand can be specified as an independent variable on associated METHOD subcommands.

METHOD Subcommand

The required METHOD subcommand specifies a variable selection method and names a block of variables to be evaluated using that method.

- The minimum specification is a method keyword and, for some methods, a list of variables. The actual keyword METHOD may be omitted.
- The default variable list for methods FORWARD, BACKWARD, STEPWISE, and ENTER consists of all variables named on the VARIABLES subcommand that are not named on the preceding DEPENDENT subcommand.
- There is no default variable list for the REMOVE and TEST methods.
- If VARIABLES=(COLLECT) is specified, you cannot use the default variable list and must name the variables.
- The keyword TO in a variable block named on METHOD refers to the order variables are named on the VARIABLES subcommand. If VARIABLES=(COLLECT) was specified, TO refers to the order of variables on the active file.
- At least one METHOD subcommand must follow each DEPENDENT subcommand.
- When you specify more than one method for a single DEPENDENT subcommand, the methods are cumulative.

The available stepwise methods are

BACKWARD *Backward elimination.* Variables in the block are considered for removal. At each step, the variable with the largest probability-of-*F* value is removed, provided that the value is larger than POUT (see CRITERIA Subcommand). If no variables are in the equation when BACKWARD is specified, all independent variables are first entered.

FORWARD *Forward entry.* Variables in the block are added to the equation one at a time. At each step, the variable not in the equation with the smallest probability of *F* is entered if the value is smaller than PIN (see CRITERIA Subcommand).

STEPWISE *Stepwise selection.* If there are independent variables already in the equation, the variable with the largest probability of *F* is removed if this value is larger than POUT. The equation is recomputed omitting the removed variable and the evaluation process is repeated until no more independent variables can be removed. Then, the independent variable not in the equation with the smallest probability of *F* is entered if this value is smaller than PIN. Then all variables in the equation are again examined for removal. This process continues until no variables in the equation need to be removed and no variables not in the equation are eligible for entry, or until the maximum number of steps has been reached (see CRITERIA Subcommand).

The methods that enter or remove the entire variable block in a single step are

ENTER *Forced entry.* All variables in the block are entered in a single step in order of decreasing tolerance. If the order in which variables are entered is important, use multiple METHOD=ENTER subcommands.

REMOVE *Forced removal.* All variables are removed in a single step. REMOVE requires a variable list.

TEST *Test indicated subsets of independent variables using* R^2 *change and its test of significance.* This method first builds a model that includes all variables named on that METHOD=TEST subcommand, by adding any variables that are not already in the equation. Variables already in the equation are not removed. Test statistics are printed for the subsets, usually in addition to any other statistics that may have been requested. Specify test subsets in parentheses. A variable can be used in more than one subset, and each subset can include any number of variables. Variables named on TEST remain in the equation when the METHOD is completed.

Example

```
REGRESSION VARIABLES=X1 TO X7,Y
/DEPENDENT=Y
/METHOD=STEPWISE
/METHOD=ENTER.
```

- STEPWISE applies the stepwise procedure to variables X1 to X7.
- Any variables not in the equation when the STEPWISE method is complete will be forced into the equation with ENTER.

Example

```
REGRESSION VARIABLES=(COLLECT)
/DEPENDENT=ZVAR
/METHOD=TEST(QVAR15 TO QVAR20)(QVAR15,PVAR)
/METHOD=ENTER XVAR.
```

- The VARIABLES=(COLLECT) specification assembles a correlation matrix that includes all variables named on the DEPENDENT and METHOD subcommands. QVAR15 TO QVAR20 refers to all variables between and including QVAR15 and QVAR20 on the active file.
- REGRESSION first builds the full equation of all the variables named on the first METHOD subcommand: ZVAR regressed on QVAR15 to QVAR20 and PVAR. For each set of test variables, the R^2 change, F, probability, sums of squares, and degrees of freedom are displayed.
- XVAR is added to the equation by the second METHOD subcommand. Variables QVAR15 to QVAR20 and PVAR are still in the equation when this subcommand is executed.

STATISTICS Subcommand

The optional STATISTICS subcommand controls the display of statistics for the equation and for the independent variables.

- The minimum specification is simply the subcommand keyword.
- If the STATISTICS subcommand is omitted or if it is included with no keywords, R, ANOVA, COEFF, and OUTS are displayed (see below).
- If any statistics are specified on STATISTICS, only statistics specifically requested are displayed.
- The statistical displays for stepwise methods (BACKWARD, FORWARD, and STEPWISE) are sometimes different from those for methods that enter and remove blocks of variables. In particular, some statistical displays are not produced for method TEST if the equation has not changed since the last variable block.
- Method TEST always produces its own display in addition to other statistics.
- A STATISTICS subcommand affects any equations that are subsequently defined and remains in effect until overridden by another STATISTICS subcommand.

- A STATISTICS subcommand may not be placed between the DEPENDENT and METHOD subcommands.
- If the width is set to less than 132, some requested statistics may not be displayed.

Global Statistics

DEFAULTS *R, ANOVA, COEFF, and OUTS.* These are displayed if the STATISTICS subcommand is omitted or if it is specified without keywords.

ALL *Display all statistics except F, LINE, and END.*

Equation Statistics

R *Multiple* R. Includes R^2, adjusted R^2, and standard error of the estimate.

ANOVA *Analysis of variance table.* Includes regression and residual sum of squares, mean square, *F*, and probability of *F*.

CHA *Change in* R^2. Includes the change in R^2 between steps, *F* at the end of each step and its probability, and *F* for the equation and its probability. For stepwise methods (BACKWARD, FORWARD, and STEPWISE), these statistics are displayed at the end of each step. For other methods, the statistics are displayed for the variable block.

BCOV *Variance-covariance matrix for unstandardized regression coefficients.* Matrix has covariances below the diagonal, correlations above the diagonal, and variances on the diagonal.

XTX *Sweep matrix.*

COND *Condition number bounds.* Displays the upper and lower bounds for the condition number of the submatrix of the sweep matrix, which contains independent variables already entered. (See Berk, 1977.)

Statistics for the Independent Variables

COEFF *Regression coefficients.* Includes regression coefficients (B), standard errors of the coefficients, standardized regression coefficients (beta), *t*, and two-tailed probability of *t*.

OUTS *Statistics for variables not yet in the equation* that have been named on METHOD subcommands for the equation. Displays beta, *t*, two-tailed probability of *t*, and minimum tolerance of the variable if it were the only variable entered next.

ZPP *Zero-order, part, and partial correlation.*

CI *95% confidence interval for the unstandardized regression coefficient.*

SES *Approximate standard error of the standardized regression coefficients.* (See Meyer & Younger, 1976.)

TOL *Tolerance.* Displays tolerance for variables in the equation and, for variables not in the equation, the tolerance each variable would have if it were the only variable entered next.

F F *value for* B *and its probability.* Displayed instead of the *t* value.

Step Summary Statistics

The full summary line displayed by keywords LINE, END, and HISTORY includes *R*, R^2, adjusted R^2, *F*, probability of *F*, R^2 change, *F* of the change, probability of R^2 change, and statistics on variables added or removed. For stepwise methods (BACKWARD, FORWARD, and STEPWISE), the statistics refer to each step. For other methods (ENTER, REMOVE, and TEST), the statistics refer to the entire variable block. If other statistics are requested, the summary line may not be produced for a block that does not entail steps.

LINE *Display a single summary line for each step, for stepwise methods only.* Does not affect direct methods. The default or requested statistics are displayed at the end of each method block for all methods.

END *Display the same summary line produced by LINE after each step for stepwise methods, and after each variable block for other methods.* For TEST, the summary line is displayed only if the equation changes. Other default or requested statistics are displayed at the completion of the last METHOD subcommand for the equation.

HISTORY *Display a final summary report.* HISTORY can be requested in addition to LINE or END. For stepwise methods, the report includes a summary line for each step. For ENTER and REMOVE, the report includes a summary line for each method; for TEST, the summary line is displayed only if the equation changes. If HISTORY is the only statistic requested, COEFF is displayed for the final equation.

CRITERIA Subcommand

The optional CRITERIA subcommand controls the statistical criteria used in building the regression equations. The way in which these criteria are used depends on the method specified on the METHOD subcommand. The default criteria are noted in the description of each CRITERIA keyword below.

- The minimum specification is a criterion keyword and its arguments, if any.
- If the CRITERIA subcommand is omitted or included with no specifications, the default criteria are in effect.
- Only default criteria that are changed explicitly on the CRITERIA subcommand are altered.
- A CRITERIA subcommand affects any subsequent DEPENDENT and METHOD subcommands and remains in effect until overridden by another CRITERIA subcommand.
- The CRITERIA subcommand may not be placed between the DEPENDENT subcommand and its METHOD subcommands.

Tolerance and Minimum Tolerance Criteria

Variables must pass both tolerance and minimum tolerance tests in order to enter and remain in a regression equation. Tolerance is the proportion of the variance of a variable in the equation that is not accounted for by other independent variables in the equation. The minimum tolerance of a variable not in the equation is the smallest tolerance any variable already in the equation would have if the variable being considered were included in the analysis.

If a variable passes the tolerance criteria, it is further tested according to the method in effect. These criteria are controlled by the CRITERIA subcommand or defaults in effect for the equations that the DEPENDENT subcommand builds.

Testing Independent Variables

The ENTER, REMOVE, and TEST methods use only the TOLERANCE criterion. The stepwise methods (BACKWARD, FORWARD, and STEPWISE) differ in the way they use criteria.

- BACKWARD selects variables according to the probability of *F*-to-remove (keyword POUT). Specify FOUT to use *F*-to-remove instead.
- FORWARD selects variables according to the probability of *F*-to-enter (keyword PIN). Specify FIN to use *F*-to-enter instead.
- STEPWISE uses both PIN and POUT (or FIN and FOUT) as criteria. If the criterion for entry (PIN or FIN) is less stringent than the criterion for removal (POUT or FOUT), the same variable may cycle in and out until the maximum number of steps is reached. Therefore, if PIN is larger than POUT or FIN is smaller than FOUT, REGRESSION adjusts POUT or FOUT and issues a warning.

DEFAULTS *PIN(0.05), POUT(0.10), and TOLERANCE(0.01).* These are the defaults if the CRITERIA subcommand has not been specified. If criteria have been changed, keyword DEFAULTS restores these defaults.

PIN(value) *Probability of* F*-to-enter.* The default value is 0.05. Keywords PIN and FIN are alternatives. If more than one is used, the last mentioned is in effect.

FIN(value) F*-to-enter.* If no value is specified, the value defaults to 3.84. Keywords PIN and FIN are alternatives. If more than one is used, the last mentioned is in effect.

POUT(value) *Probability of* F-*to-remove*. The default value is 0.10. Keywords POUT and FOUT are alternatives. If more than one is used, the last mentioned is in effect.

FOUT(value) F-*to-remove*. If no value is specified, the value defaults to 2.71. Keywords POUT and FOUT are alternatives. If more than one is used, the last mentioned is in effect.

TOLERANCE(value) *Tolerance*. The default value is 0.01. If the tolerance chosen is very low, REGRESSION issues a warning.

MAXSTEPS(n) *Maximum number of steps*. The value of MAXSTEPS is the sum of the maximum number of steps over each method for the equation. The default values are

BACKWARD or FORWARD methods: the number of variables meeting PIN/POUT or FIN/FOUT criteria.

STEPWISE method: twice the number of independent variables.

Example

```
REGRESSION VARIABLES=X1 TO X7,Y
/CRITERIA=PIN(.1) POUT(.15) TOL(.001)
/DEPENDENT=Y
/METHOD=FORWARD
/CRITERIA=DEFAULTS
/DEPENDENT=Y
/METHOD=STEPWISE.
```

- The first CRITERIA subcommand relaxes the default criteria for entry and removal while the FORWARD method is used. Note that the specified PIN is less than POUT.
- The second CRITERIA subcommand reestablishes the defaults for the second equation.

ORIGIN and NOORIGIN Subcommands

The optional ORIGIN and NOORIGIN subcommands control whether or not the constant is suppressed.

- The minimum specification is simply the ORIGIN or NOORIGIN subcommand. There are no additional specifications.
- ORIGIN and NOORIGIN must be specified before the DEPENDENT and METHOD subcommands they modify.
- ORIGIN and NOORIGIN are mutually exclusive.
- If neither ORIGIN nor NOORIGIN is specified, NOORIGIN is the default and all equations include a constant term (intercept).
- ORIGIN requests regression through the origin. The constant term is suppressed.
- Once specified, ORIGIN remains in effect until NOORIGIN is requested.
- If you specify ORIGIN, statistics requested on the DESCRIPTIVES subcommand are computed as if the mean were zero.

Example

```
REGRESSION VAR=(COL)
/DEP=A
/METHOD=ENTER B
/ORIGIN
/DEP=A
/METHOD=ENTER B
/NOORIGIN
/DEP=B
/METHOD=ENTER C.
```

- The subcommand VAR=(COL) builds a correlation matrix that includes A, B, and C.
- The REGRESSION command requests three equations. The first regresses A on B and includes a constant term because the default (NOORIGIN) is in effect. The second regresses A on B and suppresses the constant (ORIGIN). The third regresses B on C and includes a constant term because NOORIGIN has been specified.
- This example takes advantage of spelling permitted by three-character truncation of keywords.

DESCRIPTIVES Subcommand

By default, descriptive statistics are not displayed. Use the optional DESCRIPTIVES subcommand to request the display of correlation and descriptive statistics for variables in the correlation matrix.

- The minimum specification is simply the subcommand keyword.
- If DESCRIPTIVES is specified without keywords, MEAN, STDDEV, and CORR are displayed.
- If DESCRIPTIVES is included and any keywords are specified, only those statistics specifically requested are displayed.
- Descriptive statistics are displayed only once for all variables named or implied on VARIABLES.
- Descriptive statistics are based on all valid cases for each variable if PAIRWISE or MEANSUBSTITUTION has been specified on MISSING. Otherwise, only cases included in the computation of the correlation matrix are included in the calculation of the descriptive statistics.
- If regression through the origin has been requested (subcommand ORIGIN), statistics are computed as if the mean were zero.

NONE *Turn off all descriptive statistics.* This is the default if the subcommand is omitted.

DEFAULTS *MEAN, STDDEV, and CORR.* Same as specifying DESCRIPTIVES without specifications.

MEAN *Variable means.*

STDDEV *Variable standard deviations.*

VARIANCE *Variable variances.*

CORR *Correlation matrix.*

SIG *One-tailed probabilities of the correlation coefficients.*

BADCORR *Display the correlation matrix only if some coefficients cannot be computed.*

COV *Covariance matrix.*

XPROD *Cross-product deviations from the mean.*

N *Numbers of cases used to compute correlation coefficients.*

ALL *Display all descriptive statistics.*

Example

```
REGRESSION DESCRIPTIVES=DEFAULTS SIG COV
/VARIABLES=X1 TO X7,Y
/DEPENDENT=Y
/METHOD=ENTER X1
/METHOD=ENTER X2.
```

- The variable means, variable standard deviations, correlation matrix, one-tailed probabilities of the correlation coefficients, and covariance matrix are displayed.
- Statistics are displayed for all variables named on VARIABLES, even though only variables Y, X1, and X2 are used to build the equations.
- X1 is entered into the equation by the first METHOD subcommand; X2 is entered by the second METHOD subcommand.

SELECT Subcommand

By default, all cases on the active file are considered for inclusion in REGRESSION. Use the optional SELECT subcommand to include a subset of cases in the correlation matrix and resulting regression statistics.

- The minimum specification is a logical expression. The default is to include all cases if the subcommand is omitted.
- Do not include the variable named on SELECT on the VARIABLES subcommand.
- The logical expression on SELECT is of the form

  ```
  /SELECT=varname relation value
  ```

  where the relation can be EQ, NE, LT, LE, GT, or GE.

- Only cases for which the logical expression on SELECT is true are included in the calculation of the correlation matrix and regression statistics. All other cases, including those with missing values for the variable named on SELECT, are not selected.
- SELECT displays additional statistics describing the fit of the model estimated among the unselected cases.
- By default, residuals and predicted values are calculated and reported separately for both selected and unselected cases (see REGRESSION: Residuals).
- Cases deleted from the active file with the SELECT IF, PROCESS IF, or SAMPLE commands are not passed to REGRESSION. Such cases are not reviewed by the SELECT subcommand and are not included among either the selected or unselected cases.
- The display of the values of the variable named on SELECT is controlled by that variable's format (see DATA LIST).

Example

```
REGRESSION SELECT SEX EQ 'M'
/VARIABLES=X1 TO X7,Y
/DEPENDENT=Y
/METHOD=STEP
/RESIDUALS=NORMPROB.
```

- Only cases with the value M for SEX are included in the correlation matrix calculated by REGRESSION.
- Separate normal probability plots are displayed for cases with SEX equal to M and for other cases (see REGRESSION: Residuals).

MISSING Subcommand

By default, a case that has a user- or system-missing value for any variable named or implied on the VARIABLES subcommand is omitted from the computation of the correlation matrix on which all analyses are based. Use the optional MISSING subcommand to change the treatment of cases with missing values.

- The minimum specification is a keyword specifying a missing-value treatment.

The available keywords are

LISTWISE *Delete cases with missing values listwise.* This is the default. Only cases with valid values on all variables named on the current VARIABLES subcommand are used. If INCLUDE is also specified, only cases with system-missing values are deleted listwise. LISTWISE is the default if the MISSING subcommand is omitted.

PAIRWISE *Delete cases with missing values pairwise.* Each correlation coefficient is computed using cases with complete data for the pair of variables correlated. If INCLUDE is also specified, only cases with system-missing values are deleted pairwise.

MEANSUBSTITUTION *Replace missing values with the variable mean.* All cases are included and the substitutions are treated as valid observations. If INCLUDE is also specified, user-missing values are included in the computation of the means. Mean substitution affects the computation of the correlation matrix and the calculation of predicted values and residuals.

INCLUDE *Include cases with user-missing values.* All user-missing values are treated as valid values. This keyword can be specified along with the methods LISTWISE, PAIRWISE, or MEANSUBSTITUTION. Including user-missing values affects the computation of the correlation matrix and the calculation of predicted values and residuals.

Example

```
REGRESSION  VARIABLES=X1 TO X7,Y
/DEPENDENT=Y
/METHOD=STEP
/MISSING=MEANS.
```

- Missing values are replaced with the means of the variables when the correlation matrix is calculated.
- The MEANS keyword takes advantage of spelling permitted by three-character truncation.

WIDTH Subcommand

The optional WIDTH subcommand controls the width of the display within the REGRESSION procedure.

- The minimum specification is an integer between 72 and 132.
- The default display uses the width specified on SET. The width specified on the WIDTH subcommand within REGRESSION overrides the width on SET for the REGRESSION display only.
- The WIDTH subcommand can appear anywhere.
- If more than one WIDTH subcommand is included, the last WIDTH specified will be in effect for the display.
- If the width is less than 132, some statistics may not be displayed.

References

Berk, K. N. 1977. Tolerance and condition in regression computation. *Journal of the American Statistical Association* 72:863-66.

Meyer, L. S., and M. S. Younger. 1976. Estimation of standardized coefficients. *Journal of the American Statistical Association* 71:154-57.

REGRESSION: Matrix Materials

```
REGRESSION [READ=[DEFAULTS] [MEAN] [STDDEV]

           [VARIANCE] {CORR} [N]]
                      {COV }

  [/WRITE=[DEFAULTS] [MEAN] [STDDEV]

          [VARIANCE] [CORR] [COV]

          [N] [NONE**]]

 /VARIABLES=varlist/DEPENDENT=varlist/METHOD=method
```

**Default if subcommand is omitted.

Example:

```
REGRESSION  VARIABLES=AGE TO SUICIDE
/DESCRIPTIVES
/WRITE
/DEP=SUICIDE
/METHOD=ENTER
/RESIDUALS.
```

Overview

Procedure REGRESSION can both read and write matrix materials. It accepts matrix materials written by REGRESSION or other SPSS/PC+ procedures such as CORRELATION, or entered directly from other sources provided that the appropriate keyword specifications are used. Such matrix materials can be processed more quickly than cases. Matrix materials to be read can be in either fixed or freefield format but must conform to certain format and record specifications (see DATA LIST: Matrix Materials).

The subcommands for writing and reading matrix materials can be used in addition to the REGRESSION subcommands described in REGRESSION.

WRITE Subcommand

Use the WRITE subcommand to write the matrix materials used in the REGRESSION computations to an external file.

- The WRITE subcommand can be specified anywhere except between the DEPENDENT subcommand and the METHOD subcommands that define an equation.
- If the WRITE subcommand is included without specifications, REGRESSION writes a vector of means, a vector of standard deviations, a correlation matrix, and the number of cases.
- If any keyword specifications are given, only those materials specified are written.
- Matrix materials are written for all variables named on the VARIABLES subcommand.
- The order of variables in vectors and matrices is the same as the order in which variables are named on the VARIABLES subcommand.
- If (COLLECT) is used on the VARIABLES subcommand, the order of the variables in the matrix materials is the order in which they are first named on REGRESSION. Use keyword CORRELATION on the DESCRIPTIVES subcommand to display a listing of the matrix you write.
- Only one WRITE subcommand should be used. If more than one WRITE appears, the last one encountered will be in effect.
- Each type of matrix material written begins on a new record. Eight 10-column fields are used on each record.
- REGRESSION displays a format table describing the contents and format of the file it has written.

- The VARIABLES, SELECT, ORIGIN or NOORIGIN, and MISSING subcommands in effect for the REGRESSION procedure also affect the materials that are written.
- Matrix materials are written to the results file named on the SET command (by default, SPSS.PRC).
- If the results file named on SET is not empty when the WRITE subcommand is executed, its contents will be overwritten.

The following keywords can be specified on WRITE. These keywords should be specified in the order they are listed below.

DEFAULTS *Includes MEAN, STDDEV, CORR, and N.* This is the default if the WRITE subcommand is used without specifications.

MEAN *Write a vector of means.*

STDDEV *Write a vector of standard deviations.*

VARIANCE *Write a vector of variances.*

CORR *Write a correlation matrix.*

COV *Write a covariance matrix.*

N *Write out the* n*'s of cases used to compute correlation coefficients.* When the MISSING=LISTWISE is in effect, the number of cases is written as one item on the last record. When PAIRWISE or MEANSUBSTITUTION is specified on MISSING, a matrix of *n*'s is written.

NONE *No matrix output* (default).

Example

```
SET RESULTS='GSS82.MAT'.
DATA LIST FILE='GSS82.DAT'/AGE 5-6 INCOME 7-13
ANOMIE1 TO ANOMIE7 14-20 SUICIDE 21.
REGRESSION  VARIABLES=AGE TO SUICIDE
/DESCRIPTIVES
/WRITE
/DEP=SUICIDE
/METHOD=ENTER
/RESIDUALS.
```

- The SET command sets the results file to the file GSS82.MAT in the current directory.
- The DATA LIST command specifies variable names and column locations for raw data in file GSS82.DAT.
- The VARIABLES subcommand on REGRESSION requests that all variables in the active file between and including AGE and SUICIDE be included in the computation of the correlation matrix for the regression.
- The DESCRIPTIVES subcommand requests descriptive statistics for all variables in the analysis.
- The WRITE subcommand requests that default matrix materials be written to the file specified on the SET RESULTS command. GSS82.MAT will contain a vector of means, a vector of standard deviations, a correlation matrix, and the number of cases, in that order. All variables from AGE to SUICIDE will be included in the matrix materials.
- A regression of SUICIDE on all other variables implied by the VARIABLES subcommand will be computed and the default display produced (see REGRESSION).
- The RESIDUALS subcommand displays a histogram of the standardized residuals, a normal probability plot of the standardized residuals, the Durbin-Watson test statistic, and a listing of the 10 worst outliers based on the absolute value of the standardzied residuals (see REGRESSION: Residuals).

READ Subcommand

Use the READ subcommand to read matrix materials.

- There can be only one READ subcommand.
- The READ subcommand cannot be specified between the DEPENDENT and METHOD subcommands.

- If the READ subcommand is used without specifications, REGRESSION assumes that the matrix materials have the default structure: a vector of means, a vector of standard deviations, a correlation matrix, and the number of cases.
- If any keywords are specified on READ, only those matrix materials specified will be expected.
- The matrix materials you read must permit regression statistics to be calculated. You must read either a correlation or covariance matrix.
- When you specify READ on REGRESSION, you must first specify a DATA LIST MATRIX command that points to the file containing the matrix materials and names the variables that will be read (see DATA LIST: Matrix Materials).
- All matrix materials read must be in a single input file.
- The order in which variables are named on the VARIABLES subcommand of REGRESSION must be the order of the variables in each vector or matrix that is read.
- If a correlation matrix is the only matrix material to be read, an N command must be entered prior to REGRESSION to specify the number of cases. Only standardized coefficients will be available.
- If more than one kind of matrix material is present, the matrix materials must be arranged in the input file in the following order: the vector of means, the vector of standard deviations, the vector of variances, the correlation or covariance matrix, and the *n*'s of cases.
- The specification on the MISSING and ORIGIN or NOORIGIN subcommands should agree with the options in effect when the matrix was written.
- The descriptive statistics available with the DESCRIPTIVES subcommand depend on which matrix materials are read.
- The RESIDUALS, CASEWISE, SCATTERPLOT, and PARTIALPLOT subcommands are not available when matrix materials are read.
- The (COLLECT) keyword on the VARIABLES subcommand is not allowed if the READ subcommand is used.

DEFAULTS *MEAN, STDDEV, CORR, and N.* If you specify READ without specifications, the input file must contain these materials in this order. Matrix materials written by REGRESSION are in this default format.

MEAN *The matrix is preceded by a vector of means.*

STDDEV *The matrix is preceded by a vector of standard deviations.*

VARIANCE *The matrix is preceded by a vector of variances.*

CORR *Correlation matrix.* Alternative to keyword COV.

COV *Covariance matrix.* Alternative to keyword CORR. A covariance matrix is not allowed if pairwise deletion of missing values is specified.

N *The number of cases used to compute correlation coefficients follows the matrix.* If the MISSING subcommand specifies MEANSUBSTITUTION or PAIRWISE, a symmetric matrix of *n*'s is expected. If the MISSING subcommand specifies LISTWISE or INCLUDE, all coefficients are based on the same number of cases and a single number is expected. If a single number of cases is expected, it will be read from the first 10 columns of the last record of the matrix file. If the keyword N is not specified, the *n* specified on the N command is used. If an *n* is read from the matrix file and an *n* is specified on the N command, the *n* read from the matrix file is used.

Example

```
DATA LIST FIXED MATRIX FILE='GSS82.MAT'
/AGE INCOME ANOMIE1 TO ANOMIE7 SUICIDE.
REGRESSION  READ
/VARIABLES=AGE INCOME ANOMIE1 TO ANOMIE7 SUICIDE
/DEPENDENT=SUICIDE
/METHOD=ENTER ANOMIE1 TO ANOMIE7.
```

- The DATA LIST command specifies that the matrix input should be read from the file GSS82.MAT and names the variables in the file. The names of the variables on the DATA LIST command will be entered in the dictionary of the active file.
- The READ subcommand on REGRESSION requests that matrix materials be read and used for the procedure. Because no keyword specifications are given and the default listwise treatment of missing values is in effect, REGRESSION expects a vector of means, a vector of standard deviations, a correlation matrix, and a single *n* of cases.
- The VARIABLES subcommand names the variables in the order in which they appear in the vectors and matrix to be read.
- The DEPENDENT subcommand defines an equation in which SUICIDE is the dependent variable.
- The METHOD subcommand requests that the variables ANOMIE1 to ANOMIE7 be entered into the equation using the ENTER method.
- The variables AGE and INCOME are not used in the equation but must be named on the VARIABLES subcommand so that the locations of all variables in the matrix file are identified accurately.

REGRESSION: Residuals

```
REGRESSION  VARIABLES=varlist/ DEPENDENT=varname/ METHOD=method
  [/RESIDUALS=[DEFAULTS] [DURBIN]

     [OUTLIERS({ZRESID      })] [ID (varname)]
               {tempvarlist}

     [NORMPROB({ZRESID      })] [HISTOGRAM({ZRESID      })]
               {tempvarlist}               {tempvarlist}

     [SIZE({SMALL})] [{SEPARATE}]]
           {LARGE}    {POOLED  }

  [/CASEWISE=[DEFAULTS] [{OUTLIERS({3    })}]
                         {         {value} }
                         {ALL              }

        [PLOT({ZRESID })]  [{DEPENDENT PRED RESID}]]
              {tempvar}     {tempvarlist         }

  [/SCATTERPLOT=(varname,*tempvarname)... [SIZE({SMALL})]]
                                                {LARGE}

  [/PARTIALPLOT=[{ALL                   }] [SIZE({SMALL})]]
                 {varname,varname,...}           {LARGE}

  [/SAVE=tempvar(newname) [tempvar(newname)... ]]
```

Temporary residual variables are:

```
PRED ADJPRED SRESID MAHAL RESID ZPRED SDRESID
COOK DRESID ZRESID SEPRED LEVER
```

Example:

```
REGRESSION VARIABLES=SAVINGS INCOME POP15 POP75
/WIDTH=132
/DEPENDENT=SAVINGS
/METHOD=ENTER
/RESIDUALS
/CASEWISE
/SCATTERPLOT (*ZRESID *ZPRED)
/PARTIALPLOT
/SAVE ZRESID(STDRES) ZPRED(STDPRED).
```

Overview

REGRESSION automatically calculates predicted values and residuals and several statistics based on these. The entire distribution of these statistics can be examined and outliers can be identified. The automatic variables are available for analysis within REGRESSION by means of casewise plots, scatterplots, and partial plots of the variables with independent variables in the analysis. In addition, you can specify any of the residuals subcommands to obtain statistics on the predicted values, residuals, and their standardized versions. You can request that any of the automatic residuals variables be added to your active file.

The residuals subcommands follow the last METHOD subcommand of any equation for which residuals analysis is requested. Statistics are based on this final equation.

Defaults

All residuals analysis subcommands are optional but most have active defaults that can be requested by including the subcommand without any further specifications. These active defaults are described by subcommand below.

Tailoring

You can specify which residuals statistics are presented, which variables are added to the active file, which are plotted, and the size of plots.

Syntax

- The optional residuals subcommands RESIDUALS, CASEWISE, SCATTERPLOT, and PARTIALPLOT must follow the last METHOD subcommand for an equation. If you specify more than one equation (more than one DEPENDENT subcommand), you can request residual analysis after each one.

- Residuals subcommands can be specified in any order.
- Residuals subcommands affect only the equation they follow.
- The residuals subcommands cannot be specified if matrix input is used.

Operations

- All calculations and plots requested on the RESIDUALS subcommand are based on the regression equation produced as a result of the last method specified.
- The temporary variables PRED, RESID, ZPRED, and ZRESID are calculated and descriptive statistics are printed for these variables whenever any residuals subcommand is specified. Referring to any of the other eight temporary variables (or specifying DEPENDENT on CASEWISE) causes these other eight temporary variables to be calculated.
- Predicted values and statistics based on predicted values are calculated for every observation that has valid values for all variables in the equation.
- Residuals and statistics based on residuals are calculated for all observations that have a valid predicted value and a valid value for the dependent variable.
- The missing-values option therefore affects the calculation of residuals and predicted values.
- The amount of information displayed in a casewise plot is limited by the display width (see REGRESSION).
- The widest page allows a maximum of eight variables in a casewise plot.
- No residuals or predictors are generated for cases deleted from the active file with the PROCESS IF, SELECT IF, or SAMPLE commands.
- All variables are standardized before plotting. If the unstandardized version of a variable is requested, the standardized version is plotted.
- For each analysis, REGRESSION can calculate 12 temporary variables:

PRED *Unstandardized predicted values.*

RESID *Unstandardized residuals.*

DRESID *Deleted residuals.*

ADJPRED *Adjusted predicted values.*

ZPRED *Standardized predicted values.*

ZRESID *Standardized residuals.*

SRESID *Studentized residuals.*

SDRESID *Studentized deleted residuals.* (See Hoaglin & Welsch, 1978.)

SEPRED *Standard errors of the predicted values.*

MAHAL *Mahalanobis' distances.*

COOK *Cook's distances.* (See Cook, 1977.)

LEVER *Leverage values.* (See Velleman & Welsch, 1981.)

Limitations

- If there is not enough storage available to assemble the requested plots, a warning is printed. Small plots are displayed and some plots may be deleted.

Example

```
REGRESSION VARIABLES=SAVINGS INCOME POP15 POP75
/WIDTH=132
/DEPENDENT=SAVINGS
/METHOD=ENTER
/RESIDUALS
/CASEWISE
/SCATTERPLOT (*ZRESID *ZPRED)
/PARTIALPLOT
/SAVE ZRESID(STDRES) ZPRED(STDPRED).
```

- This REGRESSION command requests a single equation in which SAVINGS is the dependent variable and INCOME, POP15, and POP75 are independent variables.
- The RESIDUALS subcommand requests the default residuals output.

- Because residuals processing has been requested, statistics for predicted values, residuals, and standardized versions of predicted values and residuals are displayed.
- The CASEWISE subcommand requests a casewise plot of ZRESID of cases for which the absolute value of ZRESID is greater than 3. Values of the dependent variable, predicted value, and residual are listed for each case.
- The SCATTERPLOT subcommand requests a small plot of the standardized predicted value by the standardized residual.
- The PARTIALPLOT subcommand requests small partial residual plots of SAVINGG by POP75, SAVINGS by POP15, and SAVINGS by INCOME.
- The SAVE subcommand requests that the standardized residual and the standardized predicted value be added to the active file, as new variables named STDRES and STDPRED.

RESIDUALS Subcommand

The RESIDUALS subcommand controls the display and labeling of summary information on outliers as well as the display of the Durbin-Watson statistic and histograms and normal probability plots for the temporary variables.

- The RESIDUALS subcommand without specifications displays a histogram of the standardized residuals, a normal probability plot of the standardized residuals, the values of $CASENUM and ZRESID for the 10 cases with the largest absolute value of ZRESID, and the Durbin-Watson test statistic. The default SIZE of both plots is LARGE when no specifications are given.
- If any keyword specifications are given for RESIDUALS, only the displays requested are produced.

DEFAULTS *SIZE(LARGE), DURBIN, NORMPROB(ZRESID), HISTOGRAM(ZRESID), OUTLIERS(ZRESID).* These are the defaults if the RESIDUALS subcommand is included without specifications.

SIZE(plotsize) *Plot sizes.* The plot size can be SMALL or LARGE. The default is large plots if the display width is at least 120 (see REGRESSION) and the page length is at least 57 (see SET). Four small histograms or normal probability plots can be displayed on a single page if the width is 132 (see REGRESSION) and the page length is 59.

HISTOGRAM(tempvars) *Histogram of the standardized temporary variable or variables.* The default is ZRESID. The other temporary variables for which histograms are available are PRED, RESID, ZPRED, DRESID, ADJPRED, SRESID, and SDRESID.

NORMPROB(tempvars) *Normal probability (P-P) plot of standardized values.* The default is ZRESID. The other temporary variables available for normal probability plots are PRED, RESID, ZPRED, DRESID, ADJPRED, SRESID, and SDRESID.

OUTLIERS(tempvars) *The 10 worst outliers based on absolute values of the specified temporary variables.* The default is ZRESID. The listing includes the value of $CASENUM and of the temporary variables for the 10 cases. The other temporary variables available for OUTLIERS are RESID, SRESID, SDRESID, DRESID, MAHAL, and COOK.

DURBIN *Durbin-Watson test statistic.*

ID(varname) *The case identifier on outlier plots.* Any variable on the active file can be named. ID also labels the list of cases produced by CASEWISE.

POOLED *Display pooled plots and statistics using all cases in the active file when the SELECT subcommand is in effect* (see REGRESSION). The alternative to POOLED is the default keyword SEPARATE, which requests separate reporting of residuals statistics and plots for selected and unselected cases.

Example

```
/RESID=DEFAULT ID(SVAR)
```

- DEFAULT implies the default residuals statistics: Durbin-Watson statistic, a normal probability plot and histogram of ZRESID, and an outlier plot of ZRESID.
- Descriptive statistics for ZRESID, RESID, PRED, and ZPRED are automatically displayed.
- ID(SVAR) names SVAR the case identifier on outlier plots. If the CASEWISE subcommand is also included, SVAR is used to label cases in the casewise plot.
- This example takes advantage of spelling permitted by three-character truncation of keywords.

CASEWISE Subcommand

The CASEWISE subcommand requests a casewise plot of residuals. On it you specify a temporary residual variable for casewise plotting (PLOT) and control the selection of cases for plotting (OUTLIERS or ALL). CASEWISE can also be used to specify variables to be listed for each case next to the plot.

- The CASEWISE subcommand without specifications displays a casewise plot of ZRESID for cases for which the absolute value of ZRESID is at least 3. By default the values of the case sequence number, DEPENDENT, PRED, and RESID are listed next to the plot entry for each case.
- Only those defaults specifically altered are changed.

DEFAULTS *OUTLIERS(3), PLOT(ZRESID), DEPENDENT, PRED, and RESID.* These are the defaults if the CASEWISE subcommand is included without specifications.

OUTLIERS(value) *Plot only cases for which the absolute standardized value of the plotted variable is at least as large as the value given.* The default value is 3. The alternative to a casewise plot of outliers is a plot of all cases (keyword ALL). The keyword OUTLIERS is ignored if the keyword ALL is also present.

ALL *Include all cases in the casewise plot.* Alternative to the OUTLIERS keyword.

PLOT(tempvar) *Plot the standardized values of the temporary variable in the casewise plot.* The default temporary variable is ZRESID. Other variables that can be plotted are RESID, DRESID, SRESID, and SDRESID.

temp varlist *Display the values of these variables for each case next to its casewise plot entry.* The default variables are DEPENDENT (the dependent variable), PRED, and RESID. Other variables that can be named are DRESID, ADJPRED, ZPRED, ZRESID, SRESID, SDRESID, SEPRED, MAHALANOBIS, COOK, and LEVER. If an ID variable is specified on the RESIDUALS subcommand, the ID variable is also listed if the width is sufficient.

Example

```
/CASEWISE=DEFAULT ALL SRE MAH COOK SDR
```

- This example requests a casewise plot of the standardized residuals for all cases.
- The dependent variable and the temporary variables PRED, RESID, SRESID, MAHAL, COOK, and SDRESID are also listed for all cases.
- This example takes advantage of spelling permitted by three-character truncation of keywords.

SCATTERPLOT Subcommand

The SCATTERPLOT subcommand names pairs of variables for scatterplots and controls the size of the plots.

- The minimum specification for SCATTERPLOT is a pair of variables in parentheses.
- There are no default specifications for SCATTERPLOT.
- The first variable named in each set of parentheses is plotted along the vertical axis, and the second variable is plotted along the horizontal axis.
- Specify as many pairs of variables in parentheses as you want.
- Plotting symbols are used to represent multiple points occurring at the same print position.
- Specify the temporary variable names with a asterisk prefix to distinguish temporary from user-defined variables.
- All scatterplots are standardized. That is, specifying *RESID is the same as specifying *ZRESID, and *PRED is the same as *ZPRED.

(varname,varname) *Plot the variables specified.* Available temporary variables are PRED, RESID, ZPRED, ZRESID, DRESID, ADJPRED, SRESID, and SDRESID, along with any variable on the VARIABLES subcommand.

SIZE(plotsize) *Plot sizes.* The plot size can be either SMALL or LARGE. The default is always small. Four small scatterplots can be displayed on a single page if the width is at least 120 (see REGRESSION) and the page length is at least 57 (see SET).

Example

```
/SCATTERPLOT (*RES,*PRE)(*RES,YVAR)
```

- This example specifies two scatterplots: residuals against predicted values and residuals against the values of the variable YVAR.
- This example takes advantage of spelling permitted by three-character truncation of keywords.

PARTIALPLOT Subcommand

Use the PARTIALPLOT subcommand to request partial residual plots and to control the size of the plots. Partial residual plots are scatterplots of the residuals of the dependent variable and an independent variable when both of these variables are regressed on the rest of the independent variables.

- If the PARTIALPLOT subcommand is included without specifications, a partial residual plot is produced for every independent variable in the equation.
- All plots are standardized.
- Plots are displayed in descending order of the standard error of *B*.

varlist *List of variables to be plotted.* Any variable entered into the equation can be named. At least two independent variables must be in the equation for partial residual plots to be produced. The default is every independent variable in the equation. You can request the defaults with keyword ALL.

SIZE(plotsize) *Plot sizes.* The plotsize can be either SMALL or LARGE. The default is always small. Four small partialplots can be displayed on a single page if the width is at least 120 (see REGRESSION) and the page length is at least 57 (see SET).

Example

```
REGRESSION VARS=YVAR TO ZVAR
/DEP=YVAR
/METH=ENTER
/RESID=DEFAULTS
/PARTIAL.
```

- A partial residual plot is produced for every variable from YVAR to ZVAR that is entered into the equation.
- This example takes advantage of spelling permitted by three-character truncation of keywords.

SAVE Subcommand

Use the SAVE subcommand to add new variables to the active file containing one or more of the temporary residual variables.

- If the SAVE subcommand causes the number of variables in the active file to exceed the limit of 200, it will generate an error and the command will not be executed.
- Specifications for SAVE consist of one or more of the 12 temporary variable names (see Operations, above), each followed by a new variable name in parentheses.
- New variable names must be unique.

Example

```
/SAVE=PRED(PREDVAL) RESID(RESIDUAL) COOK(CDISTANC)
```

- This subcommand adds three variables to the end of the file: PREDVAL, containing the unstandardized predicted value for the case; RESIDUAL, containing the unstandardized residual for the case; and CDISTANC, containing the Cook's distance for the case.
- The new variables are saved onto the active file. If you do not save the file (using the SAVE *command*), they exist only for the duration of your session, or until you replace the active file.

References

Cook, R. D. Detection of influential observations in linear regression. *Technometrics* 19 (1977) 15–18.

Hoaglin, D. C., and R. E. Welsch. The hat matrix in regression and ANOVA. *American Statistician* 32 (1978) 17–22.

Velleman, P. F., and R., E. Welsch. Efficient computing of regression diagnostics. *American Statistician* 35 (1981) 234–42.

REPORT

```
REPORT

   [/FORMAT=[{MANUAL    }] [{NOLIST    }] [ALIGN({LEFT  })] [TSPACE({1})]
             {AUTOMATIC}     {LIST[(n)]}          {CENTER}            {n}
                                                  {RIGHT }

       [CHDSPACE({1})] [FTSPACE({1})] [SUMSPACE({1})] [COLSPACE({4})]
                 {n}            {n}             {n}             {n}

       [BRKSPACE({1  })] [LENGTH({1,length})] [MARGINS({1,width})]
                 {n  }           {n,n     }            {n,n    }
                 {-1†}           {*,*     }            {*,*    }

       [CHALIGN({TOP    })] [UNDERSCORE({OFF})] [PAGE1({1})] [MISSING {'.'}]]
                {BOTTOM†}               {ON†}          {n}            {'s'}

   [/OUTFILE=file]

   [/STRING=stringname (varname[(width)] [(BLANK)] ['literal...'])]

   /VARIABLES=varname ({VALUE}) [+ varname({VALUE})] ['col head'] [(option lis
                       {LABEL}              {LABEL}
                       {DUMMY}              {DUMMY}

   where option list can contain any of the following:

          width     OFFSET({0      })     {LEFT  }
                           {n      }      {CENTER}
                           {CENTER†}      {RIGHT }

   [/MISSING={VAR             }]
             {NONE            }
             {LIST(varlist{1})}
                          {n}

   [/TITLE='line1' 'line2'...      ]   [/FOOTNOTE='line1' 'line2'...     ]
              or                                  or
   [/TITLE=LEFT 'line1' 'line2'...]   [/FOOTNOTE=LEFT 'line1' 'line2'...]
   [     CENTER 'line1' 'line2'...]   [         CENTER 'line1' 'line2'...]
   [      RIGHT 'line1' 'line2'...]   [          RIGHT 'line1' 'line2'...]

               [)PAGE]  [)DATE]  [)var]

   [/BREAK=varlist [(TOTAL)] ['col head'] [(option list)]]

  where option list can contain any of the following:

         width   {VALUE }   {NOTOTAL}   SKIP({1})   PAGE[(RESET)]
                 {LABEL†}   {TOTAL  }        {n}

         OFFSET({0      })   UNDERSCORE[(varlist)]   {LEFT  }   {NONAME}
                {n      }                            {CENTER}   {NAME  }
                {CENTER†}                            {RIGHT }

   [/SUMMARY=function...['summary title'][(break col #1)] [SKIP({0})]]
                                                               {n}
   or

    [/SUMMARY=PREVIOUS[({1})]]
                        {n}

   where function is

    aggregate [(varname[(d)][({PLAIN })][varname...])]
                              {DOLLAR}
                              {COMMA }

   or

    composite(agg(varname)...)[(report col[(d)][({PLAIN })])]
                                                 {DOLLAR}
                                                 {COMMA }
```

†Default if FORMAT=AUTOMATIC.

Aggregate Functions:

VALIDN	VARIANCE	PCLT(n)
SUM	KURTOSIS	PCIN(min,max)
MIN	SKEWNESS	FREQUENCY(min,max)
MAX	MEDIAN(min,max)	PERCENT(min,max)
MEAN	MODE(min,max)	
STDDEV	PCGT(n)	

Composite Functions

DIVIDE(agg(varname) agg(varname)[factor])
PCT(agg(varname) agg(varname))
SUBTRACT(agg(varname) agg(varname))
ADD(agg(varname) agg(varname)...)
GREAT(agg(varname) agg(varname)...)
LEAST(agg(varname) agg(varname)...)
AVERAGE(agg(varname) agg(varname)...)
MULTIPLY(agg(varname) agg(varname)...)

Example:

```
REPORT FORMAT=AUTOMATIC LIST
/VARIABLES=PRODUCT (LABEL) ' ' 'Retail' 'Products'
           SALES 'Annual' 'Sales' '1981'
/BREAK=DEPT 'Department'
/SUMMARY=VALIDN (PRODUCT) MEAN (SALES).
```

Overview

The REPORT procedure produces both case listings and summary statistics and gives you considerable control over the appearance of the output. REPORT calculates all the univariate statistics available in DESCRIPTIVES and the statistics and subpopulation means available in MEANS. In addition, REPORT calculates statistics not directly available in any other SPSS/PC+ procedure, such as computations involving aggregated statistics.

REPORT provides complete format defaults or lets you customize column widths, titles, footnotes, spacing, and other elements. Because REPORT is so flexible and its output can have so many components, it is often efficient to preview REPORT output using a small number of cases until you find the format that best meets your needs.

Defaults

A listing report without subgroup classification requires the FORMAT and VARIABLES subcommands. A listing report with subgroup classification requires the FORMAT, VARIABLES, and BREAK subcommands. A report with summary statistics requires the VARIABLES, BREAK, and SUMMARY subcommands.

By default, column heads use variable labels, or variable names if no variable labels have been specified. Default column widths are determined by REPORT using the maximum of the following for each variable:

- The widest print format in the column, whether it is a variable print format or a summary print format.
- The width of any temporary variable defined with the STRING subcommand.
- If a column heading is assigned, the length of the longest title line in that heading.
- If you specify (LABEL), the length of the variable's longest value label. (If MANUAL is in effect, REPORT uses the length of the variable's longest label, up to a maximum of 20 characters.)

Intercolumn spacing adjusts automatically, using a minimum of 1 and a maximum of 4 spaces between columns. By default, cases with user-missing values are excluded from the calculation of report statistics, and missing-value indicators are ignored for variables named on BREAK.

Many default settings depend on whether FORMAT=AUTOMATIC or MANUAL. Table A below shows the defaults according to both specifications.

Table A Default print formats for functions

Subcommand	Keyword	Default for Automatic	Default for Manual
FORMAT	ALIGN	left	left
	BRKSPACE		
	summary report	1	1
	listing report	-1	1
	CHALIGN	bottom	top
	CHDSPACE	1	1
	COLSPACE	4	4
	FTSPACE	1	1
	LENGTH	1,system length	1,system length
	LIST|NOLIST	NOLIST	NOLIST
	MARGINS	1,system length	1,system length
	MISSING	'.'	'.'
	PAGE1	1	1
	SUMSPACE	1	1
	TSPACE	1	1
	UNDERSCORE	on	off
VARIABLES	LABEL|VALUE|DUMMY	VALUE	VALUE
	LEFT|CENTER|RIGHT	CENTER	RIGHT for numbers LEFT for strings
	OFFSET	CENTER	0
BREAK	LABEL|VALUE	LABEL	VALUE
	LEFT|CENTER|RIGHT	CENTER	RIGHT for numbers LEFT for strings
	NAME|NONAME	NONAME	NONAME
	OFFSET	CENTER	0
	PAGE	off	off
	SKIP	1	1
	TOTAL|NOTOTAL	NOTOTAL	NOTOTAL
	UNDERSCORE	off	off
SUMMARY	PREVIOUS	1	1
	SKIP	0	0

Tailoring

Display Format. REPORT provides full report format defaults and offers you optional control over page length, vertical spacing, margins and column widths, page titles and footnotes, and labels for statistics. The maximum width and length of the report are controlled by specifications on the SET command.

The FORMAT subcommand controls how the report is laid out on a page and whether case listings are displayed. The STRING subcommand concatenates variables to create temporary variables that can be referenced on the VARIABLES or BREAK subcommands. The VARIABLES subcommand names the report variables used to compute statistics and controls the titles, width, and contents of report columns. The BREAK subcommand specifies the variables that define groups and controls the titles, width, and contents of break columns. The SUMMARY subcommand specifies statistics and controls the titles and spacing of summary lines. Additional subcommands control the specification and placement of multiple-line titles and footnotes.

Statistical Display. The statistical display is controlled by the SUMMARY subcommand. Statistics can be calculated for each category of a break variable and for the group as a whole. Available statistics include mean, variance, standard deviation, skewness, kurtosis, sum, minimum and maximum value, mode, median, and percentages. Composite functions perform arithmetic operations using two or more summary statistics calculated on single variables.

Missing Values. You can override the default to include user-missing values in report statistics and listings with the MISSING subcommand. You can also define a missing-value symbol to represent missing data on the FORMAT subcommand.

Syntax

- The minimum specification for REPORT is the VARIABLES subcommand, plus either FORMAT=LIST or the BREAK and SUMMARY subcommands.
- Only one each of the FORMAT, STRING, VARIABLES, and MISSING subcommands are allowed. If used, they must be entered in this order and must precede all BREAK and SUMMARY subcommands.
- Multiple BREAK subcommands are allowed, and multiple SUMMARY subcommands are allowed per BREAK.
- Each BREAK subcommand immediately precedes the SUMMARY subcommands that request statistics for that BREAK.
- To suppress subgroup classification when LIST is specified on FORMAT, omit the BREAK subcommand.
- Title and footnote subcommands can appear anywhere after FORMAT except among the BREAK and SUMMARY subcommands.
- Keywords on REPORT subcommands have default specifications that are in effect if the keyword is not specified. Specify keywords only when you wish to change a default.
- Keywords are enclosed in parentheses if the subcommand takes variable names as arguments.
- Subcommands are separated from each other by slashes.

Operations

- REPORT causes the data to be read.
- REPORT processes cases sequentially. When the value of a break variable changes, REPORT displays a statistical summary for cases processed since the last set of summary statistics was displayed.
- The file must be sorted in order on the break variable or variables.
- The maximum width and page length of the report are the width and page length specified on the SET command.
- The format used to display values in case listings is controlled by the dictionary format of the variable. Each statistical function in REPORT has a default format.

Limitations

- The length of titles and footnotes cannot exceed the report width.
- The length of string variables created on the STRING subcommand cannot exceed the page width.
- There is no fixed limit on the number of BREAK and SUMMARY subcommands. However, the page width limits the number of variables displayed and thereby limits the number of break variables. The limit of 10 variables on SORT CASES often effectively acts as a limit on the number of breaks.
- The number of report variables that can be specified depends upon the width of the report, the width of the variable columns, and the number of BREAK subcommands.
- Maximum 10 dummy variables per VARIABLES subcommand.
- Maximum 50 strings per STRING subcommand.
- Maximum 20 MODE and MEDIAN requests per SUMMARY subcommand.
- Maximum 20 PCGT, PCLT, and PCIN requests per SUMMARY subcommand.
- Workspace is required to store all labeling information, frequency counts if summaries request MEDIAN or MODE, strings, and computed summary statistics.
- Memory requirements significantly increase if MEDIAN or MODE is requested for variables with a wide range of values. The amount of workspace required is 20 + 8*(max −min +1) bytes per variable per function per break.

- If TOTAL is in effect, workspace requirements are almost doubled.
- Memory requirements also increase if value labels are displayed for variables with many value labels. The amount of workspace required is 4 + 24*(number of labels) per variable.

Example

```
SORT CASES BY DEPT.
REPORT FORMAT=AUTOMATIC LIST
/VARIABLES=PRODUCT (LABEL) ' ' 'Retail' 'Products'
           SALES 'Annual' 'Sales' '1981'
/BREAK=DEPT 'Department'
/SUMMARY=VALIDN (PRODUCT) MEAN (SALES) 'No.Sold,Mean Sales'.
```

- This report is a listing of products and sales by department. A summary of the total number of products sold and the average sales by department is also produced.
- Cases are first sorted by DEPT to ensure that cases are appropriately grouped for the calculation of statistics.
- FORMAT requests AUTOMATIC format and a report that includes a listing of cases within each break group.
- VARIABLES specifies PRODUCT and SALES as the report variables and requests that the value labels identifying products be displayed. Three-line column headings are provided for each report column. The first line of the column head is blank for the variable PRODUCT.
- BREAK identifies DEPT as the break variable and provides a one-line column title for the break column.
- SUMMARY requests the calculation of the valid number of cases for PRODUCT and the mean of SALES for each value of DEPT. A title is provided for the summary line to override the default title, VALIDN.

FORMAT Subcommand

```
FORMAT=[AUTOMATIC|MANUAL]  [NOLIST|LIST[(n)]]
       [PAGE(n)]  [LENGTH(t,b)]  [MARGINS(l,r)]
       [ALIGN(LEFT|CENTER|RIGHT)]  [COLSPACE(n)]
       [CHALIGN(TOP|BOTTOM)]  [UNDERSCORE(ON|OFF)]
       [TSPACE(n)]  [CHDSPACE(n)]  [BRKSPACE(n)]
       [SUMSPACE(n)]  [FTSPACE(n)]  [MISSING 's']
```

The optional FORMAT subcommand controls the overall width and length of the report and vertical spacing.

- Keyword specifications and their arguments can be named in any order.

The following can be specified on FORMAT:

AUTOMATIC
MANUAL *The default settings.* AUTOMATIC facilitates report design by doing the following: displays labels for break variables; centers all data; centers column headings but left-justifies column headings if value labels or string values exceed the width of the longest word in the heading; bottom aligns and underscores column headings; extends column widths to accommodate the longest word in a variable label or the variable's longest value label; shrinks a report that is too wide for its margins. MANUAL does the following: displays values for break variables; right-justifies numeric values and their column headings; left-justifies value labels and string values and their column headings; top aligns and does not underscore column headings; extends column widths to accommodate the variable's longest value label (but not the longest word in the variable label), up to a width of 20; generates an error message when a report is too wide for its margins. MANUAL is the default.

NOLIST
LIST[(n)] *Listing of individual cases.* List the values of all variables named on the VARIABLES subcommand for each case. The optional *n* indicates that a blank line be inserted after each *n* cases; the default is not to insert blank lines. Values for cases are listed using the default formats for the variables. The default is the alternative NOLIST, which requests that no case listing be produced.

PAGE(n) *The number for the first page of the report.* The default is 1.

LENGTH(t,b) *The top and bottom lines of the report.* The value for the bottom line cannot be greater than the system page length. The system page length is controlled by SET. By default, the top of the report begins at line 1 and the bottom of the report is the last line of the system page length. You can use an asterisk to indicate a default value.

MARGINS(l,r) *The columns for the left and right margins.* By default, the left margin is print column 1 and the right margin is the rightmost print column of the system page width, which is controlled by the SET WIDTH command. The right column cannot be beyond the width specified on SET. You can use an asterisk to indicate a default value.

ALIGN *The report's placement relative to its margins.* The specification is either (LEFT), (CENTER), or (RIGHT). (LEFT) left justifies the report. (CENTER) centers the report between its margins. (RIGHT) right justifies the report. The default is (LEFT).

COLSPACE(n) *The number of spaces between each column.* The default is the lesser of either 4 or the result obtained by first subtracting the combined column widths of the break and report variables from the REPORT margins and then dividing the difference by the number of columns minus one. When AUTOMATIC is in effect, REPORT overrides the specified column spacing if necessary to fit the report between its margins.

CHALIGN *Alignment of column headings.* The specification is either (TOP) or (BOTTOM). (TOP) aligns all column headings with the first, or top, line of multi-line headings. (BOTTOM) aligns headings with the last, or bottom, line of multi-line headings. The default when AUTOMATIC is in effect is (BOTTOM); when MANUAL is in effect, it is (TOP).

UNDERSCORE *Heading underscores.* The specification is either (ON) or (OFF). (ON) underscores the bottom line of each column heading for the full width of the column. (OFF) does not underscore column headings. The default when AUTOMATIC is in effect is (ON); when MANUAL is in effect, it is (OFF).

TSPACE(n) *The number of blank lines between the report title and the column heads.* The default is 1.

CHDSPACE(n) *The number of blank lines beneath the longest column head.* The default is 1.

BRKSPACE(n) *The number of blank lines between the break head and the next line.* The next line is a case if LIST is in effect or the first summary line if NOLIST is in effect. BRKSPACE(−1) places the first summary statistic or the first case listing on the same line as the break value. When a summary line is placed on the same line as the break value, the summary title is suppressed. When AUTOMATIC is in effect, the default is -1; when MANUAL is in effect, it is 1.

SUMSPACE(n) *The number of blank lines between the last summary line at the lower break and the first summary line at the higher break when they break simultaneously.* SUMSPACE also controls spacing between the last case listed and the first summary line if LIST is in effect. The default is 1.

FTSPACE(n) *The minimum number of blank lines between the last listing on the page and the footnote.* The default is 1.

MISSING 's' *Missing-value symbol.* The symbol can be only 1 character and is used to represent both system- and user-missing values. The default is a period (.).

Example

```
FORMAT=AUTOMATIC LIST MARGINS(1,60) LENGTH(5,30) MISSING ('*')
```

- This FORMAT subcommand requests a case listing, defines a new page size smaller than the system page size, and specifies an asterisk as the missing-value symbol.

Page Layout

Figure A displays the complete page layout and subcommand specifications used to control the basic structure of the report.

Figure A Page layout for REPORT

```
------------------------------- top of page -------------------------------
                    ***************** TITLE ****************           <---- LENGTH
                                                                      <---- TSPACE
BREAK HEAD          BREAK HEAD          COLUMN  COLUMN  COLUMN  COLUMN
                                        HEAD    HEAD    HEAD    HEAD
                                        [VAR]   [VAR]   [VAR]   [VAR]
                                                                      <---- CHDSPACE
BREAK A VALUE 1     BREAK B VALUE 1
                                                                      <---- BRKSPACE
                                        VALUE   VALUE   VALUE   VALUE
                                        VALUE   VALUE   VALUE   VALUE
                                                                      <---- LIST
                                        VALUE   VALUE   VALUE   VALUE
                                        VALUE   VALUE   VALUE   VALUE
                                                                      <---- SUMSPACE
                    SUMMARY TITLE        AGG.    AGG.    AGG.    AGG.
                                                                      <---- SKIP with SUMMARY
                    SUMMARY TITLE        AGG.    AGG.    AGG.    AGG.
                                                                      <---- SKIP with BREAK
                    BREAK B VALUE 2
                                                                      <---- BRKSPACE
                                        VALUE   VALUE   VALUE   VALUE
                                        VALUE   VALUE   VALUE   VALUE
                                                                      <---- LIST
                                        VALUE   VALUE   VALUE   VALUE
                                        VALUE   VALUE   VALUE   VALUE
                                                                      <---- SUMSPACE
                    SUMMARY TITLE        AGG.    AGG.    AGG.    AGG. <---- stats for B=2, A=1

                    SUMMARY TITLE        AGG.    AGG.    AGG.
                                                                      <---- SUMSPACE
SUMMARY TITLE                            AGG.    AGG.    AGG.    AGG. <---- stats for A=1

SUMMARY TITLE                            AGG.    AGG.    AGG.    AGG.
                                                                      <---- SKIP with BREAK
BREAK A VALUE 2     BREAK B VALUE 1
                                                                      <---- BRKSPACE
                                        VALUE   VALUE   VALUE   VALUE
                                        VALUE   VALUE   VALUE   VALUE
                                                                      <---- LIST
                                        VALUE   VALUE   VALUE   VALUE
                                        VALUE   VALUE   VALUE   VALUE
                                                                      <---- SUMSPACE
                    SUMMARY TITLE        AGG.    AGG.    AGG.    AGG.
                                                                      <---- SKIP
                    SUMMARY TITLE        AGG.    AGG.    AGG.    AGG.
                                                                      <---- SKIP with BREAK
                    BREAK B VALUE 2
                                                                      <---- BRKSPACE
                                        VALUE   VALUE   VALUE   VALUE
                                        VALUE   VALUE   VALUE   VALUE
                                                                      <---- LIST
                                        VALUE   VALUE   VALUE   VALUE
                                        VALUE   VALUE   VALUE   VALUE

                    SUMMARY TITLE        AGG.    AGG.    AGG.    AGG.

                    SUMMARY TITLE        AGG.    AGG.    AGG.    AGG.
                                                                      <--- SUMSPACE
SUMMARY TITLE                            AGG.    AGG.    AGG.    AGG.

SUMMARY TITLE                            AGG.    AGG.    AGG.    AGG.
                                                                      <---- FTSPACE
                    **************** FOOTNOTE ***************          <---- LENGTH
------------------------------ bottom of page -----------------------------
|                                                                         |
left margin                                                    right margin
```

OUTFILE Subcommand

The OUTFILE subcommand allows you to direct the report to a disk file other than the SPSS/PC+ listing file. You can later print this file from DOS, without having to delete the extraneous material that would be present in the listing file.

- OUTFILE must follow the FORMAT subcommand and must precede the BREAK subcommand.

- Specifications on OUTFILE consist of a filename in apostrophes.
- You can send the report to a directory or drive other than your current defaults by specifying an appropriate DOS file specification.
- You can print the report with the DOS *PRINT* command at the end of your session.
- The report does not appear on your screen when you specify OUTFILE.

Example

```
REPORT FORMAT = LIST
/OUTFILE = 'B:SALES.RPT'
/VARIABLES = SALES,QUOTA
/BREAK = REGION
/SUMMARY = MEAN.
```

- The report is written to SALES.RPT on the floppy diskette in drive B:.

VARIABLES Subcommand

```
/VARIABLES=var|var TO var [(VALUE|LABEL|DUMMY)] ['col head']
             [(LEFT|CENTER|RIGHT)] [(width)]
             [(OFFSET(n|CENTER))] [varname...]
```

The required VARIABLES subcommand names the variables to be listed and summarized in the report. Optionally, you can use VARIABLES to control column titles, column widths, and the contents of report columns.

- The minimum VARIABLES specification is a list of variables. These are the report variables. The number of variables that can be named is limited by the system page width.
- Variables named on the BREAK subcommand can also be named on VARIABLES.
- Each report variable defines a report column. Each report column can be thought of as having the name of the variable that defines it.
- The value of the variable or an aggregate statistic calculated on the variable is displayed in that variable's report column.
- Variables are assigned to columns in the order in which they are named on the VARIABLES subcommand.
- When FORMAT=LIST, variables can be stacked in a single column by linking them with + signs on the VARIABLES subcommand. If no column heading is specified, REPORT uses the default heading from the first variable on the list. Only values from the first variable in the stacked list are used to calculate summaries.
- Optional specifications can be given in any order following the variable name to which they apply.
- Optional specifications apply only to the immediately preceding variable or list of variables implied by the TO keyword.

The following options can be specified:

(VALUE)
(LABEL)
(DUMMY)

Contents of the report column assigned to the variable. If no specification is given, the keyword (VALUE) is in effect. (VALUE) specifies that values of the variable be displayed in the column. The alternative keyword (LABEL) displays value labels if value labels are defined, and values otherwise. (VALUE) and (LABEL) have no effect unless LIST has been specified on the FORMAT subcommand.

(DUMMY) defines a report column for a variable that does not exist in the active file. Such dummy variables are used to control spacing or to reserve space for statistics computed upon other variables. Do not name an existing SPSS/PC+ variable as a dummy variable.

When AUTOMATIC is in effect, value labels or string values are centered in the column based on the length of the longest string or label; numeric values are centered based on the width of the widest value or summary format. When MANUAL is in effect, value labels or string values are left-justified in the column; numeric values are right-justified.

'column title' *Title used for the report column assigned to the variable.* Specify multiple-line titles by enclosing each line in a set of apostrophes or quotes, using the conventions for strings (see Universals: Strings). Separate the specifications for title lines with at least one blank.

If no column title is specified, the default column title is the variable label or the variable name if no variable label has been specified.

Default column titles wrap for as many lines as are required to display the entire label. If AUTOMATIC is in effect, user-specified column titles appear exactly as specified, even if the column width must be extended. If MANUAL is in effect, user-specified titles wrap to fit within the column width.

(LEFT)
(CENTER)
(RIGHT) *Alignment of the column heading.* If AUTOMATIC is in effect, column headings are centered within their columns; if value labels or string values exceed the width of the longest word in the heading, the heading is left-justified. If MANUAL is in effect, column headings are left-justified for value labels or string values, and right-justified for numeric values.

(width) *Width for the report column.* If no width is specified for a variable, REPORT determines a default width using the criteria described under Defaults, above. If you specify a width that is not wide enough to display numeric values, REPORT first rounds decimal digits, then converts to scientific notation if possible, and then displays asterisks. Value labels or string values that exceed the width are truncated.

(OFFSET) *Adjust the position of the report column contents.* The specification is either (*n*) or (CENTER). (OFFSET(*n*)) indicates the number of spaces to offset. Contents are offset from the left for value labels or string values, and from the right for numeric values. (OFFSET(CENTER)) centers contents within the center of the column. If AUTOMATIC is in effect, the default is CENTER. (However, entering a number on OFFSET offsets the contents from the justified position, not from the center.) if MANUAL is in effect, the default is 0.

Example

```
/VARIABLES=V1 TO V3(LABEL) (15)
   V4 V5 (LABEL)(OFFSET (2))(10)
   SEP1 (DUMMY) (2) ''
   V6 'Results using' "Lieben's Method" 'of Calculation'
```

- The variables from V1 through V3 have report columns with a width of 15 each. Values are listed in the case listing. Value labels are displayed for these variables in the case listing.
- Variable V4 has a report column with the default width. Values are listed in the case listing.
- Value labels are displayed for variable V5. The column has a width of 10 columns. Column contents are offset two spaces from the left.
- SEP1 is a dummy variable. The column width is 2, and there is at least one space column on each side of SEP1. Thus, there are at least 4 blanks between the columns for V5 and V6. SEP1 is given a null title to override the default column title SEP1.
- V6 is given a three-line title. Its column has the default width, and values are listed in the case listing.

STRING Subcommand

```
/STRING=stringname (varname [(width)] [(BLANK)] ['literal'])
        [stringname...]
```

The optional STRING subcommand concatenates variables and user-specified strings into temporary string variables that exist only within REPORT.

- The minimum specification is a name for the string variable followed by a variable name or a user-specified string enclosed in parentheses.

- The name assigned to the string variable must be unique.
- Any combination of string variables, numeric variables, and user-specified strings can be used enclosed in parentheses to define the string.
- The keyword TO cannot be used within the parentheses to imply a variable list.
- More than one string variable can be defined on the STRING subcommand.
- If a variable within the parentheses has a missing value, the string has a system-missing value, and the missing-value symbol is used for that value in case listings.
- A string variable defined in REPORT cannot exceed the system page width.
- String variables defined on the STRING subcommand can be used on the VARIABLES or BREAK subcommand.

The following options can be specified:

(width) *Column width within the string of the preceding variable.* The default width is the dictionary width of the variable.

If the width specified is less than required by the value, asterisks are displayed for numeric values, and string values are truncated on the right. If the width exceeds the width of a value, values of numeric variables are padded with zeros on the left and values of string variables are padded with blanks on the right.

The maximum width for numeric variables within the string definition is 16. The maximum width for a string variable is the system page width.

(BLANK) *Left-pad values of the preceding numeric variable with blanks.* If the specification is omitted, the default is to left-pad values of numeric variables with zeros. If a numeric variable has a DOLLAR or COMMA format, it is automatically left-padded with blanks.

'literal' *A user-specified string.* Any combination of characters can be specified, enclosed in apostrophes or quotes.

Example

```
/STRING=JOB1(AVAR VARN)
        JOB2(AVAR(2) VARN(3))
        JOB3(AVAR(2) VARN(BLANK) (4))
```

- This STRING subcommand defines three string variables to be used within the report.
- Assume that AVAR is a string variable read from a four-column field using the FIXED keyword on DATA LIST, and that VARN is a computed numeric variable with the default format of 8 columns with 2 implied decimal places.
- If a case has the value 'KJ ' for AVAR and the value 241 for VARN, JOB1 displays the value KJ 00241.00, JOB2 the value KJ241, and JOB3 the value KJ 241.

Example

```
/STRING=SOCSEC(S1 '-' S2 '-' S3)
```

- This STRING subcommand concatenates the three variables S1, S2, and S3, which each contain a portion of the social security number.
- Hyphens are inserted between the portions when the values of SOCSEC are displayed.
- This example assumes that the variables S1, S2, and S3 were read from three-column, two-column, and four-column fields respectively, using the FIXED keyword on DATA LIST. These variables have format widths of 3, 2, and 4 columns, and are not left-padded with zeros.

BREAK Subcommand

```
/BREAK=[varlist] [(VALUE|LABEL)] ['col title']
       [(LEFT|CENTER|RIGHT)] [(width)] [(OFFSET(n|CENTER))]
       [(UNDERSCORE(ON|OFF))] [(NOTOTAL|TOTAL)]
       [(NONAME|NAME)] [(SKIP(n|PAGE))]
```

The BREAK subcommand specifies the variables that define the subgroups for the report display, or specifies summary totals for reports with no subgroups. BREAK also allows you to control the titles, width, and contents of break columns and to begin a new page for each level of the break variable.

- A break occurs when any one of the break variables named on BREAK changes value. Cases must be presorted by the values of all BREAK variables on all BREAK subcommands.
- The keyword TO can be used to specify an implied variable list.
- To obtain summary totals without any break levels, use the keyword (TOTAL) on BREAK without listing any variables.
- Optional specifications can be given in any order following the last variable named.
- Multiple BREAK subcommands can be used.
- When you use more than one BREAK subcommand, specify (TOTAL) on only one BREAK to get a summary for the entire file. Otherwise, redundant statistics may be displayed. (Use as many SUMMARY subcommands as you need to specify the statistics to be totaled.)
- Missing-value specifications are ignored for variables named on the BREAK subcommand. There is one break category for system-missing values and one for user-missing values. The values are displayed using the missing-value symbol controlled by the FORMAT subcommand.
- The BREAK subcommand must precede the SUMMARY subcommand that defines the summary line for the break.
- A break column is reserved for each BREAK subcommand.
- If more than one variable is specified on a BREAK subcommand, a single break column is used. The value or value label for each variable is displayed on a separate line, in the order in which the variables are named on BREAK. The first variable named changes most slowly. The default column width is the longest of the default widths for any of the break variables.
- Optional specifications apply to all variables in the break column and to the break column as a whole.

The following can be specified on BREAK:

(VALUE)
(LABEL)
Contents of the break column. (VALUE) specifies that values of the break variables be displayed in the column. The alternative keyword (LABEL) displays value labels if value labels have been defined, and values otherwise. The value is displayed only once for each break change and is not repeated at the top of the page in a multiple-page break group. When AUTOMATIC is in effect, the default is (LABEL); when MANUAL is in effect, it is (VALUE).

When AUTOMATIC is in effect, value labels and string values are centered in the column based on the length of the longest string or label; numeric values are centered based on the width of the widest value or summary format. When MANUAL is in effect, value labels and string values are left-justified in the column; numeric values are right-justified.

'column head'
Title used for the break column. Specify multiple-line titles by enclosing each line in a set of apostrophes or quotes, following the conventions for strings (see Universals: Strings). Separate the specifications for title lines with at least one blank.

The default title is the variable label of the break variable or the variable name if no label has been defined. If the break column is defined by more than one variable, the label or name of the first variable is used.

Default column titles wrap for as many lines as are required to display the entire label. If AUTOMATIC is in effect, user-specified column titles appear exactly as specified, even if the column width must be extended. If MANUAL is in effect, user-specified titles wrap to fit within the column width.

(LEFT)
(CENTER)
(RIGHT)
Alignment of the column heading. When AUTOMATIC is in effect, column headings are centered within their columns. However, if value labels or string values exceed the width of the longest word in the heading, the heading is left-justified.

When MANUAL is in effect, column headings are left-justified for value labels or string values, and right-justified for numeric values.

(width) *Column width for the break column.* If no width is specified for a variable, REPORT determines a default width using the criteria described under Defaults, above. If you specify a width that is not wide enough to display numeric values, REPORT first rounds decimal digits, then converts to scientific notation if possible, and then displays asterisks. Value labels or string values that exceed the width are truncated.

(OFFSET) *Adjust the position of the break column contents.* The specification is either (*n*) or (CENTER). (OFFSET(*n*)) indicates the number of spaces to offset. Contents are offset from the left for value labels or string values, and from the right for numeric values. (OFFSET(CENTER)) centers contents within the center of the column. If AUTOMATIC is in effect, the default is CENTER. (However, entering a number on OFFSET offsets the contents from the justified position, not from the center.) If MANUAL is in effect, the default is 0.

(UNDERSCORE) *Heading underscores.* The specification is either (ON) or (OFF). (UNDERSCORE(ON)) underscores the bottom line of each column heading for the full width of the column. (UNDERSCORE(OFF)) does not underscore column headings. The default when AUTOMATIC is in effect is (ON); when MANUAL is in effect, it is (OFF).

(TOTAL)
(NOTOTAL) *(TOTAL) calculates summary statistics specified on the next SUMMARY subcommand for all the cases on the report.* (NOTOTAL) is the default and displays summary statistics only for each break.

(SKIP)
(PAGE) *The vertical spacing between the last summary line for a break and the next break.* The specification is either (*n*) or (PAGE). If (SKIP(*n*)) is specified, each break begins following *n* blank lines. The default is 1. If (PAGE) is specified, each break begins on a new page. If (RESET) is specified on (PAGE), the page counter resets to the PAGE1 setting on the FORMAT subcommand every time the break value changes for the specified variable. (PAGE(RESET)) is not allowed on listing reports with no break levels.

(NAME)
(NONAME) *Display the name of the break variable alongside each value or value label of the break variable.* (NAME) requires 10 spaces (the maximum eight-character length of SPSS/PC+ variable names plus two parentheses) in addition to the space needed to display break values or value labels. (NAME) is ignored if the break-column width is insufficient. If the default keyword (NONAME) is specified, the name of the break variable is omitted.

Example

```
SORT DIVISION BRANCH DEPT.
REPORT FORMAT=AUTOMATIC MARGINS (1,70) BRKSPACE(-1)

/VARIABLES=SPACE(DUMMY) ' ' (4)
            SALES 'Annual' 'Sales' '1981' (15) (OFFSET(2))
            EXPENSES 'Annual' 'Expenses' '1981' (15) (OFFSET(2))

/BREAK=DIVISION
        BRANCH (10) (TOTAL) (OFFSET(1))
  /SUMMARY=MEAN

/BREAK=DEPT 'Department' (10)
  /SUMMARY=MEAN.
```

- This example creates a report that breaks on three variables. BRANCH breaks within values of DIVISION, and DEPT breaks within values of BRANCH.
- FORMAT sets margins to a maximum of 70 columns and requests that the summary line be displayed on the same report line as the break values. Because LIST is not specified on FORMAT, only summary statistics are displayed.
- VARIABLES defines three report columns, each occupied by a report variable: SPACE, SALES, and EXPENSES.
- The variable SPACE is a dummy variable that exists only within REPORT. It has a null title and a width of 4. It is used as a space holder to separate the break columns from the report columns.
- SALES is given a three-line title and a width of 15. The values of SALES are offset 2 spaces from the right.
- EXPENSES is the third report variable and has the same width and offset specifications as SALES.
- The leftmost column in the report is reserved for the first two break variables, DIVISION and BRANCH. The break column has a width of 10, and the value labels are offset 1 space from the left. Any value label more than 9 characters long is truncated. The default column title is used. (TOTAL) requests that the BRANCH summary line be displayed when all values of DIVISION and BRANCH have been cycled through, that is, at the end of the report.
- The summary line for the first BREAK subcommand consists of the mean of each report variable, displayed in its own report column. This line is displayed each time the value of DIVISION or BRANCH changes.
- The third break variable, DEPT, occupies the second column from the left in the report. The break column has a width of 10 and has a one-line title. The first 10 characters of the value labels are displayed in the break column.
- The second SUMMARY subcommand displays the mean for each report variable when the value of DEPT changes.

SUMMARY Subcommand

```
/SUMMARY={function...['summary title'][(break col #)] [SKIP(n)]}
         {PREVIOUS[(n)]                                         }
```

where function is

```
aggregate [(varname[(d)][(PLAIN|DOLLAR|COMMA)][varname...])]
```

or

```
composite(agg(varname)...)[(report col[(d)][(PLAIN|DOLLAR|
COMMA)])]
```

The SUMMARY subcommand calculates a wide range of aggregate and composite statistics.

- A SUMMARY subcommand must be specified unless LIST is specified on the FORMAT subcommand.
- Each SUMMARY subcommand following a BREAK subcommand specifies a new summary line.
- The minimum specification is an aggregate function or a composite function and its arguments. This must be the first specification on SUMMARY.
- The default format can be altered for any function. Format specifications in an aggregate modify only the immediately preceding result.
- The default location of the summary title is the column of the break variable to which the summary applies.
- When more than one function is named on the SUMMARY subcommand, the default summary title is that of the function named first.
- Both the title and the default column location of the title can be altered.
- SUMMARY subcommands apply only to the preceding BREAK subcommand. If there is no SUMMARY subcommand after a BREAK subcommand, no statistics are displayed for that break level.

- Specify the keyword PREVIOUS on SUMMARY to use the summary specifications from a previous BREAK subcommand for the current BREAK subcommand.
- More than one function can be specified on SUMMARY as long as you do not attempt to place two results in the same column.
- Multiple summary statistics requested on one SUMMARY subcommand are all displayed on the same summary report line.
- Use the SKIP keyword to insert blank lines when more than one summary line is requested for a break.
- An implicit or explicit attempt to place the result of two or more functions in the same report column stops execution of REPORT. Use multiple SUMMARY subcommands to place results of more than one function in the same report column.
- Summary lines can combine any composite functions and aggregrate functions (except FREQUENCY and PERCENT).

Aggregate Functions

Use the aggregate functions to request descriptive statistics on report variables.

- If no variable names are given as arguments to an aggregate function, the statistic is calculated for all variables named on the VARIABLES subcommand, that is, for all report variables.
- To request an aggregate function for a subset of the report variables, specify the list of report variables in parentheses after the function keyword.
- All variables specified on an aggregate function must have been named on the VARIABLES subcommand.
- The keyword TO cannot be used to specify a list of variables for an aggregate function.
- The result of an aggregate function is always displayed in the report column reserved for the variable on which the function was calculated.
- To use several aggregate functions on the same report variable, specify multiple SUMMARY subcommands. The results are displayed on different summary lines.
- The aggregate functions FREQUENCY and PERCENT have special display formats and cannot be placed on the same summary line with other aggregate or composite functions.
- Aggregate functions use only cases with valid values.

The following aggregate functions are available:

VALIDN *Valid number of cases.* This is the only function that operates on string variables.

SUM *Sum of values.*

MIN *Minimum value encountered.*

MAX *Maximum value encountered.*

MEAN *Mean.*

STDDEV *Standard deviation.* Aliases are SD and STDEV.

VARIANCE *Variance.*

KURTOSIS *Kurtosis.*

SKEWNESS *Skewness.*

MEDIAN(min,max) *Median value for values within the range.* MEDIAN sets up integer-valued bins for counting all values in the specified range. Noninteger values are truncated when the median is calculated.

MODE(min,max) *Modal value for values within the range.* MODE sets up integer-valued bins for counting all values in the specified range. Noninteger values are truncated when the mode is calculated.

PCGT(n) *Percentage of cases with values greater than specified value.*

PCLT(n) *Percentage of cases with values less than specified value.*

PCIN(min,max) *Percentage of cases within the inclusive value range specified.*

FREQUENCY(min,max) *Frequency counts for values within the inclusive range.* FREQUENCY sets up integer-valued bins for counting all values in the specified range. Noninteger values are truncated when the frequency is computed. FREQUENCY cannot be mixed with other aggregate statistics on a summary line.

PERCENT(min,max) *Percentages for values within the inclusive range.* PERCENT sets up integer-valued bins for counting all values in the specified range. Noninteger values are truncated when the frequency is computed. PERCENT cannot be mixed with other aggregate statistics on a summary line.

Example

```
SORT CASES BY BVAR AVAR.
REPORT FORMAT=AUTOMATIC LIST/ VARIABLES=XVAR YVAR ZVAR

/BREAK=BVAR
  /SUMMARY=SUM
  /SUMMARY=MEAN (XVAR YVAR ZVAR)
  /SUMMARY=VALIDN(XVAR)

/BREAK=AVAR
  /SUMMARY=PREVIOUS.
```

- The FORMAT subcommand requests a case listing, and the VARIABLES subcommand establishes a report column for variables XVAR, YVAR, and ZVAR. The report columns have default widths and titles.
- Both break variables, BVAR and AVAR, have default widths and titles.
- Every time the value of BVAR changes, three summary lines are displayed. The first line contains the sums for variables XVAR, YVAR, and ZVAR. The second line contains the means of all three variables. The third line displays the number of valid cases for XVAR in the report column for XVAR.
- Every time the value of AVAR changes within each value of BVAR, the three summary lines requested for BVAR are displayed. These summary lines are based on cases with the current value of BVAR that also have the current value of AVAR.

Example

```
SORT CASES BY DEPT.
REPORT FORMAT=AUTOMATIC
/VARIABLES=WAGE BONUS TENURE
/BREAK=DEPT (23)
/SUM=SUM(WAGE BONUS) MEAN(TENURE) 'Sum Income: Mean Tenure'.
```

- The SUMMARY subcommand defines a summary line consisting of the sums of WAGE and BONUS and the mean of TENURE. The result of each aggregate function is displayed in the report column of the variable on which the function is calculated.
- A title is assigned to the summary line. A width of 23 is defined for the break column to accommodate the right-justified summary-line title.

Composite Functions

Use composite functions to obtain statistics based on aggregated statistics, to place a summary statistic in a column other than that of the report variable on which it was calculated, or to manipulate variables not named on the VARIABLES subcommand.

- Composite functions can be computed upon constants and any variable in the active file.
- The following aggregate functions can also be arguments to composite functions: VALIDN, SUM, MIN, MAX, MEAN, STDEV, VARIANCE, KURTOSIS, and SKEWNESS. When used within composite functions, aggregate functions can have only one variable as an argument.
- A composite function and its arguments cannot be separated from each other by other SUMMARY specifications.

- By default, the results of a composite function are placed in the report column of the first variable named on the composite function that is also named on the VARIABLES subcommand.
- The result of a composite function can be placed in any report column, including columns of dummy or string variables, by specifying a target column. To specify a target column, enclose the variable name of a report column in parentheses after the composite function and its arguments.
- The format for the result of a composite function can be specified in parentheses after the name of the column location and within the parentheses that enclose the column-location specification.

The following composite functions are available:

DIVIDE(agg() agg() [factor]) *Divide the first argument by the second and multiply by the optional factor.*

MULTIPLY(agg() ... agg()) *Multiply the arguments.*

PCT(agg() agg()) *Percentage of the first argument over the second.*

SUBTRACT(agg()agg()) *Subtract the second argument from the first argument.*

ADD(agg() ... agg()) *Add the arguments.*

GREAT(agg() ... agg()) *Give the maximum of the arguments.*

LEAST(agg() ... agg()) *Give the minimum of the arguments.*

AVERAGE(agg() ... agg()) *Give the average of the arguments.*

Example

```
SORT CASES BY DEPT.
REPORT FORMAT=AUTOMATIC BRKSPACE(-1)
/VARIABLES=WAGE BONUS SPACE1 (DUMMY) '' BNFT1 BNFT2 SPACE2
(DUMMY)''
/BREAK=DEPT
  /SUMMARY=MEAN(WAGE BONUS BNFT1 BNFT2)
     ADD(VALIDN(WAGE)) (SPACE2)

  /SUMMARY=ADD(SUM(WAGE) SUM(BONUS))
      ADD(SUM(BNFT1) SUM(BNFT2)) 'Totals' SKIP(1)

  /SUMMARY=DIVIDE(MEAN(WAGE) MEAN(BONUS)) (SPACE1 (COMMA)(2))
      DIVIDE(MEAN(BNFT1) MEAN(BNFT2)) (SPACE2 (COMMA)(2)) 'Ratios'
      SKIP(1).
```

- The VARIABLES subcommand defines six report columns. The columns called WAGE, BONUS, BNFT1, and BNFT2 contain aggregate statistics based on those variables. The variables SPACE1 and SPACE2 are dummy variables that are created for use as space holders; each is given a blank title to suppress the default column head.
- The first SUMMARY computes the means of the variables WAGE, BONUS, BNFT1, and BNFT2. Because BRKSPACE=−1, this summary line will be placed on the same report line as the break value and will have no summary title. The means are displayed in the report column for each variable. SUMMARY also computes the valid number of cases for WAGE inside a composite function and places the result in SPACE2 column.
- The second SUMMARY adds the sum of WAGE to the sum of BONUS. Since no location is specified, the result is displayed in the WAGE column. In addition, the sum of BNFT1 is added to the sum of BNFT2, and the result is placed in the BNFT1 column. One line is skipped before the summary line requested by this SUMMARY subcommand is displayed.
- The third summary line divides the mean of WAGE by the mean of BONUS and places the result in SPACE1. The ratio of the mean of BNFT1 to the mean of BNFT2 is displayed in the SPACE2 column. Because locations are specified, formats can also be given. The results are displayed with commas and 2 decimal places. One line is skipped before the summary line requested by this SUMMARY subcommand is displayed.

Summary Titles

The default titles for summary lines are listed below:

Keyword	Title
VALIDN	N
VARIANCE	Variance
SUM	Sum
MEAN	Mean
STDDEV	StdDev
MIN	Minimum
MAX	Maximum
SKEWNESS	Skewness
KURTOSIS	Kurtosis
PGT(n)	>n
PLT(n)	<n
PIN(n1,n2)	In n1 to n2
FREQUENCY(min,max)	Total
PERCENT(min,max)	Total
MEDIAN(min,max)	Median
MODE(min,max)	Mode

- You can specify a summary title enclosed in apostrophes or quotes, following the conventions for strings (see Universals: Strings).
- The summary title must be specified after the first function and its arguments. It cannot separate any function from its arguments.
- A summary title can be only one line long.
- A summary title wider than the break column extends into the next break column to the right. If the title is wider than the all of the available break columns, it is truncated.
- Only one summary title applies per summary line. If more than one is specified, the last is used.
- The summary title is left- or right-justified depending upon whether the break title is left- or right-justified.
- The default location for the summary title is the column of the BREAK variable to which the summary applies.
- With multiple breaks, you can override the default placement of the title by specifying the break-column number in which you want the summary title to be displayed in parentheses following the title.
- In a report with no break levels, REPORT displays the summary title above the summary line at the left margin.

Summary Print Formats

All functions have default formats that are used to print results (see Table B). You can override these defaults by specifying a format keyword and/or the number of decimal places.

- A format specification must be enclosed in parentheses.
- For aggregate functions, one or both format specifications are placed after the variable name, within the parentheses that enclose the variable name. The variable must be explicitly named as an argument.
- For composite functions, one or both format specifications are placed after the variable name of the column location, within the parentheses that enclose the variable name. The column location must be explicitly specified.
- If the report column is wide enough, SUM, MEAN, STDDEV, MIN, MAX, MEDIAN, and MODE use DOLLAR or COMMA format if one has been declared for the variable on the FORMATS command.

- If the column is not wide enough to display the decimal digits for a given function, REPORT displays fewer decimal places. If the column is not wide enough to display the integer portion of the number, REPORT adopts scientific notation and then displays asterisks.
- An exact value of zero is displayed with one zero to the left of the decimal point and as many 0 digits to the right as specified by the format. A number less than 1 in absolute value is displayed without a zero to the left of the decimal point, except with DOLLAR and COMMA formats.

The following fomrat keywords are available:

(DOLLAR) *Display the value using DOLLAR format.*

(COMMA) *Display the value using COMMA format.*

(PLAIN) *Override DOLLAR or COMMA dictionary formats.* PLAIN is the default for all functions except MEAN, STDDEV, MIN, MAX, MEDIAN, and MODE. For these functions, the default is the dictionary format.

Example

```
/SUMMARY=MEAN(INCOME (DOLLAR)(2))
         ADD(SUM(INCOME)SUM(WEALTH) (WEALTH(DOLLAR)(2))
```

- SUMMARY displays the mean of INCOME with dollar format and two decimal places. The format can be specified because INCOME is specified as an argument to the MEAN function. The result is displayed in the INCOME column.
- The sums of INCOME and WEALTH are added and the result is displayed with dollar format and two decimal places. The format can be specified because an explicit location is given for the results of ADD. The result is displayed in the WEALTH column.

Table B Default print formats for functions

Function	Width	Decimal places
VALIDN	5	0
SUM	Dictionary print format + 2	Dictionary print format
MEAN	Dictionary print format	Dictionary print format
STDDEV	Dictionary print format	Dictionary print format
VARIANCE	Dictionary print format	Dictionary print format
MIN	Dictionary print format	Dictionary print format
MAX	Dictionary print format	Dictionary print format
SKEWNESS	5	2
KURTOSIS	5	2
PGT	6	1
PLT	6	1
PIN	6	1
MEDIAN	Dictionary print format	Dictionary print format
MODE	Dictionary print format	Dictionary print format
PERCENT	6	1
FREQUENCY	5	0
DIVIDE	Dictionary print format	0
PCT	6	2
SUBTRACT	Dictionary print format	0
ADD	Dictionary print format	0
GREAT	Dictionary print format	0
LEAST	Dictionary print format	0
AVERAGE	Dictionary print format	0
MULTIPLY	Dictionary print format	0

Other SUMMARY Keywords

Spacing between multiple summary lines for a single break and references to previously defined summary lines are controlled by the following keywords:

SKIP(n) *Blank lines before the summary line.* SKIP is not enclosed in parentheses. The default is 0. SKIP on the first SUMMARY subcommand for a BREAK skips the specified lines after skipping the number of lines specified for BRKSPACE on FORMAT.

PREVIOUS(n) *Use the SUMMARY subcommands for the* nth *BREAK.* If no specification is given in parentheses, PREVIOUS points to the set of SUMMARY subcommands for the previous BREAK. If an integer specification is given, the SUMMARY subcommands from the *n*th BREAK are used.

No other specification can be used on SUMMARY with PREVIOUS. For a multiple-break report for which you want the same sets of summaries, specify SUMMARY subcommands for the higher BREAK subcommand and keyword PREVIOUS for lower breaks.

Title and Footnote Subcommands

```
/TITLE ='title' (centered head)
/FOOTNOTE='title' (centered foot)
```

or

```
/TITLE=LEFT 'title' (left-justified head)
/TITLE=CENTER 'title' (centered head)
/TITLE=RIGHT 'title' (right-justified head)
/FOOTNOTE=LEFT 'title' (left-justified foot)
/FOOTNOTE=CENTER 'title' (centered foot)
/FOOTNOTE=RIGHT 'title' (right-justified foot)

                [)PAGE]   [)DATE]   [)var]
```

- The title and footnote subcommands are optional and can be placed anywhere after the FORMAT subcommand except among the BREAK and SUMMARY subcommands.
- Title subcommands should be specified before any footnote subcommands.
- The default REPORT title is the title specified on the TITLE command. If there is no TITLE command specified in your SPSS/PC+ session, the default REPORT title is the first line of the SPSS/PC+ header.
- A title or footnote is specified by providing a string in apostrophes or quotes on one of the title or footnote subcommands.
- If the title or footnote is more than one line, enclose each line in apostrophes or quotes and separate the specifications for each line by at least one blank.
- The positional keywords LEFT, CENTER, and RIGHT can each be specified only once.
- Titles are displayed beginning in the first line of the system page.
- Footnotes end in the last line of the system page.
- Centered titles and footnotes are centered within the report page width.
- Titles and footnotes are repeated on each page of a multiple-page report.
- If the total width requested for the combined titles or foonotes for a line exceeds the page width, REPORT generates an error message.

Three keywords can be used in a title or footnote. *You must specify)PAGE and)DATE in upper case, as shown here.*

)PAGE *Display the page number right-justified in a five-character field.*

)DATE *Display the current date in the form* dd/mmm/yy *right-justified in a nine-character field.*

)var *Print this variable's value label in this relative position.* If you specify a variable that has no value label, the value itself will print, formatted according to its print format. You cannot specify a scratch or system variable, nor can you specify a variable you create with the STRING subcommand. In addition, you cannot use variables named DATE or PAGE in the *)var* argument because they will only print the current date or a page number. If you want to use a variable named DATE or PAGE, change the variable's name with the RENAME VARIABLES command before you use it in the *)var* argument.

Each variable you specify with *)var* must be one you've defined in the active file, though it does not need to be a variable you've included as a column on your report. One label or value from each variable specified in a *)var* argument prints on every page of the report. The label printed for each varies from page to page and is chosen from cases determined as follows:

- If a new page starts with a case listing, REPORT takes the value label from the first case listed.
- If a new page starts with a BREAK line, REPORT takes the value label from the first case of the new break group.
- If a new page starts with a summary line, REPORT takes the value label from the last case of the break group being summarized.
- If you specify the same variable in both a title and a footnote, REPORT takes the title value from the top case on the page, and the footnote value from the last case on the page.

Example

```
/TITLE=LEFT 'Personnel Report' 'Prepared on )DATE'
/TITLE=RIGHT 'Page )PAGE'
```

- The TITLE=LEFT subcommand provides a two-line title. The second line of this title contains the date on which the report was processed. This title is displayed left-justified at the top of each page of the report.
- The TITLE=RIGHT subcommand displays the page number following the string Page, right-justified at the top of each page.
- The keywords)DATE and)PAGE are entered in upper case.

MISSING Subcommand

```
/MISSING=VAR|NONE|LIST [([varlist][n])]
```

Use the optional MISSING subcommand to control the handling of cases with missing values on the VARIABLES and SUMMARY subcommands. By default, cases with missing values are included in case listings but are excluded from the calculation of functions on a function-by-function basis.

- MISSING specifications apply to variables named on VARIABLES and SUMMARY as well as to strings created with the STRING subcommand.
- The character used to indicate missing values is controlled by the FORMAT subcommand.

The following keywords are available for the MISSING subcommand:

VAR *Missing values are treated separately for each variable.* Missing values are displayed in case listings but are not included in summary statistics. This is the default.

NONE *User-missing-value indicators are ignored.* Applies to all variables named on the VARIABLES subcommand.

LIST[([varlist][n])] *Eliminates any case with the specified number of missing values among the specified list of variables.* If no *n* is specified, the default is 1. If no variables are specified, all variables named on the VARIABLES subcommand are assumed.

Example

```
/MISSING= LIST (XVAR,YVAR,ZVAR 2)
```

- Any case with a missing value for two or more of the variables XVAR, YVAR, and ZVAR is omitted from the report.

REVIEW

```
REVIEW [{SCRATCH                    }]
        {LISTING                    }
        {LOG                        }
        {BOTH                       }
        {'filename' ['filename']}
```

Example:

```
REVIEW.
```

Overview

The REVIEW command initiates a REVIEW editing session. REVIEW is a full-screen editor specifically designed for editing SPSS/PC+ scratch pad and listing files. It allows you to edit two files at once in separate windows. It also allows you to submit revised command lines directly to SPSS/PC+ for execution.

The REVIEW command places you in a full-screen editing environment, complete with the Menu and Help system. In REVIEW, you can build your SPSS/PC+ commands in three ways: by selecting from the Menu and Help windows, by browsing the menus and then typing commands yourself, or by clearing the menus and typing complete commands in the scratch pad. REVIEW's editing commands are issued by selecting options from mini-menus called up with the function keys on your keyboard. For example, pressing [F4] calls up a mini-menu from which you can choose to insert, delete, or undelete lines.

Defaults

If no keywords are specified, REVIEW initiates an editing session of the current scratch pad and listing files. The Menu and Help windows initially cover up the listing file in the upper window. The scratch pad is displayed in the bottom window. On color monitors, REVIEW uses a default color scheme of (1,2,4) for lower-window (or only window) background color, upper-window background color, and frame color, respectively. See the manual for your monitor for the actual colors that correspond to these numbers. The default selection for each mini-menu is highlighted when the menu is called up.

Tailoring

You can edit any file or pair of files that consist of legible characters (see "Limitations" below). On color monitors, you can choose colors for the two editing windows and frame. You can also turn color off.

Syntax

- The minimum specification is the word REVIEW followed by the command terminator.
- Only one keyword (SCRATCH, LISTING, LOG, or BOTH) can be specified.
- Use the SCRATCH keyword to edit your SPSS/PC+ scratch pad file.
- Use the LISTING keyword to edit your SPSS/PC+ listing file.
- Use the LOG keyword to edit your SPSS/PC+ log file.
- Use the BOTH keyword to edit both the log and listing files simultaneously.
- If you do not use a keyword, you can specify one or two filenames within apostrophes to edit files other than your scratch pad, listing, and log files.
- At least one space must separate the keyword REVIEW and its specifications.
- When you enter REVIEW, the Menu and Help system is displayed in the upper window, concealing the listing file. [Alt] [M] reveals the listing file.
- Issue the SET RCOLOR command before entering REVIEW to set nondefault colors (see SET).
- Specify SET COLOR OFF before entering REVIEW to turn color off.
- As an alternative to using the mini-menus, you can perform some editing functions by using the [Alt] key in combination with various other keys.
- Some commands cause REVIEW to prompt for additional information, such as a filename, at the bottom of the screen. To cancel such commands when the prompt appears, press [Esc].

Operations

- You can enter REVIEW from SPSS/PC+ or directly from DOS.
- If you edit two files at once, REVIEW divides your screen into upper and lower windows. The file named first occupies the upper window and the file named second occupies the lower window. At first, the Menu and Help windows cover the upper window. You can clear the menus (or recall them) with Alt M.
- To keep the menus up while you type in the other window, press Alt E. To redirect your typing back to the menus, press Esc.
- When you first enter REVIEW and the menus are on screen, the system searches the menu for a match to every character you type. The cursor moves to the first menu entry that matches what you've typed. This is called *incremental search*. To direct your keystrokes to the other window, press Alt E.
- Editing occurs at the current line and character as determined by the position of the cursor.
- There are two editing modes: insert mode and overtype mode. In *insert mode*, entered text is inserted to the left of the current character. In *overtype mode*, entered text replaces existing text.
- A two-digit number in the lower-right corner of the screen displays the column location of the cursor.
- To enter extended ASCII characters in the Edit window, activate NumLck and use Alt with the numeric keypad. If NumLck is not activated, the keypad controls cursor movement, and when used with Alt scrolls the Help window.

Limitations

- You can edit only ASCII files with REVIEW. Do not attempt to edit formatted (non-ASCII) files used by spreadsheet, word-processing, or other software. If the listing of a file that results from a DOS *TYPE* command is meaningful, the file is probably an ASCII file.
- REVIEW displays only the first 80 characters of any line. Use the Split Line command to view lines longer than 80 characters.
- REVIEW works on files held in memory. It starts with the *last* lines in the file in memory and reads additional lines as required. You may not be able to edit very large files.

Example

```
REVIEW 'A:PLOT.INC' 'PLOT.LIS'.
```

Entering REVIEW from DOS

To enter REVIEW directly from DOS, specify

```
SPSSPC/RE filename1 [filename2] [/La] [/Ub] [/Fc]
[/B]
```

- Place the names of the files you wish to edit in place of filename1 and filename2.
- If you wish to edit only one file, omit the second filename.
- The /La, /Ub, and /Fc options allow you to change the colors used for the lower (or only) window, upper window, and frame, respectively. Specify integers from 0 to 6 in place of a, b, and c. See the manual for your monitor for the colors that correspond to numbers 0 through 6.
- To turn off color use, specify /B on the command line.

If you plan to use REVIEW frequently from outside SPSS/PC+ with nondefault colors, create a batch file (a file with a .BAT extension) that contains the following single line:

```
SPSSPC /REVIEW /La /Ub /Fc %1 %2
```

- Specify the desired color numbers in place of a, b, and c.
- The batch file should be in the directory that holds your SPSS/PC+ program files.
- After you create the batch file, you can enter REVIEW by typing the name of the batch file followed by the names of the files you wish to edit. The colors you have specified in the batch file will take effect automatically.

Setting and Information Commands

Command Name	Key Combination/Function
REVIEW help*	F1 followed by R (or Alt R) reveals the Help screen for REVIEW function keys and Menu system commands. Pressing F1 again reveals the Help screen for motion commands. When pressed in response to a prompt, F1 provides a short message explaining the prompt.
Menus*	F1 followed by M (or Alt M) invokes the SPSS/PC+ Menu and Help system.
Glossary*	F1 followed by G (or Alt G or Alt D) calls up the online glossary. You are prompted for the term to look up; press ↵ to look up the term currently under the cursor, or enter another term before pressing ↵. The glossary displays the closest alphabetic match.
File list*	F1 followed by F (or Alt F) switches the Files menu on and off. The Files menu displays all (or a specified subset) the files in your current directory. When you need a filename to execute an SPSS/PC+ command, you can select the name from this menu. Esc clears the menu.
Var list*	F1 followed by V (or Alt V) switches the Variables menu on and off. The Variables menu displays all the variables defined in your active file. You can select variable names from this menu. Esc also clears the menu.
Switch windows*	F2 followed by S (or Alt S) causes the editor to switch windows.
Change window size*	F2 followed by C allows you to change the relative sizes of the two windows.

Menu Commands

Command Name	Key Combination/Function
menu Hlp on*	F1 followed by H (or Alt H or Alt ?) switches the Help window on and off. The Help window contains an explanation of the currently highlighted menu item.
Incremental Search	When the Menu and Help windows are displayed, the cursor automatically moves to the first item on the menu that matches what you type. If you type an *s*, the cursor moves to the first item that begins with *s*. If you then type an *e*, the cursor moves to the first item that begins with *se*.
Get Typing Window	Alt T opens a narrow window at the bottom of the Menu and Help system window. You can type anything you want in this window and when you press ↵ the contents of this typing window are pasted in the Edit window.
Down Menu	↓ moves the cursor down one item on the menu.
Up Menu	↑ moves the cursor up one item on the menu.
Bottom of Menu	End moves the cursor to the bottom of the menu.
Top of Menu	Home moves the cursor to the top of the menu.
Down Help Window	Alt ↓ moves you down one line in the Help window. Alt PgDn moves you down one page in the Help window.
Up Help Window	Alt ↑ moves you up one line in the Help window. Alt PgUp moves you up one page in the Help window.
Bottom of Help Window	Alt End moves you to the bottom of the Help window.
Top of Help Window	Alt Home moves you to the top of the Help window.
Paste Menu Item, Descend Menus	↵ pastes the highlighted menu item into the Edit window and descends to a lower-level menu (if any). If the menu item is followed by empty parentheses or

* These commands are mini-menu items that can be selected by highlighting the item with the cursor arrows and pressing ↵.

apostrophes, you'll be prompted to fill in the space before descending to a lower menu, unless lower menu selections fill in the space.

Browse Menus	(→) or (Tab) selects the highlighted menu item without pasting the item in the Edit window. Therefore you can descend through the lower menus, browsing, before deciding what to paste into your Edit window. Your selections are stored in a memory buffer in case you later decide to paste them. You can climb back out of the menus, and clear the buffer, with either (←) or (Esc).
Clear Memory Buffer	(Alt) (K) clears ("kills") the memory buffer. When you browse through the menus without pasting items into the Edit window, your selections are stored in a memory buffer. If you had already typed part of a command, for example, and you wanted to paste just one keyword instead of a complete command, you would first clear the buffer, then paste the keyword you need.
Remove Window, Ascend Menus	(Esc) either returns you to the Main Menu or brings you one step closer to the Main Menu. (Esc) removes the Variables and Files menus and the typing window. If you've descended several levels into the menus, (Esc) brings you up one level at a time. If you're typing in the Edit window, (Esc) returns you to the Menu and Help window.
Jump to Main Menu	(Alt) (Esc) returns you to the Main Menu from any level in the Menu and Help system.
Standard/Extended Menus	(Alt) (X) switches you back and forth between standard and extended menus. The *standard menus* contain the procedures and operations that you will use most of the time. The *extended menus* contain keywords that you may need, but probably not as often. A note at the bottom of the screen says either **STD** or **EXT**.

File Commands

Command Name	Key Combination/Function
write Marked area*	(F9) followed by (M) writes the current marked area to disk under a name you specify. The default name is REVIEW.TMP.
write Whole file*	(F9) followed by (W) (or (Alt) (W)) saves the current file to disk under a name you specify. The default is the current name. The original copy is renamed with an extension of .BAK.
Delete file on disk*	(F9) followed by (D) allows you to delete any file from a DOS directory.
Insert File*	(F3) followed by (I) inserts the contents of a DOS file into the current file below the current line.
Edit different file*	(F3) followed by (E) replaces the file in the current window.
run marked Area*	(F10) followed by (A) (or (Alt) (A)) submits a marked area to SPSS/PC+ for execution. This command is available only if REVIEW was entered from SPSS/PC+. If unsaved changes have been made in either window, you're prompted to save or abandon the changes.
run from Cursor*	(F10) followed by (C) (or (Alt) (C)) submits the commands beginning with the one containing the cursor and continuing to the end of the file to SPSS/PC+ for execution. This command is available only if REVIEW was entered from SPSS/PC+. If unsaved changes have been made in either window, you're prompted to save or abandon the changes.

Exit to prompt*	(F10) followed by (E) exits REVIEW. If unsaved changes have been made in either window, you're prompted to save or abandon the changes.
exit Fastgraf*	(F10) followed by (F) exits Fastgraph. If unsaved changes have been made in either window, you're prompted to save or abandon the changes.

Editing Commands

Command Name	Key Combination/Function
Change Edit Mode	(Ins) changes the editing mode from overtype to insert and from insert to overtype.
Insert after*	(F4) followed by (I) (or (Alt) (I)) inserts a blank line after the current line. The blank line becomes the current line.
insert Before*	(F4) followed by (B) (or (Alt) (B)) inserts a blank line before the current line. The blank line becomes the current line.
Delete*	(F4) followed by (D) deletes the current line. The line below becomes the current line.
Undelete*	(F4) followed by (U) restores the most recently deleted line and places it immediately above the current line. It is only possible to restore the single most recently deleted line.
Split Line	(↵) in insert mode splits the current line at the current character. In overtype mode, moves the cursor to the next line.
Join Line	(Del) joins the line below to the current line if the cursor is at the end of a line.
Delete Current Character	(Del) deletes the current character. The text to the right of the cursor moves left one space.
Delete Previous Character	The backspace key deletes the character to the left of the cursor. The cursor and righthand text move left.
fOrward change*	(F5) followed by (O) replaces one character string with another following the current cursor position. At each occurrence, prompts for (C) to change the occurrence and move to the next one, (A) to change all occurrences without individual prompting, (N) to move to the next occurrence without changing the current one, (S) to stop the search, or (X) to change the occurrence and then stop the search. The previous search and replacement strings are the defaults but can be typed over. Upper and lower case are distinguished in searches. To identify only those occurrences of the string that begin in column 1, place a tilde (~) at the beginning of the search string. It is not possible to change a character string that begins with a tilde.
bAckward change*	(F5) followed by (A) works the same as forward change moving backward through the file to the earliest line in memory.

The following commands deal with marking or unmarking areas:

Command Name	Key Combination/Function
Lines*	(F7) followed by (L) marks an area of lines for subsequent action. Press (F7) followed by (L) at the first or last line of an area and (F7) again at the opposite border to mark the area. Press (F7) a third time to unmark the area. Areas of lines consist of *whole* lines.
Rectangle*	(F7) followed by (R) marks any rectangle for subsequent action. Press (F7) followed by (R) at one corner of the rectangle, and (F7) again at the other corner. A rectangle

* These commands are mini-menu items that can be selected by highlighting the item with the cursor arrows and pressing (↵).

C Command Reference

	consists of a given number of lines of the marked width, which may be shorter than the width of the longest line. When you have marked a rectangular area, place the cursor at the *top left* of the area to which you want to copy or move it.
Command*	(F7) followed by (C) marks the current command for subsequent execution. Place the cursor anywhere within the command and press (F7) followed by (C). Press (F7) again to unmark the command.

The following commands apply to marked areas:

Command Name	Key Combination/Function
Copy*	(F8) followed by (C) copies a marked area to the line below the current line; it copies a marked rectangle to the right of and below the cursor. In insert mode, a rectangle pushes existing text to the right; in overtype mode, a rectangle replaces existing text. Areas of lines, commands, and rectangles can be copied between windows.
Move*	(F8) followed by (M) moves a marked area to the line below the current line; it moves a marked rectangle to the right of and below the cursor. In insert mode, a rectangle pushes existing text to the right; in overtype mode, a rectangle replaces existing text. Areas of lines, commands, and rectangles can be moved between windows.
Delete*	(F8) followed by (D) deletes a marked area or rectangle.
Round*	(F7) followed by (R) rounds all numeric values in the marked area or rectangle to a specified number of decimal places. When you press (F7) followed by (R), REVIEW prompts you for the desired number of decimal places. There is no way to reverse this command after execution; save important files before you round them in case the results are not what you expected. To round columns of numbers that are adjacent to one another, mark and round each column separately as a rectangle. Do not try to round numbers with imbedded commas (REVIEW will round each group of three digits) or numbers in scientific notation.

Motion Commands

Command Name	Key Combination/Function
Left	(←) moves the cursor left one space.
Right	(→) moves the cursor right one space.
Up	(↑) moves the cursor up one line.
Down	(↓) moves the cursor down one line.
Start of Line	(Ctrl) (←) moves the cursor to the first character of the current line.
End of Line	(Ctrl) (→) moves the cursor to the last character of the current line.
Tab Forward	(Tab) moves the cursor eight characters to the right.
Tab Backward	(⇧) (Tab) moves the cursor eight characters left.
Previous Page	(PgUp) moves the cursor to the last line of the preceding page.
Next Page	(PgDn) moves the cursor to the first line of the following page.
Top of Page	(Home) moves the cursor to the start of the first line in the window.
Bottom of Page	(End) moves the cursor to the end of the last line in the window.
Top of File	(Ctrl) (Home) moves the cursor to the first line of the file.
Bottom of File	(Ctrl) (End) moves the cursor to the last line of the file.

Top of Memory	(Ctrl) (PgUp) moves the cursor to the first line of the portion of the file currently in memory.
Put Line at Top	(Ctrl) (PgDn) moves the current line to the top of the window.
Start of Next Line	(↵) moves the cursor to the first character of the next line in overtype mode.
Forward search*	(F5) followed by (F) moves the cursor to the first occurrence of a specified character string on or below the current line. Upper and lower case are not distinguished in searches. If you want to locate the string only when it appears at the beginning of a line, specify a tilde (~) at the beginning of the search string. The search continues to the end of the file and the search string defaults to the last specified string. It is not possible to search for a string that begins with a tilde.
Backward search*	(F5) followed by (B) works the same as Forward search, moving backward through the file to the earliest line in memory.

The following commands tell the cursor to go to the indicated location:

Command Name	**Key Combination/Function**
Marked area*	(F6) followed by (M) moves the cursor to the top left corner of the marked area. If the cursor is already at the top left corner, this command moves it to the bottom right corner. If you are editing two files, this command may switch the cursor from one window to the other.
after executed Line*	(F6) followed by (L) (or (Alt) (L)) moves the cursor to the line below the last one executed.
Error line*	(F6) followed by (E) moves the cursor to the line below the first one that caused an error when the commands where executed.
Output page*	(F6) followed by (O) moves the cursor to a specified page in the listing file. REVIEW prompts you for the page number. The cursor must be located in the listing file when you use this command. Some types of output are not saved in listing files; thus some page numbers may be missing.

* These commands are mini-menu items that can be selected by highlighting the item with the cursor arrows and pressing (↵).

SAMPLE

```
SAMPLE {sampling fraction         }
       {sample size FROM file size}
```

Example:

```
SAMPLE .25.
```

Overview

The SAMPLE transformation temporarily draws a random sample of cases for processing in the next procedure. Sampling is based on a pseudo-random-number generator.

Syntax

- A decimal value between 0 and 1 selects an approximate percentage of cases.
- A positive integer value less than the file size followed by keyword FROM and the file size selects an exact-sized random sample.
- SAMPLE can be entered anywhere in an SPSS/PC+ session, except between BEGIN DATA and END DATA.

Operations

- SAMPLE is a temporary transformation and is executed when the data are read for the next procedure.
- By default, the initial seed value is randomly assigned. Use the SEED subcommand on SET to assign a specific seed value.
- A proportional sample (a sample based on a decimal value) usually does not produce the exact proportion specified.
- If the number (*n*) following FROM is less than the actual file size, the sample is drawn only from the first *n* cases.
- If the number following FROM is greater than the actual file size, SPSS/PC+ will sample an equivalent proportion of cases from the active file (see example below).
- If SAMPLE follows SELECT IF or PROCESS IF, it samples only cases selected by SELECT IF or PROCESS IF.
- If SAMPLE precedes SELECT IF or PROCESS IF, cases are selected from the sample.
- SAMPLE does not affect SORT CASES, and is ignored if the next procedure is SORT CASES.
- If more than one SAMPLE is specified before a procedure, only the last SAMPLE command is executed.
- If the same SAMPLE specification is used for different procedures in the same session or in different sessions, the two samples may be different because of the random assignment of the seed value. To obtain the same sample for different procedures, use the SET command specifying the same seed value before entering the SAMPLE command.

Example

```
SAMPLE .25.
```

- This command samples approximately 25% of the cases in the active file.

Example

```
SAMPLE 500 FROM 3420.
```

- In this example, the active file must have 3,420 cases or more to obtain a random sample of exactly 500 cases.
- If the file contains fewer than 3420 cases, approximately 15% of the cases are sampled.
- If the file contains more than 3420 cases, a random sample of 500 cases would be drawn from the first 3420 cases.

SAVE

```
SAVE [OUTFILE={SPSS.SYS**}] [/DROP=varlist]
              {'filename'}

     [/{COMPRESSED  }]
       {UNCOMPRESSED}
       {QUICK       }
```

**Default if subcommand is omitted.

Example:

```
SAVE DROP V1 TO V20.
```


Overview

SAVE produces an SPSS/PC+ system file. The system file includes all data and a data dictionary with variable and value labels (if specified), missing-value flags, and print formats for each variable. System files are read with the GET command.

System files saved by SPSS/PC+ *cannot* be read by SPSS/PC Release 1.0 or 1.1. To transfer a system file to SPSS/PC, use the EXPORT command from SPSS/PC+ and the IMPORT command in SPSS/PC. System files from SPSS/PC *can* be read directly by SPSS/PC+.

Defaults SAVE writes all variables in your active file to the file SPSS.SYS in the current directory. To save processing time, the system file is normally written in compressed form if and only if the active file is currently in compressed form (that is, COMPRESS=ON has been specified on SET).

Tailoring You can direct the system file to a file of your choosing. You can save a subset of variables on the system file. You can specify that the system file be written in compressed or uncompressed form.

Syntax

- The minimum specification is simply the command keyword.
- Subcommands, if used, must be separated by a slash.
- Only one of the subcommands COMPRESSED, UNCOMPRESSED, or QUICK can be specified per SAVE command.

Operations

- The system file created by SAVE includes the system variables $CASENUM, $DATE, and $WEIGHT.
- The system file dictionary is arranged in the same order as the active file.
- System files are binary files designed to be read and written by SPSS/PC+ only and cannot be edited.
- The active file remains available for SPSS/PC+ transformations and procedures after a SAVE file is created.
- SAVE messages that display variable counts treat each 8-character portion of a long string variable as one variable.

Example

```
SAVE DROP=V1 TO V3.
```

- The system file is written to SPSS.SYS in the current directory.
- Any variables between and including V1 and V3 on the active file are excluded from the system file.

OUTFILE Subcommand

Use OUTFILE to direct the system file to a file other than the default SPSS.SYS.

- The only specification on OUTFILE is the name of the file.
- The file specification must be enclosed in apostrophes.
- You can save a system file in another directory or on a drive other than the default by specifying a fully qualified filename (see Universals: Files).

Example

```
SAV OUT='\NSDIR\SALDATA.SYS'.
```

- The complete active file is written to file SALDATA.SYS in directory NSDIR.

- This example takes advantage of spelling permitted by three-character truncation of keywords.

DROP Subcommand

Use the DROP subcommand to exclude a variable or list of variables from the system file. All variables not named on DROP will be in the system file. The SPSS/PC+ active file is not affected by specifying DROP on the SAVE command.

- You can specify variables in any order.
- You can use the TO keyword to specify a group of consecutive variables in the active file.
- If you drop a long string variable, SPSS/PC+ will report that it has dropped one variable for each 8-character portion of the long string. Thus if you drop a 9-character string variable, a message will report that two variables have been dropped.

Example

```
SAVE OUTFILE='NEWSUM.SYS'/DROP=DEPT79 TO DEPT81, SALARY82.
```

- The system file is written to NEWSUM.SYS in the current directory.
- All variables between and including DEPT79 and DEPT81, as well as SALARY82, are excluded from the system file.
- All other variables are saved on the system file.

COMPRESSED Subcommand

The optional COMPRESSED subcommand tells SPSS/PC+ to save the system file in compressed form. In a compressed file, small integers (from -99 to 155) are stored in one byte, instead of the usual eight bytes.

- Compressed system files occupy less disk space than do uncompressed system files.
- Compressed system files take longer to read than do uncompressed system files.
- No additional specification is required on the GET command to read a compressed system file.

Example

```
SAVE OUTFILE='EMPLOY.SYS' /COMPRESSED.
```

- The system file is written to EMPLOY.SYS in the current directory.
- The system file is saved in compressed form.

UNCOMPRESSED Subcommand

The optional UNCOMPRESSED subcommand tells SPSS/PC+ to save the system file in uncompressed form.

- Uncompressed system files are quicker to read than are compressed system files.
- Uncompressed system files occupy more disk space than do compressed system files.
- No additional specification is required on the GET command to read an uncompressed system file.

QUICK Subcommand

The optional QUICK subcommand tells SPSS/PC+ to save the system file in the same form as that of the active file. This minimizes the time required to save the system file.

- QUICK is the default.
- If the active file is uncompressed, the system file is saved in uncompressed form.
- If the active file is compressed, the system file is saved in compressed form.
- Use the SET COMPRESS command at any time during an SPSS/PC+ session to set the active file to either compressed or uncompressed form. The default for COMPRESS is OFF.
- No declaration that a system file is either compressed or uncompressed is ever required by other SPSS/PC+ commands.

SELECT IF

```
SELECT IF (logical expression)
```

Relational Operators:

EQ or =	Equal to	NE or ~= or <>	Not equal to
LT or <	Less than	LE or <=	Less than or equal to
GT or >	Greater than	GE or >=	Greater than or equal to

Logical Operators:

AND or & OR or | NOT or ~

Missing-Value Functions:

SYSMIS	Selects the case if value is system-missing
MISSING	Selects the case if value is system- or user-missing
VALUE	Selects the case if value meets stated criteria, ignoring user-missing flags

Example:

```
SELECT IF (SEX EQ 'MALE  ').
SELECT IF (AVAR GE BVAR).
SELECT IF (SYSMIS(VARA)).
```

Overview

The SELECT IF transformation permanently selects cases for analysis based upon logical conditions found in the data. These conditions are specified in a *logical expression.* The logical expression can contain relational operators, logical operators, missing-value functions, and arithmetic operations and functions allowed in COMPUTE transformations (see COMPUTE, Universals: Logical Expressions, and Universals: Functions). For temporary case selection, see PROCESS IF.

Syntax

- The logical expression must be enclosed in parentheses.
- Parentheses also may be used to specify the order of evaluation.
- Relational operators are EQ, NE, LT, LE, GT, and GE (and their symbolic equivalents).
- Each relational operator must be preceded and followed by a variable name or an expression.
- Abbreviated syntax such as (AGE GE 18 AND LE 65) is *not* allowed.
- Logical operators are AND, OR, and NOT (and their symbolic equivalents).
- At least one relation, SYSMIS function, or MISSING function must be included in the logical expression.
- A relation includes a variable name, a relational operator, and a value or variable.
- Long string variables cannot be used in the SELECT IF command.
- SELECT IF can be entered anywhere in an SPSS/PC+ job, except between BEGIN DATA and END DATA.

Operations

- SELECT IF is a transformation and is executed when the data are read for the next procedure.
- SELECT IF permanently selects cases.
- The logical expression is evaluated as true or false.
- If a logical expression is true, the case is selected; if it is false or missing, the case is not selected.
- Multiple SELECT IF commands issued prior to a procedure command must all be true for a case to be selected.
- SELECT IF should be placed before other transformations for efficiency considerations.
- Logical expressions are evaluated in the following order: first numeric functions, then exponentiation, then arithmetic operators, then relational operators, and finally logical operators.
- Use parentheses to change the order of evaluation.

Limitations

- The complexity of logical expressions is limited by available memory.

Example

```
SELECT IF (SEX EQ 'MALE  ').
```

- All subsequent procedures will use only cases in which the value of SEX is equal to MALE.
- SEX is a short string variable with a format width of six characters. The specification for value MALE includes trailing blanks to fill the field.
- Since upper and lower case are different in comparisons of string variables, cases for which SEX equals 'male ' are not selected.

Example

```
SELECT IF (INCOME GT 75000 OR INCOME LE 10000).
```

- The logical expression tests whether a case has a value greater than 75000 or less than 10000. If either relation is true, the case is used in subsequent analyses. Note that the variable name INCOME is repeated before the second relational operator.

Example

```
SELECT IF (AVAR GE BVAR).
```

- This example selects cases where variable AVAR is greater than or equal to BVAR.

Example

```
SELECT IF (SEX = 'F' & INCOME <= 10000).
```

- The logical expression tests whether short string variable SEX is equal to F and if numeric variable INCOME is less than or equal to 10000. Cases that meet both conditions are included in subsequent analyses.

Example

```
SELECT IF (SYSMIS(VARA)).
```

- The logical expression tests whether VARA is equal to the system-missing value. If the value of VARA is system-missing, the case is selected for subsequent analyses.

Example

```
SELECT IF (VALUE(VARA) GT 0).
```

- Cases are selected if VARA is greater than 0, even if the value of VARA has been declared user-missing.

```
SELECT IF (VARA GT 0).
```

- In this example cases are not selected if VARA is user-missing, even if its value is greater than 0.

Example

```
SEL IF (RECEIV GT DUE OR (REVNUS GE EXPNS AND BALNCE GT 0)).
```

- The expression uses parentheses to change the order of evaluation. First, SPSS/PC+ tests whether variable REVNUS is greater than or equal to variable EXPNS and variable BALNCE has a value greater than 0. Second, SPSS/PC+ tests whether RECEIV is greater than DUE. If either of these conditions are met, the case is included in the active file for subsequent analyses.

Example

```
SELECT IF ((V1-15) LE (V2*(-0.001))).
```

- The logical expression compares whether V1 minus 15 is less than or equal to V2 multiplied by −0.001. If the expression is true, the case is selected for subsequent analyses.

Example

```
SELECT IF ((YRMODA(84,13,0) - YRMODA(YVAR,MVAR,DVAR)) LE 30).
```

- The logical expression subtracts the number of days representing the date YVAR (year), MVAR (month), and DVAR (day) from the number of days representing the last day in 1984. If the difference is less than or equal to 30, the case is selected for subsequent analyses.

SET

```
SET [SCREEN={ON }]/  [PRINTER={OFF}]/  [LISTING={'SPSS.LIS'}]/
            {OFF}              {ON }            {ON        }
                                                {OFF       }
                                                {'filename'}

    [RUNREVIEW={AUTO  }]/  [AUTOMENU={ON }]/
               {MANUAL}              {OFF}

    [MENUS={STANDARD}]/     [HELPWINDOWS={ON }]/
           {EXTENDED}                    {OFF}

    [LENGTH={24}]/  [WIDTH={79    }]/  [EJECT={OFF}]/
            {n }           {132   }           {ON }
                           {n     }
                           {NARROW}
                           {WIDE  }

    [VIEWLENGTH={MINIMUM}]/     [ERRORBREAK={ON }/
                {MEDIUM }                   {OFF}
                {MAXIMUM}
                {number }

    [LOG={'SPSS.LOG'}]/  [RESULTS={'SPSS.PRC'}]/
         {ON        }             {'filename'}
         {OFF       }
         {'filename'}

    [HISTOGRAM={'■'   }]/  [BLOCK={'█'   }]/
               {'char'}           {'char'}

    [BOXSTRING={─│┼└┌┘┐├┤┬┴}]/ [PTRANSLATE={ON }]/
               {'11 char'  }               {OFF}
               {' 3 char'  }

    [INCLUDE={ON }]/  [ECHO={ON }]/
             {OFF}          {OFF}

    [PROMPT={'SPSS/PC:'}]/  [CPROMPT={'       :'}]/  [MORE={ON }]/
            {'string'  }             {'string'  }          {OFF}

    [ENDCMD={'.'   }]/  [NULLINE={ON }]/
            {'char'}             {OFF}

    [BEEP={ON }]/
          {OFF}

    [COLOR={(7,1,1)    }]/  [RCOLOR={(1,2,4)    }]/
           {ON         }            {(a,b[,c])}
           {OFF        }
           {(a,b[,c])}

    [COMPRESS={OFF}]/  [WORKDEV=device]/
              {ON }

    [BLANKS={'.'     }]/  [SEED={RANDOM}]
            {real num}          {number}
```

When SCREEN is OFF, the following defaults are in effect:

```
LENGTH=59  EJECT=ON
BOXSTRING='-|+++++++++'  HISTOGRAM='X'  BLOCK='X'
```

Example:

```
SET MORE=OFF/ECHO=ON/LENGTH=59/EJECT=ON.
```

Overview

The SET command changes SPSS/PC+ running options. To see the current running options, use the SHOW command.

Defaults

The default running options include the command prompt **SPSS/PC:** , the continuation prompt **:** , the period (.) and blank line as command terminators, and a prompt of **MORE** after each page of output. To remind you to move to the next screen of output, a high-pitched beep is sounded. Error messages are accompanied by a low-pitched beep. The default foreground, background, and border colors are 7, 1, and 1, respectively. On many color monitors this corresponds to white on blue. However, your monitor may interpret 7, 1, 1 differently. The default environment is the Menu and Help system within REVIEW, displaying the standard menus and their help windows.

SPSS/PC+ directs the output from statistical and reporting procedures to your screen, with a length of 24 rows and a width of 79 columns. Output does not go automatically to the printer and no blank lines are left between pages. In addition to appearing on the screen, procedure output is sent to a listing file whose default name is SPSS.LIS. Command lines are not sent to the listing file. If you write out a data set or matrix materials, SPSS/PC+ writes the results to the file SPSS.PRC.

Any commands you enter are sent to a log file. By default, this file is SPSS.LOG. Commands processed from a command file with the INCLUDE command are shown on the screen. If a command causes an error, processing stops.

Crosstabulation tables use special characters for the screen display. Histograms (available in FREQUENCIES) use a solid bar. Icicle plots (available in CLUSTER) use a solid box. By default, special characters are translated to simpler ones when output is sent directly to a printer.

The SAMPLE transformation and UNIFORM and NORMAL functions (available in COMPUTE) use a random seed obtained from the clock. Numeric blanks are interpreted as system-missing values and printed as a period (.) in output. Active files are kept in uncompressed form on the current device.

Tailoring

Output Destination. You can send your output to multiple destinations (screen, disk, and printer), and you can change destinations or turn off any of them throughout the course of an SPSS/PC+ session.

Output Form and Layout. You can change the length and width of output. Additionally, you can include printer carriage-control characters for page ejects in disk file output.

Optional Output. You can write data from the WRITE procedure or matrix materials from REGRESSION, CORRELATION, CLUSTER, ONEWAY, and FACTOR procedures to either SPSS.PRC or to a file of your choice.

Environment. Instead of beginning in the Menu and Help system environment, you can begin at the SPSS/PC+ command prompt, or in REVIEW without the Menu and Help system. If you do use the Menu and Help system, you can suppress the help windows, and/or use extended menus.

Error Processing. You can direct the system to try to continue processing commands after one causes an error.

Special Characters. You can change the characters SPSS/PC+ uses to print crosstabulation table grids. You can also specify the symbols SPSS/PC+ uses in histograms (available in FREQUENCIES) and in icicle plots (available in CLUSTER).

Character Translation. SPSS/PC+ automatically translates special characters to simpler ones for output going directly to a printer. You can suppress this translation if you have a printer that can handle special characters.

Command Printback. You can direct all your SPSS/PC+ commands to your output file and/or printer. You can also suppress the screen printback of commands included with the INCLUDE command.

Prompts. You can specify your own command prompt and continuation prompt. You can turn off the prompt that appears by default after each screen of output.

Command Terminators. You can change the default command terminator from a period to a character of your choice. You can specify either that a blank line be ignored or taken as a command terminator.

Sound and Color. You can suppress the paging and error-message beeps. On a color monitor, you can choose colors for the type, the background, and the border.

Active and Scratch Files. You can direct your active and scratch files to any disk. You can specify that the active file be kept in compressed or uncompressed form.

Treatment of Numeric Blanks. You can substitute a numeric value for blank fields in numeric variables. All numeric blanks encountered are translated to this value.

Random Number Seed. You can change the initial seed value to a particular number or have SPSS/PC+ select a random number.

Syntax

- Subcommands on SET can be named in any order.
- Subcommands can be separated by an optional slash.
- Only one keyword or argument can be specified for each subcommand.
- SET can be used more than once in an SPSS/PC+ session.
- Each time SET is entered, only the named specifications are affected. All others remain at their previous settings or the default.
- YES is accepted as an alias for keyword ON.
- NO is accepted as an alias for keyword OFF.

Operations

- SET is an operation command and takes effect immediately.
- Since SPSS/PC+ output, by default, is primarily directed to the screen, the default character settings use graphics characters that may not be available on your printer.
- SPSS.LIS and SPSS.LOG are erased at the *beginning* of each SPSS/PC+ session. If you want to preserve their contents, use the DOS *RENAME* command to rename them before reentering SPSS/PC+.
- If you use multiple log, listing, or results files, you should not start with one file, go to another file, and then return to the first file. SPSS/PC+ will *not* append new information to the first file. It will *write over* the previous contents.
- The results file can be directed only to the current directory. The log and listing files may be sent to any directory.

Limitations

- Some characters used in screen displays are not available on all printers. Additionally, some of the characters used for the screen do not have a printable equivalent.
- A screen display length of 35 or 43 lines can only be used with an EGA or VGA adapter. A length of 43 may not be compatible with an ANSI.SYS file.

Example

```
SET MORE=OFF/ECHO=ON/LENGTH=59/EJECT=ON.
```

- The specification MORE=OFF requests that SPSS/PC+ send output to the screen continuously, without pausing to give the *MORE* prompt when the screen fills. (Use REVIEW to examine output when a procedure is finished.)
- The specification ECHO=ON integrates all input commands in the output *listing* file. This is independent of whether commands are written to a log file.
- SPSS/PC+ uses a page length of 59 for both screen and disk output.
- EJECT=ON inserts a page-eject carriage-control character in the disk output file after each 59 lines of output and suppresses the dashed line between pages.

Output Destination

By default, SPSS/PC+ directs output both to the screen and to a listing file named SPSS.LIS in the current directory. You can change the name of the listing file. Additionally, you can direct output directly to a printer.

- When SCREEN=ON is in effect, the output is, by default, formatted especially for screen-by-screen displays and includes special graphics characters that are available only on the screen.
- If you turn screen display off, SPSS/PC+ uses different defaults for output. The defaults for SCREEN=ON and SCREEN=OFF are listed in Table A below. You can change these options with other SET subcommands.
- When you direct output to both the screen and printer or disk file, you must specify format options that are compatible with your printer.
- The keywords ON and OFF on subcommand LISTING apply to the current listing file, regardless of its name. You can suspend sending output to a file and then resume sending output to the same file, as long as you do not specify a different listing file in the interim.
- The output from different procedures can be directed to different disk files by naming a different file on the LISTING subcommand prior to a procedure command.
- When you use multiple disk files, you cannot send output to one file, then to another file, and then resume sending output to the first file. If you do, SPSS/PC+ writes over the first set of output. It cannot append output to a file once another file has been named.

The subcommands for controlling output destinations are

SCREEN *Screen output.* SCREEN=ON is the default. You can suppress screen display by specifying SCREEN=OFF and then restart screen display by specifying SCREEN=ON.

PRINTER *Printer output.* PRINTER=OFF is the default. With PRINTER=ON, output is sent directly to your printer in addition to any other destinations. Your printer must be turned on if you specify PRINTER=ON. You can stop sending output to the printer by specifying PRINTER=OFF.

LISTING *Disk output.* LISTING=ON is the default. You can suppress the copying of your output to a disk file by specifying LISTING=OFF. The default disk output file is SPSS.LIS in the current directory. You can direct the output to another disk file in a directory of your choice by specifying a filename on the LISTING subcommand. Enclose the filename in apostrophes. The keyword DISK is accepted as an alias for LISTING.

Example

```
GET FILE='MARCH.SYS'.
LIST VAR=V1 TO V10.
SET LISTING='MEANS.LIS'
MEANS V1 TO V5 BY V7.
SET LISTING=OFF.
FREQ VAR=V1,V3,V4/HISTOGRAM.
SET LISTING=ON/HISTOGRAM='X'.
FREQ VAR=V3/HISTOGRAM.
FINISH.
```

- Output is initially routed to the default file SPSS.LIS in the current directory.
- The GET command reads the SPSS/PC+ system file MARCH.SYS from the current directory.
- The LIST procedure output is directed to the screen and to SPSS.LIS.
- The first SET command directs output to the file MEANS.LIS in the current directory.
- The output from the MEANS procedure is directed to MEANS.LIS and to the screen.
- The next SET command suspends sending output to the file MEANS.LIS. Procedure display continues coming to the screen.

- The FREQUENCIES command requests frequency tables and histograms for three variables. The histograms use the default histogram characters for the screen.
- The next SET command with LISTING=ON resumes sending output to the file MEANS.LIS. The character used in any histogram on the screen and in the disk file is X.
- The FREQUENCIES command produces a table and histogram for the variable V3. Output is directed to both the screen and to the disk file.
- The FINISH command ends the SPSS/PC+ session.
- The file SPSS.LIS contains the results from the LIST procedure. The file MEANS.LIS contains output from the MEANS procedure and from the last FREQUENCIES procedure.

Output Form and Layout

SCREEN=ON and SCREEN=OFF set up different defaults for output format (see Table A). You can change these defaults with the LENGTH, VIEWLENGTH, WIDTH, and EJECT subcommands.

- The LENGTH subcommand affects the page length for the screen, the printer, and for the listing file.
- On the screen, the **MORE** prompt appears after the specified number of lines have been displayed.
- For printer and listing-file output, a dotted line is printed or a page eject occurs at the end of the specified number of lines.
- The VIEWLENGTH subcommand determines the display of lines on the screen. Computers with special adapters can display more than 25 lines at a time, by making the lines appear smaller.
- The PC must have a special graphics adapter (EGA or VGA) to use any VIEWLENGTH other than MINIMUM (25 lines).
- If the CONFIG.SYS file contains the specification DEVICE=ANSI.SYS, a VIEWLENGTH of 43 may not work.
- The WIDTH subcommand affects screen, printer, and listing-file output.
- The screen allows a width of only 79 characters. Lines longer than 79 characters are normally wrapped onto two lines. Some SPSS/PC+ output is produced by special-purpose routines that do not support wrapping of long lines. To examine the portion of such lines that is off the screen, either specify SET WIDTH 79, print the listing file, or use an editor that wraps or allows horizontal scrolling to view the listing file (REVIEW does not do this).
- Printer and disk output can contain lines longer than 79 characters.
- To print long lines (from 80 to 132 characters), use the DOS *MODE* command to reduce the size of characters on your printer.
- The EJECT subcommand affects only printer and disk file output. With LENGTH=59 and EJECT=ON, insert printer paper with perforations one-half inch above the type head; SPSS/PC+ will print each page with skips over subsequent perforations.

Table A Defaults for SCREEN settings

<table>
<tr><th>SCREEN</th><th>LENGTH</th><th>BOXSTRING</th><th>HISTOGRAM</th><th>BLOCK</th><th>EJECT</th></tr>
<tr><td>ON</td><td>24</td><td>-|+└┌┘┐├┤┬┴</td><td>▪</td><td>■</td><td>OFF</td></tr>
<tr><td>OFF</td><td>59</td><td>-|+++++++++</td><td>X</td><td>X</td><td>ON</td></tr>
</table>

The subcommands for controlling output, form, and layout are

LENGTH *The page length* (the number of lines before the MORE prompt, and before a page eject is sent to the listing file and the printer). Specify a positive integer. LENGTH is normally set to one line less than VIEWLENGTH. When you intend to print the listing file, you may want to set LENGTH to a suitable value for your printer. A standard printer page length is 59 lines.

VIEWLENGTH *The number of lines shown on the screen.* Specify MINIMUM, MEDIUM, MAXIMUM, or a number. Available settings depend upon your display adapter. MDA and CGA allow only MINIMUM (25 lines). EGA allows 25, 35, and 43 line displays. VGA likewise honors all three keywords, with 50 lines displayed for MAXIMUM. If you specify an inappropriate number, SPSS/PC+ rounds it to the nearest available number.

WIDTH *The page width.* Specify an integer or keywords NARROW (79 characters) or WIDE (130 characters). Most SPSS/PC+ output is formatted to fit in a page width of 79 characters. For procedures that let you specify your own width, the width specified in the procedure should agree with the width specified on SET.

EJECT *Carriage control for page ejects for printer and disk output.* Specify ON or OFF. With EJECT=OFF, output sent to the printer and/or to a disk file contains a dotted line indicating page breaks. With EJECT=ON, no dotted line is printed, and pages are printed with margins between each page. When SCREEN=ON, the default is EJECT=OFF.

Example

```
SET LISTING='REG.LIS'/VIEWLENGTH=MAXIMUM/LENGTH=59/WIDTH=130/
EJECT=ON.
REG VAR=SAVINGS POP15 POP75 INCOME GROWTH
/SELECT IN2 NE 1
/DEP SAVINGS /ENTER POP15/RES SEPARATE
/CASEWISE DEP OUTLIER (3) ALL PLOT (RESID)
/DEP SAVINGS /ENTER INCOME/RES POOLED.
```

- The SET command directs output to the file REG.LIS in addition to the screen; the latter will display the maximum number of lines possible for the display adapter. The output is formatted with 59 lines for each page with an eject after each page. The width of each page is 130 characters.
- The REGRESSION procedure uses 130 characters on each line of the output display.
- If REG.LIS is printed, each page includes top and bottom margins to skip over the perforations. No lines are wrapped in printed output.

Optional Output

SPSS/PC+ provides two types of additional output: a log file, which includes all commands processed in a session; and a results file, which contains matrix or casewise data produced by options on SPSS/PC+ procedures.

LOG Subcommand

By default, SPSS/PC+ copies all the commands you enter into log file SPSS.LOG in the current directory.

- You can send the log file to a file other than SPSS.LOG by specifying a different filename, in apostrophes, on the LOG subcommand.
- You can suppress the building of the log file by specifying LOG=OFF.
- You can resume sending commands to the file SPSS.LOG or to the log file last specified by specifying LOG=ON. Any existing contents of that file will be overwritten.
- The log file contains each command processed during a session.
- If you enter a command with errors, the log file contains a comment line after that command indicating the error.
- The log file can be edited to remove erroneous lines for use as a command file (see INCLUDE). REVIEW allows you to do this without leaving SPSS/PC+.
- When you use a log file from a completed session as a command file, you must initially send the log file to a file other than SPSS.LOG or rename the log file using the DOS *RENAME* command since SPSS.LOG is initialized at the start of each session.

When you use INCLUDE in your session, the commands from the included command file are preceded by a square bracket ([) in the log file. If this log file is then used as a command file in a subsequent session, only the INCLUDE command is processed, not the commands preceded by the square bracket. This avoids double processing of commands. You cannot enter the square bracket yourself in interactive sessions.

Example

```
SET LOG='DATADEF.FIL'.
DATA LIST FILE='MARCH.DAT'/OP1 TO OP7 1-14.
VARIABLE LABELS OP1 'Opinion on sex'/OP2 'Opinion on religion'/
     OP7 'Opinion on divorce'.
VALUE LABELS OP1 TO OP7 1 'Agree' 2 'Disagree'.
MISSING VALUE OP1 TO OP7(9).
INCLUDE 'DES.INC'.
FIN.
```

- The SET command directs the log file to DATADEF.FIL in the current directory.
- The DATA LIST, VARIABLE LABELS, VALUE LABELS, MISSING VALUE, and INCLUDE commands and their specifications are copied to the file. Each of the commands processed from the INCLUDE file, DES.INC, is copied to the log file prefaced with a square bracket.
- The file DATADEF.FIL can be used as a command file in a subsequent SPSS/PC+ session.

RESULTS Subcommand

Some SPSS/PC+ procedures produce data that can be used in other sessions. WRITE produces a rectangular data file of individual values. Procedures CORRELATION, ONEWAY, and REGRESSION (and CLUSTER, FACTOR, MANOVA, and QUICK CLUSTER in SPSS/PC+ Advanced Statistics) produce summary data in the form of matrix materials. SPSS/PC+ writes these data files to the default file SPSS.PRC in the current directory.

- You can direct these data files to other files in the current directory by specifying a filename in apostrophes on the RESULTS subcommand.
- The results file is always written to the current directory.
- The results file can be used as input (named on the DATA LIST or, after editing, on the INCLUDE command) at a later time or in a subsequent SPSS/PC+ session.
- SPSS/PC+ does *not* append data to the results file. If two procedures write results to the same file, the second set of data will overwrite the first set.

Example

```
SET LISTING 'NEWDAT.LIS' / RESULTS='NEWDAT.DAT'.
GET FILE='INVENT.SYS'.
COMPUTE TOTCOST=V1+V7+V9.
SELECT IF (V3 LT 7).
WRITE VARIABLES=ID TO V15, TOTCOST.
FINISH.
```

- The SET command specifies that procedure output be sent to the disk file NEWDAT.LIS and that data results be written on the file NEWDAT.DAT. Both files are written in the current directory.
- The GET command reads the system file INVENT.SYS from the current directory.
- The COMPUTE command creates (or modifies) the variable TOTCOST.
- The SELECT IF command tells SPSS/PC+ to select cases in which the value of variable V3 is less than 7.
- The WRITE command writes the values of variable TOTCOST and variables between and including ID and V15 (provided the value of V3 is less than 7). The data are written on the file NEWDAT.DAT.

Environment

SPSS/PC+ provides three environments in which to enter commands: the Menu and Help system; the REVIEW text editor; and the command prompt. The default environment when you enter SPSS/PC+ is the Menu and Help system within REVIEW. You can make one of the other environments the default; additionally you can alter the default settings of the Menu and Help system environment.

Default Environment

- To make the default environment the REVIEW text editor *without* the Menu and Help system, use the subcommand AUTOMENU=OFF. You can still call up the menus whenever you like by pressing Alt M.
- To make the default environment the SPSS/PC+ command prompt, use the subcommand RUNREVIEW=MANUAL. To get into REVIEW from the command prompt, use the REVIEW command.

Example

```
SET RUNREVIEW MANUAL.
```

- The default environment becomes the SPSS/PC+ command prompt. If commands are entered and run from REVIEW, the system returns to the command prompt upon completion of processing, not to REVIEW.

Menu Settings

- To automatically display menus with all possible keywords, use the subcommand MENUS=EXTENDED. The alternative, MENUS=STANDARD, omits seldom-used keywords. You can switch between standard and extended menus at any time by pressing Alt X.
- To display the menus without the help windows that annotate each menu selection, use the subcommand HELPWINDOWS=OFF. (This speeds up the display of menus on slow machines.) You can bring back the help windows at any time by pressing Alt H.

Example

```
SET MENUS = EXTENDED /HELPWINDOWS = OFF.
```

- When the menus display, all keywords are included, but the help windows do not appear.

Error Processing

To direct the system to continue processing commands after one causes an error, use the subcommand ERRORBREAK=OFF. This is useful when initial commands are not essential for the processing of subsequent commands. Regardless of the setting of ERRORBREAK, when the system returns to REVIEW after processing commands that caused errors, the cursor is always positioned at the first command that caused an error.

Example

```
SET ERRORBREAK = OFF.
VRIABLE LABELS OP1 'Opinion on sex'/OP2 'Opinion on religion'/
     OP7 'Opinion on divorce'.
LIST.
```

- The SET command directs the system to continue processing even if errors are encountered.
- The VRIABLE LABELS command is misspelled, and causes an error.
- The LIST command is processed even though the preceding command caused an error, because ERRORBREAK is set to OFF. However, the listing does not include variable labels.

Special Characters

SPSS/PC+ uses special graphic characters in tables and plots. These graphic characters appear on the screen but may not print correctly (depending upon the type of printer you are using) or may be interpreted as special printer escape sequences. Three subcommands, HISTOGRAM, BLOCK, and BOXSTRING, are used to specify other characters.

HISTOGRAM *The character used in histograms and barcharts.* Specify a single character in apostrophes. If output is directed to the screen, the default is a lower-half solid block (ASCII decimal 220). If output is not directed to the screen, X is the default.

BLOCK *The character used in icicle plots.* Specify a single character in apostrophes. If output is directed to the screen, the default is a full solid block (ASCII decimal 219). If output is not directed to the screen, the default is X.

BOXSTRING *Box-building characters.* Specify the characters to be used in building boxes. You can specify a set of three characters or a full set of eleven. Enclose either set in apostrophes.

A set of three characters comprises the horizontal bar, vertical bar, and intersection characters, in that order. The intersection character is used for all types of intersections.

A set of eleven characters comprises the following characters: horizontal bar, vertical bar, intersection, lower-left corner, upper-left corner, lower-right corner, upper-right corner, left T, right T, top T, bottom T. The characters must be specified in this order.

If the display is going to the screen, the defaults are the eleven single-stroke graphics characters (see Table B). If the display is not directed to the screen, '-|+++++++++' are the default BOXSTRING characters.

You can enter special characters by holding down (Alt) and entering the ASCII decimal value for the character desired on the numeric keypad on the right side of the keyboard. For example, to enter the lightest-shade block character, press (Alt) and hold it down while entering 176. Then release (Alt). Depending upon the editor you use to build command files, you may be able to enter special characters in the same manner.

A complete list of characters and their ASCII decimal values is available in the DOS BASIC manual. Table B below documents some of the more useful characters and their ASCII decimal codes.

Table B Special characters

Character	ASCII value	Character	ASCII value
Single stroke horizontal bar	196	Double stroke horizontal bar	205
Single stroke vertical bar	179	Double stroke vertical bar	186
Single stroke intersection	197	Double stroke intersection	206
Single stroke lower-left corner	192	Double stroke lower-left corner	200
Single stroke upper-left corner	218	Double stroke upper-left corner	201
Single stroke lower-right corner	217	Double stroke lower-right corner	188
Single stroke upper-right corner	191	Double stroke upper-right corner	187
Single stroke left T	195	Double stroke left T	204
Single stroke right T	180	Double stroke right T	185
Single stroke top T	194	Double stroke top T	203
Single stroke bottom T	193	Double stroke bottom T	202
Light intensity block	176	Lower-half solid block	220
Medium intensity block	177	Upper-half solid block	223
Heavy intensity block	178	Left-half solid block	221
Solid block	219	Right-half solid block	222
Small solid block	254		

Character Translation

Many printers cannot print the special characters used by SPSS/PC+ for histograms, plots, and tables. By default, SPSS/PC+ translates these characters to simpler ones when output is sent directly to a printer (PRINTER=ON). If your printer can print special characters, you may want to turn off this translation.

PTRANSLATE *Character translation for output to printers.* Specify ON or OFF. PTRANSLATE=ON (the default) translates special characters to simpler ones when PRINTER=ON. PTRANSLATE=OFF prevents this translation.

Example

```
SET PTRANSLATE=OFF.
```

- Translation of special characters to simpler ones is turned off. When PRINTER=ON, the printer will receive unsimplified characters.

Command Printback

SPSS/PC+ provides two options for printing back commands either on the screen or in output: the INCLUDE and ECHO subcommands.

- If you don't want to see SPSS/PC+ commands from an INCLUDE file on your screen, specify INCLUDE=OFF prior to using the INCLUDE command.
- If you do not want SPSS/PC+ commands (from whatever source) copied to your listing file, specify ECHO=OFF.

INCLUDE *Printback of commands from command files on the screen.* Specify ON or OFF. INCLUDE=ON (the default) displays commands on your screen from any command file that you include. Inline data from included command files is never sent to the screen.

ECHO *Printback of commands in output.* By default ECHO=ON, and SPSS/PC+ sends SPSS/PC+ commands along with results of statistical and reporting procedures to listing and printer files. Specify ECHO=OFF if you want only results of statistical and reporting procedures sent to these files. This does not affect screen output.

Example

```
SET ECHO=OFF.
GET FILE='MUFILE.SYS'.
CROSSTABS TABLES=AVAR BY BVAR BY CVAR.
SET SCREEN=OFF.
INCLUDE 'REG.INC'.
FIN.
```

- The first SET command specifies that the listing file commands should not include the command printback.
- The second SET command stops sending output display to the screen. Output is still sent to the listing file.
- The INCLUDE command processes commands from file REG.INC in the current directory. By default, all commands from REG.INC are sent to the screen as they are processed. If INCLUDE=OFF is specified in REG.INC, further commands from that file are not sent to the screen as they are processed.

Prompts

Use the PROMPT, CPROMPT, and MORE subcommands to control the characters used for prompts and to suppress page prompting.

- SPSS/PC+ uses two types of prompts for command input: a command prompt and a continuation prompt. The command prompt prompts for a new command and the continuation prompt prompts for the next line of a command.
- The continuation prompt is always issued if a command terminator has not been used on the previous input.
- You can specify special characters for both PROMPT and CPROMPT.
- By default SPSS/PC+ pauses after each screen of output and displays the **MORE** prompt. You must press a key to see the next screen of output. The MORE subcommand lets you eliminate the pause and prompt between screens of output.

PROMPT *The command prompt.* Specify a string of up to 8 characters in apostrophes. The default is SPSS/PC: .

CPROMPT *The continuation prompt.* Specify a string of up to 8 characters in apostrophes. The default is : .

MORE *The full screen prompt.* Specify ON or OFF. MORE=ON (the default) causes SPSS/PC+ to pause between screens of output and display the **MORE** prompt. MORE=OFF eliminates the prompt, and output is displayed continuously across the screen.

Example

```
SET PROMPT='===> '/CPROMPT='this it?'/MORE=OFF.
```

- The command prompt is changed to '===> '.
- The continuation prompt becomes, 'this it?'.
- Pausing and prompting between screens of output is turned off.

Command Terminators

Use the ENDCMD and NULLINE subcommands to control command termination.

- Every SPSS/PC+ command must end with a command terminator.
- By default, the terminator is a period, but you can change it with the ENDCMD. SPSS/PC+ continues to read lines as part of the current command until it encounters the command terminator.

- When NULLINE is ON (the default), a completely empty input line is accepted as an alternative command terminator. If you SET NULLINE OFF, you must enter the character specified for ENDCMD.
- The ENDCMD specification should be a printable character.
- Do not specify a command terminator that is the last character in any of your variable names, nor a character with special syntactic meaning in SPSS/PC+ (such as a slash). Such specifications will often result in syntactic ambiguity.

ENDCMD *The command terminator.* Specify a character in apostrophes. The default is a period.

NULLINE *Empty lines as an alternative command terminator.* Specify ON or OFF. The default is NULLINE=ON.

Example

```
SET ENDCMD='!' / NULLINE=OFF.
```

- The command terminator is set to '!'.
- A blank line will not be accepted as a command terminator.

Sound and Color

Use the BEEP subcommand to suppress the beep that signals the next screen of output and errors. Use the COLOR subcommand to control the text, background, and border colors on color monitors. On some monochrome monitors, you can use the COLOR subcommand to control the patterns and shading that may appear in place of colors. The RCOLOR subcommand allows you to set the background colors in REVIEW, the SPSS/PC+ text editor.

- A high-pitched beep reminds you to page to the next screen of output.
- A low-pitched beep is sounded when an error occurs.
- On most color displays, it is possible to assign one color to text, another color to the background on which the text appears, and a third color to a border surrounding the central text area.
- In REVIEW, you can assign colors to the primary and secondary text windows, and to the frame surrounding them.

BEEP *Beep.* To suppress the beep, specify OFF. To turn it on, specify ON. The default is ON.

COLOR *Color.* Specify a set of two or three integers in parentheses or specify ON or OFF. A set of three integers represents text color, background color, and border color, in that order. An asterisk in place of an integer retains the existing color. If you specify the first two integers only, the border color is taken to be the same as the background color. The default color scheme is COLOR=(7,1,1), which on many color monitors corresponds to white text on a blue background with a blue border. Check your monitor manual for the colors available on your monitor and the integers that represent these colors. Colors for the standard IBM color monitor are shown in Table C.

RCOLOR *Color in REVIEW.* Specify a set of three integers in parentheses, representing background color in the lower (or only) window, background color in the upper window (when present), and frame color, in that order. The three integers must be between 0 and 6. An asterisk in place of an integer retains the existing color. The default color scheme is RCOLOR=(1,2,4). Text is always white in REVIEW. The keywords OFF and ON are not recognized for RCOLOR, but if you SET COLOR OFF, REVIEW will use a monochrome display.

- The SET RCOLOR command must be issued from the SPSS/PC+ environment; you cannot change colors from within REVIEW itself.
- On some monochrome monitors, colors may be interpreted as patterns and shading. Your monitor manual will tell you how to control these patterns.
- You can specify COLOR=OFF (on color or monochrome monitors) to eliminate all color representation.

- COLOR=OFF affects REVIEW as well as the SPSS/PC+ command environment; however, the specific colors used by REVIEW when COLOR is on are determined by RCOLOR rather than COLOR.
- Specify COLOR=ON to return to the color patterns that were in effect before color was turned off.
- Put the command SET COLOR=OFF in your automatic profile SPSSPROF.INI to force a monochrome display if you prefer it to the default display.
- The high-intensity colors 8–15 can only be used for the first (text) color in SPSS/PC+. Only colors 0–6 are used in REVIEW.

Table C Color selection for standard IBM color monitor

0	Black	8	Gray
1	Blue	9	Light Blue
2	Green	10	Light Green
3	Cyan	11	Light Cyan
4	Red	12	Light Red
5	Magenta	13	Light Magenta
6	Brown	14	Yellow
7	White	15	High Intensity White

Example

```
SET BEEP=OFF.
GET FILE='AUTO.SYS'.
SET COLOR=(6,1,3).
FREQUENCIES VARIABLES=COSTCAT,SIZE,POWER.
SET COLOR OFF.
CROSSTABS TABLES=MANUFACT BY POWER.
SET COLOR ON.
FREQUENCIES VARIABLES=COSTCAT.
FINISH.
```

- The first SET command turns off beeps for this session.
- Prior to the first SET COLOR command, the default color pattern (7,1,1) is in effect.
- The first SET COLOR command sets the text color to 6, the background color to 1, and the border color to 3.
- The second SET COLOR command turns color off, creating a monochrome display.
- The third SET COLOR command restores the text, background, and border colors to their previous values of 6, 1, and 3, respectively.

Active and Scratch Files

Use COMPRESS subcommand to indicate whether the active file for an SPSS/PC+ session should be kept in compressed or uncompressed form. Use WORKDEV to direct the active and scratch files to a specific block device (disk drive).

- A compressed active file occupies less space in your directory than does an uncompressed active file.
- A compressed active file takes longer to access when you run a procedure than does an uncompressed active file.
- When you perform transformations during a session, SPSS/PC+ creates a new active file but both the new and old active files exist simultaneously. Thus, the space you save by keeping your active file compressed doubles if you do any transformations during an SPSS/PC+ session.
- If you save a system file from a compressed active file, the system file is saved in compressed form unless you have issued commands that modify the active file before it is saved.
- SPSS/PC+ can write the active files for a session to any disk drive.

COMPRESS *Compression of the active file.* Specify ON or OFF. COMPRESS=ON causes SPSS/PC+ to keep the active file in compressed form. COMPRESS=OFF causes SPSS/PC+ to keep the active file in uncompressed form. COMPRESS=OFF is the default.

WORKDEV *Location of active files.* Specify a device letter that corresponds to one of your disk drives. The active file and most scratch files for the session will be kept on that device. The default device is that of the current directory. Once an active file has been written during the session, you cannot change the WORKDEV.

Example

```
SET COMPRESS=ON / WORKDEV=B.
```

- The active file for this session will be kept in compressed form.
- The active file and other scratch files for the session will be kept on device B. If an active file has already been written during the session, WORKDEV will generate an error message.

Treatment of Numeric Blanks

By default, SPSS/PC+ translates entirely blank fields read with a numeric format to the system-missing value. You can use the BLANKS subcommand to specify some other value.

- BLANKS controls only the translation of numeric fields.
- If a blank field is read with an A format, the resulting value is a blank.
- The BLANKS specification controls all numeric-variable blanks. You cannot have different specifications for different variables.
- BLANKS must be specified before data are read, since blanks in numeric fields are converted to system-missing as they are read.

BLANKS *Blanks for numeric fields.* Specify any number. Blanks are translated to this number. The number is not automatically defined as missing. The default, however, is the system-missing value.

Example

```
SET BLANKS=-1.
```

- The SET command translates all numeric-variable blanks to the value −1.

Random Number Seed

SPSS/PC+ has a random number generator used by the SAMPLE command and the NORMAL and UNIFORM functions. By default, the seed is a random seed generated from the clock. Optionally, you can specify a seed on the SEED subcommand.

- You can change the random number seed any number of times within a session.
- To replicate samples across sessions or procedures, specify the same seed each time.

SEED *The random number seed.* Specify a positive integer or keyword RANDOM. The default is RANDOM. If you specify an integer, it must be no larger than 2147483647.

Example

```
SET SEED=200000000.
```

- The random number seed is set to the value 200000000.

SHOW

```
SHOW
```

Overview

The SHOW command displays a table of all the current specifications on the SET command.

Syntax

- The minimum specification is simply the command keyword. SHOW has no additional specifications.

Operations

- SHOW is an operation command and is executed immediately.
- SHOW lists every current SET specification, including the default settings.

Limitations

- BOXSTR and HISTOGRAM graphics characters that are displayed on the screen can be printed on graphics printers but may not be printable on nongraphics printers.

Example

```
SHOW.
```

- The SHOW command produces a table similar to the one below of the current specifications on the SET command.
- In this example, the settings shown are the default settings, except the graphics characters (BOX, BLOCK, and HIST).

```
SPSS/PC+                              Workspace:  171.9K
Machine:  IBM                         Free disk space:  6401K
Coprocessor not installed             Work Device C:    6401K
Current directory:  C:\myfiles
SPSS/PC+ directory:  c:\spss

LISTING  SPSS.LIS          SCREEN   ON        INCLUDE  ON
LOG      SPSS.LOG          PRINTER  OFF       BEEP     ON
RESULTS  SPSS.PRC          PTRANSL  ON        MORE     ON
NULLINE  ON                ECHO     ON        EJECT    OFF

PROMPT   SPSS/PC:          LENGTH    24       WIDTH      79
CPROMPT         :          BLOCK    X         BOX       -|+++++++++
ENDCMD   .                 HIST     X         SEED        564738291
COLOR    ( 7, 1, 1)        COMPRESS OFF       BLANKS
WEIGHT OFF                 ERRORBREAK ON      VIEWLENGTH 25

-----------------------------REVIEW SETTINGS --------------

AUTOMENU ON                HELPWINDOWS ON     MENUS STANDARD
RCOLOR ( 1, 2, 4)                             RUNREVIEW AUTO
```

SORT CASES

```
SORT CASES [BY] varlist [({A})] [varlist...]
                          {D}
```

Example:

```
SORT CASES BY XVAR (A) YVAR (D).
```

Overview

The SORT CASES procedure reorders the sequence of cases in the active file based on the values of one or more variables.

Defaults

By default, cases are sorted in ascending order for each variable, starting with the first variable named. For each subsequent variable, cases are sorted in ascending order within categories of previously named variables.

Tailoring

Optionally, you can sort cases in descending order or use combinations of ascending and descending order for different variables.

Syntax

- The minimum specification is a list of variables that are used as sort keys.
- Variables can be numeric or string.
- Keyword BY is optional.
- You can explicitly request the default sort order (ascending) by specifying A in parentheses after the variable name.
- To sort cases in descending order, specify (D).
- An order specification (A or D) applies to all variables to its left until it reaches a previous order specification. Thus, if you combine ascending and descending order on the same specification, you may need to specify the default (A) explicitly.
- If the first sort variable begins with the letters CAS, you must either use the keyword BY or enter both words of the command SORT CASES.

Operations

- SORT CASES causes the data to be read.
- SORT CASES begins by sorting the file according to the first variable named. For subsequent variables, cases are sorted within categories of previously named variables.
- For string variables, the sort order is the ASCII collating sequence (see the IBM *BASIC* manual). An ascending (A) sort order places numeric characters first, then upper-case letters in alphabetical order, then lower-case letters in alphabetical order; a descending (D) sort order does the reverse.
- The values of system variable $CASENUM are not changed from their original values as the file is sorted. You can use the variable $CASENUM as a sorting key to restore the original order.
- The temporary transformation commands SAMPLE and PROCESS IF are ignored by SORT CASES.

Limitations

- Up to 10 variables can be used as sort keys.
- SORT CASES uses intermediate scratch files for which sufficient disk space must be available on the default device. SET WORKDEV does not control the drive to which scratch files for SORT CASES are written.

Example

```
SORT CASES BY XVAR (A) YVAR (D).
```

- Cases are sorted in ascending order of variable XVAR.
- Cases are further sorted in descending order of YVAR within categories of XVAR.

Example

```
SORT XVAR YVAR (A) ZVAR (D).
```

- Cases are sorted in ascending order of XVAR.
- Cases are further sorted in ascending order of YVAR within values of XVAR. Specification A applies to both XVAR and YVAR.
- Cases are further sorted in descending order of ZVAR within values of YVAR and XVAR.
- This example takes advantage of the ability to omit the keyword BY and the second word of two-word commands.

SPSS MANAGER

```
SPSS MANAGER

 {REMOVE  =procedure-name [;procedure-name...] }
 {INSTALL =procedure-name [;procedure-name...] }
 {STATUS [=procedure-name [;procedure-name...]]}

 [/{CONFIRM  }]
   {NOCONFIRM}

 [/FROM={'device or path specification'}]
        {'A:'                         }
```

Example:

```
SPSS MANAGER REMOVE=TABLES;MANOVA.
```

Overview

SPSS MANAGER allows you to reduce the disk space requirements of the SPSS/PC+ system by removing seldom-used procedures from the hard disk. If you need a procedure that is not on the hard disk, use SPSS MANAGER to install it from either the original distribution diskette or a backup copy.

- You can install and remove a module as many times as you wish.
- You can make backup copies of the distribution diskettes in case the originals are damaged. You are strongly advised to do so if you are not going to keep the entire system on your hard disk.
- SPSS MANAGER removes and installs modules in the SPSS/PC+ system directory. By default this is C:\SPSS. If you have issued the DOS environment command *SET SPSS,* the system is assumed to reside in the directory named on that command.

Many of the program modules that hold the SPSS/PC+ system contain more than one procedure. For such modules, you cannot remove one of these procedures without removing all of them. As of the initial release of SPSS/PC+, procedures are grouped into modules as follows:

- *The central system* contains DATA LIST, GET, SAVE, JOIN, all transformation commands, plus CORRELATION, CROSSTABS, DESCRIPTIVES, FREQUENCIES, LIST, MEANS, MODIFY VARS, REVIEW, SPSS MANAGER, SYSFILE INFO, TTEST, and WRITE. This module must be installed at all times.
- AGGREGATE, PLOT, REPORT, and SORT.
- ONEWAY and REGRESSION.
- ANOVA, EXPORT, IMPORT, and NPAR TESTS.
- CLUSTER, FACTOR, HILOGLINEAR, and QUICK CLUSTER (SPSS/PC+ Advanced Statistics only).
- DSCRIMINANT and RELIABILITY (SPSS/PC+ Advanced Statistics only).
- MANOVA (SPSS/PC+ Advanced Statistics only).
- TABLES and PRINT TABLES (SPSS/PC+ Tables only).
- CREATE, RMV, SPECTRA, WLS, and 2SLS (SPSS/PC+ Trends only).
- SEASON, EXSMOOTH, TSPLOT, CASEPLOT, NPPLOT, and CURVEFIT (SPSS/PC+ Trends only).
- ARIMA, AREG, ACF, PACF, CCF (SPSS/PC+ Trends only).

It is *not* guaranteed that this arrangement of procedures into modules will remain constant. To find out which procedures are grouped with a given procedure, for example, MEANS, use the STATUS subcommand (see below).

Syntax

- One and only one of the subcommands INSTALL, REMOVE, or STATUS can be specified. Invoke SPSS MANAGER once for each type of action you need.
- Unless the NOCONFIRM subcommand is entered, SPSS MANAGER will request confirmation before installing or removing any modules.

- Subcommands are separated by slashes.
- Semicolons are required to separate procedure names on the INSTALL, REMOVE, and STATUS subcommands.
- Drive or path specifications on the FROM subcommand should be enclosed in apostrophes and may not contain embedded blanks.
- A colon is required after a drive specification.

Operations

- SPSS MANAGER always reports the names of all procedures affected by an INSTALL or REMOVE operation before taking action, unless you have specified NOCONFIRM.
- SPSS MANAGER prints the name of each file as it installs or removes it.
- The Ctrl C and Ctrl Break interrupts are not accepted during the processing of an INSTALL or REMOVE operation.

Example

```
SPSS MANAGER REMOVE=MANOVA.
SPSS MANAGER INSTALL=TABLES.
```

- The first SPSS MANAGER command removes the MANOVA procedure from the hard disk.
- The second SPSS MANAGER command installs the TABLES procedure on the hard disk. You will be prompted to insert the necessary diskettes into the A: drive.

REMOVE Subcommand

The REMOVE subcommand deletes modules from the directory in which the system is stored on the hard disk.

- Specifications for REMOVE consist of the names of one or more SPSS/PC+ procedures separated by semicolons. Procedure names can be abbreviated to three characters.
- REMOVE cannot be specified along with INSTALL or STATUS. Invoke the SPSS MANAGER once for each type of action you need.
- You cannot remove the central system, nor can you remove procedures which are in the same module as the central system.
- SPSS MANAGER first tells you if any procedures you specify are already absent from the hard disk. It then displays the following: the names of all remaining procedures requested and of other procedures that are packaged in the same module(s); the disk space required to hold these modules; and the available disk space.
- After confirmation by the user (unless NOCONFIRM is specified), SPSS MANAGER simply deletes the appropriate modules from disk. If you attempt to use a procedure from a deleted module, you will receive a message stating that the procedure must be installed before use.

INSTALL Subcommand

The INSTALL subcommand places the module(s) containing one or more procedures requested by the user into the directory with the rest of the system on the hard disk.

- Specifications for INSTALL consist of the names of one or more SPSS/PC+ procedures separated by semicolons. Procedure names can be abbreviated to three characters.
- INSTALL cannot be specified along with REMOVE or STATUS. Invoke the SPSS MANAGER once for each type of action you need.
- SPSS MANAGER first tells you if any procedures you specify are already on the hard disk. It then displays the following: the names of all remaining procedures requested and of other procedures that are packaged in the same module(s); the disk space required to hold these modules; and the available disk space.
- After confirmation by the user (unless NOCONFIRM has been specified), SPSS MANAGER will prompt for the insertion of the necessary diskettes into

the A: drive (unless you have used the FROM subcommand described below). Diskettes are requested by number.

- By default, SPSS MANAGER expects to install procedures from diskettes in the A: drive. The FROM subcommand, described below, allows you to install from a different drive, which can be another hard disk or similar device.

STATUS Subcommand

The STATUS subcommand reports on the presence or absence of requested procedures and the other procedures grouped together in modules with them.

- Specifications for STATUS consist of the names of one or more SPSS/PC+ procedures separated by semicolons. Procedure names can be abbreviated to three characters.
- If no procedure is listed on the STATUS subcommand, all procedure names are listed in alphabetical order.
- To find what is grouped with any particular procedure in your system, specify STATUS and the name of that procedure.
- STATUS cannot be specified along with INSTALL or REMOVE. Invoke the SPSS MANAGER once for each type of action you need.
- For every procedure named, SPSS MANAGER displays whether that procedure is installed and the names of other procedures that are packaged in the same module.
- Files are never added to or deleted from the hard disk when the STATUS subcommand is specified. The CONFIRM and NOCONFIRM subcommands are therefore ignored with STATUS.

CONFIRM and NOCONFIRM Subcommands

CONFIRM and NOCONFIRM control whether or not the user must confirm a REMOVE or INSTALL specification before it is executed.

- CONFIRM requests that no action be taken without confirmation from the user. CONFIRM is the default.
- Confirmation is requested after the SPSS MANAGER has reported on the procedures involved, their disk space requirements, and the available space on the hard disk.
- NOCONFIRM requests that the specified action be carried out without further prompting, aside from prompts for the insertion of diskettes during installation.
- The NOCONFIRM subcommand is intended for batch processing.
- You will have to press ↵ once before installing procedures, even if you have specified NOCONFIRM.

FROM Subcommand

The FROM subcommand allows you to install procedures from a drive other than A:, which is the default. For example, you can install procedures from backup copies kept on high-density floppy diskettes or a removable-cartridge storage device.

- Specifications on FROM consist of a device or path specification in apostrophes. The apostrophes and the colon following a device specification are required.
- FROM is ignored unless the INSTALL subcommand is entered.
- If the modules containing the requested procedures cannot be found at the location specified on the FROM subcommand, the SPSS MANAGER will display a message explaining the problem and terminate. After determining the correct drive or directory, enter the SPSS MANAGER command again.

- The specified device and directory should contain copies of all files on the distribution diskettes. For example, use the DOS command

```
COPY  A:*.*  D:
```

when copying the distribution diskettes to a cartridge system on the D: drive.

- When making backup copies of the distribution diskettes, back up the diskettes in numerical order and do not divide the files distributed on a single diskette. All files on diskette B4, for example, should be copied to the same high-density diskette or removable cartridge.

Example

```
SPSS MANAGER INSTALL FACTOR / FROM 'D:\SPSSBACK'.
```

- The module containing the FACTOR procedure will be installed from directory \SPSSBACK on the D: drive.
- CONFIRM, the default, is in effect.

SUBTITLE

```
SUBTITLE [{'}] text [{'}]
          {"}        {"}
```

Example:

```
SUBTITLE FREQUENCIES FOR SEX=FEMALE..
```

Overview

The SUBTITLE command inserts a left-justified subtitle on the second line from the top of each page of the display file. The default subtitle is a blank line.

Syntax

- The only specification on SUBTITLE is the subtitle itself.
- The subtitle can include any characters.
- The subtitle can be up to 64 characters long. Subtitles longer than 64 characters are truncated.
- You can specify the subtitle as a string by enclosing it within either apostrophes or quotation marks (see Universals: Strings). This allows you to include quotation marks, apostrophes, and lowercase text within the subtitle.
- A subtitle entered without apostrophes or quotation marks is converted to uppercase.
- More than one SUBTITLE command is allowed in a single session.

Operations

- Each SUBTITLE command overrides the previous one and takes effect on the next display page.
- SUBTITLE is independent of TITLE and can be changed separately.

Example

```
TITLE 'Preliminary analysis of Santa Survey'.
SUBTITLE FREQUENCIES FOR SEX=FEMALE..
```

- This SUBTITLE command includes a period for the subtitle as well as the required command terminator.

Example

```
SUB "Crosstabs of Respondent's Age by Gift Preference".
```

- This subtitle is specified within quotation marks to allow the use of an apostrophe in the subtitle and to preserve upper and lower case.
- This example takes advantage of spelling permitted by three-character truncation of keywords.

SYSFILE INFO

This command available as of SPSS/PC+ V2.0.

```
SYSFILE INFO [FILE =] 'file specifications'
```

Example:

```
SYSFILE INFO '\DATA\MYSTERY.SYS'.
```

Overview

The SYSFILE INFO command allows you to determine the contents of an SPSS/PC+ system file without having to use GET, which replaces your active file. SYSFILE INFO works only with SPSS/PC+ system files that were created with the SAVE command in SPSS/PC+, in SPSS/PC+ Data Entry or SPSS Data Entry II, or by an SPSS/PC+ procedure such as AGGREGATE.

Syntax

The sole specification on SYSFILE INFO is a DOS file specification in apostrophes.

- The file specification can include a drive and/or a path.

Operations

SPSS/PC+ determines if the specified file exists and if it is an SPSS/PC+ system file. If so, it displays

- The filename, creation date, and creation time.
- The file title, if any.
- The number of cases and variables.
- The use of case weights.
- Whether the file is compressed or not.
- Whether the file contains special information for use by SPSS Data Entry II.
- The names and variable labels of all user-defined variables in the file.

The variable names in the file are particularly useful prior to a JOIN command.
Error conditions:

- If SPSS/PC+ cannot find or obtain access to the specified file, it displays a message indicating that an error occurred while referencing the file.
- If the file is found but is not an SPSS/PC+ system file, SPSS/PC+ displays a message indicating this.

TITLE

```
TITLE [{'}] text [{'}]
       {"}        {"}
```

Example:

```
TITLE This run will contain preliminary analyses.
```

Overview

The default page heading on SPSS/PC+ output contains the date, SPSS/PC+, and the display page number. The TITLE command replaces the centered title portion (SPSS/PC+) of the default display heading with user-supplied text. The date and page number remain.

Syntax

- The only specification on TITLE is the title itself.
- The title can include any characters.
- The title can be up to 58 characters long. Titles longer than 58 characters are truncated.
- You can specify the title as a string by enclosing it within either apostrophes or quotation marks (see Universals: Strings). This allows you to include quotation marks, apostrophes, and lower-case text within the title.
- A subtitle entered without apostrophes or quotation marks is converted to upper case.
- More than one TITLE command is allowed in a single session.

Operations

- Each TITLE command overrides the previous one and takes effect on the next display page.
- TITLE does not cause a new page.

Example

```
TITLE This run will contain preliminary analyses.
```

- This title will replace the default title on subsequent output pages.
- Since the title is not enclosed in apostrophes or quotation marks, it is converted to upper case.
- Note that the period (.) is the command terminator and will not appear as part of the title.

Example

```
TITLE "Regression on Nora's Dissertation Data".
```

- This title is specified within quotation marks to allow the use of an apostrophe in the title and to preserve upper and lower case.

TRANSLATE

```
TRANSLATE {FROM = 'filename.ext'}[/RANGE={name        }]*
          {TO   = 'filename.ext'}        {start..stop}
                                         {start:stop }

[/TYPE={WKS}]
       {WK1}
       {WRK}
       {SLK}
       {DBF}
       {DB2}
       {DB3}

[/DROP=variable list]

[/KEEP=variable list]

[/FIELDNAMES]**

[/MAP]
```

*RANGE is available only for translating from spreadsheet files.
**FIELDNAMES is available only for spreadsheet files.

Keyword	Foreign file
WK1	1-2-3 Release 2.0
WKS	1-2-3 Release 1A
WRK	Symphony
SLK	Multiplan (symbolic format only)
DBF	all dBase (with FROM)
DB2	dBase-II (with TO)
DB3	dBase-III, dBase-III PLUS (with TO)

Examples:

```
TRANSLATE FROM='PROJECT.WKS'
/FIELDNAMES
/RANGE=D3..J279.

TRANSLATE FROM='ANNUAL.DBF'
/DROP=JANUARY TO MARCH/MAP.

TRANSLATE TO='SALESREP.SLK'
/KEEP=SALES, UNITS, MONTHS, PRICE1 TO PRICE20
/FIELDNAMES
/MAP.
```

Overview

Procedure TRANSLATE either creates an active file from a foreign file (subcommand FROM) or translates the SPSS/PC+ active file into a foreign file (subcommand TO). Supported formats are 1-2-3, Symphony, Multiplan, dBase-II, dBase-III, and dBase-III PLUS.

Defaults

By default, all data are translated from the input file into the resulting file. The format of the input or output file is inferred from the file extension (except for translating to database files; see TO subcommand, databases).

Tailoring

You can use the DROP and KEEP subcommands to specify variables to omit or retain in the resulting active file. You can use the RANGE subcommand to translate a subset of cells from a spreadsheet file. You can use the FIELD-NAMES subcommand to translate field names in the spreadsheet file to SPSS/PC+ variable names or to translate SPSS/PC+ variable names to field names. The optional MAP subcommand displays the names of variables retrieved from a foreign file.

Syntax

- The minimum specification is subcommand FROM or TO with an associated filename. FROM or TO must precede any other specifications.
- The TYPE subcommand is required for translating from the SPSS/PC+ active file to databases.
- The file specification is enclosed in apostrophes and can be fully qualified, including drive, directory, filename, and extension.
- The RANGE subcommand can be used only for translating spreadsheet files into SPSS/PC+ active files.

- The FIELDNAMES subcommand is available for spreadsheet files only.
- Subcommands are separated by slashes.

Operations

- TRANSLATE FROM replaces an existing active file. If TRANSLATE FROM is not successful, the current active file is not changed and is still available.
- After TRANSLATE TO is executed, the active file remains available for SPSS/PC+ transformations and procedures.
- TRANSLATE TO will *not* overwrite an existing file with the same name.
- Commands N, SAMPLE, PROCESS IF, SELECT IF and any transformations are executed prior to TRANSLATE TO.

Limitations

- Maximum 200 variables can be translated into the active file. Thus, you cannot retrieve more than 200 columns from a .WKS, .WK1, .WRK, or .SLK file.
- Maximum 2048 cases can be translated to 1-2-3 Release 1A, maximum 8192 cases to 1-2-3 Release 2.0 or Symphony files, and maximum 4095 cases to Multiplan files.
- Maximum 32 variables and 65,535 cases can be translated to a dBase-II file, and maximum 128 variables and 1 billion cases (subject to disk space availability) to a a dBase-III file.

FROM Subcommand

TRANSLATE FROM reads a foreign file and translates it into the active file. The only required specification on FROM is the name of the foreign file enclosed in apostrophes.

Spreadsheets

A spreadsheet file suitable for SPSS/PC+ should be arranged so that each row represents a case and columns indicate variables.

- By default, the new active file contains all rows and up to 200 columns from 1-2-3, Symphony, and Multiplan files.
- By default, TRANSLATE FROM uses the column letters as variable names in the active file.
- The first row of a spreadsheet or specified range may contain field labels immediately followed by rows of data. These names can be transferred as SPSS/PC+ variable names (see FIELDNAMES Subcommand).
- The current value of a formula is translated to the active file.
- Blank, ERR, and NA values in 1-2-3 and Symphony are translated as system-missing in the SPSS/PC+ active file.
- Hidden columns and cells in 1-2-3 Release 2 and Symphony files *are* translated into the SPSS/PC+ active file.
- Column width and format type are transferred to the active file dictionary.
- The format type is assigned from values in the first data row. By default, the first data row is row 1. If RANGE is specified, the first data row is the first row in the range. If FIELDNAMES is specified, the first data row follows immediately after the single row containing field names.
- If a cell in the first data row is empty, the variable is assigned the global default format from the spreadsheet.
- The formats from 1-2-3, Symphony, and Multiplan are translated by default as follows:

1-2-3/Symphony	Multiplan	SPSS/PC+
Fixed	Fixed	Number
	Integer	Number
Scientific	Exponent	Number
Currency	$ (dollar)	Dollar
,(comma)		Comma
General	General	Number
+/−	* (bargraph)	Number
Percent	Percent	Number
Date		String
Time		String
Text		Number
Label	Alpha	String

- If a string is encountered in a column with numeric format, the active file contains the system-missing value.
- If a numeric value is encountered in a column with string format, the active file contains a blank.
- TRANSLATE FROM creates string variables for 1-2-3 and Symphony variables with DATE format. The SPSS/PC+ variable is in the form MM/DD/YY if the worksheet column has a width of 9 or less. If the width of the worksheet column is greater than 9, the SPSS/PC+ variable is in the form MM/DD/YYYY.
- Blank lines are translated as cases containing the system-missing value for numeric variables and blanks for string variables.
- 1-2-3 and Symphony date and time indicators (shown at the bottom of the screen) are not transferred from WKS, WK1, or WRK files. The values for system variables $CASENUM, $WEIGHT, and $DATE are assigned by TRANSLATE as each case is written to the active file.

Example

```
TRANSLATE FROM='PROJECT.WKS'.
```

- TRANSLATE creates an active file from the 1-2-3 Release 1.0 worksheet with the name PROJECT.WKS.
- The active file contains all rows and uses the column letters as variable names.
- The format for each variable is determined by the format of the value in the first row of each column.

Databases

Database files are logically very similar to SPSS/PC+ data files.

- By default, all fields and records from dBase-II or dBase-III files are included in the SPSS/PC+ active file.
- Field names are automatically translated to SPSS/PC+ variable names.
- If the FIELDNAMES subcommand is used with database files, it is ignored.
- Field names to be translated should comply with SPSS/PC+ variable-naming conventions (see Command Reference: Universals).
- Field names longer than eight characters are truncated.
- Field names must be unique in the first eight characters. If a field name is not unique, the field will be dropped.
- Colons used in dBase-II field names are translated to underscores.
- If you have marked records for deletion in dBase-II or dBase-III but not actually purged them, the records *are* included in the active file. To differentiate these cases, TRANSLATE creates a new string variable D_R, which contains an asterisk for cases marked for deletion. Other cases contain a blank for D_R.
- Character and numeric fields are transferred directly to SPSS/PC+ variables. Date and logical fields are converted into string variables. Memo fields are ignored. The following table shows how dBase formats are translated to SPSS/PC+:

dBase	SPSS/PC+
Character	String
Logical	String
Date	String
Numeric	Number
Memo	Ignored

Example

```
TRANSLATE FROM='ANNUAL.DBF'.
```

- TRANSLATE creates an active file from the database file ANNUAL.DBF.
- All cases are translated and each field becomes a variable in the active file.

TO Subcommand

Procedure TRANSLATE with the TO subcommand converts an SPSS/PC+ active file to a foreign file.

- User-missing values are transferred as the actual value.
- The three system variables, $DATE, $WEIGHT, and $CASENUM are translated only if they are specified explicitly on the KEEP subcommand.

$DATE is passed as a short string value with an A8 format; $WEIGHT and $CASENUM are passed as numeric values with formats of F8.2 and F8.0.

Spreadsheets Variables in the active file become columns and cases become rows in the spreadsheet file.

- If you specify FIELDNAMES, variable names become the first row and indicate field names.
- String variable values are left-justified and numeric variable values are right-justified.
- The resulting spreadsheet file is given the range name of SPSSPC.
- System-missing values are translated to NA in spreadsheet files.
- SPSS/PC+ formats are translated as follows:

SPSS/PC+	1-2-3/Symphony	Multiplan
Number	Fixed	Fixed
Comma	Comma	Fixed
Dollar	Currency	$ (dollar)
String	Label	Alpha
$DATE	Date	Alpha

Example

```
TRANSLATE TO='STAFF.WRK'.
```

- TRANSLATE TO creates a Symphony worksheet file containing all variables in the active file. The variable names are *not* transferred to the Symphony file.

Databases Variables in the active file become fields and cases become records in the database file.

- You must use the TYPE subcommand with keyword DB2 or DB3 when translating from SPSS/PC+ to a dBase file.
- Characters that are allowed in SPSS/PC+ variable names but are not allowed in dBase field names are translated to colons in dBase-II and underscores in dBase-III. If requested, the system variables $CASENUM, $WEIGHT, and $DATE are translated to fields PC:CASENUM, etc. for dBase-II and PC_CASENUM, etc. for dBASE-III.
- SPSS/PC+ numeric variables containing the system-missing value are translated to **** in dBase III or 0 (zero) for dBase II.
- SPSS/PC+ variable formats are translated to dBase formats as follows:

SPSS/PC+	dBase
Number	Numeric
String	Character
Dollar	Numeric
Comma	Numeric
$DATE	Date

Example

```
TRANSLATE TO='STAFF.DBF'/TYPE=DB3.
```

- TRANSLATE TO creates a dBase-III file called STAFF.DBF. The TYPE subcommand is required.

TYPE Subcommand

The TYPE subcommand indicates the format of a foreign file.

- You can use the TYPE subcommand with both FROM and TO subcommands.
- The TYPE subcommand can be omitted for translating to and from spreadsheet files if the file extension named on the TO or FROM subcommand corresponds to the format keyword.
- The TYPE subcommand with keyword DB2 or DB3 is required for translating the SPSS/PC+ active file into a dBase file. TYPE is optional for translating dBase files into SPSS/PC+ if the file extension named on FROM corresponds to the format keyword.
- The TYPE subcommand takes precedence over the file extension.
- You can create a Lotus format file in Multiplan and translate it to an SPSS/PC+ active file with TYPE=WKS.

- Available keywords on TYPE are

WK1 *1-2-3 Release 2.0.*

WKS *1-2-3 Release 1A.*

WRK *Symphony.*

SLK *Multiplan (symbolic format).*

DBF *dBase files.* Use DBF as the file extension or on the TYPE subcommand when translating any dBase file into an SPSS/PC+ active file.

DB2 *dBase-II.* Specify DB2 on TYPE when translating to a dBase-II file.

DB3 *dBase-III.* Specify DB3 on TYPE when translating to a dBase-III or dBase-III PLUS file.

Example

```
TRANSLATE FROM='PROJECT.OCT'/TYPE=SLK.
```

- TRANSLATE creates an active file from the Multiplan spreadsheet PROJECT.OCT.

FIELDNAMES Subcommand

When used with the FROM subcommand, the optional FIELDNAMES subcommand translates spreadsheet field names into SPSS/PC+ variable names. When used with TO, FIELDNAMES translates SPSS/PC+ variable names into field names in the spreadsheet.

- FIELDNAMES can be used with spreadsheets only. FIELDNAMES is ignored when used with database files.
- SPSS/PC+ variable names are transferred to the first row of the spreadsheet file.
- For translating spreadsheet field names to SPSS/PC+, each cell in the first row of the spreadsheet file or range must contain a field name. If a column does not contain a name, the column is dropped.
- Field names to be translated into SPSS/PC+ should conform to the SPSS/PC+ variable-naming conventions. They must be unique in the first eight characters and cannot have leading blanks. For other variable-naming conventions, refer to Command Reference: Universals.
- Variable names that exceed eight characters are truncated.
- If two or more columns in the spreadsheet have the same field name, only the first occurrence is translated to the active file.
- Illegal characters in field names are changed to underscores in SPSS/PC+.
- If you use SPSS/PC+ reserved words (ALL, AND, BY, EQ, GE, GT, LE, LT, NE, NOT, OR, TO, or WITH) as field names in a spreadsheet file, TRANSLATE FROM appends a dollar sign ($) to the variable name. For example, 1-2-3 columns named GE, GT, EQ, and BY are named GE$, GT$, EQ$, and BY$ in the SPSS/PC+ active file.

Example

```
TRANSLATE TO='STAFF.WRK'/FIELDNAMES.
```

- TRANSLATE TO creates a Symphony worksheet file containing all variables in the active file. The variable names *are* transferred to the Symphony file.

Example

```
TRANSLATE FROM='MONTHLY.WRK'/FIELDNAMES.
```

- TRANSLATE creates an active file from a Symphony spreadsheet. The first row in the spreadsheet contains field names that are used as variable names in the active file.

RANGE Subcommand

The RANGE subcommand translates a specified set of cells from a spreadsheet file.

- The RANGE subcommand cannot be used with subcommand TO or for translating from database files.
- For 1-2-3 or Symphony, specify the beginning of the range with a column letter and row number, two periods, and the end of the range with a column letter and row number, as in A1..K14.

- For Multiplan spreadsheets, specify the beginning and ending cells of the range separated by a colon, as in R1C1:R14C11.
- You can also specify the range using range names supplied in Symphony, 1-2-3, or Multiplan.
- If you specify FIELDNAMES with RANGE, the first row of the range must contain field names.

Example

```
TRANSLATE FROM='PROJECT.WKS'/FIELDNAMES/RANGE=D3..J279.
```

- TRANSLATE FROM creates an active file from the 1-2-3 Release 1A file PROJECT.WKS.
- Variable names are assigned from the field names in the first row of the range, in this case row 3.
- Data from cells D4 through J279 are transferred to the active file.

DROP and KEEP Subcommands

Use DROP or KEEP to include only a subset of variables in the resulting file. DROP specifies a set of variables to exclude, and KEEP specifies a set of variables to retain.

- Use DROP or KEEP with TO or FROM file translation.
- Specifications consist of a list of variable, column, or field names separated by commas or spaces.
- You can use TO to refer to adjacent variables or columns.
- You can use keyword ALL to specify all user-defined (not system) variables.
- KEEP=ALL is equivalent to omitting the KEEP subcommand.
- KEEP does not affect the order of variables in the resulting file. Variables are kept in their original order.
- If FIELDNAMES is specified when translating from a spreadsheet, the DROP and KEEP subcommands must refer to the field names, not the default column letters.
- If you specify both RANGE and KEEP, the resulting file contains only variables that are both within the range and specified on KEEP.
- If you specify both RANGE and DROP, the resulting file contains only variables within the range and excludes those mentioned on the DROP subcommand, even if they are within the range.

Example

```
TRANSLATE FROM='ADDRESS.DBF'/DROP=PHONENO, ENTRY.
```

- TRANSLATE creates an active file from the dBase file ADDRESS.DBF, omitting the fields named PHONENO and ENTRY.

Example

```
TRANSLATE FROM='PROJECT.OCT'/TYPE=WK1/FIELDNAMES
/KEEP=NETINC, REP, QUANTITY, REGION, MONTH, DAY, YEAR.
```

- TRANSLATE creates an active file from the 1-2-3 Release 2.0 file called PROJECT.OCT.
- Subcommand FIELDNAMES indicates that the first row of the spreadsheet contains field names, which will be translated into variable names in the active file.
- Subcommand KEEP translates columns with the field names NETINC, REP, QUANTITY, REGION, MONTH, DAY, and YEAR to the active file.

MAP Subcommand

Subcommand MAP produces a listing of the variables transferred from or to a foreign file.

- MAP displays a listing of the names of the variables, the variable type (string or numeric), the width of the variable, and the number of decimal places.
- MAP can be specified anywhere in the TRANSLATE command.

Example

```
TRANSLATE TO='STAFF.DBF'/TYPE=DB3/MAP.
```

- The keyword MAP requests a listing of the variables translated to the dBase III database.

T-TEST

Independent Samples:

```
T-TEST GROUPS=varname [({1,2        })] /VARIABLES=varlist
                        {value      }
                        {value,value}

      [/OPTIONS=option numbers]
```

Paired Samples:

```
T-TEST PAIRS=varlist [WITH varlist] [/[PAIRS=] varlist ...]

      [/OPTIONS=option numbers]
```

Options:

1 Include user-missing values
2 Exclude missing values listwise
3 Suppress variable labels
5 Special pairing for WITH

Examples:

```
T-TEST GROUPS=XVAR(1,3)/VARIABLES=YVAR1 YVAR2 YVAR3.

T-TEST PAIRS=ZVAR1 ZVAR2 ZVAR3/ZVAR4 ZVAR5.
```

Overview

T-TEST compares sample means by calculating Student's *t* and displays the two-tailed probability of the difference between the means. Statistics are available for either independent samples (different groups of cases) or paired samples (different variables). Other procedures that compare group means are ANOVA and ONEWAY.

Defaults

In addition to Student's *t*, degrees of freedom, and two-tailed probabilities, T-TEST produces the mean, standard deviation, standard error, and count for each group or variable. The default display includes the variable label and always uses narrow format. Cases with missing values for the variables that specify a test are excluded from that test.

Independent-Samples Test. For independent-samples tests, both pooled- and separate-variance estimates are calculated, along with the *F* value used to test homogeneity of variance and its probability. The two-tailed probability is displayed for the *t* value.

Paired-Samples Test. For paired-samples tests, the default output includes the difference between the means, the two-tailed probability level for a test of the difference, the correlation coefficient for the two variables, and the two-tailed probability level for a test of the coefficient.

Tailoring

Display Format. You can suppress the printing of variable labels.

Statistical Display. You can control which variables are paired in paired-samples tests. There are no optional statistics. All statistics available are displayed by default.

Missing Values. You can include cases with user-missing values in the calculation of the statistics for each test. Alternatively, you can exclude cases with missing values for any of the variables named on the command from all analyses.

Syntax

- To request both independent- and paired-samples tests, specify GROUPS and VARIABLES first, then PAIRS.
- Subcommands can each be specified only once and must be separated by a slash.

Operations

- T-TEST is a procedure and causes the data to be read.
- If a GROUPS variable is a long string, only the short-string portion is used to identify groups in the analysis.
- Probability levels are two-tailed.
- The T-TEST display always uses narrow format regardless of the width defined on SET.

- The BOXSTRING subcommand controls the characters used in the table display (see SET).

Limitations

- Maximum 1 grouping variable per independent-samples test (the GROUPS subcommand).
- Maximum 50 analysis variables per independent-samples test (the VARIABLES subcommand).
- Maximum 1 each GROUPS, VARIABLES, and OPTIONS subcommands.
- The maximum number of variables for paired-sample tests is the same as the system limit.

Example

```
T-TEST GROUPS=XVAR(1,3)/VARIABLES=YVAR1 YVAR2 YVAR3.
```

- This independent-samples example compares the means of the two groups defined by XVAR for variables YVAR1, YVAR2, and YVAR3.

Example

```
T-T PAIRS=ZVAR1 ZVAR2 ZVAR3/ZVAR4 ZVAR5.
```

- The first analysis list in this paired-samples example compares the means of ZVAR1 and ZVAR2, ZVAR1 and ZVAR3, and ZVAR2 and ZVAR3. The second analysis list compares the means of ZVAR4 and ZVAR5.
- This example takes advantage of spelling permitted by three-character truncation of keywords.

GROUPS and VARIABLES Subcommands

Independent samples t-tests are requested with the GROUPS and VARIABLES subcommands. The GROUPS subcommand names a variable used to group cases. The VARIABLES subcommand names the dependent variables.

- The minimum specification for an independent-samples test is the GROUPS subcommand with a variable name followed by the VARIABLES subcommand with a variable name.
- GROUPS can name only one variable, which can be numeric or string.
- The GROUPS variable can be followed by a range specification in parentheses.
- A single GROUPS value in parentheses groups all cases with a code equal to or greater than the value into one group and the remaining cases into the other group.
- Two GROUPS values in parentheses include cases with the first value in one group and cases with the second value in the other group. Cases with other values are excluded.
- If no GROUPS values are specified, (1,2) is assumed for numeric variables and (DEFAULT1, DEFAULT2) is assumed for string variables.
- VARIABLES names the variables being analyzed.
- VARIABLES can name only numeric variables.

PAIRS Subcommand

Use the PAIRS subcommand to request paired-samples tests.

- The minimum specification for a paired-samples test is the PAIRS subcommand with a single analysis list.
- The minimum analysis list is two variables.
- Analysis lists can name only numeric variables.
- An analysis list without keyword WITH compares each variable with every other variable.
- An analysis list with keyword WITH compares every variable to the left of the keyword with every variable to the right of the keyword.
- Multiple PAIRS subcommands can be used. The keyword PAIRS is required only on the first PAIRS subcommand.
- Use a slash to separate multiple PAIRS subcommands.

Display Format

The default display includes the variable label. You can change this default with the following option on the OPTIONS subcommand:

Option 3 *Suppress variable labels.*

Statistical Display

All statistics available with T-TEST are displayed automatically. However, you can control the pairing of variables in the analysis list with the following option on the OPTIONS subcommand:

Option 5 *Special pairing for paired-samples test.* Must be used with the keyword WITH. The first variable before WITH is compared to the first variable after WITH, the second variable before WITH is compared to the second variable after WITH, and so forth. The same number of variables should be specified before and after WITH; unmatched variables are ignored. Option 5 is ignored with independent samples or when keyword WITH is not specified in the PAIRS subcommand.

Example

```
T-TEST PAIRS=SCALE1 TO SCALE3 WITH SCALE4 TO SCALE6
/OPTIONS=5.
```

- This analysis compares SCALE1 and SCALE4, SCALE2 and SCALE5, and SCALE3 and SCALE6.

Missing Values

For independent-samples tests, cases missing on either the grouping variable or the analysis variable are excluded from the analysis of that variable. For paired-samples tests, a case missing on either of the variables in a given pair is excluded from the analysis of that pair. You alter the handling of cases with missing values by specifying the following on the OPTIONS subcommand:

Option 1 *Include user-missing values.* Cases with user-missing values are included in the calculation of test statistics.

Option 2 *Exclude missing values listwise.* A case with a missing value for any variable named on the command is excluded from all analyses the command requests.

VALUE LABELS

```
VALUE LABELS  varlist value 'label' value 'label' ...

                  [/varlist ...]
```

Example:

```
VALUE LABELS XVAR TO ZVAR  0 'NORTH' 1 'CENTRAL' 2 'WEST'
      3 'SOUTH' 4 'EAST'
  /RESIDE 'URBAN    ' "Within city proper" 'SUBURBAN' "Metro area"
      'EXURBAN ' "Outside SMSA".
```

Overview

The VALUE LABELS command provides descriptive labels for individual values of a variable.

Syntax

- The VALUE LABELS command is followed by a variable name (or variable list) and the individual values with associated labels.
- The variable must have been previously defined, either on a DATA LIST, GET, JOIN, AGGREGATE, or IMPORT command, or on a transformation command.
- Value labels must be specified as strings enclosed in apostrophes or quotation marks (see Universals: Strings).
- The values of short string variables as well as their labels are specified as strings.
- You cannot assign value labels to long string variables.
- The same labels can be assigned to the same values of several variables by listing all of the variables followed by the values and associated labels.
- The value labels for one variable or variable list must be separated from the next variable or variable list by a slash.
- More than one VALUE LABELS command is allowed.

Operations

- Value labels are stored in the active file dictionary.
- If any value label is assigned to a variable that already has value labels, the new assignment completely replaces the old assignment. This is true even if the new label is assigned to a previously unlabeled value.

Limitations

- Each value label can be up to 60 characters long, although most procedures only print 20 characters.
- Some procedures print fewer than 20 characters in labels.
- Only the TABLES procedure (available in SPSS/PC+ Tables) will print all 60 characters of a label.

Example

```
VALUE LABELS XVAR TO ZVAR  0 'NORTH' 1 'CENTRAL' 2 'WEST'
                     3 'SOUTH' 4 'EAST'/
  AVAR 'X' 'SALES STAFF'  'S' 'SUPERVISORY STAFF'
       'M' 'MANAGERIAL STAFF'.
```

- Labels are assigned to the values 0, 1, 2, 3, and 4 of the variables between and including XVAR and ZVAR on the active file.
- Following the required slash, labels for values X, S, and M of variable AVAR are specified. X, S, and M are string values and must be enclosed in apostrophes or quotes.

Example

```
VAL LAB SOCROLE 1 'MOTHER' 2 'FATHER' 3 'SISTER' 4 'BROTHER'
   5 'COUSIN' 10.0 'NEIGHBOR' 97 'REFUSED TO ANSWER' 98 "DON'T KNOW"
   /OPIN1 TO OPIN5 1 'STRONGLY AGREE' 3 'NO OPINION'
      5 'STRONGLY DISAGREE'.
```

- The DATA LIST command defines ID as a three-column variable, SOCROLE as a two-column variable, and each of the five OPIN variables as one-column variables.
- The VALUE LABELS command assigns labels to variables SOCROLE, OPIN1, OPIN2, OPIN3, OPIN4, and OPIN5.
- The value label for value 98 is enclosed in quotation marks to allow the embedded apostrophe.
- Three values of the OPIN variables are assigned labels.
- This example takes advantage of spelling permitted by three character truncation of keywords.

String Value Labels

- Both the values and the labels for short string variables must be enclosed in apostrophes or quotation marks.
- The *values* of a short string variable must be specified with the correct width of that variable, including any leading or trailing blanks that are in the data.

Example

```
DATA LIST / REGION 1 RESIDE 2-9 (A) V1 TO V15 10-24.
VAL LAB RESIDE "URBAN   " "Within city proper"
    "SUBURBAN" "Metro area" "EXURBAN " "Outside SMSA".
```

- The DATA LIST command defines one short string variable, RESIDE, which is eight characters wide.
- The VALUE LABELS command assigns labels to three values of RESIDE. Each value and each label is specified in quotation marks. The value URBAN contains three trailing blanks in order to match the format width of eight characters. The value EXURBAN contains one trailing blank in order to match data values.

VARIABLE LABELS

```
VARIABLE LABELS  varlist 'label' [[/]varname ...]
```

Example:

```
VARIABLE LABELS  XVAR 'SALES TERRITORY X'
/YVAR "John's Territory"
/ZVAR 'SALES TERRITORY Z'.
```

Overview

The VARIABLE LABELS command assigns a descriptive label of up to 60 characters to a variable.

Syntax

- The variable name or list of variables is followed by at least one blank or comma and the associated label.
- Variable labels must be specified within apostrophes or quotation marks (see Universals: Strings).
- The variable must have been previously defined, either on a DATA LIST, GET, JOIN, AGGREGATE, or IMPORT command, or on a transformation command.
- The slash separating a variable name and its label from subsequent names and labels is optional.
- More than one VARIABLE LABELS command is allowed per session.
- Individual variable labels cannot be specified across multiple lines.

Operations

- If a label is assigned to a variable that already has a label assigned to it, the new label completely replaces the old label.
- Variable labels are stored in the active file dictionary.
- If a list of variables precedes a label specification, the label is assigned to each of them.

Limitations

- Variable labels are limited to 60 characters, although most procedures only print 40 characters.
- Some procedures show fewer than 40 characters of a label.
- Only the TABLES procedure (available in SPSS/PC+ Tables) will print all 60 characters of a variable label.

Example

```
VARIABLE LABELS  XVAR 'SALES TERRITORY X'
/YVAR "John's Territory"
/ZVAR 'SALES TERRITORY Z'.
```

- This command assigns labels to variables XVAR, YVAR, and ZVAR.
- The label for YVAR is enclosed in quotation marks to allow the inclusion of an apostrophe in the label.

WEIGHT

```
WEIGHT  {BY varname}
        {OFF       }
```

Example:

```
WEIGHT BY AVAR.
FREQ VAR=VAR1.
```

Overview

The WEIGHT transformation reassigns the value of the system variable $WEIGHT for each case (see Universals: System Variables). With WEIGHT, you can arithmetically alter the sample size or its distribution.

Syntax

- Only one variable can be specified.
- The WEIGHT variable must be numeric.
- If the keyword BY is omitted, the first word following WEIGHT is taken to be the weighting variable.
- WEIGHT OFF turns weighting off. You cannot weight the file by a variable named OFF.

Operations

- WEIGHT is a transformation and is executed when data are read for the next procedure.
- The default value of $WEIGHT for each case is 1.0.
- To list the values of $WEIGHT or any other system variable, you must request it explicitly on the LIST command.
- Weighting is permanent during a session unless it is changed or turned off with the WEIGHT OFF specification.
- Each WEIGHT command overrides the previous one.
- WEIGHT tells SPSS/PC+ to use the value of the specified variable to arithmetically replicate cases for subsequent procedures.
- Weight values do not need to be integer.
- Cases are *not* physically replicated.
- System-missing, user-missing, and negative values for the weighting variable are treated as 0 in the computation of weights.
- A file saved with EXPORT or SAVE when weighting is in effect maintains the weighting in the system variable $WEIGHT. When the file is reread, weighting is maintained.
- If the weighted number of cases exceeds the sample size, tests of significance are inflated; if it is smaller, they are deflated.

Example

```
WEIGHT BY AVAR.
FREQ VAR=VAR1.
```

- The frequency counts for the values of variable VAR1 will be weighted (multiplied) by the values of variable AVAR.

Example

```
COMPUTE WVAR=1.
IF (GROUP EQ 1) WVAR=.5.
WEIGHT BY WVAR.
```

- Variable WVAR is initialized to 1 with the COMPUTE command. The value of WVAR is changed to 0.5 with the IF command for cases where GROUP equals 1.
- Subsequent procedures will use a case base in which cases from Group 1 only count half as much as other cases.

WRITE

```
WRITE [VARIABLES={ALL**   }]
                 {varlist}

  [/CASES=[FROM {1**}] [TO {eof**}]
                {n  }      {n    }

        [BY {1**}]]
            {n  }

  [/FORMAT=[{UNNUMBERED**}] [{WRAP**}] [WEIGHT]]
            {NUMBERED    }   {SINGLE}
```

**Default if subcommand is omitted.

Example:

```
WRITE VARIABLES=XVAR AVAR /CASES=FROM 10 TO 100 BY 2.
```

Overview

The WRITE procedure writes cases to an ASCII file. The cases are written to the results file specified on SET (by default, SPSS.PRC). SPSS/PC+ automatically specifies the format of the case file that is written using the dictionary formats and a record length of 80. In addition to writing the file, WRITE displays a table that shows the names and the record and column locations of the variables on the written file.

Defaults

By default, all user-defined variables are written for all cases in the file. Each case is written on as many records as are needed.

Tailoring

You can limit the written file to specified variables or a specified sequence of cases. In addition, you can include each case's weight or sequence number in the file. You can also limit each case to a single record.

Syntax

- The minimum specification is simply the command keyword.
- All subcommands are optional.
- A subcommand or its abbreviation may not be recognized as such if a variable name is identical to it, unless the subcommand is followed by an equals sign.
- Subcommands can be specified in any order and must be separated by slashes.

Operations

- WRITE is a procedure and causes the data to be read.
- WRITE uses the dictionary print formats (see Universals: Formats). If a value is greater than the format allows, asterisks (*) are written.
- Records are written with 80 characters regardless of the width defined on SET.
- If a long string variable cannot be written within the record length of 80, it is truncated.
- Records are always written with a single blank between variables.
- System-missing values are written as blanks.
- System variables are only written when explicitly requested.
- Records are written to the results file named on SET. The default is SPSS.PRC.
- If you write to the same file with more than one procedure in the same session, the file will contain only the results of the last procedure.
- Use the SET command to change the results file if you write output more than once in a session.

Limitations

- The VARIABLES, FORMAT, and CASES subcommands can each be specified only once.

Example

```
DATA LIST FILE='EMPLOYEE.DAT'
  /XVAR 1 YVAR 10-15 ZVAR 3-9(2) / AVAR 25-30(A).
WRITE VARIABLES=XVAR AVAR /CASES=FROM 10 TO 100 BY 2.
```

- This example writes a file containing every second case starting with Case 10 and stopping at 100 for variables XVAR and AVAR.
- The records are written to the default results file SPSS.PRC.
- A display table describing the format of the written file is sent to the screen and to the default listing file SPSS.LIS.

VARIABLES Subcommand

By default, all user-defined variables in the active file are included. Use the optional VARIABLES subcommand to control the variables to be written for each case.

- If you use the VARIABLES subcommand, a list of variables must be specified and only specified variables are written.
- Variables named on VARIABLES must already exist.
- Variables are written in the order named on VARIABLES.
- If a variable is named more than once it is written more than once.
- The actual keyword VARIABLES is optional.

Optionally, you can specify keyword ALL on the VARIABLES subcommand:

ALL *Write all user-defined variables.* Variables are written in the order in which they appear in the active file. This is the default if the VARIABLES subcommand is omitted.

- ALL does not include the system variables $CASENUM, $WEIGHT, and $DATE. You must request them explicitly to write them.

CASES Subcommand

By default, all cases in the active file are written. Use the CASES subcommand to limit the number and pattern of cases written.

- If you omit CASES or use it without specifications, all cases in the active file are written.
- Only defaults that you change are altered.

The following can be specified on CASES:

FROM n *The case number of the first case to be written.* The default is 1.

TO n *Upper limit on the cases to be written.* The default is the end of the file. CASES 100 is interpreted as CASES TO 100.

BY n *Increment used to choose cases for writing.* The default is 1.

FORMAT Subcommand

By default, the written file does not include the case weight or sequence number for each case, and each case uses as many records as it needs. Use the FORMAT subcommand to specify format options for the written file.

- If the FORMAT subcommand is omitted or if it is included without specifications, the file is written with the default format.
- Only defaults that you change are altered.

The following may be specified on FORMAT:

NUMBERED *Include the sequence number of the case.* The sequence number is the first item written for each case and is labeled Case# in the display table. The default is UNNUMBERED.

SINGLE *Limit each case to one record.* If variables requested cannot be written on a single record, WRITE is not executed. The default is to WRAP the case, using as many records as necessary.

WEIGHT *Write the value of the case's weight in the active file.*

Examples

Contents

EXAMPLES

Examples

AGGREGATE and JOIN

This example creates a file that can be used to produce a report of product sales by different sales representatives. The file contains data on the individual sales of each product by each representative, the average and total sales for each product, and the performance of each representative relative to the product average and total. On two system files for the first and second quarters of a year, the original input variables are

- SALESREP—the name of the sales representative who closed a sale.
- PRODUCT—the product, coded 1=Primo, 2=Mocha, 3=Blend, and 4=Special.
- NUMBER—the quantity sold of each product.
- MONTH—the month of a sale, coded as a number with a value label.
- DAY—the day of a sale, coded as a number.
- YEAR—the year of a sale.

The two system files, containing data on individual sales by each representative, are combined. The resulting active file is sorted and aggregated to produce variables summarizing the sales of each product. The summary variables are appended to the individual sales data so that variables can be computed for the sales of each representative in comparison to the summarized sales. Finally, the REPORT command produces a report displaying all the desired information. The SPSS/PC+ commands in the command file named on the INCLUDE command are

```
GET FILE='FSTQTR.SF'.
SORT CASES BY PRODUCT MONTH.
SAVE OUTFILE='FSTQTRSO.SF'.

GET FILE='SECQTR.SF'.
SORT CASES BY PRODUCT MONTH.

JOIN ADD FILE='FSTQTRSO.SF'/FILE=*
   /BY PRODUCT MONTH/MAP.

AGGREGATE OUTFILE='AGGSALES.SF'/BREAK=PRODUCT
   /TOTAL=SUM(NUMBER)
   /AVSALES=MEAN(NUMBER).

JOIN MATCH FILE=*/TABLE='AGGSALES.SF'/BY PRODUCT/MAP.

SORT BY PRODUCT SALESREP.

COMPUTE CONTRIB=NUMBER/TOTAL*100.
COMPUTE DIFF=RND(NUMBER-AVSALES).
FORMATS DIFF(F4.0).

REPORT FORMAT=LIST BRKSPACE(-1)
   /TITLE 'Mid Year Sales Report'
   /VARIABLES=NUMBER 'Number of' 'Units Sold' (10)
              MONTH (label) 'Month' (5)
              DIFF 'Deviation' (9)
              CONTRIB '% of Total' (10)
   /BREAK=PRODUCT (label) 'Product' 'Name' (12)
   /SUMMARY=SUM '   Total' (NUMBER CONTRIB)
   /SUMMARY=MEAN '  Average' (NUMBER)
   /BREAK=SALESREP 'Sales' 'Rep' (SKIP(0)).
```

- The first GET command retrieves the system file FSTQTR.SF. The cases are ordered according to months within product categories, as specified on the SORT command. The SAVE command saves a copy of the sorted system file, named FSTQTRSO.SF.
- The second GET command retrieves the system file SECQTR.SF, and SORT again orders the cases by months within product categories. The sorted system file is the current active file.
- ADD (synonym for JOIN ADD) combines cases from FSTQTRSO.SF and the active file, interleaving the cases according to values of the key variables, PRODUCT and MONTH. The MAP subcommand produces a listing of the variables contained on the resulting active file (Figure A).
- The AGGREGATE command writes an aggregated system file called AGGSALES.SF. The file contains one case for each category of the variable PRODUCT, named on the BREAK subcommand. Each case contains a value for PRODUCT; a value for TOTAL, representing the sum of variable NUMBER for each break group; and a value for AVSALES, representing the mean of NUMBER for each break group. Figure B shows the variables in the system file AGGSALES.SF.
- MATCH (a synonym for JOIN MATCH) combines variables from the active file and from the aggregated file, which has been declared a TABLE. Cases with matching values of PRODUCT are joined. The MAP subcommand requests a listing of the variables that are in the resulting active file (Figure B).
- The SORT command orders the file according to SALESREP within categories of PRODUCT. The subsequent REPORT command requires that the input file be sorted on each variable named on a BREAK subcommand.
- The two COMPUTE commands create two new variables, CONTRIB and DIFF, based on variables from the combined files. The FORMATS command assigns four columns for DIFF.
- The REPORT command requests a case-listing report of the variables SALESREP, NUMBER, MONTH, DIFF, and CONTRIB for each category of product. For each category, two summary measures are computed. Figure C shows the report.

The exact appearance of the following printed displays will depend on the characters available on the printer used.

A ADD MAP display

```
RESULT          FSTQTRSO.SF     *
------------    ------------    ------------
SALESREP        SALESREP        SALESREP
PRODUCT         PRODUCT         PRODUCT
NUMBER          NUMBER          NUMBER
MONTH           MONTH           MONTH
DAY             DAY             DAY
YEAR            YEAR            YEAR
```

B Table look-up MAP display

```
RESULT          *               AGGSALES.SF
------------    ------------    ------------
SALESREP        SALESREP
PRODUCT         PRODUCT         PRODUCT
NUMBER          NUMBER
MONTH           MONTH
DAY             DAY
YEAR            YEAR
TOTAL                           TOTAL
AVSALES                         AVSALES
```

C REPORT display

Mid Year Sales Report

Product Name	Sales Rep	Number of Units Sold	Month	Deviation	% of Total
Primo	Brennan	6	Feb	-3	23.08
		1	Jun	-8	3.85
	Walsh	19	Feb	10	73.08
Total		26			100.00
Average		8.67			
Mocha	Curtis	6	Feb	-9	5.08
		6	Mar	-9	5.08
		14	May	-1	11.86
		26	Jun	11	22.03
	Feyerherd	11	Feb	-4	9.32
		3	Jun	-12	2.54
	Sullivan	24	Mar	9	20.34
	Walsh	28	May	13	23.73
Total		118			100.00
Average		14.75			
Blend	Brennan	5	Feb	-6	6.58
		13	Jun	2	17.11
	Curtis	11	Feb	0	14.47
		10	May	-1	13.16
	Feyerherd	1	Jun	-10	1.32
	James	18	Jan	7	23.68
	Schultz	18	Jan	7	23.68
Total		76			100.00
Average		10.86			
Special	Brennan	6	Apr	-6	7.06
	James	10	Mar	-2	11.76
		9	Jun	-3	10.59
	Schultz	5	Jan	-7	5.88
		12	May	0	14.12
	Sullivan	29	Jun	17	34.12
	Walsh	14	Mar	2	16.47
Total		85			100.00
Average		12.14			

ANOVA

This example, illustrating the use of ANOVA, is a three-way analysis of variance with one covariate. The data are 500 cases from the 1980 General Social Survey. The variables are

- PRESTIGE—the respondent's occupational prestige scale score. PRESTIGE is the dependent variable.
- EDUC—the respondent's education in years.
- RACE—the respondent's race, coded 1=WHITE, 2=BLACK, and 3=OTHER.
- SEX—the respondent's sex, coded 1=MALE and 2=FEMALE.
- REGION—the respondent's residence, coded as one of nine regions.

The task is twofold: determine the degree to which the American occupational structure differs across race, sex, and region; and measure the effect of the respondent's educational level, since education might prove to be a concomitant influence. The data are in an external file named AANOVA.DAT. The SPSS/PC+ commands in the command file named on the INCLUDE command are

```
DATA LIST FILE='AANOVA.DAT'
 / PRESTIGE 1-2 EDUC 3-4 RACE 5 SEX 6 REGION 7.
VARIABLE LABELS PRESTIGE "Resp's Occupational Prestige Score"
                EDUC 'Highest Year School Completed'
                REGION 'Region of Interview'.
ANOVA PRESTIGE BY REGION(1,9) SEX,RACE(1,2) WITH EDUC
      /STATISTICS 2
      /OPTIONS 10,11.
FINISH.
```

- The DATA LIST command assigns variable names and gives column locations for the variables in the analysis (see DATA LIST).
- The VARIABLE LABELS command completes the file definition (see VARIABLE LABELS).
- The ANOVA command names PRESTIGE as the dependent variable; REGION, SEX, and RACE as the factors; and EDUC as the covariate. The minimum and maximum values for REGION are 1 and 9, and the minimum and maximum values for both SEX and RACE are 1 and 2. Since variable RACE actually has values 1, 2, and 3, cases with value 3 are eliminated from the model.
- Statistic 2 requests the regression coefficient for the covariate EDUC.
- Option 10 requests the hierarchical approach for decomposing sums of squares. The covariate EDUC is assessed first to establish statistical control. Then the effect of REGION is assessed; next, the effect of SEX adjusted for REGION; and next, the effect of RACE adjusted for REGION and SEX. Finally, each of the interaction effects is assessed.
- Option 11 tells SPSS/PC+ to display the results within a narrow width.

The display produced by ANOVA is on the following page. The exact appearance of the printed output depends on the characters available on the printer used.

ANOVA display

```
              PRESTIGE Resp's Occupational Prestige Score
         BY   REGION   Region of Interview
              SEX
              RACE
         WITH EDUC     Highest Year School Completed


                                    SUM OF                 MEAN              SIGNIF
SOURCE OF VARIATION                SQUARES      DF        SQUARE        F      OF F

COVARIATES                       23715.522       1     23715.522  191.701   0.000
  EDUC                           23715.522       1     23715.522  191.701   0.000

MAIN EFFECTS                      2708.380      10       270.838    2.189   0.018
  REGION                          1260.552       8       157.569    1.274   0.255
  SEX                               22.413       1        22.413    0.181   0.671
  RACE                            1425.415       1      1425.415   11.522   0.001

2-WAY INTERACTIONS                3144.833      17       184.990    1.495   0.092
  REGION   SEX                    1349.220       8       168.653    1.363   0.211
  REGION   RACE                   1138.839       8       142.355    1.151   0.328
  SEX      RACE                    534.154       1       534.154    4.318   0.038

3-WAY INTERACTIONS                1663.399       6       277.233    2.241   0.039
  REGION   SEX      RACE          1663.399       6       277.233    2.241   0.039

EXPLAINED                        31232.135      34       918.592    7.425   0.0

RESIDUAL                         52205.957     422       123.711

TOTAL                            83438.092     456       182.978



COVARIATE   RAW REGRESSION COEFFICIENT

 EDUC            2.331

    500 CASES WERE PROCESSED.
     43 CASES (  8.6 PCT) WERE MISSING.
```

CORRELATION

This example analyzes 1979 prices and earnings in 45 cities around the world, compiled by the Union Bank of Switzerland. The variables are

- FOOD—the average net cost of 39 different food and beverage items in the city, expressed as a percentage above or below that of Zurich, where Zurich equals 100%.
- RENT—the average gross monthly rent in the city, expressed as a percentage above or below that of Zurich, where Zurich equals 100%.
- SERVICE—the average cost of 28 different goods and services in the city, expressed as a percentage above or below that of Zurich, where Zurich equals 100%.
- PUBTRANS—the average cost of a three-mile taxi ride within city limits, expressed as a percentage above or below that of Zurich, where Zurich equals 100%.
- TEACHER, COOK, ENGINEER, MECHANIC, BUS—the average gross annual earnings of primary-grade teachers in public schools, cooks, electrical engineers, automobile mechanics, and municipal bus drivers; working from 5 to 10 years in their respective occupations. Each of these variables is expressed as a percentage above or below those of Zurich, where Zurich equals 100%.

This example analyzes the degree to which variation in the costs of goods and services in a city is related to variation in earnings in several occupations. CORRELATION is used to compute correlations between the average costs of various goods and services and the average gross earnings in five different occupations. The data are in an external file ACORR.DAT. The SPSS/PC+ commands in the file named on the INCLUDE command are

```
DATA LIST FILE='ACORR.DAT'
 /FOOD 2-4 RENT 6-8 PUBTRANS 10-12 TEACHER 14-16 COOK 18-20
  ENGINEER 22-24 SERVICE 26-28 MECHANIC 30-32 BUS 34-36.
CORRELATION VARIABLES=FOOD RENT PUBTRANS TEACHER COOK ENGINEER/
            SERVICE PUBTRANS WITH MECHANIC BUS
            /STATISTICS=1,2.
FINISH.
```

- The DATA LIST command assigns variable names and gives column locations for the variables in the analysis (see DATA LIST).
- The CORRELATION command requests two correlation matrices. The first variable list produces correlation coefficients for each variable with every other variable. The second variable list produces four coefficients, pairing SERVICE with MECHANIC and BUS, and PUBTRANS with MECHANIC and BUS.
- The STATISTICS subcommand requests the mean, standard deviation, and number of nonmissing cases for each variable, and the cross-product deviations and covariance for each pair of variables. The statistics for all the variable lists precede all the correlation matrices in the CORRELATION display.

The display produced by CORRELATION is on the following page. The exact appearance of the printed display will depend on the characters available on the printer used.

A CORRELATION statistics

Variable	Cases	Mean	Std Dev
FOOD	43	70.9767	18.8319
RENT	43	122.2558	95.7054
PUBTRANS	43	48.8837	24.9258
TEACHER	43	38.3023	25.4797
COOK	43	64.9767	30.5626
ENGINEER	43	60.1163	26.4802
SERVICE	43	73.5116	18.9892
PUBTRANS	43	47.8372	24.0445
MECHANIC	43	50.4884	31.0762
BUS	43	42.9535	27.3652

Variables		Cases	Cross-Prod Dev	Variance-Covar
FOOD	RENT	43	18459.2558	439.5061
FOOD	PUBTRANS	43	11121.8837	264.8068
FOOD	TEACHER	43	11244.3023	267.7215
FOOD	COOK	43	9574.9767	227.9756
FOOD	ENGINEER	43	10106.1163	240.6218
RENT	PUBTRANS	43	-3527.7209	-83.9934
RENT	TEACHER	43	-2515.3256	-59.8887
RENT	COOK	43	17443.2558	415.3156
RENT	ENGINEER	43	26859.7209	639.5172
PUBTRANS	TEACHER	43	18659.5116	444.2741
PUBTRANS	COOK	43	19394.8837	461.7829
PUBTRANS	ENGINEER	43	17621.5814	419.5615
TEACHER	COOK	43	20326.3023	483.9596
TEACHER	ENGINEER	43	18627.4884	443.5116
COOK	ENGINEER	43	25524.1163	607.7171

Variables		Cases	Cross-Prod Dev	Variance-Covar
SERVICE	MECHANIC	43	11965.2558	284.8870
SERVICE	BUS	43	12806.0233	304.9053
PUBTRANS	MECHANIC	43	23496.4186	559.4385
PUBTRANS	BUS	43	21561.6744	513.3732

B CORRELATION matrices

Correlations:	FOOD	RENT	PUBTRANS	TEACHER	COOK	ENGINEER
FOOD	1.0000	.2439	.5641**	.5579**	.3961*	.4825**
RENT	.2439	1.0000	-.0352	-.0246	.1420	.2523
PUBTRANS	.5641**	-.0352	1.0000	.6995**	.6062**	.6357**
TEACHER	.5579**	-.0246	.6995**	1.0000	.6215**	.6573**
COOK	.3961*	.1420	.6062**	.6215**	1.0000	.7509**
ENGINEER	.4825**	.2523	.6357**	.6573**	.7509**	1.0000

N of cases: 43   1-tailed Signif: * - .01 ** - .001

" . " is printed if a coefficient cannot be computed

Correlations:	MECHANIC	BUS
SERVICE	.4828**	.5868**
PUBTRANS	.7487**	.7802**

N of cases: 43   1-tailed Signif: * - .01 ** - .001

" . " is printed if a coefficient cannot be computed

CROSSTABS

This example uses CROSSTABS to examine how respondents in different age groups answer a question on alcohol-drinking habits. The data are drawn from a 500-case sample from the 1980 General Social Survey. The variables are

- AGE—the respondent's age recoded to four categories.
- DRUNK—the response to the question, Did you ever drink too much?

The raw data are stored in an external file named AXTABS.DAT. The SPSS/PC+ commands in the command file named on the INCLUDE command are

```
DATA LIST FILE='AXTABS.DAT'
 /DRUNK 1 AGE 2-3.
RECODE AGE (LOW THRU 29=1) (29 THRU 40=2) (40 THRU 58=3)
           (58 THRU HI=4) /DRUNK (1=1) (2=2) (ELSE=8).
MISSING VALUE DRUNK(8).
VARIABLE LABELS AGE 'Age in Four Categories'/
                 DRUNK 'Ever Drink Too Much'.
VALUE LABELS AGE 1 'YoungestQuarter' 4 'Oldest  Quarter'/
              DRUNK 1 'Yes' 2 'No' 8 "Don't Drink/NA".
CROSSTABS TABLES=DRUNK BY AGE
          /OPTIONS=3,4
          /STATISTICS=1,4,6,7,8,9.
FINISH.
```

- The DATA LIST command identifies the data file, and gives the names and locations of the variables to be analyzed (see DATA LIST).
- The RECODE command redefines the variable AGE into four categories and recodes all missing values for the variable DRUNK to one missing value (see RECODE).
- The VARIABLE LABELS, VALUE LABELS, and MISSING VALUE commands complete the data definition for AGE and DRUNK (see VARIABLE LABELS, VALUE LABELS, and MISSING VALUE). The value labels for AGE are formatted to print appropriate labels for a column variable. Note that the label "Youngest-Quarter" has no blanks between the words and the label "Oldest Quarter" has two blanks separating the words.
- The CROSSTABS command sets up the table. It requests a table with DRUNK as the row variable and AGE as the column variable.
- Options 3 and 4 request row and column percents.
- The STATISTICS subcommand requests chi-square, lambda, Kendall's tau-*b*, Kendall's tau-*c*, gamma, and Somers' *d*.

The results produced by CROSSTABS are on the following page. The exact appearance of a printed display will depend on the characters available on the printer used.

CROSSTABS display

```
Crosstabulation:      DRUNK       Ever Drink Too Much
                   By AGE         Age in Four Categories

             Count  |Youngest |         |         |Oldest   |
   AGE->     Row Pct|Quarter  |         |         |Quarter  |   Row
             Col Pct|       1 |       2 |       3 |       4 |  Total
DRUNK       --------+---------+---------+---------+---------+
                 1  |     62  |     33  |     36  |     16  |    147
  Yes               |   42.2  |   22.4  |   24.5  |   10.9  |   38.5
                    |   57.9  |   34.7  |   37.9  |   18.8  |
                    +---------+---------+---------+---------+
                 2  |     45  |     62  |     59  |     69  |    235
  No                |   19.1  |   26.4  |   25.1  |   29.4  |   61.5
                    |   42.1  |   65.3  |   62.1  |   81.2  |
                    +---------+---------+---------+---------+
            Column       107        95        95        85       382
             Total      28.0      24.9      24.9      22.3     100.0

 Chi-Square     D.F.       Significance        Min E.F.       Cells with E.F.| 5
 ----------     ----       ------------        --------       -----------------

  31.57228        3           0.0000            32.709           None

         Statistic                      Symmetric     With DRUNK        With AGE
         ---------                      ---------     Dependent         Dependent
                                                      ----------        ---------

Lambda                                   0.09716        0.11565          0.08727
Somers' D                                0.23546        0.19222          0.30381
         Statistic                        Value        Significance
         ---------                        -----        ------------

Kendall's Tau B                          0.24165         0.0000
Kendall's Tau C                          0.28768         0.0000
Gamma                                    0.39632

Number of Missing Observations =      118
```

DESCRIPTIVES

This example analyzes 1979 prices and earnings in 45 cities around the world, compiled by the Union Bank of Switzerland. The variables are

- NTCPUR—the city's net purchasing-power level, calculated as the ratio of labor expended (measured in number of working hours) to the cost of more than 100 goods and services, weighted by consumer habits. NTCPUR is expressed as a percentage above or below that of Zurich, where Zurich equals 100%.
- FOOD—the average net cost of 39 different food and beverage items in the city, expressed as a percentage above or below that of Zurich, where Zurich equals 100%.
- RENT—the average gross monthly rent in the city, expressed as a percentage above or below that of Zurich, where Zurich equals 100%.
- APPL—the average cost of six different household appliances, expressed as a percentage above or below that of Zurich, where Zurich equals 100%.
- SERVICE—the average cost of 28 different goods and services in the city, expressed as a percentage above or below that of Zurich, where Zurich equals 100%.
- WCLOTHES—the cost of medium-priced women's clothes, expressed as a percentage above or below that of Zurich, where Zurich equals 100%.
- MCLOTHES—the cost of medium-priced men's clothes, expressed as a percentage above or below that of Zurich, where Zurich equals 100%.
- CLOTHES—the average cost of medium-priced women's and men's clothes, expressed as a percentage above or below that of Zurich, where Zurich equals 100%.

In this example, we obtain univariate summary statistics about purchasing power and the costs of various goods and services in cities. The data are in an external file named ADESC.DAT. The command file specified on the INCLUDE command contains the following SPSS/PC+ commands:

```
DATA LIST FILE='ADESC.DAT'
 /  NTCPUR 1-3 FOOD 12-14 RENT 23-25 APPL 34-36
    SERVICE 45-47 WCLOTHES 56-58 MCLOTHES 67-69.
VARIABLE LABLES NTCPUR 'Net Purchasing Level'/
                FOOD 'Avg Food Prices'/
                RENT 'Normal Rent'/
                APPL 'Price of Appliances'/
                SERVICE 'Price for Services'/
                WCLOTHES "Medium-Priced Woman's Clothes"/
                MCLOTHES "Medium-Priced Men's Clothes"/.
COMPUTE CLOTHES=(WCLOTHES + MCLOTHES)/2.
VARIABLE LABELS CLOTHES 'Ave. Cost of W and M Clothes'.
DESCRIPTIVES VARIABLES=NTCPUR, FOOD, RENT TO SERVICE, WCLOTHES,
             MCLOTHES, CLOTHES
             /STATISTICS=1,5,9
             /OPTIONS=5.
FINISH.
```

- The DATA LIST command defines the names and locations of the variables to be analyzed (see DATA LIST).
- The VARIABLE LABELS command completes the definition of the variables (see VARIABLE LABELS).
- The COMPUTE command creates the variable CLOTHES by adding the values for WCLOTHES and MCLOTHES and dividing by 2 (see COMPUTE).
- The VARIABLE LABELS command assigns a label to the new variable CLOTHES (see VARIABLE LABELS).
- The DESCRIPTIVES command requests statistics for the variables named.
- The associated STATISTICS subcommand requests the mean, standard deviation, and range for each variable in the analysis.
- Option 5 specifies listwise deletion of missing values. A case missing on any variable specified on the DESCRIPTIVES command is excluded from the computation of statistics for all variables.

- Because the default width of 79 is used, DESCRIPTIVES displays the statistics and variable labels for each variable on one line using only 79 columns.

The display produced by DESCRIPTIVES is shown below. The exact appearance of the printed display will depend on the characters available on the printer used.

DESCRIPTIVES display

```
Number of Valid Observations (Listwise) =      44.00

Variable       Mean    Std Dev      Range  Label

NTCPUR        58.70      28.81     100.00  Net Purchasing Level
FOOD          71.00      18.61      90.00  Avg Food Prices
RENT         121.75      94.65     413.00  Normal Rent
APPL          78.70      22.23     111.00  Price of Appliances
SERVICE       73.68      18.80      71.00  Price for Services
WCLOTHES      81.20      30.36     153.00  Medium-Priced Woman's Clothes
MCLOTHES      87.86      25.91     125.00  Medium-Priced Men's Clothes
CLOTHES       84.53      26.75     139.00  Ave. Cost of W and M Clothes
```

FREQUENCIES

The following example demonstrates the use of FREQUENCIES to do some preliminary checks on a newly defined file. The file is based on employment data from Hubbard Consultants Inc. Variables include: date employee was hired, employee's department, salary, job category, name, age, and sex, as well as salary increases from 1980 to 1982. The data are in an external file named AFREQ.DAT. The SPSS/PC+ commands in the command file named on the INCLUDE command are

```
DATA LIST FILE='AFREQ.DAT'
              /MOHIRED YRHIRED 12-15 DEPT79 TO DEPT82 SEX 16-20
              /SALARY79 TO SALARY82 6-25
               AGE 54-55 RAISE80 TO RAISE82 56-70
              /JOBCAT 6 EMPNAME 25-48 (A).
MISSING VALUE   SALARY79 TO SALARY82 AGE (0)
                JOBCAT (9).
VARIABLE LABELS  SALARY79 'Salary in 1979'
                 SALARY80 'Salary in 1980'
                 SALARY81 'Salary in 1981'
                 SALARY82 'Salary in 1982'
                 JOBCAT 'Job Categories'.
VALUE LABELS  SEX 1 'Male' 2 'Female'/
              JOBCAT 1 'Officials & Managers' 2 'Professionals'
              3 'Technicians' 4 'Office & Clerical' 5 'Craftsmen'
              6 'Service Workers'.
FREQUENCIES  VARIABLES=SALARY79 TO SALARY82 SEX AGE JOBCAT
             /FORMAT=LIMIT(10)
             /STATISTICS=DEFAULT MEDIAN.
FINISH.
```

- The DATA LIST command assigns variable names and gives column locations for the variables in the analysis (see DATA LIST).
- MISSING VALUE, VARIABLE LABELS, and VALUE LABELS complete the file definition (see MISSING VALUE, VARIABLE LABELS, and VALUE LABELS).
- FREQUENCIES displays frequency tables for variables having 10 or fewer categories and statistics for all the variables named. The default statistics are the mean, standard deviation, minimum, and maximum.
- The output uses the default format of 79 characters.

The FREQUENCIES display produced by this job is on the following page. The exact appearance of the printed display will depend on the characters available on the printer used.

FREQUENCIES display

```
SALARY79  Salary in 1979

Mean       12247.323        Median     10140.000        Std Dev     6665.182
Minimum     6337.000        Maximum    45500.000


Valid Cases      158        Missing Cases    117

- - - - - - - - - - - - - - - - - - - - - - - - - - - - - - - - - - - - - -

SALARY80  Salary in 1980

Mean       12123.725        Median     10400.000        Std Dev     6316.356
Minimum     5720.000        Maximum    48100.000


Valid Cases      273        Missing Cases      2

- - - - - - - - - - - - - - - - - - - - - - - - - - - - - - - - - - - - - -

SALARY81  Salary in 1981

Mean       15096.212        Median     12359.500        Std Dev     8074.387
Minimum     7605.000        Maximum    52000.000


Valid Cases      160        Missing Cases    115

- - - - - - - - - - - - - - - - - - - - - - - - - - - - - - - - - - - - - -

SALARY82  Salary in 1982

Mean       17161.552        Median     15132.000        Std Dev     8695.734
Minimum     5830.000        Maximum    50700.000


Valid Cases      145        Missing Cases    130

- - - - - - - - - - - - - - - - - - - - - - - - - - - - - - - - - - - - - -

SEX
                                                             Valid     Cum
  Value Label                      Value  Frequency  Percent  Percent  Percent

Male                                   1        83     30.2     30.2     30.2
Female                                 2       192     69.8     69.8    100.0
                                            -------  -------  -------
                                   TOTAL       275    100.0    100.0

Mean           1.698        Median         2.000        Std Dev         .460
Minimum        1.000        Maximum        2.000


Valid Cases      275        Missing Cases      0

- - - - - - - - - - - - - - - - - - - - - - - - - - - - - - - - - - - - - -

AGE

Mean          37.158        Median        34.000        Std Dev       11.335
Minimum       20.000        Maximum       69.000


Valid Cases      272        Missing Cases      3

- - - - - - - - - - - - - - - - - - - - - - - - - - - - - - - - - - - - - -

JOBCAT     Job Categories

                                                             Valid     Cum
  Value Label                      Value  Frequency  Percent  Percent  Percent

Officials & Managers                   1        48     17.5     17.5     17.5
Professionals                          2        62     22.5     22.5     40.0
Technicians                            3        98     35.6     35.6     75.6
Office & Clerical                      4        67     24.4     24.4    100.0
                                            -------  -------  -------
                                   TOTAL       275    100.0    100.0

Mean           2.669        Median         3.000        Std Dev        1.030
Minimum        1.000        Maximum        4.000


Valid Cases      275        Missing Cases      0
```

D Examples

LIST

This example demonstrates the use of LIST to display values of selected cases in a newly defined file. The file is based on employment data from Hubbard Consultants Inc. Variables include: date employee was hired, employee's department, salary, job category, name, age, and sex, as well as salary increases from 1979 to 1982. The data are in an external file named ALIST.DAT. The SPSS/PC+ commands in the command file named on the INCLUDE command are

```
SET WIDTH=WIDE.
DATA LIST FILE='ALIST.DAT'
  / EMPLOYID 1-5 MOHIRED YRHIRED 12-15 DEPT79 TO DEPT82 SEX 16-20
  / SALARY79 TO SALARY82 6-25 HOURLY81 HOURLY82 40-53(2) PROMO81 72
     AGE 54-55 RAISE82 66-70
  / JOBCAT 6 NAME 25-48 (A).
LIST VARIABLES=MOHIRED YRHIRED DEPT82 SALARY79 TO SALARY82 NAME/
  CASES FROM 50 TO 100 BY 5/ FORMAT=SINGLE,NUMBERED.
FINISH.
```

- The SET command requests a width of 132 (WIDE) so the column titles will all be horizontal (see SET command).
- The DATA LIST command assigns variable names and gives column locations for the variables in the analysis. The variable NAME is defined as a string variable (see DATA LIST).
- LIST displays values for the variables MOHIRED, YRHIRED, DEPT82, SALARY79 to SALARY 82, and NAME. Beginning with the 50th case, every 5th case will be listed. The display is forced to have only line per case, and the sequence number for each case will be displayed.

The LIST display produced by this job is shown below. The exact appearance of the printed display will depend on the characters available on the printer used.

LIST display

Cas	MOHIRED	YRHIRED	DEPT82	SALARY79	SALARY80	SALARY81	SALARY82	NAME
50	1	79	0	14300	14300	15730	0	EVA ELDER
55	6	79	0	0	15600	0	0	EDWARD GREEN
60	12	79	0	0	8840	9503	0	LOVEY E. HUDSON
65	5	80	0	0	13520	0	0	PATRICIA SMITH
70	8	79	0	0	8255	0	0	HELEN D. SMITH
75	10	70	4	14300	18850	21450	26182	MONICA C. RIVERS
80	1	79	0	0	7442	0	0	THOMAS P. JOHNSON
85	4	80	3	0	18200	18395	19682	ANN JOHNSON
90	10	79	0	0	5720	0	0	CHRISTINA P. NORRIS
95	5	79	0	7670	9490	0	0	M. ELLIOT KRAFT
100	2	70	3	11830	12545	13799	18083	FANNIE SMITH

Number of cases read = 100 Number of cases listed = 11

MEANS

This example uses MEANS to analyze personnel data from Hubbard Consultants Inc. 1981 salaries are analyzed by sex, within departments and grades. The data are in an external file named AMEANS.DAT. The SPSS/PC+ commands in the command file named on the INCLUDE command are

```
DATA LIST FILE='AMEANS.DAT'
 /SALARY81 1-5 DEPT81 6 GRADE81 7-8 SEX 9.
VARIABLE LABELS SALARY81 'Yearly Salary in 1981'/
                DEPT81 'Department Code in 1981'/
                SEX "Employee's Sex".
VALUE LABELS DEPT81 1 'Admin' 2 'Project Directors'
             3 'Chicago Operations' 4 'St Louis Operations'/
             SEX 1 'Male' 2 'Female'.
COMPUTE GRADE81S=GRADE81.
RECODE GRADE81S (1 THRU 4=1) (5 THRU 7=2) (8 THRU 15=3) (ELSE=0).
VALUE LABELS GRADE81S 1 'Grades 1-4' 2 'Grades 5-7' 3 'Grades 8-15'.
MISSING VALUE GRADE81S(0).
MEANS TABLES=SALARY81 BY DEPT81 BY GRADE81S BY SEX
 /OPTIONS=6,9,10,12.
FINISH.
```

- The DATA LIST command identifies the column locations of the four variables used in the analysis (see DATA LIST).
- The VARIABLE LABELS and VALUE LABELS commands complete the file definition (see VARIABLE LABELS and VALUE LABELS).
- The COMPUTE command creates the variable GRADE81S as a copy of GRADE81 (see COMPUTE).
- The RECODE command recodes the variable GRADE81S into three values which contain the 15 valid values of GRADE81. Other values of GRADE81S are recoded into the value 0 (see RECODE).
- The VALUE LABELS and MISSING VALUE commands provide dictionary definitions for the new variable GRADE81S (see VALUE LABELS and MISSING VALUE).
- The MEANS command specifies a three-way breakdown of salaries with SALARY81 as the dependent variable.
- Since no missing-value option is specified, MEANS deletes cases with missing values on a tablewide basis.
- The OPTIONS subcommand requests that group sums and variances be printed. OPTIONS are used to suppress printing of the independent variable names and values, so that the output will fit within the default width of 79 columns.

The display produced by MEANS is on the following page. The exact appearance of the printed display depends on the characters available on the printer used.

MEANS display

```
Summaries of    SALARY81    Yearly Salary in 1981
By levels of    DEPT81      Department Code in 1981
                GRADE81S
                SEX         Employee's Sex

Label                           Sum       Mean    Std Dev   Variance      Cases

For Entire Population    2415394.00 15096.2125  8074.3872 65195729.2        160

Admin                    590438.000 15537.8421  9810.5522 96246934.7         38
  Grades 1-4             120922.000 10076.8333  1685.2658 2840120.70         12
    Male                 111172.000 10106.5455  1764.2221 3112479.67         11
    Female                9750.0000  9750.0000     0.0        0.0             1

  Grades 5-7             179291.000 11952.7333  2019.7453 4079371.07         15
    Male                 13910.0000 13910.0000     0.0        0.0             1
    Female               165381.000 11812.9286  2019.2662 4077435.92         14

  Grades 8-15            290225.000 26384.0909 12759.5664  162806534         11
    Male                 170625.000 34125.0000 15498.1047  240191250          5
    Female               119600.000 19933.3333  4858.3605 23603666.7          6

Project Directors        428804.000 15314.4286  8146.9522 66372830.5         28
  Grades 1-4             181449.000 11340.5625  1999.6042 3998416.80         16
    Male                 105839.000 10583.9000  1143.2161 1306942.99         10
    Female               75610.0000 12601.6667  2566.9469 6589216.27          6

  Grades 5-7             38480.0000 12826.6667  2015.3494 4061633.33          3
    Female               38480.0000 12826.6667  2015.3494 4061633.33          3

  Grades 8-15            208875.000 23208.3333 10558.8272  111488831          9
    Male                 143065.000 28613.0000 11587.9159  134279795          5
    Female               65810.0000 16452.5000  2953.7674 8724741.67          4

Chicago Operations       940294.000 14925.3016  7705.3167 59371905.9         63
  Grades 1-4             168677.000  9922.1765  1536.2349 2360017.78         17
    Male                 20917.0000 10458.5000   836.5073 699744.500          2
    Female               147760.000  9850.6667  1612.6409 2600610.67         15

  Grades 5-7             197356.000 12334.7500  2190.5735 4798612.47         16
    Male                 40924.0000 13641.3333  3333.0827 11109440.3          3
    Female               156432.000 12033.2308  1903.0008 3621412.19         13

  Grades 8-15            574261.000 19142.0333  9294.0232 86378866.9         30
    Male                 142090.000 28418.0000 15680.5949  245881058          5
    Female               432171.000 17286.8400  6471.7384 41883398.6         25

St Louis Operations      455858.000 14705.0968  6624.5319 43884422.6         31
  Grades 1-4             66118.0000  9445.4286   680.5620 463164.619          7
    Male                 18395.0000  9197.5000   873.2769 762612.500          2
    Female               47723.0000  9544.6000   679.0183 461065.800          5

  Grades 5-7             160420.000 12340.0000  1925.6254 3708033.33         13
    Male                 35100.0000 11700.0000  1357.2398 1842100.00          3
    Female               125320.000 12532.0000  2087.3897 4357195.56         10

  Grades 8-15            229320.000 20847.2727  7667.4707 58790106.8         11
    Male                 21775.0000 21775.0000     0.0        0.0             1
    Female               207545.000 20754.5000  8075.7134 65217146.9         10

  Total Cases =      275
Missing Cases =      115 OR  41.8 PCT.
```

ONEWAY

This example analyzes a 500-case sample from the 1980 General Social Survey. The variables are

- WELL—the respondent's score on a scale measuring sense of well-being. WELL is the dependent variable, computed from measures of happiness, health, life, helpfulness of others, trust of others, and satisfaction with city, hobbies, family life and friendships.
- EDUC—the respondent's education in six categories, where the original codes are years of education completed.

In this example we determine the degree to which sense of well-being differs across educational levels. The data are in an external file named AONE.DAT. The SPSS/PC+ commands in the command file named on the INCLUDE command are

```
DATA LIST FILE='AONE.DAT'
 / EDUC 1-2 HAPPY 3 HEALTH 4 LIFE 5 HELPFUL 6 TRUST 7 SATCITY 8
    SATHOBBY 9 SATFAM 10 SATFRND 11.
COUNT  X1=HAPPY HEALTH LIFE HELPFUL TRUST SATCITY SATHOBBY
            SATFAM SATFRND(1).
COUNT  X2=HAPPY HEALTH SATCITY SATHOBBY SATFAM SATFRND(2).
COUNT  X3=HEALTH HELPFUL TRUST (3).
COUNT  X4=SATCITY SATHOBBY SATFAM SATFRND(6).
COUNT  X5=HAPPY LIFE (3).
COUNT  X6=SATCITY SATHOBBY SATFAM SATFRND(7).
COMPUTE WELL=X1 + X2* .5 - X3* .5 - X4* .5 - X5 - X6.
VAR LABELS  WELL 'Sense of Well-Being Scale'.
COMPUTE EDUC6=EDUC.
RECODE  EDUC6 (0 THRU 8=1)(9,10,11=2)(12=3)(13,14,15=4)
             (16=5)(17,18,19,20=6).
VAR LABELS  EDUC6 'Education in 6 Categories'.
VALUE LABELS  EDUC6 1 'Grade School or Less' 2 'Some High School'
                    3 'High Sch Grad' 4 'Some College' 5 'College Grad'
                    6 'Grad Sch'.
ONEWAY  VARIABLES=WELL BY EDUC6 (1,6)/
  POLYNOMIAL=2/
  CONTRAST= 2* -1, 2* 1/
  CONTRAST=2* 0,2* -1,2* 1/
  CONTRAST=2* -1,2* 0,2* 1/
  RANGES=SNK/
  RANGES=SCHEFFE(.01)/
  STATISTICS=ALL.
FINISH.
```

- The DATA LIST names the file that contains the data and gives the column locations of the variables in the analysis (see DATA LIST).
- The COUNT and COMPUTE commands create variable WELL by counting the number of "satisfied" responses for each variable on the scale and computing a weighted sum of these responses (see COUNT and COMPUTE).
- A copy of EDUC is created with COMPUTE and then recoded into six categories with RECODE (see COMPUTE and RECODE).
- The VARIABLE LABELS and VALUE LABELS commands assign labels to the new variables, WELL and EDUC6 (see VARIABLE LABELS and VALUE LABELS).
- The ONEWAY command names WELL as the dependent variable and EDUC6 as the independent variable. The minimum and maximum values for EDUC6 are 1 and 6.
- The POLYNOMIAL subcommand specifies second-order polynomial contrasts. The sum of squares using the unweighted polynomial contrasts is calculated because the analysis design is unbalanced. (See Figure A.)
- The CONTRAST subcommands request three different contrasts. (See Figure B.)
- The RANGES subcommands calculate multiple comparisons between means using the Student-Newman-Keuls and Scheffe tests. (See Figure C.)
- The STATISTICS subcommand requests all optional statistics. (See Figure D.)

The display produced by ONEWAY begins on the following page. The exact appearance of the printed display will depend on the printer used and the LENGTH and WIDTH that govern the display (see SET). This example uses the default settings.

A ONEWAY polynomial contrasts

```
- - - - - -  - - - - - - - - - - - O N E W A Y - - - - - - - - - - - - - - - -

     Variable  WELL        Sense of Well-Being Scale
  By Variable  EDUC6       Education in 6 Categories

                                  Analysis of Variance

                                 Sum of          Mean              F       F
       Source           D.F.    Squares         Squares          Ratio   Prob.

Between Groups             5    361.3217        72.2643        11.5255   .0000

  Unweighted  Linear Term  1       257.3422       257.3422       41.0439  .0000
    Weighted  Linear Term  1       307.2051       307.2051       48.9966  .0000
   Deviation from  Linear  4        54.1166        13.5291        2.1578  .0727

  Unweighted   Quad. Term  1         6.6073         6.6073        1.0538  .3051
    Weighted   Quad. Term  1        16.6406        16.6406        2.6540  .1039
   Deviation from   Quad.  3        37.4759        12.4920        1.9924  .1142

Within Groups            494    3097.3463         6.2699

Total                    499    3458.6680
```

B ONEWAY contrasts

```
- - - - - -  - - - - - - - - - - - O N E W A Y - - - - - - - - - - - - - - - -

     Variable  WELL        Sense of Well-Being Scale
  By Variable  EDUC6       Education in 6 Categories

  Contrast Coefficient Matrix

           Grp 1       Grp 3       Grp 5
                 Grp 2       Grp 4       Grp 6

Contrast  1  -1.0  -1.0   1.0   1.0   0.0   0.0

Contrast  2   0.0   0.0  -1.0  -1.0   1.0   1.0

Contrast  3  -1.0  -1.0   0.0   0.0   1.0   1.0

                                        Pooled Variance Estimate
                  Value       S. Error       T Value    D.F.        T Prob.

Contrast  1      3.3207         0.5230         6.349   494.0          0.000

Contrast  2      1.1517         0.6613         1.742   494.0          0.082

Contrast  3      4.4724         0.6990         6.398   494.0          0.000

                                      Separate Variance Estimate
                  Value       S. Error       T Value    D.F.        T Prob.

Contrast  1      3.3207         0.5401         6.148   252.5          0.000

Contrast  2      1.1517         0.6108         1.886   123.2          0.062

Contrast  3      4.4724         0.6984         6.404   172.7          0.000
```

C ONEWAY multiple comparisons

```
------ ----------- O N E W A Y ---------------

     Variable  WELL       Sense of Well-Being Scale
  By Variable  EDUC6      Education in 6 Categories


Multiple Range Test


Student-Newman-Keuls Procedure
Ranges for the 0.050 level -

          2.81    3.34    3.65    3.88    4.05

The ranges above are table ranges.
The value actually compared with Mean(J)-Mean(I) is..
        1.7706 * Range * Sqrt(1/N(I) + 1/N(J))


  (*) Denotes pairs of groups significantly different at the 0.050 level

      Variable  WELL       Sense of Well-Being Scale
      (Continued)


                           G G G G G G
                           r r r r r r
                           p p p p p p

     Mean       Group      1 2 3 4 5 6

     2.6462     Grp 1
     2.7737     Grp 2
     4.1796     Grp 3      * *
     4.5610     Grp 4      * *
     4.6625     Grp 5      * *
     5.2297     Grp 6      * *
------ ----------- O N E W A Y ---------------

     Variable  WELL       Sense of Well-Being Scale
  By Variable  EDUC6      Education in 6 Categories


Multiple Range Test


Scheffe Procedure
Ranges for the 0.010 level -

          5.53    5.53    5.53    5.53    5.53

The ranges above are table ranges.
The value actually compared with Mean(J)-Mean(I) is..
        1.7706 * Range * Sqrt(1/N(I) + 1/N(J))


  (*) Denotes pairs of groups significantly different at the 0.010 level

      Variable  WELL       Sense of Well-Being Scale
      (Continued)


                           G G G G G G
                           r r r r r r
                           p p p p p p

     Mean       Group      1 2 3 4 5 6

     2.6462     Grp 1
     2.7737     Grp 2
     4.1796     Grp 3      * *
     4.5610     Grp 4      * *
     4.6625     Grp 5      * *
     5.2297     Grp 6      * *
```

D ONEWAY statistics

Group	Count	Mean	Standard Deviation	Standard Error	95 Pct Conf Int for Mean		
Grp 1	65	2.6462	2.7539	.3416	1.9638	To	3.3285
Grp 2	95	2.7737	2.8674	.2942	2.1896	To	3.3578
Grp 3	181	4.1796	2.4220	.1800	3.8243	To	4.5348
Grp 4	82	4.5610	2.1450	.2369	4.0897	To	5.0323
Grp 5	40	4.6625	2.3490	.3714	3.9113	To	5.4137
Grp 6	37	5.2297	2.3291	.3829	4.4532	To	6.0063
Total	500	3.8920	2.6327	.1177	3.6607	To	4.1233
Fixed Effects Model			2.5040	.1120	3.6720	To	4.1120
Random Effects Model				.4492	2.7374	To	5.0466

Random Effects Model - Estimate of Between Component Variance 0.8491

Group	Minimum	Maximum
Grp 1	-4.0000	8.5000
Grp 2	-5.0000	8.5000
Grp 3	-4.0000	9.0000
Grp 4	-.5000	9.0000
Grp 5	-1.0000	8.0000
Grp 6	-1.5000	9.0000
Total	-5.0000	9.0000

Tests for Homogeneity of Variances

```
Cochrans C = Max. Variance/Sum(Variances) =  .2209, P =  .093 (Approx.)
Bartlett-Box F =                            1.905 , P =  .090
Maximum Variance / Minimum Variance         1.787
```

PLOT

Example 1: An Overlay Plot

Overlay plots are useful when several variables represent the same type of measurement, or the same variable is measured at different times. This example overlays two time series: marriage and divorce rates, 1900–1981. The data are drawn from the 1983 *Information Please Almanac*. Rates are specified for five-year periods for 1900–1940 and annually after 1943. The variables are

- MARRATE—Marriage rate per 1,000 population, excluding armed forces overseas.
- DIVRATE—Divorce rate (including annulments) per 1,000 population. (The rates for 1941-1946 include armed forces overseas.)

The data are in an external file named APLOT1.DAT. The SPSS/PC+ commands in the command file named on the INCLUDE command are

```
DATA LIST FILE='APLOT1.DAT'
         / YEAR 1-4 MARRATE 6-9 (1) DIVRATE 11-13 (1).
PLOT SYMBOLS='MD'/
     VSIZE=30 /HSIZE=70/
     FORMAT=OVERLAY/
     TITLE 'MARRIAGE AND DIVORCE RATES  1900-1981'/
     VERTICAL='RATES PER 1000 POPULATION'/
     HORIZONTAL='YEAR' REFERENCE (1918,1945) MIN (1900) MAX (1983)/
     PLOT=MARRATE DIVRATE WITH YEAR.
FINISH.
```

- The DATA LIST command defines the variables to be used in the overlay plot.
- The PLOT subcommand, placed last within the PLOT command, requests two bivariate plots, marriage rate by year and divorce rate by year.
- The SYMBOLS subcommand specifies the symbol M for the plot of marriage rate with year. The divorce rate by year plot is represented by the symbol D.
- The VSIZE and HSIZE subcommands establish the plot frame size of 30 lines high and 70 columns wide.
- The FORMAT subcommand requests an overlay plot of the variables specified on the PLOT subcommand.
- The VERTICAL subcommand supplies a label for the vertical axis.
- The HORIZONTAL subcommand supplies a label for the horizontal axis. The REFERENCE keyword requests reference lines to be drawn at the dates on which World War I and World War II ended. These lines point out the time period following the wars in which a sharp increase in marriage and divorce rates occurred. The MIN and MAX keywords specify the horizontal scale. SPSS/PC+ automatically divides the scale into equal-width intervals.

The PLOT display produced by this job follows. The exact appearance of the printed display will depend on the characters available on the printer used.

Overlay plot

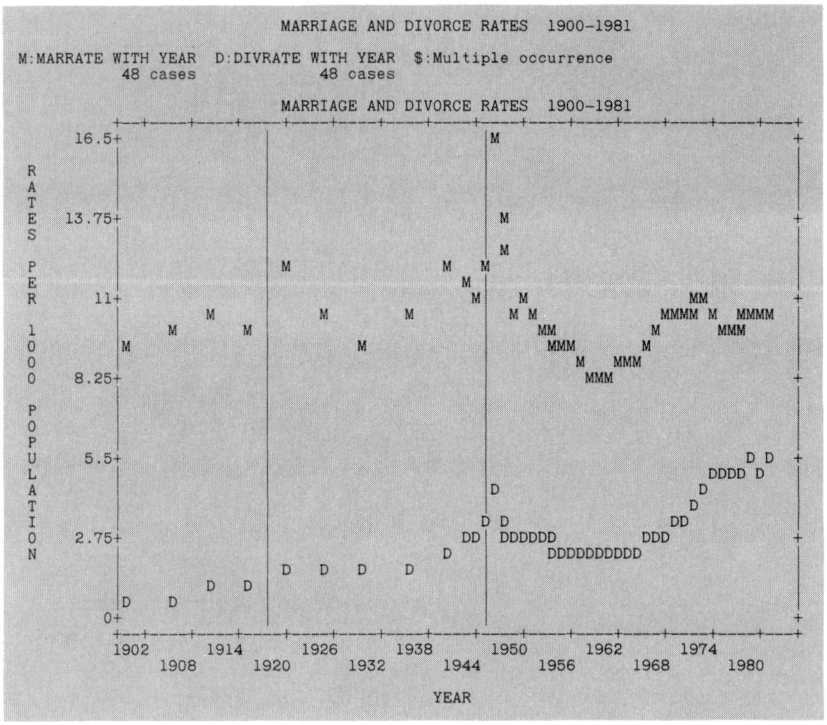

Example 2: A Regression Plot

Regression plots print a set of basic regression statistics below your plot and mark the regression-line intercepts on each axis. The statistics include the correlation coefficient, slope and intercept values, standard errors, and significance level.

This example examines the relationship of beginning salary to current salary as a first step in an analysis of salary differentials between race and sex groups. The data are drawn from bank employees hired between 1969 and 1971. The variables are

- SALBEG—annual starting salary of bank employees.
- SALNOW—current annual salary.

The data are in an external file named APLOT2.DAT. The SPSS/PC+ commands in the command file named on the INCLUDE command are

```
DATA LIST FILE='APLOT2.DAT'
         / SALNOW 1-5 SALBEG 6-10.
PLOT   HSIZE=35/ VSIZE=40/
       CUTPOINTS=EVERY (4)/
       SYMBOLS='+X*'/
       TITLE='SALARY REGRESSION'/
       VERTICAL='CURRENT ANNUAL SALARY'/
       HORIZONTAL= 'ANNUAL STARTING SALARY'/
       FORMAT=REGRESSION/
       PLOT=SALNOW WITH SALBEG.
FINISH.
```

- The DATA LIST command defines the variables to be used in the regression plot.
- The PLOT subcommand, the last specification on the PLOT command, names the variables to be plotted.
- The HSIZE subcommand requests a horizontal size of 35 print positions. VSIZE requests a vertical size of 40.

- The CUTPOINTS subcommand requests that each successive symbol represent accumulated frequency intervals of 4.
- The SYMBOLS subcommand defines the symbols to be plotted in accordance with the frequency cutpoints specified on the CUTPOINTS command. Thus, the symbol + represents positions with 4 or fewer cases; X, positions with 5 to 8 cases; and *, positions with 9 or more cases.
- The TITLE subcommand supplies a plot title. Note that the title is less than 35 print positions because of the HSIZE specification.
- The VERTICAL subcommand supplies an extended label for the vertical axis. The HORIZONTAL subcommand supplies an extended label for the horizontal axis.
- The FORMAT subcommand requests a regression plot. The following display shows the regression statistics generated by the FORMAT subcommand. The regression intercepts for the regression of SALNOW on SALBEG are printed on the top and bottom horizontal axis lines.

The display produced by the above commands is shown below. The exact appearance of the printed display will depend on the characters available on the printer used.

Regression plot

```
Frequencies and symbols used (not applicable for control or overlay plots)

        4 - +
        8 - X
       12 - *

                     SALARY REGRESSION

474 cases plotted. Regression statistics of SALNOW on SALBEG:
Correlation   .88012 R Squared   .77461   S.E. of Est 3246.14226 Sig.   .0000
Intercept(S.E.)  771.28230(355.47192)  Slope(S.E.)     1.90945(   .04741)

                     SALARY REGRESSION
```

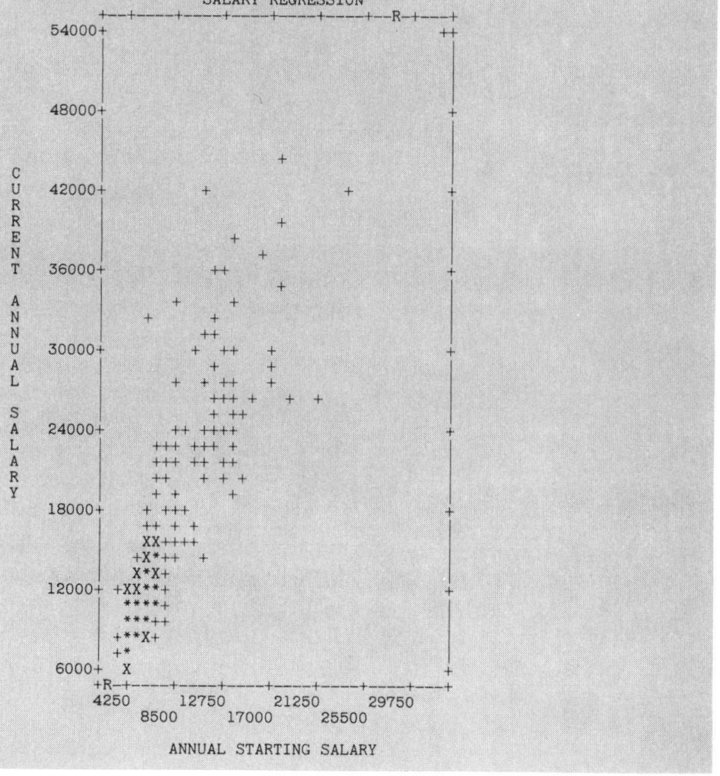

REGRESSION

The example attempts to predict the average aggregate personal savings rate of a country as a function of the age distribution of the population, the average level of real per capita disposable income, and the average percentage growth rate of real per capita disposable income. The data are 50 cases taken from an example in Belsley, Kuh, and Welsch (1980).
The variables are

- COUNTRY—the country in question.
- SAVINGS—the average aggregate personal savings rate in a country over the period 1960–1970.
- POP15—the average percentage of the population under 15 years of age over the period 1960–1970.
- POP75—the average percentage of the population over 75 years of age over the period 1960–1970.
- INCOME—the average level of real per capita disposable income in a country over the period 1960–1970, measured in United States dollars.
- GROWTH—the average percentage growth rate of INCOME over the period 1960–1970.

The data are in an external file named AREG.DAT. The SPSS/PC+ commands in the command file named on the INCLUDE command are

```
DATA LIST FILE='AREG.DAT'
 /   COUNTRY 1-8(A) SAVINGS POP15 POP75
     INCOME GROWTH 11-60.
VAR LABELS       SAVINGS 'Avg Agg Personal Savings Rate'
                 POP15 'Avg % Pop Under 15 Years Old'
                 POP75 'Avg % Pop Over 75 Years Old'
                 INCOME 'Avg Level Real Per-Cap Disposable Inc'
                 GROWTH 'Avg % Growth Rate of DPI'.
REGRESSION   VARIABLES=SAVINGS TO GROWTH/DEP=SAVINGS/ENTER/
             RESID=DEFAULT SIZE(SMALL) ID(COUNTRY)/
             SCATTERPLOT (*RES,*PRE)/PARTIALPLOT.
FINISH.
```

- The DATA LIST command names the file that contains the data, names the variables, and gives their column locations.
- The VAR LABELS command assigns labels to the variables (see VARIABLE LABELS).
- The REGRESSION command requests a direct-entry regression analysis with variable SAVINGS as the dependent variable.
- The RESIDUALS subcommand requests the default residuals results. In addition, the SIZE(SMALL) keyword overrides the default plot sizes so that small plots are displayed. The ID(COUNTRY) keyword specifies that the values for variable COUNTRY are to be used to label outlier plots. Figure A shows the residuals statistics and outlier plots. Figure B displays the histogram of the standardized residual and the normal probability plot.
- The SCATTERPLOT subcommand requests a plot of the residuals against the predicted values. Since *RES is specified first, it is plotted along the vertical axis (see Figure C).

The display produced by REGRESSION begins on the following page. The exact appearance of the printed display depends on the printer used and the LENGTH and WIDTH that govern the display (see SET). This example uses the default settings.

A REGRESSION residuals statistics and outliers

```
Residuals Statistics:

                 MIN          MAX      MEAN   STD DEV   N

*PRED          5.5874      15.8185    9.6710   2.6066   50
*RESID        -8.2422       9.7509     .0000   3.6441   50
*ZPRED        -1.5666       2.3584    -.0000   1.0000   50
*ZRESID       -2.1675       2.5642     .0000    .9583   50

Total Cases =        50

Durbin-Watson Test =    1.68579

Outliers - Standardized Residual

  Case #  COUNTRY          *ZRESID

      50  Zambia           2.56423
       7  Chile           -2.16749
      36  Philippi         1.75534
      35  Peru             1.71969
      18  Iceland         -1.63321
      34  Paraguay        -1.61093
      24  Korea           -1.60598
      10  Denmark          1.42014
      23  Japan            1.38890
       9  Costa Ri         1.34776
```

B REGRESSION histograms and normal probability plots

```
Histogram - Standardized Residual

NExp N       (* = 1 Cases,     . : = Normal Curve)
0  .04   Out
0  .08  3.00
1  .20  2.67 *
0  .45  2.33
0  .91  2.00 .
2 1.67  1.67 *:
3 2.74  1.33 **:
3 4.03  1.00 ***.
7 5.31   .67 ****:**
4 6.26   .33 **** .
6 6.62  0.0  ******.
8 6.26  -.33 *****:**
9 5.31  -.67 ****:****
3 4.03 -1.00 ***.
0 2.74 -1.33    .
3 1.67 -1.67 *:*
0  .91 -2.00 .
1  .45 -2.33 *
0  .20 -2.67
0  .08 -3.00
0  .04   Out

Normal Probability (P-P) Plot
Standardized Residual
    1.0 +---------+---------+---------+--------**
        I                                    ** I
        I                                 ***   I
        I                                *.     I
        I                             ***       I
    .75 +                             .         +
        I                         ****          I
O       I                      ***.             I
b       I                       .               I
s       I                     *                 I
e    .5 +                 ****                  +
r       I                *.                     I
v       I              **                       I
e       I             *                         I
d       I         ****                          I
    .25 +         .                             +
        I       .*                              I
        I     . *                               I
        I  *****                                I
        I .                                     I
        +**-------+---------+---------+---------+ Expected
                .25        .5       .75       1.0
```

C REGRESSION fit against residual plots

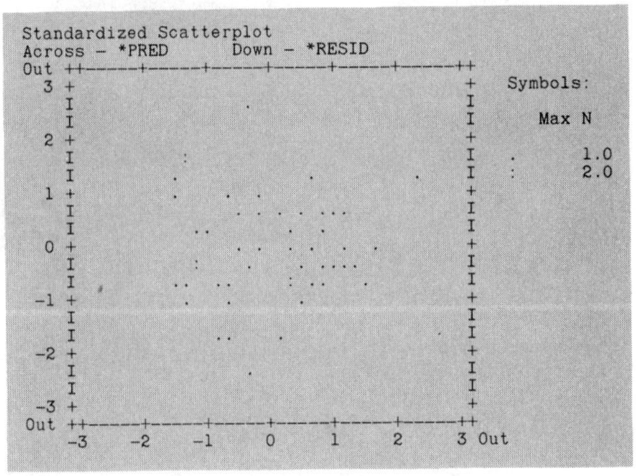

REPORT

Example 1: A Report with Summaries

This example produces a report that summarizes information from a retail company's personnel file. It reports summary statistics for employees in each division of the company within each store. The variables are

- AGE—age of employee in years.
- TENURE—length of employment at the company in months.
- JTENURE—length of employment in job grade in months.
- SALARY—annual salary in dollars.

The data are in an external file named AREPORT.DAT. The SPSS/PC+ commands are

```
DATA LIST FILE='AREPORT.DAT' /
   AGE 4-8 TENURE 13-16 JTENURE 21-24 SALARY 25-29
   STORE 30 DIVISION 31.
VARIABLE LABELS TENURE 'Tenure in Company' JTENURE 'Tenure in Grade'
                DIVISION 'Product Division' SALARY 'Annual Salary'
                STORE 'BRANCH STORE' age 'Age'.
VALUE LABELS DIVISION  1  'CARPETING'  2  'APPLIANCES'  3  'FURNITURE'
                       4  'HARDWARE' / STORE 1 'SUBURBAN' 2 'DOWNTOWN'.
FORMATS AGE (F2.0) TENURE JTENURE (F3.1).
SORT CASES  BY STORE DIVISION.
REPORT FORMAT=AUTOMATIC MARGINS(1,72) LENGTH(1,24) BRKSPACE(-1)/
       VARIABLES=AGE TENURE JTENURE SPACE(DUMMY)' '(4) SALARY /
       TITLE='Chicago Home Furnishing'/
       FOOTNOTE=LEFT 'Tenure measured in months'/
       BREAK=STORE /
          SUMMARY=MEAN 'AVERAGE:' (2) /
          SUMMARY=VALIDN (AGE) '  Count:' (2) /
       BREAK=DIVISION (SKIP(0))/
       SUMMARY=MEAN.
FINISH.
```

- The DATA LIST command assigns variable names and gives column locations for the variables, which are in the file AREPORT.DAT (see DATA LIST).
- The VARIABLE LABELS command defines labels for the variables; these appear on the report as column headings.
- The VALUE LABELS command defines labels for the break variables' values. These labels will be displayed in upper case since they are entered in upper case on this command (see VALUE LABELS).
- The FORMATS command overrides the default print formats (see FORMATS).
- The SORT CASES command sorts the file into the major and minor breaks required for REPORT (see SORT CASES).
- The FORMAT subcommand implements the AUTOMATIC settings; sets the left margin at column 1 and right margin at column 72; sets the top of the report page on the first line and defines the last line of the page as line 24; and places the break-group label on the first line of summary statistics.
- The VARIABLES subcommand defines five columns in the body of the report. AGE, TENURE, JTENURE, and SALARY are SPSS/PC+ variables. SPACE defines a 4-column wide dummy column, for spacing purposes, with a blank column heading.
- The TITLE subcommand defines a one-line centered title.
- The FOOTNOTE subcommand defines a left-justified one-line footnote.
- The first BREAK subcommand defines the major break in this two-break report. Variable STORE breaks the file into two categories: the downtown store and the suburban store.
- The first two SUMMARY subcommands print two lines of summary statistics for each store. The first SUMMARY subcommand computes means for AGE, TENURE, JTENURE, and SALARY. The second SUMMARY subcommand computes the number of employees in each store. Titles are specified for each summary line.
- The second BREAK subcommand breaks the file into divisions within each store. The SKIP specification suppresses blank lines between the summary for each division.

- The last SUMMARY subcommand computes means for the report variables for each division.

The display produced by REPORT is on the next page.

Summary report

```
                          Chicago Home Furnishing

                                              Tenure      Tenure
BRANCH       Product                            in          in                   Annual
STORE        Division           Age          Company      Grade                  Salary
--------     ----------         -----        -------      ------                 ------

SUBURBAN     CARPETING            40           2.4          2.4                   20869
             APPLIANCES           35           2.3          2.3                   16105
             FURNITURE            38           2.4          2.4                   18821
             HARDWARE             35           2.3          2.4                   15234

             AVERAGE:             36           2.3          2.4                   17011
               Count:             97

DOWNTOWN     CARPETING            37           2.3          2.4                   14207
             APPLIANCES           37           2.3          2.4                   14130
             FURNITURE            38           2.3          2.4                   14403
             HARDWARE             37           2.3          2.4                   14469

             AVERAGE:             37           2.3          2.4                   14307
               Count:            153

Tenure measured in months
```

Example 2: A Report Using LIST

This example produces a report using data from the October 1980 issue of *Runner's World* magazine. It lists the top-rated shoes in the survey, organized by manufacturer. Measures used by the raters to determine an overall evaluation for each shoe are reported. The data are in an external file named AREPT.DAT. The SPSS/PC+ commands are

```
SET WIDTH=132/LENGTH=45.
DATA LIST FILE='AREPT.DAT' FIXED /
             TYPE 1 MAKER 2-3 QUALITY 5-9
             REARIMP FOREIMP FLEX SOLEWEAR 10-29
             REARCONT SOLETRAC 31-40 WEIGHT 42-46 LASTYEAR 48
             PREFER 50-53 STARS 55 NAME 57-72 (A).
VARIABLE LABELS NAME 'SHOE' STARS 'RATING'
                REARIMP 'REARFOOT IMPACT'
                FOREIMP 'FOREFOOT IMPACT'
                SOLEWEAR 'SOLE WEAR' REARCONT 'REARFOOT CONTROL'
                SOLETRAC 'SOLE TRACTION' LASTYEAR '1979 STARS'
                PREFER 'READER PREFERENCE'.
VALUE LABELS MAKER 1 'ADIDAS'  2 'AUTRY'  3 'BROOKFIELD'  4 'BROOKS'
              5 'CONVERSE'  6 'REEBOK'  7 'NEW BALANCE'  8 'PUMA'
              9 'OSAG'
              10 'PONY'  11 'ETONIC'  12 'NIKE'  13 'SAUCONY'
              14 'WILSON-BATA'  15 'VOL SHOE CORP'
              16 'SPECS INTERNATIONAL'  17 'POWER SPORT'
              18 'THOM MCAN JOX'  19 'REGAL SHOES'  20 'SHOE CORP'
              21 'ASICS'  22 'INTL FOOTWEAR'  23 'EB SPORT INTL'
              24 'VAN DOREN'/
             TYPE 1 'MALE'  2 'FEMALE'/
             STARS 6 '******'  5 '*****'/.
FORMATS  QUALITY (F5.3)/REARIMP FOREIMP SOLEWEAR (F4.1)/
               FLEX SOLETRAC (F4.2)/REARCONT WEIGHT (F5.1)/
               PREFER (F4.3).
SELECT IF (STARS GE 5).
SORT CASES MAKER STARS(D).
REPORT   FORMAT=AUTOMATIC LIST MISSING ' '  /
               VARIABLES=TYPE(LABEL) NAME STARS(LABEL)
                         SEP1(DUMMY)(1)' '
                         REARIMP FOREIMP
                         FLEX 'FLEXI-' 'BILITY'
                         SOLEWEAR REARCONT SOLETRAC WEIGHT
                         LASTYEAR PREFER /
               TITLE='RATINGS OF TRAINING SHOES'
                      "RUNNER'S WORLD MAGAZINE - OCTOBER, 1980"/
               FOOTNOTE=LEFT '****** HIGHLY RECOMMENDED'
                             '***** RECOMMENDED'/
                        RIGHT ' ' 'PAGE )PAGE'/
               BREAK=MAKER 'MANUFACTURER'.
FINISH.
```

- Because there are many variables, the SET command sets the WIDTH to 132 columns, the maximum, and the LENGTH to 45 lines (see SET).
- The DATA LIST command assigns variable names and gives column locations for the variables in the analysis (see DATA LIST).
- The VARIABLE LABELS command defines labels for the variables; these appear on the report as column headings.
- The VALUE LABELS command supplies value labels for the manufacturer, type of shoe, and rating. These labels are used in the report (see VALUE LABELS).
- The FORMATS command overrides the default print formats (see FORMATS).
- The SELECT IF command selects shoes with the top two ratings (see SELECT IF).
- The SORT CASES command sorts cases in descending order of ranking for each manufacturer. They are sorted by manufacturer so that the report can group them by manufacturer. They are sorted by descending order of ranking so that the top-rated shoes for the manufacturer are listed first (see SORT).
- In REPORT, the FORMAT subcommand implements the AUTOMATIC settings and specifies a case listing. In a listing report, AUTOMATIC places the first case for each break on the same line as the break value, so BRKSPACE(−1) is not needed. The MISSING keyword prints a blank in place of the period for variables with missing values.
- The VARIABLES subcommand names all the variables to be listed as well as a dummy column (SEP1) to separate the measurements from the rating. Value labels are printed in place of values for variables TYPE and STAR. The column heading specified for FLEXIBILITY splits the variable name so that the column is narrower.
- The TITLE subcommand prints a two-line centered title.
- The FOOTNOTE subcommand prints a two-line left-justified footnote, and a two-line right-justified footnote in which the first line is blank while the second line uses the special keyword)PAGE to print page numbers.
- The BREAK subcommand groups the shoes by manufacturer.

The report produced by these commands is shown below and on the following page.

Report on running-shoe data—page 1

RATINGS OF TRAINING SHOES
RUNNER'S WORLD MAGAZINE - OCTOBER, 1980

MANUFACTURER	TYPE	SHOE	RATING	REARFOOT IMPACT	FOREFOOT IMPACT	FLEXI- BILITY	SOLE WEAR	REARFOOT CONTROL	SOLE TRACTION	WEIGHT	1979 STARS	READER PREFERENCE
SAUCONY	MALE	TC84	******	9.3	15.1	1.56	6.5	5.2	.85	278.0	0	.028
	MALE	HORNET 84	******	9.9	13.1	2.65	7.6	3.0	.68	265.0	4	.097
	FEMALE	MS TRAINER	******	10.2	13.3	1.58	6.4	22.4	.86	237.7	5	.053
	MALE	JAZZ	*****	8.9	12.7	2.04	7.6	-7.0	.64	270.8	0	
	MALE	TRAINER 80	*****	10.5	14.5	2.18	4.1	11.5	.82	307.6	5	.232
	FEMALE	JAZZ	*****	9.0	12.2	1.86	6.1	-7.5	.63	223.0	0	.013
	FEMALE	TC 84	*****	9.3	14.6	1.46	7.5	1.3	.77	231.1	0	
	FEMALE	MS HORNET	*****	9.8	13.2	2.59	6.4	6.5	.67	224.0	4	.046
NIKE	MALE	DAYBREAK	******	10.8	15.4	2.17	3.7	7.8	.54	304.2	5	.6
	MALE	YANKEE	*****	10.9	13.7	1.93	2.0	9.8	.66	276.6	0	
	FEMALE	LIBERATOR	*****	10.6	14.7	2.20	5.8	6.5	.52	254.2	5	.503
ETONIC	MALE	ECLIPSE TRAINER	******	10.0	12.9	1.65	10.0	-2.6	.51	237.4	0	
	FEMALE	ECLIPSE TRAINER	******	9.6	12.8	1.78	10.0	1.4	.57	204.1	0	
	MALE	STABILIZER	*****	10.3	15.5	2.25	1.2	-.6	.53	283.1	4	.232
	MALE	STREETFIGHTER	*****	10.8	15.5	2.28	1.4	-.4	.61	266.1	4	.222
	FEMALE	STREETFIGHTER	*****	10.7	15.5	1.66	.7	-7.7	.70	214.1	4	.3
	FEMALE	STABILIZER	*****	10.8	14.4	2.09	2.6	-6.9	.67	235.3	4	.298
PONY	MALE	TARGA FLEX	******	9.6	14.3	1.32	2.5	-22.7	.86	253.0	3	
	MALE	SHADOW	*****	9.9	13.8	1.53	2.5	-17.9	.77	270.2	0	
	FEMALE	LADY SHADOW	*****	10.6	17.4	.91	3.0	-7.1	.90	211.8	0	
OSAG	MALE	FAST RIDER	*****	10.5	14.0	2.48	4.9	1.9	.66	296.7	5	.025
	FEMALE	KT-26	*****	10.7	17.3	1.66	5.5	8.1	.60	223.1	2	
NEW BALANCE	MALE	420	*****	9.8	14.8	2.09	1.8	-17.7	.46	267.9	0	.516
	MALE	620	*****	12.0	14.6	2.73	1.1	-3.5	.41	242.0	5	.475
	FEMALE	420	*****	9.9	13.9	1.94	1.6	-.7	.46	219.3	0	.411
REEBOK	MALE	AZTEC	******	10.9	12.6	2.07	2.5	3.7	.65	260.8	5	.065
	MALE	SHADOW I	*****	10.7	13.1	1.79	1.9	-8.7	.63	253.0	0	
	FEMALE	SHADOW III	*****	10.2	12.9	1.63	2.4	-24.6	.66	212.8	0	
	FEMALE	AZTEC PRINCESS	*****	10.2	12.8	2.18	5.9	-20.3	.70	221.3	5	.033

****** HIGHLY RECOMMENDED
***** RECOMMENDED

PAGE 1

Report on running-shoe data—page 2

RATINGS OF TRAINING SHOES
RUNNER'S WORLD MAGAZINE - OCTOBER, 1980

MANUFACTURER	TYPE	SHOE	RATING	REARFOOT IMPACT	FOREFOOT IMPACT	FLEXI- BILITY	SOLE WEAR	REARFOOT CONTROL	SOLE TRACTION	WEIGHT	1979 STARS	READER PREFERENCE
CONVERSE	MALE	ARIZONA 84	*****	10.1	13.6	1.90	6.6	-5.1	.55	302.9	4	.006
	FEMALE	WORLD CLASS 84	*****	9.4	14.0	2.19	4.3	-.3	.65	234.7	3	.020
BROOKS	MALE	VANTAGE	******	8.3	11.0	1.33	10.0	-13.6	.55	232.4	5	.531
	MALE	VANTAGE SUPREME	******	8.5	10.9	1.31	10.0	-16.5	.58	239.1	5	
	MALE	HUGGER GT	******	8.5	11.2	1.32	9.4	-11.7	.60	234.5	5	.488
	MALE	NIGHTHAWK	******	8.7	13.5	1.57	3.1	-8.6	.45	216.7	0	
	MALE	SUPER VILLANOVA	******	10.0	14.1	1.07	10.0	14.4	.61	238.7	5	.155
	FEMALE	VANTAGE	******	8.1	11.0	1.27	10.0	-13.1	.58	199.9	5	.563
	FEMALE	HUGGER GT	******	8.2	11.1	1.28	10.0	-12.7	.60	203.8	0	.126
	FEMALE	VANTAGE SUPREME	******	8.2	11.1	1.34	10.0	.6	.62	201.4	3	.205
	FEMALE	SUPER VILLANOVA	******	9.0	13.4	1.01	10.0	11.9	.62	195.1	5	.298
	FEMALE	NIGHTHAWK	*****	8.6	13.1	1.54	2.4	-9.3	.45	189.0	0	
BROOKFIELD	MALE	COLT	*****	12.4	17.4	2.31	3.5	21.5	1.13	289.3	4	
AUTRY	MALE	MACH III	*****	8.7	13.0	2.13	3.0	-37.6	.66	250.2	4	
	MALE	NEW JET	*****	9.1	14.5	1.88	4.0	-37.9	.69	242.4	4	
	MALE	CONCORDE	*****	9.2	13.2	2.41	2.0	-33.9	.61	261.7	5	.023
	FEMALE	CLOUD 9	*****	9.4	17.6	1.79	2.3	-27.3	.63	198.6	3	
ADIDAS	MALE	TRX TRAINER	*****	10.5	16.8	2.07	2.1	-.6	.72	309.0	5	.143
	MALE	MARATHON TRAINER	*****	13.0	17.2	2.75	10.0	14.5	.63	302.3	5	.315
	FEMALE	MARATHON TRAINER	*****	11.7	16.5	2.14	10.0	17.5	.58	243.6	5	.298

****** HIGHLY RECOMMENDED
***** RECOMMENDED

PAGE 2

T-TEST

This example uses T-TEST to analyze 1979 prices and earnings in 45 cities around the world, compiled by the Union Bank of Switzerland. The variables are

- WORLD—the economic class of the country in which the city is located. The 45 cities are divided into three groups: cities in economically developed nations such as the United States and most European nations; cities in nations that are members of the Organization for Petroleum Exporting Countries (OPEC); and cities in underdeveloped countries. These groups are coded from 1 to 3 and are labeled 1ST WORLD, PETRO WORLD, and 3RD WORLD, respectively.
- NTCPRI—the city's net price level, based on more than 100 goods and services weighted by consumer habits. NTCPRI is expressed as the percentage above or below that of Zurich, where Zurich equals 100%.
- NTCSAL—the city's net salary level, calculated from average net hourly earnings in 12 occupations. NTCSAL is expressed as a percentage above or below that of Zurich, where Zurich equals 100%.
- NTCPUR—the city's net purchasing power level, calculated as the ratio of labor expended (measured in number of working hours) to the cost of more than 100 goods and services, weighted by consumer habits. NTCPUR is expressed as a percentage above or below that of Zurich, where Zurich equals 100%.
- WCLOTHES—the cost of medium-priced women's clothes, expressed as the percentage above or below that of Zurich, where Zurich equals 100%.
- MCLOTHES—the cost of medium-priced men's clothes, expressed as the percentage above or below that of Zurich, where Zurich equals 100%.

This example compares mean price, salary and purchasing power for cities grouped by economic class. It also compares the mean costs of women's and men's clothes. The data are in an external file named ATTEST.DAT. The SPSS/PC+ commands in the command file specified on the INCLUDE command are

```
DATA LIST FILE='ATTEST.DAT'
 /NTCPRI 9-11 NTCSAL 20-22 NTCPUR 31-33 WCLOTHES 42-44
    MCLOTHES 53-55 WORLD 66.
VARIABLE LABELS NTCPRI 'Net Price Level'/
                NTCSAL 'Net Salary Level'/
                NTCPUR 'Net Purchasing Level'/
                WCLOTHES "Medium-Priced Women's Clothes"/
                MCLOTHES "Medium-Priced Men's Clothes"/.
T-TEST GROUPS=WORLD(1,3)/VARIABLES=NTCPRI NTCSAL NTCPUR/
       PAIRS=WCLOTHES MCLOTHES/NTCPRI WITH NTCPUR NTCSAL.
FINISH.
```

- The DATA LIST command assigns variable names and gives the column locations of the variables to be analyzed (see DATA LIST).
- The VARIABLE LABELS command completes the data definition of these variables (see VARIABLE LABELS).
- The T-TEST command requests an independent-samples test and a paired-samples test. For the independent-samples test, the variable WORLD specifies a grouping criterion that compares cities in first-world countries to cities in third-world countries. Cities in petro-world countries are not included.
- By default, the display is formatted within 79 columns.

The results produced by T-TEST are on the next page. The exact appearance of the printed display depends on the characters available on the printer used.

T-TEST display

```
Independent samples of  WORLD

Group 1:  WORLD  EQ 1          Group 2:  WORLD  EQ 3

t-test for:  NTCPRI    Net Price Level

                 Number               Standard      Standard
                of Cases      Mean    Deviation       Error

      Group 1      25       83.8400     13.309        2.662
      Group 2      13       67.3077     14.773        4.097

                 | Pooled Variance Estimate | Separate Variance Estimate
                 |                          |
   F    2-Tail   |    t   Degrees of 2-Tail |    t    Degrees of  2-Tail
 Value  Prob.    |  Value   Freedom   Prob. |  Value   Freedom    Prob.
                 |                          |
  1.23  0.637    |  3.50     36      0.001  |  3.38     22.28     0.003
Independent samples of  WORLD

Group 1:  WORLD   EQ 1          Group 2:  WORLD   EQ 3

t-test for:  NTCSAL    Net Salary Level

                 Number               Standard      Standard
                of Cases      Mean    Deviation       Error

      Group 1      25       64.4000     19.026        3.805
      Group 2      12       25.6667     13.241        3.822

                 | Pooled Variance Estimate | Separate Variance Estimate
                 |                          |
   F    2-Tail   |    t   Degrees of 2-Tail |    t    Degrees of  2-Tail
 Value  Prob.    |  Value   Freedom   Prob. |  Value   Freedom    Prob.
                 |                          |
  2.06  0.210    |  6.33     35      0.000  |  7.18     30.07     0.000
Independent samples of  WORLD

Group 1:  WORLD   EQ 1          Group 2:  WORLD   EQ 3

t-test for:  NTCPUR    Net Purchasing Level

                 Number               Standard      Standard
                of Cases      Mean    Deviation       Error

      Group 1      25       76.7600     21.491        4.298
      Group 2      12       31.9167     17.573        5.073

                 | Pooled Variance Estimate | Separate Variance Estimate
                 |                          |
   F    2-Tail   |    t   Degrees of 2-Tail |    t    Degrees of  2-Tail
 Value  Prob.    |  Value   Freedom   Prob. |  Value   Freedom    Prob.
                 |                          |
  1.50  0.493    |  6.28     35      0.000  |  6.74     26.26     0.000

Paired samples t-test:  WCLOTHES  Medium-Priced Woman's Clothes
                        MCLOTHES  Medium-Priced Men's Clothes

Variable      Number              Standard    Standard
             of Cases     Mean    Deviation     Error

WCLOTHES        45      80.7111     30.195      4.501
MCLOTHES        45      87.0444     26.192      3.905

(Difference) Standard    Standard  |        2-Tail |    t    Degrees of   2-Tail
    Mean     Deviation     Error   |  Corr.  Prob. |  Value    Freedom     Prob.
                                   |               |
  -6.3333     17.916       2.671   |  0.807  0.000 |  -2.37     44        0.022

Paired samples t-test:  NTCPRI    Net Price Level
                        NTCPUR    Net Purchasing Level

Variable      Number              Standard    Standard
             of Cases     Mean    Deviation     Error

NTCPRI          44      82.1591     19.773      2.981
NTCPUR          44      58.7045     28.806      4.343

(Difference) Standard    Standard  |        2-Tail |    t    Degrees of   2-Tail
    Mean     Deviation     Error   |  Corr.  Prob. |  Value    Freedom     Prob.
                                   |               |
  23.4545     33.310       5.022   |  0.098  0.528 |   4.67     43        0.000

Paired samples t-test:  NTCPRI    Net Price Level
                        NTCSAL    Net Salary Level

Variable      Number              Standard    Standard
             of Cases     Mean    Deviation     Error

NTCPRI          44      82.1591     19.773      2.981
NTCSAL          44      50.3409     24.295      3.663

(Difference) Standard    Standard  |        2-Tail |    t    Degrees of   2-Tail
    Mean     Deviation     Error   |  Corr.  Prob. |  Value    Freedom     Prob.
                                   |               |
  31.8182     22.753       3.430   |  0.482  0.001 |   9.28     43        0.000
```

Glossary

Glossary

active default Options that are in effect when an optional subcommand is included but is not followed by any keywords. For example, in REGRESSION, including the subcommand DESCRIPTIVES without any modifying keywords requests the active defaults MEAN, STDDEV, and CORR. See also *passive default.*

active file The file defined in a session by a DATA LIST command or called by a GET or IMPORT command. Consists of the data and a dictionary and is the file that you modify using transformations, and analyze using any of the procedures.

active window In REVIEW, the portion of the screen that accepts file modifications. Two windows of different sizes can be accessed, but only the active window can be modified.

alias A keyword that is a synonym for another SPSS/PC+ keyword. For example, XTABS is an alias for CROSSTABS.

alphanumeric variable A variable whose values are stored as characters and are not available for arithmetic operations. The values may contain letters or special characters as well as numbers. Also known as "string" variable. Compare *numeric variable.*

analysis list Portion of an SPSS/PC+ command that requests an analysis (such as a correlation matrix) or a set of related analyses (such as a set of crosstabs that use the same variables).

argument An expression (usually within parentheses) following a function or another SPSS/PC+ keyword that is the object of the operation. For arithmetic function SQRT(X), variable X is the argument. For REPORT function PCGT(150), the value 150 is the argument.

arithmetic operators Graphic symbols used in a variable transformation to represent arithmetic functions such as addition (+), subtraction (−), multiplication (*), division (/), and exponentiation (**).

ASCII file A data or command file containing characters conforming to the American Standard Code for Information Interchange standards. These files contain a carriage control line-feed sequence at the end of each line and a CTRL−Z at the end of the file. SPSS/PC+ reads and writes ASCII files.

assignment expression The expression following the equals sign in the COMPUTE and IF commands that assigns a value for each case to the target variable. See also *target variable.*

batch processing Using the INCLUDE command to process SPSS/PC+ commands stored in a file. While SPSS/PC+ is processing the commands from the file, you cannot change any of the specifications given on the commands in the file. See also *interactive operation.*

case identifier Unique code assigned to each case.

case, observation Basic unit of analysis for which measurements are taken.

categorical variable A variable that has a limited number of values that form categories of cases. See also *continuous variable.*

code A numeric or alphabetical value that represents the status of a case on a variable.

codebook Document that describes each variable, its name, label, values and value labels, formats, and missing values. Also known as a coding form.

command A specific instruction that controls the execution of SPSS/PC+.

command file A file that contains SPSS/PC+ commands and sometimes includes data. Use the INCLUDE command to process command files. See also *batch processing*.

command line A line of specifications for one SPSS/PC+ command. It may take more than one line to enter all of the instructions for a single command (see *continuation line*).

conditional transformation Changing the values of a variable contingent upon logical conditions found in the data; e.g., altering variables one way for one subset of cases and other ways for other subsets. See the IF command.

constant A value that is the same for every case. For example, in the expression B=A+12, 12 is a constant. Constants can be either numeric or string values.

contingency table A table containing the joint frequency distribution of two or more variables. See the CROSSTABS command.

continuation line The second or any subsequent line of specifications for one SPSS/PC+ command.

continuous variable A variable that does not have a fixed number of values. For example, the variable INCOME, measured in dollars, can take on many different values.

control variable The variable whose values are used to separate cases into subgroups. In a crosstabulation, the variable whose values form subtables.

CPU The abbreviation for central processing unit. The CPU performs tasks based on instructions from SPSS/PC+ commands.

current character The character highlighted by the cursor.

current line The line containing the cursor.

cursor The highlighted symbol on the screen indicating the current location.

data Information organized for analysis.

data line A line of information for a single case recorded on a disk. It may take more than one line to enter all the data for a case.

default Instruction assumed when no other specification is stated.

delimiter A symbol or a blank used by SPSS/PC+ to detect when one specification on a command ends and another one begins (blanks between keywords, commas between arguments, slashes between subcommands, and so on).

dictionary Descriptive information about variables on the active file. Includes variable names and labels, print formats, value labels, missing-value flags, and a positional index.

directory (DOS) The type of structure used for storing files. Your personal computer uses a tree structure of directories. The primary level is the root directory. SPSS/PC+ is loaded into the directory \SPSS. This is a subdirectory under the root. See your *DOS* manual, "Using Tree-Structured Directories."

discrete variable A variable that has a limited number of values. The values can have nominal or ordinal properties. For example, INCOME with values high, medium, and low is a discrete variable. Compare *continuous variable*.

disk Oxide-coated plastic disk that stores data magnetically. See also *hard disk, diskette*.

diskette A flexible disk (5 1/4 inches in diameter for the IBM PC) that can be removed from the drive.

drive A device for reading data or commands from a disk or diskette. The floppy disk drive is usually called the A: drive. The hard disk drive is usually called the C: drive.

editor A program which allows you to enter text or data from your keyboard into the computer, to edit text or data, and to save files on disk. A text editor named REVIEW is integrated into the SPSS/PC+ system. A simpler text editor, EDLIN, is supplied with DOS.

external file A file residing on the hard disk that can be called in during an SPSS/PC+ session. SPSS/PC+ can read external files that contain data or commands. The term is generally not used in this manual to refer to system files or portable files.

file A physical organization of records, usually stored on a disk. Refer to Command Reference: Universals for files used by SPSS/PC+.

file definition Description of file characteristics. Points SPSS/PC+ to the data file and indicates the format of the file and the number of records SPSS/PC+ should read per case from fixed-format data files. See also *variable definition*.

floppy disk A flexible disk.

format The way values of a variable are represented to the computer and the way these values are displayed. The DATA LIST and FORMATS commands assign formats that SPSS/PC+ uses to write or display values. The components of a format are the variable type (string or numeric), the variable width, and the number of decimal places.

fully qualified name A file name that includes the drive, directory, and name of a file.

hard disk Rigid, oxide-coated plastic disk that stores data magnetically. Generally holds more information than a floppy.

implied decimal places The number of digits that are placed to the right of a decimal point in the stored value for a value that contains no decimal places when it is read. Implied decimal places are declared on the DATA LIST command when fixed-format data are read.

include file A file of SPSS/PC+ commands and/or data to be processed in an SPSS/PC+ session via the INCLUDE command. See also *batch processing*.

initialization The assignment of a value to a new variable before a data transformation assigns a computed value. New numeric variables are initialized with the system-missing value. New string variables are initialized with blanks. See the IF and COMPUTE commands.

inline data Data included as lines in a command file or entered from the terminal during a session. The BEGIN DATA command precedes the first data line and the END DATA command follows the last data line.

input Information entered into the computer.

input data file Contains data in almost any format stored on a disk. This file can be included within an SPSS/PC+ command file as inline data, or it can be a separate file defined on the DATA LIST command.

interactive operation Entering commands (and sometimes data) directly into SPSS/PC+, as opposed to entering a file of SPSS/PC+ commands using the INCLUDE command. See also *batch processing*.

key diskette A floppy disk inserted in the A: drive to unlock SPSS/PC+. The key diskette is required to operate SPSS/PC+.

keyword A word already defined by SPSS/PC+ to identify a command, subcommand, or specification. See Command Reference: Universals for rules for using keywords.

label A string that contains an extended description associated with a file, a variable, or a value of a variable. See also *dictionary*.

line A line of SPSS/PC+ command text or a line of data. See *command line*, *continuation line*, *data line* (or *record*).

listing file A file that contains the statistical and tabular output from SPSS/PC+ procedures, diagnostic information, and messages about the session. The default listing file is SPSS.LIS. This default can be changed with the SET command.

listwise deletion Cases that have missing values for any of the variables named are omitted from the analysis.

literal The value of a string variable or a label, enclosed within apostrophes or quotation marks. See also *string*.

log file A file that contains all commands entered and processed by SPSS/PC+. The default log file is SPSS.LOG. This default can be changed using the SET command.

logical expression Expression composed of logical operators and relations comparing two or more variables, or a compound expression. Results in assignment as true or false (or missing).

logical operator Symbol that joins two or more relations logically in a conditional transformation. The SPSS/PC+ logical operators are AND (both relations must be true) and OR (either relation can be true). The NOT logical operator reverses the outcome of the expression that immediately follows.

logical variable A numeric variable whose values are 0 (false), 1 (true), or missing.

long string variable A string variable that has more than 8 and up to 255 alphanumeric characters.

map Lists variable names from the active-file dictionary, showing the result of the IMPORT or EXPORT command.

matrix materials Summary data used as input to procedures CLUSTER, FACTOR, ONEWAY, and REGRESSION. The data can contain correlation coefficients, covariance coefficients, standard deviations, distance measures, counts, means, standard deviations, as well as other summary measures. See these commands and DATA LIST: Matrix Materials in the Command Reference for more information.

memory The internal storage area of the CPU where SPSS/PC+ operates and where calculations are performed. See also *workspace*.

missing value A code that indicates that the true value of the variable could not be obtained, or a code that you would like to have ignored during statistical calculations. These values are defined on the MISSING VALUE command and are then flagged in the dictionary of the active file. The SPSS/PC+ statistical procedures and transformation commands recognize this flag. See also *user-missing value*, *system-missing value*.

module A portion of SPSS/PC+ software. SPSS/PC+ is stored on the hard disk as a set of modules. Each module contains different parts of the whole system.

narrow format The default width, 79 characters, of SPSS/PC+ display. Narrow format is designed to fit on your screen. Compare *wide format*.

numeric expression An expression consisting of numeric variables, constants, and operators.

numeric function A function that operates on a numeric expression and returns a number or system-missing value. The expression to be transformed by a function is called an *argument* and usually consists of a variable name or a list of variable names.

numeric variable A variable whose values are numbers. Compare *string* (or *alphanumeric*) *variable*.

operating system A collection of programs that supervise the operation of the computer and handle files and jobs. SPSS/PC+ uses MSDOS.

option numbers Numbers used on the OPTIONS subcommand to request that options be in effect for a procedure.

output Results produced by a computer from specific input. SPSS/PC+ produces several types of output (statistics, tables, reports, plots, data, matrices, and so forth). The output destination is controlled by the SET command.

output file Contains data formatted to be read by a computer. Some procedures create output files containing matrix or other materials. The WRITE command produces a rectangular file to your specifications. SAVE produces a system file that is read by SPSS/PC+. EXPORT produces a portable file that can be read by SPSS/PC+ or $SPSS^X$.

pairwise deletion Cases that have valid values on both variables used in a calculation are included in the calculation. Other cases are deleted.

passive default Options that are in effect when an optional subcommand is omitted. Passive defaults are indicated by two asterisks (**) in the syntax diagrams. For example, in REGRESSION, when the optional subcommand STATISTICS is omitted, the passive defaults R, COEFF, ANOVA, ZPP, and OUTS are requested. See also *active default*.

portable file An ASCII file produced by SPSS/PC+ or SPSS^{X} EXPORT, or by SAS procedure TOSPSS. Portable files contain data and a dictionary of labels and variable formats. See also *system file*.

print format Format used for displaying and printing the values of a variable. Controlled by the DATA LIST and FORMATS commands. See also *dictionary*.

procedure Any SPSS/PC+ command that reads data from the active file. Procedures cause preceding transformations to be executed. Most SPSS/PC+ procedures produce statistical results.

prompt The string that SPSS/PC+ displays to indicate that it is ready to receive your commands. The default prompt is SPSS/PC: for the first line of a command and : for a continuation line. You can change the prompts by using the SET command.

record Line of machine-readable information.

rectangular data file A file in which each case contains one and only one value for each variable and in which each case is the same type of unit of analysis throughout the entire file. The variables are in the same order for each case.

relational operator Symbol that compares two values of a logical expression in a conditional transformation such as EQ (equal to), LT (less than), LE (less than or equal to), and so forth.

reserved keyword Keywords that cannot be used as variable names since they can appear in variable lists (e.g., TO, EQ, WITH, BY).

scientific notation Expression of a number as a fractional part and a power of ten. 350,000,000 can be expressed .3500E9 or 3.5000E8.

scratch file A file used by SPSS/PC+ to hold intermediate results from transformation and procedure commands. You cannot directly access this file.

scrolling The phenomenon of watching your results fly by on your screen when you set the page length too long.

session An SPSS/PC+ session begins with the SPSSPC command given in response to the DOS prompt, and ends with the SPSS/PC+ FINISH command.

short string variable A string variable with a width of 8 or fewer alphanumeric characters.

specifications Instructions added to a command. May include subcommands, keywords, numbers, strings, arithmetic operators, variable names, special delimiters, and spacing as needed to separate these elements. Specifications begin at least one space after the command keyword and continue for as many lines as necessary.

statistic number A number that is included on the STATISTICS subcommand to request optional statistics from a procedure.

status area The area in the upper right-hand corner of your screen that contains messages about what SPSS/PC+ is doing. The status area informs you when modules are swapped and when data is read into memory.

string One or more alphanumeric characters. Strings are specified in apostrophes or quotation marks in SPSS/PC+ commands. See also *alphanumeric variable*, *string variable*, *literal*.

string variable A variable whose values contain letters or special characters (as well as numbers). Also known as *alphanumeric variable*. Compare *numeric variable*.

subcommand Additional instructions that further specify SPSS/PC+ commands. A command may contain one or more subcommands, each with specifications to that subcommand. Defined in Command Reference: Universals.

syntax General rules for the structure of the SPSS/PC+ language.

system-missing value Value assigned by SPSS/PC+ when a value in your data is undefined according to the format type that you have specified, when a numeric field is blank for the default format type, or when a value resulting from a transformation command is undefined. See also *user-missing value*.

system file A binary file specially formatted for use by SPSS/PC+, containing both data and the dictionary that is written by the SPSS/PC+ SAVE command. See also *portable file*.

system variable Special variables assigned by SPSS/PC+ that determine the number of cases read by the system, compare the system-missing value, obtain the current date, and so forth. System variable names begin with a dollar sign. You cannot modify system variables or use them in most procedures, but you can use them in transformations.

table list The portion of an SPSS/PC+ command that requests tables using a subset of variables.

tabulation Counting, arranging, or listing values and cases in table format.

target variable A variable that contains the result of a transformation. See COMPUTE, IF, and COUNT commands in Command Reference. See also *assignment expression*.

terminator The character used to indicate the end of an SPSS/PC+ command. The default terminator is a period.

transformation Changing the values of a variable to correct coding errors, modify the coding scheme, create new variables, or construct an index.

truncation The convention that allows you to enter only the first three characters of an SPSS/PC+ keyword.

unary operator The algebraic operation of using the minus sign before a numeric variable or value to reverse the sign or using the plus sign, which does not produce a reversal.

user-missing value A value that indicates missing information. Defined on the MISSING VALUE command. See also *system-missing value*.

valid Not missing (having neither the system-missing value nor a value defined as missing).

value A numeric or alphabetical code that represents the status of a case on a variable.

variable Observable entity which can take on more than one value or characteristic.

variable definition The portion of the DATA LIST command that assigns a name to each variable you intend to analyze and provides information about the location and format of the variables in the data file.

variable name A name assigned to a variable. The name can be up to 8 characters in length. It must begin with a letter (A–Z). The remaining characters in the name can be any letter, any digit, a period (.), an underscore (_), and the symbols @ or $. See also *reserved keyword*.

wide format The display is 132 columns wide. Usually for printed output. See also *narrow format*.

window A portion of the screen used for displaying parts of a file. REVIEW allows access to two windows for displaying two files.

workspace The amount of memory available to SPSS/PC+ to perform procedures and transformations.

Communications

Contents

Communications

If you use both an IBM PC and another computer, particularly one of the many computers on which SPSS^X is available, you will probably want to transfer files between them. Both SPSS/PC+ and SPSS^X read and write portable system files (see IMPORT and EXPORT in the Command Reference). You might also want to transfer a command file (see INCLUDE in the Command Reference) or a data file. In any case, you need a reliable mechanism for sending and receiving files.

Kermit is an error-correcting file transfer program that is available on many different computers and operating systems. It was originally designed at Columbia University to transfer files between their mainframes and various microcomputers. Columbia has given SPSS Inc. permission to distribute Kermit. This documentation is adapted from that supplied by Columbia.

Kermit must exist on both machines between which you want to transfer files. Therefore this document describes fully the IBM PC version (Kermit-MS) and provides one example of a mainframe implementation—that for VAX/VMS operating systems. Mainframe sites usually have documentation available for their own version of Kermit.

Kermit is widely available on mainframe computers. SPSS Inc. distributes Kermit for IBM CMS, IBM OS, Digital VAX/VMS, Prime, and Honeywell GCOS. These versions are available at no charge to licensed SPSS^X installations upon written request from the local SPSS coordinator. If you need complete documentation for various implementations of Kermit or need a mainframe version of Kermit not available from SPSS Inc., write to:

Kermit Distribution
Columbia University Center for Computing Activities
7th Floor, Watson Laboratory
612 West 115th Street
New York, NY 10025

WHY USE KERMIT?

Everyone wants to get computers talking to one another. There are many ways to do this, and most of them are very expensive. But there is one way that is cheap and relatively easy: connect the two computers through their terminal (TTY) ports, tricking one computer (or both) into believing that the other is a terminal. Once two computers are connected in this way, cooperating programs can be run on each host to achieve the desired communication by means of a communication protocol.

Why is a protocol necessary at all? Two major problems occur when you try to connect two computers via a TTY line. First, noise on the line can garble data, sometimes in subtle ways which you may not notice until it's too late. This is especially true of the SPSS portable system file, which is not designed to be conveniently read by humans. One minor transmission error can make the file entirely unusable. Second, there is the problem of synchronization. If one computer can send data faster than the other can accept it, then some kind of flow control mechanism, either in the data stream (XON/XOFF) or in parallel to the data (modem control signals), can be employed to prevent data overruns; but no two computers can be assumed to honor the same conventions for flow control, or indeed to do it at all.

To prevent corruption of data and to synchronize communication, cooperating computers can send control information to one another at the same time that they are transferring data. This intermingling of control information with data, and the resulting actions, constitute a "protocol."

The Kermit protocol is specifically designed for transfer of files over ordinary serial communication lines. Kermit transfers data by encapsulating it in "packets" of control information. This information includes a synchronization marker, a packet number to allow detection of lost packets, a length indicator, and a "checksum" to allow verification of the data. Lost or corrupt packets are detected, and retransmission is requested. In addition, various special control packets allow cooperating Kermits to connect and disconnect from each other and to exchange various kinds of information. Very few assumptions are made about the capabilities of either computer, so the Kermit protocol can work between many different kinds of systems.

USING KERMIT

Kermit embodies a set of rules for transferring files reliably between computers. In general, one computer is a large system (a host, often with many terminals), and the other is a personal computer (PC). Host-to-host and PC-to-PC connections are also possible. The host believes that the PC is an ordinary terminal. In order for the Kermit protocol to occur, a Kermit program must be running on each end of the communication line—one on the host, one on the PC.

Talking to Two Computers at Once

Your task is just to get the two Kermits started. To do this, you have to use a single keyboard and screen to talk to two different computers and two different programs. Let's consider the specific case of transferring files between your PC and a mainframe. You are sitting at your PC, which has a serial communication port. The serial port is probably connected to a host computer using a telephone modem. The actual means of connection isn't important in this case—it also could be a direct line to the host, or some kind of switched line, etc.

Normally, when you use your PC, you are "talking" directly to it; your commands are interpreted directly by the PC's operating system or by some program that runs on the PC such as an editor, a text formatter, or SPSS/PC+. The version of Kermit on your PC is a program like any other, but it has a special ability either to interpret your commands directly, like other PC programs, or to pass everything you type through to the host. When you tell Kermit to CONNECT, it sends every character you type out the serial port, and it will put every character that comes in the serial port on the screen. This is called *virtual terminal service*—one computer acts "virtually" as though it were a terminal on another. You are now "talking" to the host, and the PC is ignoring you.

Kermit has a prompt which it types on the left margin to indicate that it is ready for you to type a command. Kermit's prompt is normally Kermit-xx>. The xx identifies the implementation of Kermit: the Kermit that runs on the DEC-20 is called *Kermit-20*, and its prompt is Kermit-20>; the Kermit on the IBM PC is Kermit-MS, and so forth. If you become confused about who you are talking to, the prompt should provide a clue. In addition, most Kermits print an informative message like

```
[Connecting to remote host, type CTRL-]C to return]
```

when you CONNECT to the remote host, and type another message like

```
[Connection closed, back at PC]
```

when you disconnect from the remote host and return to the PC.

Having connected to the host, you need a way to get back to the PC. This is accomplished by an escape sequence. As Kermit passes your characters through to the host, it checks each one to see if it's a special predefined escape character. When the PC sees this character, it stops ignoring you—you are once again "talking" to the PC, not the host. The escape character is chosen to be one that you will not need to type while talking to the host, and one that is hard to type by accident. It's usually a control character, such as Control-], which is accomplished by holding down the key marked CTRL and typing the indicated character—in this case, a right bracket (]). The CTRL key works just like a SHIFT key. Control characters are written either as CTRL-A or ∧A, where A is the character to be typed while holding down CTRL.

Transferring a File

To transfer a file, you must first get the attention of the PC's operating system. This is normally done by starting the PC. Once you are at DOS command level on your PC, run Kermit. Tell Kermit to CONNECT you to the host. Dial the host system. If you are using an external modem, place the receiver in the acoustic coupler (if required) and hit the data switch. If you are using an internal modem board, type in its commands to dial the remote host. The remote host acknowledges. Now you are talking to the host—at this point you must log on and then run Kermit on the host.

Now you have a Kermit on each end of the wire. The next step is to tell each Kermit what to do. Suppose you want to transfer a file from the host to your PC; you would first tell the host Kermit to SEND the file, then "escape" back to the PC Kermit and tell it to receive the file. The transfer begins—all you have to do now is sit back and watch. The PC Kermit will continuously show packet and retry counts on your screen and will notify you when the transfer is complete.

The desired file is now on your PC's disk in the current directory. The Kermit protocol has ensured that the file arrived correctly and completely. Now you must clean up after yourself: CONNECT back to the host, exit from Kermit on the host, log off the host, "escape" back to PC Kermit, hang up the phone or reset the modem, and exit from Kermit-MS.

Kermit works best with "printable" files—files composed only of letters, digits, punctuation marks, carriage returns, tabs, and so forth—since these can be represented on almost any kind of computer. Kermit is also able to transfer some binary files—files such as executable programs—composed of arbitrary bit patterns, but binary files normally are meaningful only to the kind of computer on which they are generated. This document does not describe the transfer of binary files.

Basic Kermit Commands

These are generic descriptions of the most basic Kermit commands. Detailed descriptions for specific implementations of Kermit will come later. In these descriptions, *local* refers to the system that you are using directly, *remote* refers to the system to which you are CONNECTed via Kermit. Commands may take one or more operands on the same line. Press the return key at the end of each command.

SEND Send the specified file or file group from this Kermit to the other. The name of each file is passed to the other Kermit in a special control packet so it can be stored there with the same name. A file group is usually specified by including "wildcard" characters like "*" in the file specification. Examples:

```
send gss82.def
send *.dat
```

Some implementations of Kermit do not support transfer of file groups; these versions require a separate SEND command for each file to be transferred. Kermit-MS does permit the specification of file groups.

RECEIVE Receive a file or file group from the other Kermit. If an incoming file name is not legal, Kermit attempts to transform it to a similar legal name, most likely by deleting illegal or excessive characters. The name thus formed cannot be guaranteed to be unique, in which case previously existing files can be overwritten. Some versions of Kermit attempt to prevent this by warning you of filename collisions and taking, or allowing for, evasive action.

CONNECT Make a "virtual terminal" connection to the remote system. On a PC, this usually means to send all keyboard input out the serial port and display all input from the serial port on the screen. To "escape" from a virtual terminal connection, type Kermit's escape character (for example, CTRL-]), followed by the letter "C" for "Close Connection."

SET Establish various nonstandard settings, such as CONNECT escape character, file characteristics, communication line number, speed, parity, or flow control.

SHOW Display the values of SET options.

HELP Type a summary of Kermit commands and what they do.

EXIT Exit from Kermit back to the system's operating system.

? Typed anywhere within a Kermit command: List the commands, options or operands that are possible at this point. This command may or may not require a carriage return, depending on the system's operating system.

An Example: PC to HOST

In this example, you are sitting at an IBM PC, which is connected through its serial port to a VAX VMS host computer. The PC is the local system, the VAX is the remote host.

You have started up your PC and have the Kermit program on your disk. Begin by running Kermit on the PC. Use Kermit's CONNECT command to connect to the VAX. The PC will now behave as a terminal to the VAX. In fact, the PC emulates the Heath-19, which most closely resembles the popular DEC VT52 terminal, so it is desirable to tell the VAX that your terminal is a VT52. Log in to the VAX and run Kermit there.

```
C>kermit                    Run Kermit on the PC.

IBM-PC Kermit-MS V2.27
TYPE ? for help
Kermit-MS>                  This is the Kermit prompt for the PC

Kermit-MS>connect           Connect to the VAX
[Connecting to host.  Type Control -]C to return to PC]

                            Dial into the remote host.
                            You are now connected to the VAX.
VAX-11/780                  The system prints its herald.
Username:                   Log in using normal login method.
```

Now you are logged in to the VAX.

```
$ kermit                    Run Kermit on the VAX.
Kermit-32>                  This is Kermit-32's prompt.
```

You are now ready to transfer files between the two machines.

```
Kermit-32>send *.inc          Send all INCLUDE files.
/\]c                          Now return back to the PC by
                              typing the escape sequence,
                              /\]C (CRTL-] followed by C).

[Back at PC.]

Kermit-MS>receive             Tell the PC that files are coming.
```

If you take more than about 5 seconds to get back to Kermit-MS and issue the RECEIVE command, the first packets from Kermit-32 may arrive prematurely and appear as garbage on your screen, but no harm will be done because the packet will be retransmitted by the VAX until the PC acknowledges that it has received it correctly.

Once the connection is established, the PC shows you what is happening. It first clears the screen and waits for incoming packets; as packets arrive, the current file name and packet number will be continuously displayed on the screen. When the PC's Kermit-MS> prompt returns to your screen, the entire transfer is done.

When the transfer is complete, you must CONNECT back to the VAX host, EXIT from Kermit there, log off the VAX, and return to the PC.

```
Kermit-MS>connect             Get back to the VAX.
[Connecting to host. Type CTRL-]C to return to PC.]

Kermit-32>                    Here we are.
Kermit-32>exit                Get out of Kermit-32.
$ log                         Log off the VAX.

LARSON  logged out at ...     The system messages appear.
                              Hang up or reset the modem.
/\]c                          Now 'escape' back to the PC,

[Back at PC.]
Kermit-MS>exit                and exit from the PC's Kermit.

C>
```

The files you transferred should now be on your PC disk.

Sending to the Remote Host. To send files from the PC to the VAX, follow a similar procedure. First follow the instructions in the previous section to access the VAX through the PC and start the remote Kermit. Then in response to Kermit's Kermit-32> prompt, type RECEIVE rather than SEND. Now escape back to the PC and use the SEND command specifying a file name or group to send the local PC files to VAX host. The PC will show you the progress of the transmission on its screen.

When the Kermit-MS> prompt indicates that the transmission is complete, follow the procedure shown above to log off the VAX host, except that you may first wish to confirm that the files have been stored correctly in your directory on the VAX.

When Things Go Wrong

Connecting between computers can be complicated, and many things can go wrong. Before you can transfer files at all, you must first establish terminal communication. If the CONNECT command doesn't seem to work, try the following:

1 Make sure the required physical connections have been made and have not worked loose. If you have an external modem, check the carrier light.

2 If you have more than one connector, make sure you are using the right one. Kermit's SET PORT command can be used to specify either COM1 or COM2.

3 Make sure the baud rate is correct. It defaults to 1200 in Kermit-MS.

4 Make sure the other communication line parameters, such as parity, are set correctly.

If all settings and connections appear to be correct, and communication still does not take place, the problem may lie in the modem. Kermit normally expects to have full control of the communication port. However, it is sometimes the case that some communications equipment controls the line between the two computers on either end. Examples include modems (particularly "smart" modems), port contention or selection units, multiplexers, local networks, and wide-area networks. Such equipment can impose *parity* upon the communication line which will cause the packet checksums to appear incorrect to the receiver. If terminal connection works but file transfer does not, parity is the most likely culprit. Find out what parity is being used and inform both Kermits (with the SET command) so that they can compose and interpret the checksums correctly.

Communications equipment can also interpret certain characters in the data stream as commands rather than passing them along to the other side. For instance, you might find your "smart" modem suddenly disconnecting you and placing a call to somewhere. The only way to work around such problems is to put the device into transparent mode.

There are various ways in which Kermit can become stuck, especially when packets get "lost." Many hosts are capable of generating timeout interrupts when some process doesn't complete quickly enough and thus can usually resend or "NAK" (negatively acknowledge) lost packets. Nevertheless, if a transfer seems to be stuck, you can type RETURN on the keyboard of most PCs to wake up Kermit and have it retransmit the last packet. The IBM VM/CMS Kermit cannot time out its "virtual console" (i.e. the user's terminal), so when using Kermit from a PC to an IBM VM/CMS remote host, occasional manual wakeups may be necessary.

Here are a few symptoms and what to do about them.

The Transfer is Stuck. The following sections discuss various reasons why a transfer in progress could become stuck. Before examining these, first make sure that you really have a Kermit on the other end of the line and that you have issued the appropriate command: SEND or RECEIVE. Remember that you must connect back between each transfer and issue a new SEND or RECEIVE command.

The PC is Hung. The PC itself sometimes becomes hung for reasons beyond Kermit's control, such as power fluctuations. If the PC's screen has not been updated for a long time, then the PC may be hung. Try these steps (in the following order):

1. Check the connection. Make sure no connectors have worked loose from their sockets. If you're using a modem, make sure you still have a carrier signal. Reestablish your connection if you have to.
2. Press RETURN to wake the PC up. This should clear up any protocol deadlock.
3. If the problem was not a deadlock, restart the PC and then restart Kermit, CONNECT back to the host, get back to your job or log in again, and restart the transfer. You may have to stop and restart Kermit on the remote host.

The Remote Host Crashed. If your local system is working but the transfer is hung, maybe the remote host crashed. "Escape" back to local Kermit command level and issue the CONNECT command so that you can see what happened. If the remote system has crashed, then you will have to wait for it to come back and restart the file that was being transferred at the time.

The Disk is Full. If your disk fills up, Kermit-MS informs you and then terminates the transfer. You can continue the transfer by repeating the whole procedure after inserting a fresh diskette or cleaning up your disk.

Message Interference. You may find that file transfers fail only occasionally and unpredictably. One explanation could be that terminal messages are being mixed with your file packet data. These could include system broadcast messages and messages from other users. Most Kermits attempt to disable intrusive messages automatically, but not all can be guaranteed to do so. You may be required to turn off such messages on the remote host.

Host Errors. Various error conditions can occur on the remote host that could affect file transmission. Whenever any such error occurs, the remote Kermit normally attempts to send an informative error message to the local one and then stops transmission, putting you back at Kermit-MS command level.

File is Garbage. There are certain conditions under which Kermit can believe it transferred a file correctly when in fact, it did not. One way this can happen is if errors occurred during transmission that did not affect the byte count or checksum. For instance, if in a single packet one character's bit n was received as a zero when it should have been a one, and in another character, bit n became a one when it should have been a zero, then the checksum would be the same as if no error had occurred. The only remedies here are to edit the file after it is sent to fix the bad characters or send the file again.

A more likely cause of garbage files has to do with the tricky business of the bytes. Most computers store files as sequences of 8-bit bytes, but there are some exceptions. To complicate matters further, different systems handle telecommunication differently: most systems transmit ASCII (a 7-bit code) in 8-bit bytes, but some use the 8th bit for data—thus allowing the transmission of binary files in 8-bit bytes. Others insist upon using the 8th bit as a parity check. These complications come into play only in attempts to transfer binary files; when transferring ordinary text files, there's never a problem about byte sizes.

Kermit Servers

The previous discussion applies to Version 1 of the Kermit protocol. Version 2 of the protocol has been enhanced to include the concept of the *Kermit server.* A Kermit server is a Kermit program that does not interact directly with you but only with another Kermit. You do not type commands to a Kermit server, you merely start it at one end of the connection and then type all further commands at the other end.

Not all implementations of Kermit have servers. Some of the ones that do are VAX/VMS and Honeywell GCOS. Some PC Kermit implementations can also act as servers, notably MS-DOS. You must still connect to that computer to log in and start the Kermit server, but you do not have to tell one side to SEND and the other to RECEIVE, nor must you connect back to the remote host to exit Kermit and log off. Using the server, you can do as many send and receive operations as you like without ever having to connect back to the remote host.

A Kermit server acts much as ordinary Kermit does after you give it a RECEIVE command — it waits for a message from the other Kermit, but in this case the message is a command saying what to do, normally to send or to receive a file or group of files. After escaping back to the local system, you can give as many SEND and GET commands as you like, and when you are finished transferring files, you can give the BYE command, which sends a message to the remote Kermit server to log itself off. You do not have to connect back to the remote host and clean up. However, if you want to connect back to the host, you can use the FINISH command instead of BYE to shut down the Kermit server on the remote host without logging it off, allowing you to CONNECT back to your job there.

Basic Server Commands

These are descriptions of the most basic server commands. Detailed descriptions will come later. In this section, *local* refers to the system that you are using directly, and *remote* refers to the system to which you are CONNECTed via Kermit.

SEND Send the specified file or file group from the local host to the remote system.

GET Ask the remote host to send a file or file group. Example:

```
get *.c
```

This command is exactly equivalent to typing "send *.c" at the remote host followed by "receive" on the local system. Note that the local Kermit does not attempt to validate the filespec. If the server cannot parse it, or cannot access the specified file(s), it will send back an appropriate error message.

REMOTE Ask the server to perform the specified command, and send the results to your screen. Not all servers are capable of performing REMOTE commands; those that can most commonly provide REMOTE DIRECTORY, REMOTE DELETE, REMOTE SPACE, and similar file management services.

BYE Shut down the remote server and exit from Kermit. This will cause the job at the remote end to log itself off. You need not connect back and clean up unless you get an error message in response to this command (for example, if your logged-off disk quota is exceeded on the remote host).

FINISH Shut down the server without having it log itself off, and don't exit from Kermit. A subsequent CONNECT command will put you back at your job on the remote host, at system command level.

An Example: PC to Host Using a Server

In this example, you are sitting at an IBM PC, which is connected through a serial port to a VAX host computer. The PC is the local system, and the VAX is the remote host.

```
C>kermit                        Run Kermit on the PC.

IBM-PC Kermit-MS V2.27
TYPE ? for help
Kermit-MS>                      This is the Kermit prompt for the PC

Kermit-MS>connect               Connect to the VAX
[Connecting to host.  Type CTRL-]C to return to PC]

                                Dial into the remote host.
                                You are now connected to the VAX.
VAX-11/780                      The system prints its herald.
Username:                       Log in using normal login method.
```

Now you are logged in to the VAX.

```
$ kermit                        Run Kermit on the VAX.
Kermit-32>                      This is Kermit-32's prompt.
```

You are now ready to transfer files between the two machines.

```
Kermit-32>server                Tell Kermit-32 to be a server

Kermit Server running on VAX/VMS.  Please type your escape sequence
to return to your local machine.  Shut down the Server by typing
the Kermit BYE command on your local machine.

^]c                             Now return to the PC by
                                typing the escape sequence,
                                ^]C (CTRL-] followed by C).
[Back at PC.]

Kermit-MS>get *.spssx           Get all my VAX/VMS SPSS-X portable files.
Kermit-MS>exit                  Exit from Kermit back to MS-DOS
```

Here you can do some work on the PC, edit files, etc.

```
C>Kermit                        Run Kermit-MS
Kermit-MS>send file.inc         Send an SPSS/PC+ INCLUDE file.
Kermit-MS>bye                   Shut down the Kermit server and log off
                                the mainframe

C>                              Back to MS-DOS automatically.
```

KERMIT IMPLEMENTATIONS

Kermit has been written for a wide variety of systems, including mainframes, minis, and microcomputers. Kermit is not written in a portable language; rather, each implementation is written in the language best suited for the particular machine. The specification, given in the Kermit Protocol Manual (available from Columbia University), is quite general and allows implementation on almost any machine.

Command Interface

Most implementations (the PRIMOS version is one exception) attempt to imitate the command parsing style of the DEC-20, which is roughly as follows: In response to the Kermit prompt you can type a keyword, such as SEND, RECEIVE, or EXIT, possibly followed by additional keywords or operands, each of which is called a *field.* You can abbreviate keywords to any length that makes them distinguishable from any other keyword valid for that field. You can terminate a field with ESC and receive a special short prompt, called a *guide word,* in parentheses, telling you what the next field should be. You can type a question mark at any time to get information about what's expected or valid at that point. The ESC and ? features work best on full duplex systems (all but the IBM mainframe, so far), where the program can "wake up" immediately and perform the required function. On half duplex or record-oriented systems, the ESC feature is not available, and the ? must be followed by a carriage return.

The following example demonstrates the parsing style of Kermit-MS. First a question mark is used to obtain items that can be set. Then the question mark is used to obtain settings for HANDSHAKE.

```
Kermit-MS>SET ?

BAUD rate
BELL
BLOCK-CHECK-TYPE
DEBUG
DEFAULT-DISK
DESTINATION
End-of-line character
EOF CTRL-Z or NOCTRL-Z
ESCAPE character change
FLOW-CONTROL
HANDSHAKE
HEATH-19
KEY
LOCAL-echo echoing (half duplex)
PARITY type
PORT for connection
PROMPT
RECEIVE parameter
REMOTE on/off
SEND parameter
TAKE-ECHO
TIMER
WARNING

Kermit-MS>SET HANDSHAKE ?
BELL  CR  ESC  LF  NONE  XOFF  XON
Kermit-MS>SET HANDSHAKE NONE
```

MS-DOS Kermit

Kermit-MS is always in control of the screen—it is always *local.* Thus it keeps the screen updated with the file name and the packet number, whether sending or receiving. Kermit-MS can't time out, so if the transmission gets stuck (the screen fails to change for a while), type a carriage return to restart it. If the remote host is capable of timeouts, this action should not normally be necessary, but PC-to-IBM mainframe transfers can require this kind of manual intervention.

When acting as a virtual terminal, Kermit-MS emulates the Heath H19 (or Zenith Data Systems Z19). The Kermit-MS command interpreter is modeled after the DEC-20's, allowing keyword abbreviation (and completion and guide words if you type ESC) and "?" help at any point in the command. Kermit-MS is interrupt driven; an interrupt is generated whenever data comes into the serial port. Kermit stores the characters in a buffer and attends to them later—this way data are not normally lost if the host is too fast for the PC.

Kermit-MS Commands

The following sections describe Kermit-MS commands for sending and receiving files.

CLOSE. The CLOSE command closes the log file and stops logging the remote session.

CONNECT. The CONNECT command makes a terminal connection to a remote system, and displays a line at the bottom of the screen showing the escape character, port, baud rate, parity, local echo, and what to type if you need help, as in:

```
Esc chr: ^], Port: 1, Speed: 1200, Parity: Mark,
Echo: Lcl, Type ^]? for Help
```

Type an escape sequence to get back to the PC. The escape character is CTRL-] by default (you can use the SET ESCAPE command to define a different escape character). You can send a BREAK to the host by typing CTRL-BREAK.

EXIT. Exit from Kermit-MS back to DOS. QUIT is a synonym.

HELP. HELP prints a list of all available commands.

LOCAL. The LOCAL keyword is a prefix for a number of commands. It indicates that the specified command is to be executed on the local MS-DOS system. The LOCAL prefix may be omitted. Local command options are described below.

LOG. The LOG command specifies that all characters that appear on your screen while connected to a remote system will be recorded in the file you name. This allows you to "capture" files from a remote system that does not have Kermit, as well as to record remote command typescripts. The log is closed when you issue either the EXIT or CLOSE commands.

RECEIVE. Receive a file or file group from the other system. Kermit simply waits for the file to arrive; this command is not to be used when talking to a Kermit server. Use the GET command for that. You should already have issued a SEND command to the remote Kermit and escaped back to Kermit-MS before issuing the RECEIVE command.

Incoming filenames that are illegal under DOS are transformed to legal names by deleting illegal or excessive characters, and so on. Optionally, you can provide a filename for the file being received. The incoming file is stored in the default directory. If the name already exists, and SET FILE-WARNING is ON, Kermit-MS changes the name to avoid conflicts.

If a file begins to arrive that you don't really want, a CTRL-X should cancel it. However, the remote host may not acknowledge it. Similarly, if you are sending a file group, CTRL-Z can be used to cancel the group. If the transfer is stuck, hitting the RETURN key a number of times causes the other Kermit to retransmit the current packet, until the retransmission limit is reached, and you will be back at the Kermit-MS level. To return to Kermit-MS without sending any kind of notification to the remote system, type CTRL-C.

Note: After file transfer is complete, you should inspect your file for stray characters at the end of the file. These should be edited out prior to using the file in SPSS/PC+.

SEND. Specify a file or file group on the SEND command. The file specification can contain a device designator, like "A:" and the wildcard characters "*" and/or "=." The wildcard character "*" matches a whole field or the trailing characters of a field. For example, *.FOR sends all files with extension .FOR. Alternatively, the wildcard "=" matches one or zero characters. For example, A==.TXT will match the files ABC.TXT and A.TXT, but not ABCD.TXT. The "=" wildcard is functionally equivalent to the MS-DOS "?" and is used in place of it because Kermit interprets the "?" as a request for help. The current release of Kermit-MS, however, does not allow pathnames in the SEND command file specification.

Files are sent with their filename. If the name is illegal on the remote host and the Kermit on the remote host can't convert incoming filenames, use the MS-DOS RENAME command prior to sending it.

If you notice a file being sent which you do not really want to send, you can cancel the operation by typing CTRL-X or CTRL-Z. CTRL-X causes the current file in a group to be skipped and CTRL-Z cancels the sending of the entire group. If a transfer is stuck, hitting the RETURN key a number of times causes Kermit-MS to keep retransmitting the same packet until you reach the maximum number of retrys, returning you to the Kermit-MS level. Type CTRL-C to cancel the transfer and return to Kermit-MS.

SERVER. Kermit-MS can act as a Kermit server, providing file transfer for users coming through one of the communications ports. The current version of Kermit-MS can send files, receive files, and terminate, giving control to the console.

SET. SET sets the value of an option. SET options are described below.

SHOW. Display macro definitions and key redefinitions. The SHOW MACROS command displays the definitions of all currently defined macros. The SHOW KEY command allows you to determine the scan code produced by pressing a given key so that you can construct a SET KEY command to redefine the key. If the key already has a redefinition in effect, that too will be displayed. In the following example, SHOW KEY determines the scan code for the Backspace key.

```
Kermit-MS>show key
Press a key: [BACKSPACE is pressed]
  Scan code: 14
  Definition: \177
```

STATUS. Display the values of options that can be modified via the SET command. This is equivalent to the SHOW ALL command available on other Kermits.

TAKE. The TAKE command executes Kermit commands from a specified file. If no path is specified, the value of the PATH variable is used. If PATH has no value, then the current disk and directory are searched. The command file can include TAKE commands, but cannot include characters to be sent to a remote host during terminal emulation (i.e. after a CONNECT command). For example, you can store a SET command within a TAKE file called CTRLZ.KER created with an editor, as in

```
set eof ctrl-z
```

which instructs Kermit to append a CTRL-Z to the end of incoming files. A CTRL-Z is required at the end of all data and command files in SPSS/PC+. Within Kermit-MS, type

```
take ctrlz.ker
```

to execute the file.

SET Options for Kermit-MS

The SET command sets options. The following items can be set. Note that there is no longer a "set ibm" command in Kermit-MS, as there was in Kermit-PC. IBM mainframe communication parameters can be selected by invoking the command macro, do ibm. For further discussion, see Command Macros.

BAUD n Set terminal communications port speed to 300, 1200 (the default), 1800, 2400, 4800, 9600, or some other common baud rate.

BELL ON or OFF; the default is ON. If ON, the bell rings when the transfer is complete.

BLOCK-CHECK-TYPE 1, 2, or 3. Kermit normally uses a 1-character block check, or "checksum," on each packet. The sender of the packet computes the block check based on the other characters in the packet, and the receiver recomputes it in the same way. If these quantities agree, the packet is accepted and transmission proceeds. If they disagree, the packet is rejected and retransmission is requested. You can decrease the probability that an error can slip through by using the SET BLOCK-CHECK-TYPE command to select more rigorous block check methods. Specify 2 for a 2-character, 12-bit checksum and 3 for a 3-character, 16-bit Cyclic Redundancy Check.

Note that 2- and 3-character block checks are not available in all versions of Kermit; if the other Kermit is not capable of performing the higher-precision block checks, the transfer will automatically use the standard single-character method.

DEBUG Set debug mode ON or OFF (normally should be OFF). If ON, outgoing packets will also be displayed on your screen.

DEFAULT-DISK Specify the default disk drive for file transfer, directory listings, etc. Equivalent to the DOS command for changing disks.

DESTINATION Specify the device for incoming files, either DISK, PRINTER, or SCREEN. For example, SCREEN will send output normally destined for the disk to the screen.

END-OF-LINE n Change the character used at the end of outgoing packets to something other than the default of carriage return (ASCII 13 decimal). It must be a number between 0 and 31 (decimal value of ASCII character). For example, to set it to linefeed, specify set end-of-line 10. Normally, you will not have to change the default.

EOF Controls how the end of file is handled. CTRL-Z appends a Control-Z character to the end of an incoming file unless it already ends with a Control-Z. NOCTRL-Z, which is the default, stores incoming files exactly as they are received, and similarly makes no changes to MS-DOS files which are sent.

If this option does not work on your system, you can copy the file to another with the COPY command and append a CTRL-Z to the end of it by using the /a option on the destination file, as in

```
copy oldmain.inc newmain.inc/a
```

ESCAPE c Change the escape character that gets Kermit-MS's attention when CONNECTed to a remote host. The default is CTRL-].

FLOW-CONTROL Specify the full duplex flow control to be done on the currently selected port. The options are XON/XOFF and NONE. If set to XON/XOFF, HANDSHAKE is automatically set to OFF.

HANDSHAKE Specify the desired half-duplex handshaking option, either BELL, CR, LF, NONE, XOFF, or XON. If HANDSHAKE is set to anything other than NONE, FLOW-CONTROL is automatically set to OFF.

HEATH19 ON or OFF; the default is ON. When CONNECTed to a remote, controls whether the PC emulates a Heath-19 or runs in *native mode*. This should be ON unless the host supports the IBM PC as a terminal in "native mode." The Heath-19 most closely resembles a VT52 terminal. However, Heath-19 codes are a superset of the VT52, which does not support cursor keys or function keys on IBM systems.

INCOMPLETE If a file is incompletely received (the host crashes, the link is broken, etc.), DISCARD discards the file and is the default. To save it, specify KEEP.

KEY

Specifies that when the designated key is pressed during terminal emulation, the associated character string is sent. The key-specifier is one of the keywords F1,F2,..., or SCAN followed by a scan code. The scan code of the key can be ascertained with the Kermit-MS SHOW KEY command, or from your system reference manual. Systems that have a BACKSPACE key also include BACKSPACE as a keyword. In the following example, SHOW KEY determines the scan code and the current octal code definition for the backspace key. SET KEY redefines the key to send ASCII octal code '010.

```
Kermit-MS>show key
Press a key:  Backspace key is pressed
  Scan code: 14
  Definition: \177
Kermit-MS>set key scan 14
Definition string: \010
```

LOCAL-ECHO

ON or OFF; the default is OFF. ON sets half-duplex; OFF sets full-duplex. It should be ON when CONNECTing to an IBM mainframe and OFF for CONNECTing to most other systems, including the DEC-10, DEC-20, UNIX, etc. DO IBM selects LOCAL-ECHO ON as well as other required options for IBM remote hosts (see Command Macros).

PARITY

EVEN, ODD, MARK, SPACE, or NONE. Select parity for outgoing characters, during both CONNECT and file transfer, to match the requirement of the remote host. NONE is the default.

PORT

Define a communications port (COM1 or COM2). SET PORT 1 selects COM1, and SET PORT 2 selects COM2. The default is COM1.

RECEIVE

RECEIVE settings allow you to specify to the other Kermit what the packets it sends should look like, or to inform Kermit-MS what to expect.

SET RECEIVE PACKET-LENGTH n asks the remote Kermit to use the specified maximum length for packets it sends to Kermit-MS. The range is 20 to 94 (decimal), where 94 is the default.

SET RECEIVE PADCHAR n asks the remote Kermit to use the character you supply for interpacket padding. The default value is 0 (ASCII NUL). The value you specify can be a decimal number in the range 0 to 31, or 127.

SET RECEIVE PADDING n asks the remote Kermit to insert the given number of padding characters before each packet it sends. The default value is 0, but can be set to a decimal value between 0 and 99. Normally no padding is necessary.

SET RECEIVE QUOTE n requests that the specified character be used for quoting control characters. The default value is 35 (ASCII pound sign). The value you set must be in the range 33-126, where n is the decimal ASCII value for the quoting character you select.

SET RECEIVE START-OF-PACKET n asks the remote Kermit to mark the beginning of packets with something other than the default value of 1 (ASCII CTRL-A). The value you set must be between 0 and 31 (decimal).

SET RECEIVE TIMEOUT n defines the number of seconds the remote Kermit should wait for a packet before asking for retransmission. The default value is 13 seconds.

REMOTE

ON or OFF. SET REMOTE ON suppresses the file transfer display screen so that the display will not interfere with the file transfer itself.

SEND

SEND settings allow you to specify parameters for outgoing packets.

SET SEND PACKET-LENGTH n requests that Kermit-MS use the specified maximum length for outbound packets. Normally, Kermit-MS uses whatever length the remote Kermit requests. The default value is 80, but can be set to a decimal value between 20 and 94.

SET SEND PADCHAR n requests that the specified character be used for interpacket padding. Some hosts may require a padding character (normally NUL or DEL) before a packet. You can specify a number between 0 and 31, or 127 (decimal).

SET SEND PADDING n defines the number of padding characters to use between packets. The default value is 0, but can be set to any decimal value between 0 and 99. Usually no padding will be necessary.

SET SEND QUOTE n requests that the specified character be used for prefixing (quoting) control characters and other prefix characters. The default value is 35 (ASCII pound sign). The value you set must be in the range 33-126, where n is the decimal ASCII value for the quoting character you select.

SET SEND START-OF-PACKET n marks the beginning of packets with a control character. The default value is 1 (ASCII CTRL-A). If CTRL-A causes problems for you communications hardware or software, you can use the SET SEND START-OF-PACKET command to select a different control character to mark the start of a packet. The value you select must be a decimal value in the range 0 to 31. Also, the remote host must have been given the corresponding SET RECEIVE START-OF-PACKET command (providing it has such a command).

SET SEND TIMEOUT n defines Kermit-MS's normal timeout interval. This command is effective only if TIMER is set to ON; it is normally OFF so that the other Kermit can control timeouts. The default value is 8 seconds.

TAKE-ECHO ON or OFF. Specifies whether Kermit-MS should display command processing on screen during TAKE operations on MSKERMIT.INI or other Kermit-MS command files, or during evaluation of macro definitions. This setting can be useful for finding errors in command files.

TIMER ON or OFF. Enable or disable timer that is used during file transfer to break the deadlock that occurs when an expected packet does not arrive. By default the timer is OFF, because Kermit-MS is usually used in conjunction with a mainframe that does its own timeouts. During file transfer, it is sufficient for one side to do the timing out and the mainframe is usually better equipped to adjust timeout intervals based on system load or other conditions. The timer should be set ON if you are communicating with systems which cannot do timeouts, such as IBM VM/CMS or IBM TSO Kermit.

WARNING ON or OFF; the default is OFF. If ON, Kermit will warn you when an incoming file has the same name as an existing file, and rename the incoming file to avoid destroying an existing one. If OFF, the pre-existing file is destroyed, even if the incoming file does not arrive completely.

SERVER Commands for Kermit-MS

The available Kermit-MS SERVER commands are:

BYE If you are connected to a remote host which is a server, BYE shuts down the host, logs you off the host, and exits Kermit-MS.

FINISH If you are connected to a remote server, FINISH exits the server but the server does not log off. When you CONNECT again you are still logged on to the remote host and at the system command level.

GET The GET command requests a remote Kermit server to send the specified file or file group. You can use the GET command to specify a different name for storing the incoming file. If you type GET without a file name, you will be prompted separately for the remote filespec and the local filespec, as in:

```
Kermit-MS>get
 Remote Source File:  bank.sps
 Local Destination File:  bank2.sps
```

LOGOUT The LOGOUT command is identical to the BYE command except you remain at the Kermit-MS prompt level, rather than exit to DOS.

REMOTE The keyword REMOTE is the prefix for remote file management commands. It indicates that the command is to be performed by the remote Kermit, which must be running as a server. For example, to erase a file called "bank.sps;2" on a remote VAX, in Kermit-MS type:

```
REMOTE DELETE bank.sps;2
```

It should be noted that not all Kermit servers can execute all these commands, and some Kermit servers may be able to perform functions for which Kermit-MS does not yet have the corresponding commands. If the server cannot execute a command, it will send back a message that the command is unknown.

REMOTE CWD changes the working directory on the remote host to the one you specify.

REMOTE DELETE scratches the specified file or files on the remote system.

REMOTE DIRECTORY provides a directory listing of files in the specified directory on the remote system. If no files are specified, all the files in the current working directory will be listed.

REMOTE HELP tells what server functions are available on the remote system.

REMOTE HOST sends the specified command to the remote system's command processor for execution.

REMOTE SPACE provides a summary of disk usage in the specified area on the remote host. If none is specified, the current area will be summarized.

REMOTE TYPE displays the contents of the specified remote file or files on the screen.

SEND Transfer files from the current PC directory to the remote system. SEND takes a filename as the argument, or you can send groups of files using PC wildcard conventions.

LOCAL Options for Kermit-MS

The LOCAL command is a prefix for local file management commands. You can issue the following commands from within Kermit-MS:

DELETE Deletes the specified file or files.

DIRECTORY Lists the names, sizes and creation dates of files that match the given specification.

SPACE Performs the MS-DOS CHKDSK function by running the CHKDSK program from the current path, or default disk.

RUN Runs the specified file, which must be in .EXE or .COM format, from the specified path or according to the value of the PATH variable if no path was included in the filespec. The RUN command requires MS-DOS 2.0 or higher.

PUSH Transfers you to MS-DOS level without terminating your Kermit-MS session. Here you can issue normal MS-DOS commands. When you return to Kermit-MS (by typing the MS-DOS EXIT command), you will find Kermit-MS as you left it, with all settings intact. The PUSH command only works in MS-DOS 2.0 or higher.

Command Macros

Kermit-MS provides a facility for combining commands into macros. Command macro definitions can be included in your MSKERMIT.INI file (see MSKERMIT.INI File), executed with the TAKE command, or typed interactively, and can be invoked with the DO command. The following commands are available for constructing and executing macros:

DEFINE Kermit-MS command macros are constructed with the DEFINE command. The syntax is

```
DEFINE macro-name [command [,command [, ...]]]
```

Any Kermit-MS commands can be included. Each command must be separated by commas. For example, if you use Kermit to communicate with several different kinds of systems, you can set up a macro for each, as in

```
define vax set parity none, set local off
define prime set parity mark, set key backspace,\010
```

You can then type DO VAX, DO PRIME, etc. to set all the desired parameters with a single command.

DO The DO command invokes Kermit-MS macros. For example, Kermit-MS comes with a predefined macro which allows communication with an IBM host. To invoke it, type

```
do ibm
```

The IBM macro is already defined in Kermit-MS. Its values are: parity mark, handshake xon, local-echo on, timer on. You can delete or replace this definition by adding a new definition, as in

```
define ibm parity even, handshake cr, local-echo on, timer on
```

or

```
define ibm
```

(In Kermit-PC, the predecessor to Kermit-MS, the function of DO IBM was accomplished with the command SET IBM ON).

MSKERMIT.INI File

Upon initial startup, Kermit-MS executes any commands found in the MSKERMIT.INI file in the current path. This file may contain macro definitions, communications settings, and any other Kermit commands. In the following MSKERMIT.INI file, SET KEY specifies that whenever you press the backspace key (scan code 14), octal code '010 is sent to the remote Kermit. This allows you to backspace over the previously sent character. SET BAUD sets the baud rate to 1200. DO IBM sets the correct communications protocol for a mainframe IBM. CONNECT establishes the link between the mainframe host. At this point, you can dial the number required to activate a port on the IBM host, and log on in the normal manner.

```
set key scan 14
\010
set baud 1200
do ibm
connect
```

Limitations on File Size

Kermit-MS has a limit of 32,767 packets in a single transfer. A packet contains approximately 75 characters of data. Thus, the maximum size of a file which can be transferred is 2.5MB. If you hit this limitation while sending multiple files by using a wildcard, only the file reaching this limit is truncated.

Specifying a Destination for Files

Kermit-MS only receives files onto the default disk and directory. It does not support pathnames. Therefore, before running Kermit to get files from a remote host, you must select the disk you want to put the files on, using the *A:*, *B:*, or *C:* DOS command. You may need to select the right directory on the current disk; use the DOS CHDIR command. You can send files from any disk.

Using a Modem Card

If your PC has a modem card, you must give it the instructions required to dial a system. For example, if you have a Hayes Smartmodem 1200B, follow this example:

```
Kermit-MS>CONNECT [Connecting to host. Type CTRL-]C to return to PC.]
AT Z                          Reset the modem.
OK                            The modem acknowledges the reset.
AT DT 123-4567                Dial the number using touch-tone dialing.
```

The commands must be entered in upper case. The DT command generates touch-tone signals. To obtain pulse dialing, specify DP. If you have to first access an outside line, use a comma in the number, as in: 9,123-4567. See your Hayes documentation for descriptions of modem card commands. Most other brands of modem cards use the same commands as Hayes, but follow your modem's documentation if this method doesn't work. Once you have logged off the host machine you probably will have to CONNECT again to enter the modem reset command.

Keep in mind that once you have CONNECTed with Kermit, the characters you type are sent out the PC's communications port, which in this case is on the modem card.

VAX/VMS Kermit

Kermit-32 implements the Kermit file transfer protocol for the Digital Equipment Corporation VAX series computers under the VAX/VMS operating system. Kermit-32 should run on any VAX/VMS system from version 3 on.

The file system used by VAX/VMS provides for a large number of attributes to be associated with each file. These attributes provide some indication of whether the file is a text file or a binary file. SPSS^X and SPSS/PC+ portable and command files are of the text variety. The "standard" method of storing text in a file under VAX/VMS is to store one line of text per record (variable length records), with a carriage return/line feed sequence implied by the end of the record (implied carriage return). This is the method Kermit-32 uses to store files it receives when using FILE TYPE TEXT.

File Transfer between Kermit-MS and Kermit-32

The following sequence of steps transfers files between Kermit-MS and Kermit-32:

1. Type Kermit to activate Kermit-MS on the PC. At this point you should use the STATUS command to make sure that the parity is set to NONE, local echo is set to OFF, and the baud rate is correct for the modem. If these parameters are incorrect, use the Kermit-MS SET command to change them.
2. Type C (for CONNECT) on the PC, then dial the number required to activate a port on the mainframe VAX. At this point your PC is functioning like a remote terminal.
3. Log in normally to the VAX. You may need to press the carriage return a few times to get the logon message.
4. Start up Kermit-32. Normally you type Kermit to do this. Check with your local systems staff to see how to activate Kermit at your site.
5. Set up the Kermit-32 environment to perform file transfers using the SEND and RECEIVE commands, or use the SERVER command described below. Actual file transfer begins after you escape back to Kermit-MS and issue the appropriate command, either SEND or RECEIVE in "normal" mode, or SEND or GET in "server" mode.
6. Type CTRL-] C to escape back to Kermit-MS.
7. Transfer files using the SEND or RECEIVE commands if you are in normal mode; in server mode use SEND or GET.
8. Type C (for CONNECT) to reconnect to Kermit-32.
9. Type EXIT to exit Kermit-32 and return control to VAX/VMS.
10. Log out normally from the VAX mainframe.

11 Type CTRL-] C to escape back to Kermit-MS.

12 Type EXIT to exit from Kermit-MS back to DOS.

Kermit-32 Commands

The following section summarizes the commands you will need for file transfer. For a complete list of available Kermit-32 commands at your site, type the "HELP" command.

EXIT. Exit Kermit-32. QUIT is a synonym.

HELP. You must specify the command for which help is required.

RECEIVE. The RECEIVE command tells Kermit-32 to receive a file or file group from the other system. If a file specification is not included, and only one file is being received, Kermit-32 uses the name provided by the incoming header. If the header name is not a legal VAX/VMS file name, Kermit-32 will normally replace the illegal characters with "X'. If an incoming file has the same name as an existing file, Kermit-32 will create a new version of the same name and type, for example FOO.BAR;3, FOO.BAR;4.

SEND. The SEND command causes a file or file group to be sent from the VAX to the other system. If wildcard characters are used, all matching files will be sent.

SERVER. The SERVER command puts Kermit-32 into server mode. While in server mode, Kermit-32 waits for commands from Kermit-MS. While running in local mode, Kermit-32 allows you to give a wide range of commands to a remote Kermit server, with no guarantee that the remote server can process them, since they are all optional features of the protocol. Commands for servers include the standard SEND, GET, BYE, and FINISH commands. These commands are described below.

SET. The SET command is used to set various options. You can examine their values with the SHOW command. Many different options are available. The following section lists some of the most useful options for file transfer between the PC and mainframe VAX. For a complete list of SET commands, refer to the *Kermit User's Guide,* or use the Kermit-32 HELP command.

SHOW. Allows you to show the various parameters you can set with the SET command. Type SHOW ALL to see all the current settings.

STATISTICS. Give statistics about the most recent file transfer.

STATUS. Shows the current status of Kermit-32. This includes the number of characters that have been sent and received from the other Kermit. Also included is an estimate of the effective baud rate of the transfer.

SET Options for Kermit-32

The following is a partial list of commands you can set via the SET command:

BLOCK_CHECK_TYPE	1, 2, or 3. Determines the type of block check sequence used to detect transmission errors. These are the single-character checksum (default), the two-character checksum, and the three-character CRC (cyclic redundancy check). This command does not ensure that the desired type of block check will be used, since both Kermits involved in the file transfer must agree on the block check type. If they are different, the single character checksum will be used.
DEBUGGING	ON or OFF. Display packet traffic and other debugging information at the console terminal.
DELAY n	Specify the delay, in seconds, before sending the first send-init packet. This gives you time to "escape" back to your local Kermit and issue a RECEIVE command.
INCOMPLETE_FILE _DISPOSITION	KEEP or DISCARD. The default is DISCARD. If you specify KEEP, then if the transfer fails, you can keep the incomplete part which was received.

RECEIVE

RECEIVE settings allow you to specify to the other Kermit what the packets it sends should look like, or to inform this Kermit what to expect.

SET RECEIVE END_OF_LINE n defines the octal end-of-line character Kermit-32 expects to receive from the remote Kermit. The default value is 15, the ASCII octal value for a carriage return.

SET RECEIVE PACKET_LENGTH n specifies the maximum length packet for the other Kermit to send. The default length is 80. You can specify any decimal number between 10 and 96.

SET RECEIVE PADDING n asks the other Kermit to insert the given number of padding characters before each packet it sends. The default value is 0.

SET RECEIVE QUOTE n defines the octal quoting character that Kermit-32 will expect on incoming messages. The default value is 43, the ASCII octal value for a pound sign.

SET RECEIVE PADCHAR n asks the remote Kermit to use the given octal character for interpacket padding. The default value is 0 (ASCII NUL). You can specify an octal value between 0 and 37, or 177.

SET RECEIVE START_OF_PACKET n sets the control character to mark the beginning of incoming packets. The default value is 1 (ASCII SOH, CTRL-A). This value you set must be in the range 0 to 36 where n is the ASCII value of the character you specify (octal). The value you use must be the same on both Kermits.

SET RECEIVE TIMEOUT n sets the number of seconds to wait for an incoming message before trying again. The default value is 15 seconds.

SEND

SEND settings allow you to set parameters for outgoing packets.

SET SEND END_OF_LINE n sets the octal end-of-line character Kermit-32 will send to the remote Kermit. This character terminates a packet. The default value is 15, the ASCII octal value for a carriage return.

SET SEND PACKET_LENGTH n sets the send packet length. The value for this parameter must be a decimal number between 10 and 96. The default value is 80.

SET SEND PADDING n defines the number of padding characters before each packet. The default value is 0.

SET SEND QUOTE n sets the octal printable character to use for quoting of control characters. The default is 43, the ASCII octal value for a pound sign.

SET SEND PADCHAR n defines the octal padding character that is sent to the remote host. The default value is 0 (ASCII NUL). You can specify an octal number in the range 0 to 37, or 177.

SET SEND START_OF_PACKET n sets the control character to mark the beginning of outbound packets. The default value is 1 (ASCII SOH, CTRL-A). The value you set must be in the range 0 to 36, where n is the ASCII value for the character you specify (octal).

SET SEND TIMEOUT n changes Kermit-32's normal timeout interval to the number of seconds you specify. The default value is 15 seconds.

RETRY

Specify the maximum number of retries allowed for trying to send a particular packet before giving up. Normally this is 5.

The SERVER Command

The SERVER command permits file transfer without moving back and forth between Kermit-MS and Kermit-32. Once you have established the connection between your PC and the VAX, type the SERVER command at Kermit-32, reconnect to Kermit-MS, and issue commands that transfer files.

1 Follow steps 1 through 4 above.
2 Issue the SERVER command by typing SERVER.
3 Reconnect to Kermit-MS by typing CTRL-] C.
4 Issue commands recognized by the server described below.

Kermit-32 server commands are:

BYE The BYE command shuts down the host, logs you off the host, and ends your Kermit-MS session. LOGOUT is identical to BYE, except you remain at the Kermit-MS prompt level rather than exit to DOS.

FINISH The FINISH command tells Kermit-32 to exit server mode and return control to VAX/VMS. When you CONNECT again you are still logged on to the remote host and at the system command level.

GET Transfer files from the currently attached VAX directory to the current PC directory. The GET command takes a filename as an argument, or a file group using VAX wildcard conventions.

SEND Transfer files from the current PC directory to the currently attached directory on the VAX. Specify a filename, or a file group by using PC wildcard conventions.

Appendixes

Contents

Using SPSS/PC+ with RAM Disks

If your PC has an extended or expanded memory board and RAM-disk software—software that makes some of your memory (RAM) appear to be an extremely fast disk drive—you can put the RAM disk to good use in SPSS/PC+.

- You should not use a RAM disk in regular DOS memory (below 640K). Such memory is best used directly by SPSS/PC+.
- The following examples assume that the RAM disk is drive D: and that SPSS/PC+ is installed into its default location (the \SPSS directory) on your default drive.

Use the RAM disk for your active file, for REVIEW, or for both.

THE ACTIVE FILE

Whenever you read a raw data file or transform your active file, SPSS/PC+ writes the data in binary format to a scratch file, called the *active file*. You can control the location of these scratch files with the SPSS/PC+ command SET WORKDEV. When you are going to be processing a file large enough so that reading the data occupies substantial time, begin your session with the SPSS/PC+ command

```
SET WORKDEV=D:.
```

SPSS/PC+ will write one and perhaps two copies of the active file to the RAM disk. Be certain that they will fit before giving this command. You cannot set the work device after the active file has been created in a session.

You may want to put the SET WORKDEV command in your automatic SPSS/PC+ profile, SPSSPROF.INI.

REVIEW

You can also put REVIEW onto a RAM disk. If you go in and out of REVIEW frequently (for example, using the default RUNREVIEW AUTO setting), you can speed up the automatic return to REVIEW considerably. Use the DOS command *SET SPSSR=path* and place REVIEW on a RAM disk. Here is a sample batch file to place REVIEW on drive D:.

```
ECHO OFF
IF EXIST D:SPSSRV.EXE GOTO RUN
COPY C:\SPSS\SPSSRV.EXE D:
:RUN
SET SPSSR=D:\
SPSSPC %1 %2 %3 %4 %5
SET SPSSR=
```

- The *IF* command checks whether REVIEW has already been copied to the RAM disk. (If you had executed this batch file earlier in the same session, it would already be there.)
- The *COPY* command places REVIEW on the RAM disk, if it is not already there. REVIEW occupies approximately 180,000 bytes of space on the RAM disk. *Do not* invoke REVIEW by the name of this .EXE file. To invoke REVIEW from DOS, use the command

  ```
  SPSSPC/RE file1 [file2]
  ```

G Appendixes

- The *SET SPSSR* command sets a DOS environment variable so that SPSS/PC+ will know that REVIEW is in drive D:.
- The SPSSPC command invokes SPSS/PC+, passing any parameters that you may have used when calling the batch file (such as the name of a command file).
- The second *SET SPSSR* command deletes the string SPSSR from the DOS environment area. There should be nothing—not even a blank space—after the equals sign.

CAUTIONS

- SPSS Inc. does not distribute or support any RAM-disk software.

Neither of the two applications suggested above places anything on the RAM disk that needs to be copied to a physical disk at the end of your session. Nevertheless, if you use a RAM disk

- Use reliable RAM-disk software and be aware of its limitations.
- Be sure that your RAM disk has enough room for the application. If you place your active file on the RAM disk, you need enough room for two copies of the data in binary form. If you place REVIEW on the RAM disk, you need enough room to hold the file SPSSRV.EXE.

Files Created by Other Software

SPSS/PC+ reads and writes several types of files. This section discusses the use of these files with other text editors typically used in the personal computing environment. You probably will use an editor to prepare command files and data files. The REVIEW editor that is part of SPSS/PC+ allows you to prepare these files, but many other editors are available as well.

You may also want to analyze data stored in a database or spreadsheet program. SPSS/PC+ has the ability to translate files produced by several popular spreadsheet or database packages into its own active file format—and the ability to translate files into formats used by the same packages. SPSS/PC+ can do this with one command: TRANSLATE. See Command Reference: TRANSLATE for a description of this command and a list of the database and spreadsheet packages supported.

EDITORS

An editor is software used to create or modify files containing text or data. SPSS/PC+ processes command and data files created with editors. SPSS/PC+ itself contains an editor called REVIEW. However, you can use another editor to create SPSS/PC+ command and data files, as long as you produce an ASCII file that conforms to the following criteria:

- Each line in the file must end with a carriage-return line-feed sequence. (This is typical of most DOS files.)
- The file must end with a Ctrl Z for SPSS/PC+ to read the last line correctly. (This is typical of most DOS files.)
- The file must not contain special format codes or other information specific to the program from which it was created. (An example of this is the use of control characters to indicate special document formatting options, such as Ctrl B for boldface.)
- The maximum line length for files containing SPSS/PC+ commands and inline data is 80 characters. Anything past column 80 is ignored.
- The maximum line length for data files named on the DATA LIST command is 1,024 characters.

The REVIEW Editor

The REVIEW editor (see REVIEW: The SPSS/PC+ Editor, in Part A) allows you to edit the ASCII files produced by SPSS/PC+ and to produce data and command files that can be read by SPSS/PC+. If you use REVIEW for editing related to SPSS/PC+, you don't have to worry about problems of compatibility between SPSS/PC+ and the editor.

You can use REVIEW as your general-purpose editor, creating and modifying files for other applications.

Other Editors

Some editors store files in a special format that combines text with escape sequences and/or control characters. The escape sequences and control characters are usually used to perform font shifts when printing, or to format text for display. For example, the editor might add a special character to the end of every line, paragraph, and page. Editors designed for word processing are more likely to use such special formats than are editors designed for programming. If you INCLUDE a file created with an editor that inserts escape sequences, SPSS/PC+ reads the escape sequences literally and cannot interpret them. Fortunately, these types of editors usually have a special mode or translate program to produce files readable by programs such as SPSS/PC+. For example, WordStar™ has a "non-document" mode for creating files. Other word processors typically let you specify ASCII or "text" format when you save a file.

Another potential problem is the requirement for a carriage-return line-feed sequence at the end of the line. If an editor has an automatic word-wrap procedure which automatically creates a new line while you continue typing, it may not insert the end-of-line sequence required by SPSS/PC+. To obtain the carriage-return line-feed sequence required by SPSS/PC+, you may need to use the enter key after typing in each line. Some editors may not even wrap to start a new line but appear to do so because they are displaying a line longer than 80 characters in 80-character window. In general, you should avoid using the word-wrap facilities of your editor when preparing files to read with SPSS/PC+ unless your editor explicitly documents that it inserts the carriage-return line-feed sequence. If your editor's documentation provides special instructions for preparing files for use with other applications such as language compilers, follow them.

How do you find out whether your favorite editor is going to work with SPSS/PC+? Use the DOS *TYPE* command to type a file created with your editor. If the file does not look the same when typed as when viewed within the editor, then it probably cannot be read correctly by SPSS/PC+. Beware of special symbols that are not directly available on the keyboard. The one exception to this rule is the use of tabs. If your editor permits tabs, and it places an ASCII 09 character into the file, SPSS/PC+ interprets it as *one* blank. Although the file does not look the same when echoed by SPSS/PC+ (since the interpretation of tabs is always context specific), it is interpreted correctly.

With the exception of forms control for listing files, ASCII files written by SPSS/PC+ do not contain special escape sequences, tabs, etc., and are generally readable and editable by most editors. The one exception is a file produced by SPSS/PC+ which does not contain a Ctrl Z at the end because a session was not ended normally. Some editors can read such a file (although they might issue a read error warning); other editors have great difficulty. If you terminate an SPSS/PC+ session normally, you should be able to *TYPE*, *PRINT*, or otherwise display listing and log files written by SPSS/PC+. Follow your editor's discussion for editing "foreign" files.

COMMUNICATIONS PROGRAMS

Communications programs such as CROSSTALK™ and PC-TALK III can be used to transfer ASCII files read and written by SPSS/PC+. However, they should not be used with SPSS/PC+ system files or portable system files. Kermit is distributed with SPSS/PC+ (see Communications, Part F) to provide an error-free transfer facility for ASCII files and portable system files. ASCII files containing special graphics characters used in SPSS/PC+ output cannot be transferred correctly to mainframes or to other types of personal computers.

Help for Old Friends

This appendix is designed to provide help for users of the original version of SPSS/PC+.

- The beginning sections outline differences between SPSS/PC+ and SPSS/PC+ V2.0.
- REPORT and REVIEW were revised substantially for SPSS/PC+ V2.0. The end of this appendix includes the Command Reference material for REVIEW and REPORT from the first edition of *SPSS/PC+*. If you are using this book with the older software, you should consult these sections to use REVIEW and REPORT.

The rest of this appendix contains reprints of the REPORT and REVIEW Command Reference material from the original *SPSS/PC+* manual.

DIFFERENCES BETWEEN SPSS/PC+ AND SPSS/PC+ V2.0

SPSS/PC+ V2.0 differs from its predecessor most visibly in the default user interface, which is the new Menu and Help system in REVIEW. The Menu and Help system provides an optional way of constructing commands by making menu selections in REVIEW and furnishes context-sensitive help on command syntax. Related features include:

- An online glossary of statistical and other terms that appear in SPSS/PC+ output. You can request a definition of specific terms, you can browse through the glossary, and you can paste the definition into REVIEW's editing window.
- A menu of files. You can look at directory information, or you can paste a file specification from the menu into a command you are building.
- A menu of variable names. Once you have defined your data, you can select variable names from this menu.

New Commands and Features

The principal commands and features that have been added to SPSS/PC+ V2.0 are:

REPORT	Using FORMAT AUTOMATIC, you can produce an attractive report with much less effort than before. This specification changes many default settings to reduce the number of subcommands that you need to specify.
REVIEW	REVIEW has been enhanced considerably to support the Menu and Help system, and to make it a more effective working environment.
MODIFY VARS	The new MODIFY VARS command lets you rename, reorder, or drop variables from your active file.
SYSFILE INFO	The new SYSFILE INFO command displays information about SPSS/PC+ system files on disk without replacing your current active file.
SET	Half a dozen new keywords on the SET command allow you to tailor your default operating environment, or to use the capabilities of an EGA or VGA display for showing more than 25 lines on the screen.
RELIABILITY	The new RELIABILITY command performs item analysis on additive scales. RELIABILITY is part of the Advanced Statistics option.

TABLES	The TABLES command now allows you to sort the cells in any dimension. TABLES is part of the TABLES option.
PRINT TABLES	The PRINT TABLES command offers enhanced support for the HP LaserJet printer. PRINT TABLES is part of the TABLES option.

Memory Requirements

We recommend that you have at least 512K of RAM to use SPSS/PC+ V2.0. Some commands, particularly in the Advanced Statistics, TABLES, and Trends options, require 640K. Almost all commands can handle larger problems if 640K is available.

SPSS/PC+ V2.0 does not use Expanded or Extended memory (memory beyond 640K) for its workspace. You can, however, use such memory as a RAM disk to speed processing. See the appendix on RAM disks for details.

REVIEW from DOS

REVIEW now uses some of the facilities of the SPSS/PC+ monitor (SPSSPC.COM) for enhanced performance. To use it from DOS, you must invoke it through the monitor, with the command

```
SPSSPC/RE file1 [file2]
```

If you are accustomed to using REVIEW from DOS, and you want to use the new REVIEW, you may wish to set up a REVIEW.BAT file containing the line

```
SPSSPC/RE %1 %2 %3 %4 %5
```

Within SPSS/PC+, you invoke REVIEW as before, with the command

```
REVIEW.
```

IF YOU DON'T HAVE SPSS/PC+ V2.0

If you are using this manual with the original version of SPSS/PC+, you will find that the new features listed above are not available in your system. Bearing this in mind, you should be able to use the manual with little difficulty, except in two areas:

- REVIEW. The older REVIEW used function keys in combination with ⇧, Ctrl, and Alt for most commands and lacked several of the functions available in the new REVIEW.
- REPORT. The sections on REPORT in this manual have been rewritten to exploit automatic formatting, and deemphasize the manual formatting options that you must use with the older REPORT.

Following are reprints of the REPORT and REVIEW references, taken from the original SPSS/PC+ manual.

REPORT

```
REPORT

 [FORMAT=[TSPACE({1})] [CHDSPACE({1})] [BRKSPACE({1})] [FTSPACE({1})]
                 {n}              {n}              {n}             {n}

       [LENGTH({SET length})] [MARGINS({SET width})]
               {n,n       }            {n,n      }

       [{NOLIST    }] [SUMSPACE({1})] [MISSING {'.'}]]
        {LIST[(n)]}              {n}           {'s'}

 [/STRING=stringname ([varname] [(width)] [(BLANK)]
          ['string']...)[stringname...]]

 /VARIABLES={var       } [({VALUE})] ['col head'] [(width)]
            {var TO var}   {LABEL}
                           {DUMMY}

            [(OFFSET({0}))] [var...]
                     {n}

 [/MISSING={VAR               }]
           {NONE              }
           {LIST(varlist{1}))}
                        {n}

 [/TITLE='line1' 'line2'...] [/FOOTNOTE='line1' 'line2'...]
      or                          or
 [/LTITLE='line1' 'line2'...] [/LFOOTNOTE='line1' 'line2'...]
 [/CTITLE='line1' 'line2'...] [/CFOOTNOTE='line1' 'line2'...]
 [/RTITLE='line1' 'line2'...] [/RFOOTNOTE='line1' 'line2'...]

 /BREAK=varlist [({VALUE})] ['col head'] [(width)]
                 {LABEL}

       [OFFSET({0})] [({NOTOTAL})] [({NONAME})] [({SKIP({1})})]
               {n}     {TOTAL  }     {NAME  }     {      {n} }
                                                  {PAGE      }
or
 /BREAK=(NOBREAK) [(width)] [(OFFSET({0}))] [(SKIP({1}))]
                                    {n}            {n}

 [/SUMMARY=function...['summary title'] [(break col #)] [SKIP({0})]]
                                                              {n}
or
 [/SUMMARY=PREVIOUS[(n)]]
where function is
 aggregate [(varname[(d)][({PLAIN })] [varname...])]
                           {DOLLAR}
                           {COMMA }
or
 composite(agg(varname)...)[(report col[(d)][({PLAIN })])]
                                              {DOLLAR}
                                              {COMMA }
```

Aggregate Functions:

VALIDN	VARIANCE	PCGT(n)
SUM	KURTOSIS	PCLT(n)
MIN	SKEWNESS	PCIN(min,max)
MAX	MEDIAN(min,max)	ABFREQ(min,max)
MEAN	MODE(min,max)	RELFREQ(min,max)
STDEV		

Composite Functions:

DIVIDE(agg(varname) agg(varname) [factor])
PCT(agg(varname) agg(varname))
SUBTRACT(agg(varname) agg(varname))
ADD(agg(varname) agg(varname)...)
GREAT(agg(varname) agg(varname)...)
LEAST(agg(varname) agg(varname)...)
AVERAGE(agg(varname) agg(varname)...)
MULTIPLY(agg(varname) agg(varname)...)

Example:

```
REPORT FORMAT=LIST
/VARIABLES=PRODUCT (LABEL) ' ' 'Retail' 'Products'
           SALES 'Annual' 'Sales' '1981'
/BREAK=DEPT 'Department' (LABEL)
/SUMMARY=VALIDN (PRODUCT) MEAN (SALES).
```

Overview

The REPORT procedure produces both case listings and summary statistics and gives you considerable control over the appearance of the output. REPORT calculates all the univariate statistics available in DESCRIPTIVES and the statistics and subpopulation means available in MEANS. In addition, REPORT calculates statistics not directly available in any other SPSS/PC+ procedure, such as computations involving aggregated statistics.

REPORT provides complete report format defaults or lets you customize column widths, titles, footnotes, spacing, and other elements. Because REPORT is so flexible and the output has so many components, it is often efficient to preview report output using a small number of cases until you find the format that best meets your needs.

Defaults

A listing report with or without subgroup classification requires the FORMAT, VARIABLES, and BREAK subcommands. Required subcommands for a report with summary statistics are VARIABLES, BREAK, and SUMMARY. By default, column heads are variable labels, or variable names if no variable labels have been specified. Default column widths are 8 columns for short string and numeric variables, 20 columns for display of value labels, and the dictionary width of long string variables. Intercolumn spacing adjusts automatically, using a minimum of 1 and a maximum of 4 spaces between columns. By default, the calculation of report statistics excludes cases with user-missing values, and missing-value indicators are ignored for variables named on BREAK.

Tailoring

Display Format. REPORT provides full report format defaults and offers you optional control over page length, vertical spacing between different types of information, margins and column widths, page titles and footnotes, and labels for statistics. The maximum width and length of the report are controlled by specifications on the SET command.

Use the FORMAT subcommand to control how the report is laid out on a page and whether case listings are displayed. The STRING subcommand concatenates variables to create temporary variables that can be referenced on the VARIABLES or BREAK subcommands. The VARIABLES subcommand names the report variables used to compute statistics and controls the titles, width, and contents of report columns. The BREAK subcommand specifies the variables that define groups and controls the titles, width, and contents of break columns. The SUMMARY subcommand specifies statistics and controls the titles and spacing of summary lines. Additional subcommands control the specification and placement of multiple-line titles and footnotes.

Statistical Display. The statistical display is controlled by the SUMMARY subcommand. Statistics can be calculated for each category of a break variable and for the group as a whole. Available statistics include mean, variance, standard deviation, skewness, kurtosis, sum, minimum and maximum value, mode, median, and percentages. Composite functions perform arithmetic operations using two or more summary statistics calculated on single variables.

Missing Values. You can override the default to include user-missing values in report statistics and listings with the MISSING subcommand. You can also define a missing-value symbol to represent missing data on the FORMAT subcommand.

Syntax

- The minimum specification for REPORT is the VARIABLES and BREAK subcommands, plus either FORMAT=LIST or the SUMMARY subcommand.
- Only one each of the FORMAT, STRING, VARIABLES, and MISSING subcommands are allowed. If used, they must be entered in this order and must precede all BREAK and SUMMARY subcommands.
- Multiple BREAK subcommands are allowed, and multiple SUMMARY subcommands are allowed per BREAK.
- Each BREAK subcommand immediately precedes the SUMMARY subcommands that request statistics for that BREAK.
- To suppress subgroup classification when LIST is specified on FORMAT, specify (NOBREAK) on BREAK.
- To obtain summary statistics but no case listing or subgroup classification, specify NOLIST on FORMAT and (NOBREAK) on BREAK.
- Title and footnote subcommands can appear anywhere after FORMAT except among the BREAK and SUMMARY subcommands.
- Only one TITLE or one each of RTITLE, CTITLE, and LTITLE subcommands can be entered.
- Only one FOOTNOTE or one each of RFOOTNOTE, CFOOTNOTE, and LFOOTNOTE subcommands can be entered.
- Keywords on REPORT subcommands have default specifications that are in effect if the keyword is not specified. Specify keywords only when you wish to change a default.
- Keywords are enclosed in parentheses if the subcommand takes variable names as arguments.
- Subcommands are separated from each other by slashes.

Operations

- REPORT causes the data to be read.
- REPORT processes cases sequentially. When the value of a break variable changes, REPORT displays a statistical summary for cases processed since the last set of summary statistics was displayed.
- The file must be sorted in order on the break variable or variables.
- The maximum width and page length of the report are the width and page length specified on the SET command.
- If the column is not wide enough to display numeric values, REPORT first rounds decimal digits, then converts to scientific notation if possible, and then displays asterisks. String variables that are wider than the column are truncated.
- The format used to display values in case listings is controlled by the dictionary format of the variable. Each statistical function in REPORT has a default format.

Limitations

- Titles and footnotes can be no wider than the number of characters in an SPSS/PC+ command line, minus 2 characters for the apostrophes or quotes that enclose the string.
- String variables created on the STRING subcommand can be no wider than the system page width.
- There is no fixed limit on the number of BREAK and SUMMARY subcommands. However, the page width limits the number of variables displayed and

thereby limits the number of break variables. The limit of 10 variables on SORT CASES often effectively acts as a limit on the number of breaks.

- Maximum 10 dummy variables per VARIABLES subcommand.
- Maximum 50 strings per STRING subcommand.
- Maximum 20 MODE and MEDIAN requests per SUMMARY subcommand.
- Maximum 20 PCGT, PCLT, and PCIN requests per SUMMARY subcommand.
- The number of report variables that can be specified depends upon the width of the report, the width of the variable columns, and the number of BREAK subcommands.
- The)PAGE function can occur only once in either the title or footnote.
- Workspace is required to store all labeling information, frequency counts if summaries request MEDIAN or MODE, strings, and computed summary statistics.
- Memory requirements significantly increase if MEDIAN or MODE is requested with variables having a wide range of values. The amount of workspace required is 20 + 8*(max −min +1) bytes per variable per function per break.
- If TOTAL is in effect, workspace requirements are almost doubled.
- Memory requirements also increase if value labels are displayed for variables with many value labels. The amount of workspace required is 4 + 24*n labels per variable.

Example

```
SORT CASES BY DEPT.
REPORT FORMAT=LIST
/VARIABLES=PRODUCT (LABEL) ' ' 'Retail' 'Products'
           SALES 'Annual' 'Sales' '1981'
/BREAK=DEPT 'Department' (LABEL)
/SUMMARY=VALIDN (PRODUCT) MEAN (SALES) 'No.Sold,Mean Sales'.
```

- This report is a listing of products and sales by department. A summary of the total number of products sold and the average sales by department is also produced.
- Cases are first sorted by DEPT to ensure that cases are appropriately grouped for the calculation of statistics.
- FORMAT requests a report that includes a listing of cases within each break group.
- VARIABLES specifies PRODUCT and SALES as the report variables and requests that the value labels identifying products be displayed. Three line column headings are provided for each report column. The first line of the column head is blank for the variable PRODUCT.
- BREAK identifies DEPT as the break variable and provides a one-line column title for the break column. (LABEL) specifies that the value label be displayed instead of the value itself.
- SUMMARY requests the calculation of the valid number of cases for PRODUCT and the mean of SALES for each value of DEPT. A title is provided for the summary line to override the default title, VALIDN.

FORMAT Subcommand

```
FORMAT=[NOLIST|LIST[(n)] ] [LENGTH(l,r)]
       [MARGINS(t,b)] [TSPACE(n)]
       [HDSPACE(n)] [BRKSPACE(n)]
       [SUMSPACE(n)] [FTSPACE(n)] [MISSING 's']
```

The optional FORMAT subcommand controls the overall width and length of the report and vertical spacing.

- Defaults not explicitly changed remain in effect.
- Keyword specifications and their arguments can be named in any order.

The following can be specified on FORMAT:

NOLIST *Listing of individual cases.* List the values of all variables on VARIABLES for each case. The optional *n* indicates that a blank line be inserted after each *n* cases; the default is not to insert blank lines. Values for cases are listed using the default formats for the variables. The default is the alternative NOLIST, which requests that no case listing be produced.

LENGTH(t,b) *The top and bottom lines of the report.* The value for the bottom line cannot be greater than the system page length. The system page length is controlled by SET. By default, the top of the report begins at line 1 and the bottom of the report is the last line of the system page length.

MARGINS(l,r) *The columns for the left and right margins.* The system page width is controlled by the SET WIDTH command. The right column cannot be beyond the width specified on SET. By default, the left margin is print column 1 and the right margin is the rightmost print column of the system page width.

TSPACE(n) *The number of blank lines between the report title and the column heads.* The default is 1.

CHDSPACE(n) *The number of blank lines beneath the longest column head.* The default is 1.

BRKSPACE(n) *The number of blank lines between the break head and the next line.* The next line is a case if LIST is in effect or the first summary line if NOLIST is in effect. The default is 1. BRKSPACE(−1) places the first summary statistic or the first case listing on the same line as the break value. When a summary line is placed on the same line as the break value, the summary title is suppressed.

SUMSPACE(n) *The number of blank lines between the last summary line at the lower break and the first summary line at the higher break when they break simultaneously.* SUMSPACE also controls spacing between the last case listed and the first summary line if LIST is in effect. The default is 1.

FTSPACE(n) *The minimum number of blank lines between the last listing on the page and the footnote.* The default is 1.

MISSING 's' *Missing-value symbol.* The symbol can be only 1 character and is used to represent both system- and user-missing values. The default is a period (.).

LIST[(n)] POSITION THIS WORD UNDER NOLIST ABOVE.

Example

```
FORMAT=LIST MARGINS(1,60) LENGTH(5,30) BRKSPACE(-1) MISSING ('*')
```

- This FORMAT subcommand requests a case listing, defines a new page size smaller than the system page size, requests that case listings begin on the line on which the break head is displayed, and specifies an asterisk as the missing-value symbol.

Page Layout Figure A displays the complete page layout and subcommand specifications used to control the basic structure of the report.

A Page layout for REPORT

```
--------------------------------------- top of page ----------------------------------------
                         ****************** TITLE *****************           <---- LENGTH
                                                                              <---- TSPACE
BREAK HEAD               BREAK HEAD          COLUMN    COLUMN    COLUMN    COLUMN
                                             HEAD      HEAD      HEAD      HEAD
                                             [VAR]     [VAR]     [VAR]     [VAR]
                                                                              <---- CHDSPACE
BREAK A VALUE 1          BREAK B VALUE 1
                                                                              <---- BRKSPACE
                                             VALUE     VALUE     VALUE     VALUE
                                             VALUE     VALUE     VALUE     VALUE
                                                                              <---- LIST
                                             VALUE     VALUE     VALUE     VALUE
                                             VALUE     VALUE     VALUE     VALUE
                                                                              <---- SUMSPACE
                         SUMMARY TITLE         AGG.      AGG.      AGG.      AGG.
                                                                              <---- SKIP with SUMMARY
                         SUMMARY TITLE         AGG.      AGG.      AGG.      AGG.
                                                                              <---- SKIP with BREAK
                         BREAK B VALUE 2
                                                                              <---- BRKSPACE
                                             VALUE     VALUE     VALUE     VALUE
                                             VALUE     VALUE     VALUE     VALUE
                                                                              <---- LIST
                                             VALUE     VALUE     VALUE     VALUE
                                             VALUE     VALUE     VALUE     VALUE
                                                                              <---- SUMSPACE
                         SUMMARY TITLE         AGG.      AGG.      AGG.      AGG. <---- stats for B=2, A=1

                         SUMMARY TITLE         AGG.      AGG.      AGG.
                                                                              <---- SUMSPACE
SUMMARY TITLE                                  AGG.      AGG.      AGG.      AGG. <---- stats for A=1

SUMMARY TITLE                                  AGG.      AGG.      AGG.      AGG.
                                                                              <---- SKIP with BREAK
BREAK A VALUE 2          BREAK B VALUE 1
                                                                              <---- BRKSPACE
                                             VALUE     VALUE     VALUE     VALUE
                                             VALUE     VALUE     VALUE     VALUE
                                                                              <---- LIST
                                             VALUE     VALUE     VALUE     VALUE
                                             VALUE     VALUE     VALUE     VALUE
                                                                              <---- SUMSPACE
                         SUMMARY TITLE         AGG.      AGG.      AGG.      AGG.
                                                                              <---- SKIP
                         SUMMARY TITLE         AGG.      AGG.      AGG.      AGG.
                                                                              <---- SKIP with BREAK
                         BREAK B VALUE 2
                                                                              <---- BRKSPACE
                                             VALUE     VALUE     VALUE     VALUE
                                             VALUE     VALUE     VALUE     VALUE
                                                                              <---- LIST
                                             VALUE     VALUE     VALUE     VALUE
                                             VALUE     VALUE     VALUE     VALUE

                         SUMMARY TITLE         AGG.      AGG.      AGG.      AGG.

                         SUMMARY TITLE         AGG.      AGG.      AGG.      AGG.
                                                                              <--- SUMSPACE
SUMMARY TITLE                                  AGG.      AGG.      AGG.      AGG.

SUMMARY TITLE                                  AGG.      AGG.      AGG.      AGG.
                                                                              <---- FTSPACE
                         **************** FOOTNOTE ****************           <---- LENGTH
-------------------------------------- bottom of page --------------------------------------
|                                                                                          |
left margin                                                                     right margin
```

VARIABLES Subcommand

```
/VARIABLES=var|var TO var [(VALUE|LABEL|DUMMY)] ['col head']
           [(width)] [(OFFSET(n))] [varname...]
```

The required VARIABLES subcommand names the variables to be listed and summarized in the report. Optionally, you can use VARIABLES to control column titles, column widths, and the contents of report columns.

- The minimum VARIABLES specification is a list of variables. These are the report variables.
- Each report variable defines a report column. Each report column can be thought of as having the name of the variable that defines it.
- Variables are assigned to columns in the order in which they are named on the VARIABLES subcommand.

- To refer to a report column as a location in a composite function, use the name of the variable that is assigned to that column.
- Optional specifications can be given in any order following the variable name to which they apply.
- Optional specifications apply only to the immediately preceding variable or list of variables implied by the TO keyword.
- If an optional specification is omitted, the default is in effect.
- The value of a variable or an aggregate statistic calculated on the variable is displayed in the report column with the name of the variable.
- The number of variables that can be named is limited by the system page width.
- Variables named on the BREAK subcommand may also be named on VARIABLES.

(VALUE)
(LABEL)
(DUMMY)
Contents of the report column assigned to the variable. If no specification is given, the keyword (VALUE) is in effect. (VALUE) specifies that values of the variable be displayed in the column. The alternative keyword (LABEL) displays value labels if value labels are defined, and values otherwise. (VALUE) and (LABEL) have no effect unless LIST has been specified on the FORMAT subcommand.

Use (DUMMY) to define a report column for a variable that does not exist in the active file. Such dummy variables are used to control spacing or to reserve space for statistics computed upon other variables. Do not name an existing SPSS/PC+ variable as a dummy variable.

Contents are left-justified in the column for string variables or when (LABEL) is in effect; otherwise column contents are right-justified.

'column head' *Title used for the report column assigned to the variable.* Specify multiple-line titles by enclosing each line in a set of apostrophes or quotes following the conventions for strings (see Universals: Strings). Separate the specifications for title lines with at least one blank.

If no column title is specified, the default column title is the variable label or the variable name if no variable label has been specified.

Default column titles wrap for as many lines as are required to display the entire label. User-specified column titles that exceed the column width are truncated.

Titles are left-justified for string variables or when (LABEL) is in effect; otherwise titles are right-justified.

(width) *Width for the report column.* If no width is specified, the default width is 20 if (LABEL) is in effect, 8 if (DUMMY) is in effect, or the width of the string for variables created by the STRING subcommand. The maximum width is the dictionary format width and 8 otherwise (see FORMATS and Universals: Formats).

(OFFSET(n)) *Offset the contents of the report column* n *columns.* Contents are offset from the left for string variables or when (LABEL) is in effect, and from the right otherwise. The default is 0.

Example

```
/VARIABLES=V1 TO V3(LABEL) (15)
   V4 V5 (LABEL)(OFFSET (2))(10)
   SEP1 (DUMMY) (2) ''
   V6 'Results using' "Lieben's Method" 'of Calculation'(15)
```

- The variables from V1 through V3 each have report columns that are 15 print columns wide. Value labels are displayed for these variables in the case listing.
- Variable V4 is assigned a report column with the default width and values are listed in the case listing.
- Value labels are displayed for variable V5. The column for variable V5 is 10 columns wide and column contents are offset two spaces from the left.

- SEP1 is a dummy variable. The column assigned to SEP1 is two columns wide and there is at least one blank column on each side of SEP1. Thus, there are at least 4 blank columns between the columns for V5 and V6.
- SEP1 is given a null title to override the default column title SEP1.
- V6 is given a three-line title. The column width 15 is assigned to V6 to ensure that the column title is not truncated.

STRING Subcommand

```
/STRING=stringname ([varname] [(width)] [(BLANK)]
        ['string']...[varname]...)

        [stringname...]
```

The optional STRING subcommand concatenates variables and user-specified strings into temporary string variables that exist only within REPORT.

- The minimum specification is a name for the string variable followed by a variable name or a user-specified string enclosed in parentheses.
- The name assigned to the string variable must be unique.
- Any combination of string variables, numeric variables, and user-specified strings can be used enclosed in parentheses to define the string.
- The keyword TO cannot be used within the parentheses to specify an implied variable list.
- More than one string variable can be defined on the STRING subcommand.
- If a variable within the parentheses has a missing value, the string has a system-missing value, and the missing-value symbol is used for that value in case listings.
- A string variable defined in REPORT cannot exceed the system page width.
- String variables defined on the STRING subcommand can be used on the VARIABLES or BREAK subcommand.

(width) *Column width within the string of the preceding variable.* The default width is the dictionary width of the variable.

If the width specified is less than required by the value, asterisks are displayed for numeric variables and string variables are truncated on the right. If the width exceeds the width of a value, values of numeric variables are padded with zeros on the left and values of string variables are padded with blanks on the right.

The maximum width for numeric variables within the string definition is 16. The maximum width for a string variable is the system page width.

(BLANK) *Left-pad values of the preceding numeric variable with blanks.* If the specification is omitted, the default is to left-pad values of numeric variables with zeros. If a numeric variable has a DOLLAR or COMMA format, it is automatically left-padded with blanks.

'string' *A user-specified string.* Any combination of characters can be specified enclosed in apostrophes or quotes.

Example

```
/STRING=JOB1(AVAR VARN)
        JOB2(AVAR(2) VARN(3))
        JOB3(AVAR(2) VARN(BLANK) (4))
```

- This STRING subcommand defines four string variables to be used within the report.
- Assume that AVAR is a string variable read from a four-column field using the FIXED keyword on DATA LIST, and that VARN is a computed numeric variable with the default format of 8 columns with 2 implied decimal places.
- If a case has the value 'KJ ' for AVAR and the value 241 for VARN, JOB1 displays the value KJ 00241.00, JOB2 the value KJ241, and JOB3 the value KJ 241.

Example

```
/STRING=SOCSEC(S1 '-' S2 '-' S3)
```

- This STRING subcommand concatenates the three variables S1, S2, and S3, which each contain a portion of the social security number.

- Hyphens are inserted between the portions when the values of SOCSEC are displayed.
- This example assumes that the variables S1, S2, and S3 were read from three-column, two-column, and four-column fields respectively, using the FIXED keyword on DATA LIST. These variables have format widths of 3, 2, and 4 columns, and are not left-padded with zeros.

BREAK Subcommand: Subgroup Breaks

```
/BREAK=varlist [(VALUE|LABEL)]['col title'] [(width)]
       [(OFFSET(n))] [(NOTOTAL|TOTAL)] [(NONAME|NAME)]
       [(SKIP(n)|PAGE)]
```

The required BREAK subcommand specifies the variables that define the subgroups for the report display. BREAK also allows you to control the titles, width, and contents of break columns and to begin a new page for each level of the break variable.

- The minimum specification is a variable name.
- A break occurs when any one of the break variables named on BREAK changes value. Cases must be sorted by the values of all BREAK variables on all BREAK subcommands.
- Optional specifications can be given in any order following the last variable named.
- Optional specifications apply to all variables in the break column and to the break column as a whole.
- If an optional specification is omitted, the default is in effect.
- If more than one variable is specified on a BREAK subcommand, a single break column is used. The first variable named changes most slowly. The default column width is the longest of the default widths for any of the break variables. The value or value label for each variable is displayed on a separate line in the order in which the variables are named on BREAK.
- Multiple BREAK subcommands can be used.
- A break column is reserved for each BREAK subcommand.
- When you use more than one BREAK subcommand, specify (TOTAL) only on the first BREAK to get a summary for the entire file. Otherwise, redundant statistics may be displayed.
- Missing-value specifications are ignored for variables named on the BREAK subcommand. There is one break category for system-missing values and one for user-missing values. The values are displayed by the missing-value symbol controlled by the FORMAT subcommand.
- The BREAK subcommand must precede the SUMMARY subcommand that defines the summary line for the break.
- Variables named on BREAK can also be named on VARIABLES.

The following can be specfied on BREAK:

(VALUE)
(LABEL) *Contents of the break column.* If no specification is given, (VALUE) is in effect. (VALUE) specifies that values of the break variables be displayed in the column. The alternative keyword (LABEL) displays value labels if value labels have been defined, and values otherwise. The value is displayed only once for each break change and is not repeated at the top of the page in a multiple-page break group.

Contents are left-justified in the column for string variables or when (LABEL) is in effect; otherwise column contents are right-justified.

'column head' *Title used for the break column.* Specify multiple-line titles by enclosing each line in a set of apostrophes or quotes following the conventions for strings (see Universals: Strings). Separate the specifications for title lines with at least one blank.

The default title is the variable label of the break variable if there is one and the variable name of the break variable otherwise. If the break column is defined by more than one variable, the label or name of the first variable is used.

Default column titles wrap for as many lines as required to display the entire label. User-specified column titles that exceed the column width are truncated.

Titles are left-justified for string variables when (LABEL) is in effect; otherwise titles are right-justified.

(width) *Column width for the break column.* The default width is 20 when (LABEL) is in effect or the width of the string for variables created with the STRING subcommand. Otherwise, the width is the maximum of 8 or the dictionary format width. The width is 0 if (NOBREAK) is specified.

(OFFSET(n)) *Offset the contents of the break column* n *columns.* Contents are offset from the left for string variables or when (LABEL) is in effect and from the right otherwise. The default is 0.

(TOTAL)
(NOTOTAL) *With (TOTAL), summary statistics are displayed when all values of the break variables on the subcommand have been cycled through* If (TOTAL) is specified for the first BREAK, statistics are reported for the entire report. (NOTOTAL) is the default and displays only summary statistics for each break.

(SKIP(n))
(PAGE) *The vertical spacing between the last summary line for a break and the next break.* If (SKIP(n)) is specified, each break begins following *n* blank lines. The default is 1. If (PAGE) is specified, each break begins on a new page.

(NAME)
(NONAME) *Display the name of the break variable alongside each value or value label of the break variable.* (NAME) requires 10 print columns (the maximum eight-character length of SPSS/PC+ variable names plus two parentheses) in addition to the columns needed to display break values or value labels. (NAME) is ignored if the break-column width is insufficient. If the default keyword (NONAME) is specified, the name of the break variable is omitted.

Example

```
SORT DIVISION BRANCH DEPT.
REPORT FORMAT=MARGINS (1,70) BRKSPACE(-1)

/VARIABLES=SPACE(DUMMY) ' ' (4)
           SALES 'Annual' 'Sales' '1981' (15) (OFFSET(2))
           EXPENSES 'Annual' 'Expenses' '1981' (15) (OFFSET(2))

/BREAK=DIVISION
       BRANCH (LABEL) (10) (TOTAL) (OFFSET(1))
  /SUMMARY=MEAN

/BREAK=DEPT 'Department' (LABEL) (10)
  /SUMMARY=MEAN.
```

- This example creates a report that breaks on three variables. BRANCH breaks within values of DIVISION, and DEPT breaks within values of BRANCH.
- FORMAT sets margins to a maximum of 70 columns and requests that the summary line be displayed on the same report line as the break values. Because the default NOLIST has not been overridden, only summary statistics are displayed.
- VARIABLES defines three report columns, each occupied by a report variable: SPACE, SALES, and EXPENSES.
- The variable SPACE is a dummy variable that exists only within REPORT. It has a null title and a width of 4 print columns. It is used as a space holder to separate the break columns from the report columns.
- SALES is given a three-line title and a width of 15 print columns. The values of SALES are offset 2 print columns from the right.

- EXPENSES is the third report variable and has the same width and offset specifications as SALES.
- The leftmost column in the report is reserved for the first two break variables, DIVISION and BRANCH. Value labels are displayed for both, the break column is 10 print columns wide, and the value labels are offset 1 print column from the left. Any value label more than 9 characters long is truncated. The default column title is used. (TOTAL) requests that the BRANCH summary line be displayed when all values of DIVISION and BRANCH have been cycled through, that is, at the end of the report.
- The summary line for the first BREAK subcommand consists of the mean of each report variable, displayed in its own report column. This line is displayed each time the value of DIVISION or BRANCH changes.
- The third break variable, DEPT, occupies the second column from the left in the report. The break column is 10 print columns wide and has a one-line title. The first 10 characters of the value labels are displayed in the break column.
- The second SUMMARY subcommand displays the mean for each report variable when the value of DEPT changes.

BREAK Subcommand: No Break Variables

```
/BREAK=(NOBREAK) [(width)] [(OFFSET(n))] [(SKIP(n))]
```

To obtain listing reports without break variables, specify (NOBREAK) on the BREAK subcommand. Column widths can be specified on BREAK to reserve room for summary titles.

- The minimum specification on BREAK for a report without breaks is the keyword (NOBREAK).
- Optionally, the (width), (SKIP), and (OFFSET) specifications can be used.

The defaults for optional keyword specifications depend on whether or not you have specified a column width on the BREAK subcommand with (NOBREAK). When a column width is specified, the following are in effect:

- Summary titles are left-justified in the break column. Use (OFFSET(n)) on BREAK to offset them from the left.
- BRKSPACE(n) on FORMAT is ignored.
- (SKIP(n)) on BREAK is ignored.
- SKIP(n) on SUMMARY operates as usual.

When a column width is not specified, the following are in effect:

- Summary titles are left-justified before the summary line.
- BRKSPACE(n) on FORMAT controls the spacing between the extra line for the summary title and the following summary line. BRKSPACE(−1) places the summary title and the statistic for the leftmost report variable on the same line.
- (SKIP(n)) on BREAK controls the spacing between a prior summary line and the next extra line for the summary title.
- SKIP(n) on SUMMARY is ignored.

Example

```
SORT CASES BY DEPT.
REPORT FORMAT=LIST /VAR=VAR1 TO VAR3
/BREAK=(NOBREAK) (10) (OFFSET(2))
/SUMMARY=SUM 'Totals'.
```

- The FORMAT subcommand requests a case listing.
- Because the file has been sorted by DEPT, the case listing are in order by DEPT.
- The VARIABLES subcommand defines three report variables; a report column is reserved for each, and each has default titles and spacing.
- The width specification on the BREAK subcommand reserves 10 print positions for the title of the summary line. Without this specification, there

would be no room to place the title "Totals," because (NOBREAK) is also requested. The title is offset two columns from the left in the break column.

- At the end of the case listing, the sum of each report variable is displayed at the bottom of its column.

SUMMARY Subcommand

```
/SUMMARY={function...['summary title'][(break col #)] [SKIP(n)]}
         {PREVIOUS[(n)]                                         }
```

where function is

```
aggregate [(varname[(d)][(PLAIN|DOLLAR|COMMA)][varname...])]
```

or

```
composite(agg(varname)...)[(report col[(d)][(PLAIN|DOLLAR| COMMA)])]
```

The SUMMARY subcommand calculates a wide range of aggregate and composite statistics. Each SUMMARY subcommand following a BREAK subcommand specifies a new summary line.

- The minimum specification is an aggregate function or a composite function and its arguments. This must be the first specification on SUMMARY.
- Each function has its own default report-column location for displaying its result.
- The default report-column location can be altered only for composite functions.
- The default format can be altered for any function. Format specifications in an aggregate modify only the immediately preceding result.
- The default summary title is the keyword of the first function named on the SUMMARY subcommand. The default location of the summary title is the column of the break variable to which the summary applies. Both the title and the default break-column location can be altered.
- Use the SKIP keyword to insert blank lines when more than one summary line is requested for a break.
- Summary statistics requested on one SUMMARY subcommand are displayed on the same summary report line.
- SUMMARY subcommands apply only to the preceding BREAK subcommand. If there is no SUMMARY sucommand after a BREAK subcommand, no statistics are displayed for that break level.
- Specify the keyword PREVIOUS to use the summary specifications from a previous BREAK subcommand for the current BREAK subcommand.
- A SUMMARY subcommand must be specified unless NOLIST is specified on the FORMAT subcommand.
- More than one function can be specified on SUMMARY as long as you do not attempt to place two results in the same column.
- An implicit or explicit attempt to place the result of two or more functions in the same report column stops execution of REPORT. Use multiple SUMMARY subcommands to place results of more than one function in the same report column.
- Summary lines can combine any composite functions and aggregrate functions (except ABFREQ and RELFREQ).

Aggregate Functions

Use the aggregate functions to request descriptive statistics on report variables.

- If no variable names are given as arguments to an aggregate function, the statistic is calculated for all variables on the VARIABLES subcommand, that is, for all report variables.
- To request an aggregate function for a subset of the report variables, specify the list of report variables in parentheses after the function keyword.

- All variables must have been named on the VARIABLES subcommand.
- The keyword TO cannot be used to specify a list of variables for an aggregate function.
- The result of an aggregate function is always displayed in the report column reserved for the variable on which the function was calculated.
- Specify multiple SUMMARY subcommands to use several aggregate functions on the same report variable. The results are displayed on different summary lines.
- The aggregate functions ABFREQ and RELFREQ have special display formats and cannot be placed on the same summary line with other aggregate or composite functions.
- Aggregate functions use only cases with valid values.

VALIDN *Valid number of cases.* This is the only function that operates on string variables.

SUM *Sum of values.*

MIN *Minimum value encountered.*

MAX *Maximum value encountered.*

MEAN *Mean.*

STDEV *Standard deviation.*

VARIANCE *Variance.*

KURTOSIS *Kurtosis.*

SKEWNESS *Skewness.*

MEDIAN(min,max) *Median value for values within the range.* MEDIAN sets up integer-valued bins for counting all values in the specified range. Noninteger values are truncated when the median is calculated.

MODE(min,max) *Modal value for values within the range.* MODE sets up integer-valued bins for counting all values in the specified range. Noninteger values are truncated when the mode is calculated.

PCGT(n) *Percentage of cases with values greater than specified value.*

PCLT(n) *Percentage of cases with values less than specified value.*

PCIN(min,max) *Percentage of cases within the inclusive value range specified.*

ABFREQ(min,max) *Frequency counts for values within the inclusive range.* ABFREQ sets up integer-valued bins for counting all values in the specified range. Noninteger values are truncated when the frequency is computed. ABFREQ cannot be mixed with other aggregate statistics on a summary line.

RELFREQ(min,max) *Percentages for values within the inclusive range.* RELFREQ sets up integer-valued bins for counting all values in the specified range. Noninteger values are truncated when the frequency is computed. RELFREQ cannot be mixed with other aggregate statistics on a summary line.

Example

```
SORT CASES BY BVAR AVAR.
REPORT FORMAT=LIST/ VARIABLES=XVAR YVAR ZVAR

/BREAK=BVAR
  /SUMMARY=SUM
  /SUMMARY=MEAN (XVAR YVAR ZVAR)
  /SUMMARY=VALIDN(XVAR)

/BREAK=AVAR
  /SUMMARY=PREVIOUS.
```

- The FORMAT subcommand requests a case listing, and the VARIABLES subcommand establishes a report column for variables XVAR, YVAR, and ZVAR. The report columns have default widths and titles.
- Both break variables, BVAR and AVAR, have the default width and titles.
- Every time the value of BVAR changes, three summary lines are displayed. The first line contains the sums of XVAR, YVAR, and ZVAR. The second line contains the means of all three variables. The third line displays the number of valid cases for XVAR in the report column for XVAR.
- Every time the value of AVAR changes within each value of BVAR, the three summary lines requested for BVAR are displayed. These summary lines are based on cases with the current value of BVAR that also have the current value of AVAR.

Example

```
SORT CASES BY DEPT.
REPORT VARIABLES=WAGE BONUS TENURE
/BREAK=DEPT (23)
/SUM=SUM(WAGE BONUS) MEAN(TENURE) 'Sum Income: Mean Tenure'.
```

- The SUMMARY subcommand defines a summary line consisting of the sums of WAGE and BONUS and the mean of TENURE. The result of each aggregate function is displayed in the report column of the variable on which the function is calculated.
- A title is assigned to the summary line. The break column is 23 columns wide to accommodate the right-justified summary-line title.

Composite Functions

A composite function operates on simple aggregate statistics and their arguments to produce a single result. Use composite functions to place a summary statistic in a column other than that of the report variable on which it was calculated, to manipulate variables not named on the VARIABLES subcommand, or to obtain statistics based on aggregated statistics.

- Composite functions can be computed upon constants and any variable in the active file.
- The following aggregate functions can also be arguments to composite functions: VALIDN, SUM, MIN, MAX, MEAN, STDEV, VARIANCE, KURTOSIS, and SKEWNESS. When used within composite functions, aggregate functions can have only one variable as an argument.
- By default, the results of a composite function are placed in the report column of the first variable named on the composite function that is also named on the VARIABLES subcommand.
- A composite function and its arguments cannot be separated from each other by other SUMMARY specifications.
- The result of a composite function can be placed in any report column, including columns of dummy or string variables, by specifying a target column.
- To specify a target column, enclose the variable name of a report column in parentheses after the composite function and its arguments.
- The format for the result of a composite function can be specified with a format specification in parentheses after the name of the target column and within the parentheses that enclose the target-column specification.

DIVIDE(agg() agg() [factor]) *Divide the first argument by the second and multiply by the optional factor.*

MULTIPLY (agg() ... agg()) *Multiply the arguments.*

PCT(agg() agg()) *Percentage of the first argument over the second.*

SUBTRACT(agg()agg()) *Subtract the second argument from the first argument.*

ADD(agg() ... agg()) *Add the arguments.*

GREAT(agg() ... agg()) *Give the maximum of the arguments.*

LEAST(agg() ... agg()) *Give the minimum of the arguments.*

AVERAGE(agg() ... agg()) *Give the average of the arguments.*

Example

```
SORT CASES BY DEPT.
REPORT FORMAT=BRKSPACE(-1)
/VARIABLES=WAGE BONUS SPACE1 (DUMMY) '' BNFT1 BNFT2 SPACE2
(DUMMY)''
/BREAK=DEPT (LABEL)
  /SUMMARY=MEAN(WAGE BONUS BNFT1 BNFT2)
           ADD(VALIDN(WAGE)) (SPACE2)

  /SUMMARY=ADD(SUM(WAGE) SUM(BONUS))
           ADD(SUM(BNFT1) SUM(BNFT2)) 'Totals' SKIP(1)

  /SUMMARY=DIVIDE(MEAN(WAGE) MEAN(BONUS)) (SPACE1 (COMMA)(2))
   DIVIDE(MEAN(BNFT1) MEAN(BNFT2)) (SPACE2 (COMMA)(2)) 'Ratios'
SKIP(1).
```

- The VARIABLES subcommand defines six report columns. The columns called WAGE, BONUS, BNFT1, and BNFT2 contain aggregate statistics based on those variables. The variables SPACE1 and SPACE2 are dummy variables that are created for use as space holders. Each column is given a blank title to suppress the default column head.
- The first SUMMARY computes the means of the variables WAGE, BONUS, BNFT1 and BNFT2. Because BRKSPACE=−1, this summary line will be placed on the same report line as the break value and will have no summary title. The means are displayed in the report column for each variable. It also computes the valid number of cases for WAGE inside a composite function so that the result can be placed in the column SPACE2.
- The second SUMMARY adds the sum of WAGE to the sum of BONUS. Since no location is specified, the result is displayed in the WAGE column. The sum of BNFT1 is added to the sum of BNFT2 and the result placed in the BNFT1 column. One line is skipped before the summary line requested by this SUMMARY subcommand is displayed.
- The third summary line divides the mean of WAGE by the mean of BONUS and places the result in SPACE1. The ratio of the mean of BNFT1 to the mean of BNFT2 is displayed in the SPACE2 column. Because locations are specified, formats can also be given. The results are displayed with commas and 2 decimal places. One line is skipped before the summary line requested by this SUMMARY subcommand is displayed.

SUMMARY Titles

By default, the title of a summary line is the keyword of the first function requested on the summary line.

- You can specify a summary title enclosed in apostrophes or quotes following the conventions for strings (see Universals: Strings).
- The summary title must be specified after the first function and its arguments and cannot separate any function from its arguments.
- A summary title can be only one line long.
- A summary title wider than the break column is truncated.
- Only one summary title applies per summary line. If more than one is specified, the last is used.
- The summary title is left- or right-justified depending upon whether the break title is left- or right-justified.
- The default location for the summary title is the column of the BREAK variable to which the SUMMARY applies.
- With multiple breaks, you can override the default placement of the title by specifying the break-column number in which you want the summary title to be displayed in parentheses following the title. You can override the default by specifying the break-column number in which you want the summary title to be displayed in parentheses following the title.
- In a (NOBREAK) report with no break-column width specified on BREAK, REPORT displays the summary title above the summary line at the left margin.

SUMMARY Print Formats

All functions have default formats that are used to print results (see Table A). You can override these defaults by specifying a format keyword or the number of decimal places.

- Formats can be specified for the results of aggregate functions only if the variable is explicitly named as an argument.
- Formats can be specified for the results of composite functions only if a report-column location for the result is also specified.
- If the report column is wide enough, SUM, MEAN, STDEV, MIN, MAX, MEDIAN, and MODE use DOLLAR or COMMA format if one has been declared for a variable on the FORMATS command.
- A format specification must be enclosed in parentheses.
- For aggregate functions, one or both format specifications are placed after the variable name within the parentheses that enclose the variable name.
- For composite functions, one or both format specifications are placed after the variable name of the target column within the parentheses that enclose the target column.
- If the column is not wide enough to display the decimal digits for a given function, REPORT displays fewer decimal places. If the column is not wide enough to display the integer portion of the number, REPORT adopts scientific notation and then displays asterisks.
- An exact value of 0 is displayed with one 0 to the left of the decimal point and as many 0 digits to the right as specified by the format. A very small number is displayed without a zero digit to the left of the decimal point, except with DOLLAR and COMMA formats.

(DOLLAR) *Display the value using DOLLAR format.*

(COMMA) *Display the value using COMMA format.*

(PLAIN) *Override DOLLAR or COMMA dictionary formats.* PLAIN is the default for all functions except MEAN, STDEV, MIN, MAX, MEDIAN, and MODE. For these functions, the default is the dictionary format.

Example

```
/SUMMARY=MEAN(INCOME (DOLLAR)(2))
         ADD(SUM(INCOME)SUM(WEALTH) (WEALTH(DOLLAR)(2))
```

- SUMMARY displays the mean of INCOME with a dollar sign, commas, and two decimal digits. The format can be specified because INCOME is specified as an argument to the MEAN function. The result is displayed in the INCOME column.
- The sums of INCOME and WEALTH are added and the result is displayed with a dollar sign, commas, and 2 decimal digits. The format can be specified because an explicit location is given for the results of ADD. The result is displayed in the WEALTH column.

A Default print formats for functions

Function	Format	Function	Format
SUM	Dictionary print format	PCIN	2
MEAN	Dictionary print format + 2	PCLT	2
STDEV	Dictionary print format + 2	PCGT	2
MIN	Dictionary print format	PCT	2
MAX	Dictionary print format	KURTOSIS	3
VARIANCE	Column width − 1	SKEWNESS	3
DIVIDE	Column width − 1	MODE	0
MULTIPLY	Column width − 1	MEDIAN	1
SUBTRACT	Column width − 1	VALIDN	0
ADD	Column width − 1	ABFREQ	0
GREAT	Column width − 1	RELFREQ	2
LEAST	Column width − 1		
AVERAGE	Column width − 1		

Other SUMMARY Keywords

Spacing between multiple summary lines for a single break and references to previously defined summary lines are controlled by the following keywords:

SKIP(n) *Blank lines before the summary line.* SKIP is not enclosed in parentheses. The default is 0. SKIP on the first SUMMARY subcommand for a BREAK skips the specified lines after skipping the number of lines specified for BRKSPACE on FORMAT.

PREVIOUS(n) *Use the SUMMARY subcommands for the nth BREAK.* If no specification is given in parentheses, PREVIOUS points to the set of SUMMARY subcommands for the previous BREAK. If an integer specification is given, the SUMMARY subcommands from the *n*th BREAK are used.

No other specification can be used on SUMMARY with PREVIOUS. For a multiple-break report for which you want the same sets of summaries, specify SUMMARY subcommands for the higher BREAK subcommand and keyword PREVIOUS for lower breaks.

Title and Footnote Subcommands

```
/TITLE ='title' (Centered head)
/FOOTNOTE='title' (Centered foot)
```

or

```
/LTITLE='title' (Left-justified head)
/RTITLE='title' (Right-justified head)
/CTITLE='title' (Centered head)
/LFOOTNOTE='title' (Left-justified foot)
/RFOOTNOTE='title' (Right-justified foot)
/CFOOTNOTE='title' (Centered foot)
```

- The title and footnote subcommands are optional and can be placed anywhere after the FORMAT subcommand except among the BREAK and SUMMARY subcommands.
- Title subcommands should be specified before any footnote subcommands.
- The default REPORT title is the title specified on the TITLE command. If there is no TITLE command specified in your SPSS/PC+ session, the default REPORT title is the first line of the SPSS/PC+ header.
- A title or footnote is specified by providing a string in apostrophes or quotes on one of the title or footnote subcommands.
- If the title or footnote is more than one line, enclose each line in apostrophes or quotes and separate the specifications for each line by at least one blank.
- Titles are displayed beginning in the first line of the system page.
- Footnotes end in the last line of the system page.
- Centered titles and footnotes are centered within the report page width.
- TITLE cannot be used with CTITLE, LTITLE, or RTITLE, and FOOTNOTE cannot be used with CFOOTNOTE, LFOOTNOTE, or RFOOTNOTE.
- Only one of each title and footnote subcommand is allowed.
- Titles and footnotes are repeated on each page of a multiple-page report.
- If the total width requested for the combined titles or foonotes for a line exceeds the page width, execution of REPORT stops.
- The maximum number of characters in a title or footnote is the number of characters that can fit on a command line, minus 2 characters for the apostrophes or quotes that enclose the string.

Two function keywords can be used in a title or footnote. *You must specify these keywords in upper case, as shown here.*

)PAGE *Display the page number right-justified in a five-character field.*

)DATE *Display the current date in the form* dd/mmm/yy *right-justified in a nine-character field.*

Example

```
/LTITLE='Personnel Report' 'Prepared on )DATE'
/RTITLE='Page )PAGE'
```

- This LTITLE subcommand provides a two-line title. The second line of this title contains the date on which the report was processed. This title is displayed left-justified at the top of each page of the report.
- The RTITLE subcommand displays the page number right-justified at the top of each page.
- The keywords)DATE and)PAGE are entered in upper case.

MISSING Subcommand

```
/MISSING=VAR|NONE|LIST [(varlist [n])]
```

Use the optional MISSING subcommand to control the handling of cases with missing values on the VARIABLES and SUMMARY subcommands. By default, cases with missing values are included in case listings but are excluded from the calculation of functions on a function-by-function basis.

- MISSING specifications apply to variables named on VARIABLES and SUMMARY as well as to strings created with the STRING subcommand.
- The character used to indicate missing values is controlled by the FORMAT subcommand.

The following keywords are available for the MISSING subcommand:

VAR *Missing values are treated separately for each variable.* Missing values are displayed in case listings but are not included in summary statistics. This is the default.

NONE *User-missing-value indicators are ignored.* Applies to all variables named on the VARIABLES subcommand.

LIST[(varlist[n])] *Eliminates any case with the specified number of missing values among the specified list of variables.* If no *n* is specified, the default is 1. If no variables are specified, all variables named on the VARIABLES subcommand are assumed.

Example

```
/MISSING= LIST (XVAR,YVAR,ZVAR 2)
```

- Any case with a missing value for two or more of the variables XVAR, YVAR, and ZVAR is omitted from the report.

REVIEW

```
REVIEW [{LOG                        }]
        {LISTING                    }
        {BOTH                       }
        {'filename' ['filename']}
```

Example:

```
REVIEW.
```

Overview

The REVIEW command initiates a REVIEW editing session. REVIEW is a full-screen editor specifically designed for editing SPSS/PC+ log and listing files. It allows you to edit two files at once in separate windows. It also allows you to submit revised command lines directly to SPSS/PC+ for execution.

The REVIEW command places you in a full-screen editing environment. REVIEW commands are issued with the function keys on your keyboard. Hold Alt and press F10 to return to the SPSS/PC+ environment.

Defaults

If no keywords are specified, REVIEW initiates an editing session of the current log and listing files. On color monitors, REVIEW uses a default color scheme of (1,2,4) for lower-window (or only window) background color, upper-window background color, and frame color, respectively. See the manual for your monitor for the actual colors that correspond to these numbers.

Tailoring

You can ask to edit any file or pair of files that consist of legible characters (see "Limitations" below). On color monitors, you can choose colors for the two editing windows and frame. You can also turn color off.

Syntax

- The minimum specification is the word REVIEW followed by the command terminator.
- Only one keyword (LOG, LISTING, or BOTH) can be specified.
- Use the LOG keyword to edit your SPSS/PC+ log file.
- Use the LISTING keyword to edit your SPSS/PC+ listing file.
- Use the BOTH keyword to edit both your log and listing files simultaneously. BOTH is the default.
- If you do not use a keyword, you can specify one or two filenames in single quotes to edit files other than your log and listing files.
- At least one space must separate the keyword REVIEW and its specifications.
- Issue the SET RCOLOR command before entering REVIEW to set nondefault colors (see Command Reference: SET).
- Specify SET COLOR OFF before entering REVIEW to turn color off.
- Four sets of key combinations are used to specify editing (see below). Keys are used alone or in combination with Ctrl, ⇧, or Alt.
- Some commands cause REVIEW to prompt for additional information, such as a filename, at the bottom of the screen. To cancel such commands at that time, enter a *backspace* character and press ↵.
- When you request help with ⇧ F1, REVIEW prompts for the SPSS/PC+ keyword(s) for which help is desired. In subsequent requests, you can use abbreviated forms to answer this prompt. A quotation mark followed by a keyword requests help for the one or two keywords in the previous request with a second or third keyword as supplied. An apostrophe refers to all but the last keywords in the previous request.

Operations

- You can enter REVIEW from SPSS/PC+ or directly from DOS.
- If you edit two files at once, REVIEW divides your screen into upper and lower windows. The file named first occupies the upper window and the file named second occupies the lower window.
- Editing occurs at the current line and character as determined by the position of the cursor.

- There are two editing modes: insert mode and overtype mode. In overtype mode, entered text replaces existing text. In insert mode, entered text is inserted to the left of the current character.
- A two-digit number in the lower right corner of the screen displays the column location of the cursor.
- The REVIEW help screen displays Ctrl as a caret, ⇧ as an up-arrow, and Alt as an *a*.

Limitations

- You can edit only ASCII files with REVIEW. Do not attempt to edit a non-ASCII file such as an SPSS/PC+ active or system file. If the listing of a file that results from a DOS *TYPE* command is meaningful, the file is probably an ASCII file.
- REVIEW displays only the first 80 characters of any line. Use the Split Line command to view lines longer than 80 characters.
- REVIEW works on files held in memory. It starts with the *last* lines in the file in memory and reads additional lines as required. You may not be able to edit very large files.

Example

```
REVIEW 'A:REGRESS.INC' 'REGRESS.LIS'.
```

Entering REVIEW from DOS

To enter REVIEW directly from DOS, specify

```
REVIEW filename1 [filename2] [/La] [/Ub] [/Fc] [/B]
```

- Place the names of the files you wish to edit in place of filename1 and filename2.
- If you only wish to edit one file, omit the second file name.
- The /La, /Ub, and /Fc options allow you to change the colors used for the lower (or only) window, upper window, and frame, respectively. Specify integers from 0 to 6 in place of a, b, and c. See the manual for your monitor for the colors that correspond to numbers 0 through 6.
- To turn off color use, specify /B on the command line.

If you plan to use REVIEW frequently from outside SPSS/PC+ with nondefault colors, create a batch file (a file with a .BAT extension) that contains the following single line:

```
REVIEW /La /Ub /Fc %1 %2
```

- Specify the desired color numbers in place of a, b, and c.
- The batch file should be in the directory that holds your SPSS/PC+ program files.
- After you create the batch file, you can enter REVIEW by typing the name of the batch file followed by the names of the files you wish to edit. The colors you have specified in the batch file will automatically take effect.

Setting and Information Commands

Command Name	Key Combination/Function
REVIEW Help	F1 displays a list of REVIEW commands. Pressing F1 a second time yields an organizational chart of the commands. When pressed in response to a prompt, F1 provides a short message explaining the prompt.
SPSS/PC+ Help	⇧ F1 invokes the SPSS/PC+ HELP facility.
SPSS/PC+ Variables	Ctrl F1 displays a list of the variables in the current SPSS/PC+ active file. If there is no active file, this command does not operate.
Switch Window	F2 causes the editor to switch windows.
Change Window Size	Ctrl F2 allows you to change the relative sizes of the two windows.

File Commands

Command Name	Key Combination/Function
Write Block to Disk	[F9] writes the current marked block to disk under a name you specify. The default name is REVIEW.TMP.
Write File to Disk	[⇧] [F9] saves the current file to disk under a name you specify. The default is the current name. The original copy is renamed with an extension of .BAK.
Delete Disk File	[Ctrl] [F9] allows you to delete any file from a DOS directory.
Insert File	[F3] inserts the contents of a DOS file into the current file below the current line.
Edit Different File	[⇧] [F3] replaces the file in the current window.
Exit and Run Block	[F10] submits a marked block of commands to SPSS/PC+ for execution and returns to the SPSS/PC+ environment. This command is available only if REVIEW was entered from SPSS/PC+. If unsaved changes have been made in either window, a prompt appears.
Exit from REVIEW	[Alt] [F10] exits REVIEW. If unsaved changes have been made in either window, a prompt appears.
Directory of Files	[Alt] [F1] provides a listing of any DOS directory you specify. Wildcards are accepted.

Editing Commands

Command Name	Key Combination/Function
Change Edit Mode	[Ins] changes the editing mode from overtype to insert or from insert to overtype.
Insert Line After	[F4] inserts a blank line after the current line. The blank line becomes the current line.
Insert Line Before	[⇧] [F4] inserts a blank line before the current line. The blank line becomes the current line.
Delete Line	[Ctrl] [F4] deletes the current line. The line below becomes the current line.
Undelete Line	[Alt] [F4] restores the most recently deleted line and places it immediately above the current line. It is only possible to restore the most recently deleted line.
Split Line	[↵] in insert mode splits the current line at the current character. In overtype mode, moves the cursor to the next line.
Join Line	[Del] joins the line below to the current line if the cursor is at the end of a line.
Delete Current Character	[Del] deletes the current character. The text to the right of the cursor moves left one space.
Delete Previous Character	The backspace key deletes the character to the left of the cursor. The cursor and righthand text move left.
Change Forward	[F5] replaces one character string with another following the current cursor positon. At each occurrence, prompts are displayed as follows: C to change the occurrence, A to change all occurrences without individual prompting, N to move to the next occurrence without changing the current one, S to stop the search, or X to change the occurrence and then stop the search. The previous search and replacement strings are the defaults but they can be typed over. Upper and lower case are distinguished in searches. To identify only those occurrences of the string that begin in column 1, place a tilde (~) at the beginning of the search string. It is not possible to change a character string that itself begins with a tilde.

Change Reverse	⇧ F5 works the same as Change Forward, moving backward through the file to the earliest line in memory.
Mark/Unmark Block	F7 marks a block of lines for subsequent action. Press F7 at the first or last line of a block and again at the opposite border to mark the block. Press F7 again to unmark the block. Blocks consist of whole lines.
Copy Block	F8 copies a marked block to the line below the current line. Blocks can be copied between windows.
Move Block	⇧ F8 moves a marked block to the line below the current line. Blocks can be moved between windows.
Delete Block	Ctrl F8 deletes a marked block.

Motion Commands

Command Name	**Key Combination/Function**
Left	← moves the cursor left one space.
Right	→ moves the cursor right one space.
Up	↑ moves the cursor up one line.
Down	↓ moves the cursor down one line.
Start of Line	Ctrl ← moves the cursor to the first character of the current line.
End of Line	Ctrl → moves the cursor to the last character of the current line.
Tab Forward	The tab key moves the cursor eight characters to the right.
Tab Backward	⇧ and the tab key moves the cursor eight characters left.
Previous Page	PgUp moves the cursor to the last line of the preceding page.
Next Page	PgDn moves the cursor to the first line of the following page.
Top of Page	Home moves the cursor to the start of the first line in the window.
Bottom of Page	End moves the cursor to the end of the last line in the window.
Top of File	Ctrl Home moves the cursor to the first line of the file.
Bottom of File	Ctrl End moves the cursor to the last line of the file.
Top of Memory	Ctrl PgUp moves the cursor to the first line of the portion of the file currently in memory.
Put Line at Top	Ctrl PgDn moves the current line to the top of the window.
Start of Next Line	↵ moves the cursor to the first character of the next line in overtype mode.
Search Down	F6 moves the cursor to the first occurrence of a specified character string on or below the current line. Upper and lower case are distinguished in searches. The search continues to the end of the file. The default search string is the last string specified To locate the string only when it appears at the beginning of a line, specify a tilde (~) at the beginning of the search string. It is not possible to search for a string that itself begins with a tilde.
Search Up	⇧ F6 works the same as Search Down, moving backward through the file to the earliest line in memory.
Find SPSS/PC+ Page	Ctrl F6 moves the cursor to the first line of the SPSS/PC+ page specified when you are editing an SPSS/PC+ listing file. Some types of output are not saved in listing files; thus some page numbers may be missing.

Index

Index

H Index